TWENTY-FIVE YEARS OF LOGICAL METHODOLOGY IN POLAND

SYNTHESE LIBRARY

MONOGRAPHS ON EPISTEMOLOGY,

LOGIC, METHODOLOGY, PHILOSOPHY OF SCIENCE,

SOCIOLOGY OF SCIENCE AND OF KNOWLEDGE,

AND ON THE MATHEMATICAL METHODS OF

SOCIAL AND BEHAVIORAL SCIENCES

Managing Editor:

JAAKKO HINTIKKA, *Academy of Finland and Stanford University*

Editors:

ROBERT S. COHEN, *Boston University*

DONALD DAVIDSON, *University of Chicago*

GABRIËL NUCHELMANS, *University of Leyden*

WESLEY C. SALMON, *University of Arizona*

VOLUME 87

TWENTY-FIVE YEARS
OF LOGICAL METHODOLOGY
IN POLAND

Edited by

MARIAN PRZEŁĘCKI *and* RYSZARD WÓJCICKI

D. REIDEL PUBLISHING COMPANY
DORDRECHT–HOLLAND/BOSTON–U.S.A.

PWN–POLISH SCIENTIFIC PUBLISHERS
WARSAW–POLAND

Library of Congress Cataloging in Publication Data
Main entry under title:

Twenty-five years of logical methodology in Poland.

(Synthese library; v. 87)
Bibliography: p.
1. Logic — Addresses, essays, lectures.   2.   Science — Methodology —
Addresses, essays, lectures.   3.   Semantics (Philosophy) — Addresses, essays,
lectures.
I.   Przełęcki, Marian.   II.   Wójcicki, Ryszard.
BC57.T85   1975   160′.9438   76-7064
ISBN 90-277-0601-8

Translated from the Polish by Edmund Ronowicz and Stanisław Wojnicki

Distributors for Albania, Bulgaria, Chinese People's Republic, Czechoslovakia,
Cuba, German Democratic Republic, Hungary, Korean People's Democratic
Republic, Mongolia, Poland, Rumania, Vietnam, the U.S.S.R. and Yugoslavia
ARS POLONA
Krakowskie Przedmieście 7, 00-068 Warszawa 1, Poland

Distributors for the U.S.A., Canada and Mexico
D. REIDEL PUBLISHING COMPANY, INC.
Lincoln Building, 160 Old Derby Street, Hingham, Mass. 02043, U.S.A.

Distributors for all other countries
D. REIDEL PUBLISHING COMPANY
P.O. Box 17, Dordrecht, Holland

The paper: "The Controversy: Deductivism Versus Inductivism" by Janina Kotarbińska, pp. 261–278 reprinted from Logic, Methodology and Philosophy of
Science, edited by Ernest Nagel, Patrick Suppes, and Alfred Tarski with the permis-
sion of the publishers, Stanford University Press. Copyright © 1962 by the Board
of Trustees of the Leland Stanford Junior University.

Printed in Poland by D.U.A.M

PREFACE

The anthology presents a selection of methodological writings published by Polish logicians after World War II (the first of them dated 1947). All the papers belong to what may be called Logical Methodology or Logical Theory of Science. The epithet 'logical' characterizes rather the general point of view than the particular methods employed by the authors. Apart from articles which make an essential use of different formal (logical and mathematical) methods, there are many which do not involve any formal apparatus whatsoever. The problems the papers deal with may be characterized as problems of the general methodology of empirical science. The papers do not consider the methodological problems of formal (mathematical) knowledge, and, as a rule, they are concerned with empirical science as a whole and not with some of its specific branches. The topics covered by the selected writings include the main issues and controversies discussed within the contemporary methodology of science. A considerable part of the anthology is concerned with the semantics of empirical languages and considers problems such as interpretation of observational and theoretical terms, analyticity, empirical meaningfulness, etc. Another group of papers deals with the problem of induction and examines various ways of its justification. Some articles discuss the nature and the status of methodology itself. The materials have been selected so as to make up a whole representative of what has been done in this field in Poland since 1945. The book comprises 33 articles by 20 authors. Almost all authors writing on logical methodology in Poland have been included. Most of the articles appear for the first time in English. They are all taken in their entirety.

MARIAN PRZEŁĘCKI and RYSZARD WÓJCICKI

TABLE OF CONTENTS

K. AJDUKIEWICZ

METHODOLOGY AND METASCIENCE*

1. Methodology of sciences is a science which takes science or rather particular sciences as the object of its investigations. Thus methodology is, in a sense, a science of the second order, viewing the other sciences as if from the floor above. However, we have not characterized methodology univocally yet by stating that it is a science about science. For, by indicating the object of investigation of a given science we do not determine its problematics: the questions that may be put forth with respect to the same object of investigation can be so diverse that they may yield problematics for a number of entirely different scientific disciplines. Indeed, there exist many different sciences about sciences. For instance, scientific disciplines so different from each other as history, sociology and psychology deal with them, and finally, the methodology of science, which is different from them all.

2. However, the definition of methodology as a science about science is a particularly inadequate description for one more reason. This reason is the ambiguity of the word 'science'. This word, whose meaning is not clear, can in addition have the following different meanings. Namely, speaking of science, we may, principally, have two things in mind: either the activities that constitute the cultivation of science, or what we might metaphorically call the products resulting from these activities. Science understood as activity would consist of the ideas of scientists that occur while they are engaged in scientific work, the activities they undertake while conducting experiments and probably many other functions. Science understood as a product would consist of scientific theorems.

3. This ambiguity of the term 'science' was overlooked during the period of psychologism in the fields of logic and methodology. During this period scientific theorems were identified with some ideas. A scientific theorem is—it was said—a certain proposition asserting that this

or that is the case, and this proposition, i.e., a conviction, is nothing else but someone's idea, his psychical activity. It seems that this psychologism has been entirely surmounted today. I do not think anybody would maintain today that scientific theorems are psychological phenomena. For Pythagoras' theorem is not a psychical phenomenon that occurred in the mind of the great Greek when he considered the relations between the squares of the sides of an isosceles right triangle. Probably Pythagoras found that the sum of the squares of the sides equalled the square of the hypothenuse more than once in his life, as innumerable other thinkers did after him. If we were to identify the Pythagorean theorem with any judgment stating the equality between the square of the hypotenuse and the sum of the squares of the sides, then we would have to say that there exists not only one Phytagorean theorem concerning the squares of the sides of an isosceles right triangle, but that there exist a great many of them. Since we speak of only one Pythagorean theorem, we are not calling anybody's thought, anybody's psychical phenomenon, a theorem.

What, then, is the Pythagorean theorem, since it is not anybody's thought? Contemporary science gives two answers to this question. According to some, the Pythagorean theorem is (like any other scientific theorem) a certain sentence, a verbal entity, a sequence of words. According to others, the Pythagorean theorem is not a sentence at all — it is the meaning — sometimes called, in order to avoid ambiguity, the 'ideal meaning' of some sentence. Neither of the above ways of understanding the term 'scientific theorem' is free of certain difficulties. Those who by a scientific theorem understand some sentence, i.e., some verbal expression, may meet with similar objections as those who consider scientific theorems to be psychical phenomena. There are at least as many sentences, i.e., verbal expressions, that may be called the Pythagorean theorem, as there are languages, in which Pythagoras' idea can be expressed. Also within one language, e.g., in the English language, all the utterances that sound the same as the sentence I produce by saying 'the sum of the squares of the sides of a triangle equals the square of its hypotenuse' may equally well pretend to be the Pythagorean theorem. In order to avoid this difficulty we might choose one of the following two ways. Namely, we might refrain from giving the name of a scientific theorem to one particular, *hic et nunc* uttered sentence, and

use the term to describe the whole class of sentences synonymous with that particular sentence. This concept of scientific theorem could be adopted, however, only after specifying what synonymous sentences are. And this is not an easy thing to do. In order to avoid using the term that has not yet been properly explicated we might choose another way. Namely, we might give up the general concept of a scientific theorem and speak of scientific theorems in a given language. We would then not speak of the Pythagorean theorem in general, but of the Pythagorean theorem in, e.g., the English language, which would be defined in the following way. The Pythagorean theorem in the English language is the class of sentences consonant to the sentence 'the sum of the squares of the sides of a given triangle equals the square of its hypotenuse'. It seems that the majority of contemporary methodologists employs this sort of concept of scientific theorem relativized to a given language. Not all of them, however, emphasize this relativization clearly—they even confuse the concept relativized to language with the absolute concept.

The other of the above mentioned nonpsychological conceptions of theorem, namely, the conception identifying it with the ideal meaning of a certain definite sentence, does not present the same difficulties. The difficulty lies, however, in the concept of ideal meaning itself, which has not been defined precisely so far in a way acceptable to all.

In spite of the above mentioned difficulties connected with the nonpsychological conceptions of a scientific theorem, these concepts play a basic role in the methodology of science.

4. We have distinguished two meanings of the term 'science': one in which science means the activities undertaken by the scientists as such, and the other, in which science consists of the theorems stated as a result of these activities, i.e. the products of these activities. Science understood as activity is a phenomenon occurring in time, even in space, and possessing its history. But also science understood as a product of these activities, i.e. the set of sentences proposed by scientists, is something that possesses its history, develops in time, changes its constituents. Thus science understood as the product of research activities is also a temporal product.

Apart from these two ways of understanding the term 'science' a third understanding should be distinguished. Let us take mathematics

as an example. One may speak of Greek mathematics and of contemporary mathematics, in other words – one may understand the term 'mathematics' as denoting a certain set of sentences stated by certain people at a certain time. We shall then have in mind not mathematics as activity but mathematics as the product of these activities. One may, however, attach to the word 'mathematics' such a meaning that the word will be a name of a certain set of sentences, regardless of whether stated by anybody or not. One might, e.g., understand by mathematics a set of sentences that can be logically derived from certain basic sentences, i.e. the set of sentences that can be derived from the mathematical axioms. Or one may understand by mathematics the set of all true sentences that can be expressed in the language of mathematics. Mathematics thus understood is not a thing that would have its history or develop in time. Thus it is a science understood as neither activity nor a product of this activity. Let us call this last concept of science, consisting of sentences that may never have been stated, the ideal concept of science.

5. Which of the three concepts of science did we have in mind when we stated that methodology is a science about science? If we are to answer this question in accordance with the actual state of methodology, we have to state that methodology speaks of science understood as activity as well as of science understood as product and, finally, of science understood in the ideal way. This is how things have been at least in traditional methodology. It speaks of inferring, proving, formulating and testing hypotheses, i.e. about some psychical activities, but it also speaks of proofs, hypotheses, theories, deductive systems, which are no longer activities. These two points of view are confused so much in traditional methodology that it is often difficult to realize whether the methodologist, while using a certain term, has in mind some activity or the product of this activity. When, e.g., methodologists use the term 'proof' they often refer to the activity of proving but might equally well be referring to the product of this activity, i.e. a certain sequence of sentences connected by the relation of consequence–beginning with axioms and ending with the sentence that is being proved.

6. Recently, however, a new and separate discipline has emerged differing from methodology, in that it is interested in science neither as activity nor as the historical product of this activity, but which takes

as the object of its investigation science in the ideal sense of the word. This discipline is what is known as metascience, and its branches so far developed bear the names of metamathematics and metalogic. These are theories of deductive systems understood as sets of expressions, and disregarding the fact whether these expressions have ever been uttered or written by anybody or not. Thus metascience does not concern itself with a certain science as a historical entity consisting of theorems that were once accepted by somebody — it deals with science understood in the way that we have called the ideal understanding of science. In order to define a deductive system it is enough to establish, first, what configurations of signs are meaningful expressions in this system, and in particular, what configurations of signs are sentences in this system, and, secondly, which of the sentences of the system are its theorems, i.e. theses. This is established by the application of certain rules. The rules establishing which expressions are sentences are called the rules of formation. The establishing which sentences of the system are its theses is usually done: 1° by clearly indicating that certain sentences are theses (these sentences are called the axioms of the system), 2° by indicating such transformations leading from certain sentences to another sentence, which from theses lead always to a thesis. The rules with enumerate axioms and specify transformations leading from theses always to a new thesis are called the rules of transformation or the rules of proving. Metascience when approaching a given system, for which the rules of formation and proving have been established, investigates this system, deriving various properties of this system from its rules in a purely deductive way. Thus metascience concerns itself with, e.g., the problems such as whether this or that sentence of the system is its thesis, whether, provided this or that sentences of the system are theses, some other definite sentence must be its thesis. It asks if a given system is not contradictory, i.e. if among its theses there are no two contradictory sentences, whether the system is complete, i.e. if taking any arbitrarily chosen sentence of the system either the sentence itself or its negation will be the thesis of the system. The statements of metascience are no less strict and precise in character than the strictest mathematical statements. It has also achieved very interesting and valuable results. Such for instance is Gödel's theorem who has demonstrated that no consistent deductive system, which is

rich enough to build in it the arithmetic of natural numbers, can ever be complete, i.e. that among its sentences such a sentence may be indicated which can be neither proved nor refuted in this system. Thus Gödel has indicated that among the sentences of any deductive system of arithmetic there will always be a sentence which is true, but which cannot be derived from the axioms of this system, i.e., in other words, that the whole of arithmetical truths can never be included in some deductive system.

Metascience has so far developed only as a theory of deductive systems. However, every science assumes at least some deductive system, which pervades its structure. Physics and other sciences using measurement assume mathematics and formal logic. The latter is assumed by every science in general. Thus metascientific investigations, although at the moment they are restricted to investigation of these deductive systems themselves, may be relevant to the investigations of other sciences, which are not deductive systems. Attempts to investigate the sciences which are not deductive systems, e.g., the natural sciences, in a way similar to metamathematics or metalogic are, at the moment, only at the embryonic stage.

7. Metascience is the highest achievement in the investigations of science, which may be included in methodology. The discussions of traditional methodology are far from these achievements. Traditional methodology — as it has been said — takes as its object of investigation the activities undertaken in cultivating science and science understood as the historical product of these activities. In attempting to review the problematics of traditional methodology and arrange it in a clear scheme, great difficulties are encountered. Every synthetic approach to the methodological problematics that comes to mind seems to be inadequate. It either seems to omit important matters, or it is the construction of a programme that does not reflect the real state of affairs. Thus I shall risk the arguments given below making all these reservations.

8. Methodology, in the traditional meaning of the word, first of all attempts to report on the activities of which (the cultivation of) science consists.

The activities of which science consists may be divided into two groups. The first group includes those activities that lead to construction

of the language of a given science and the conceptual apparatus connected
with it. The other group includes the activities resulting in the acceptance
of some scientific theorem. Questions, descriptive as well as normative
in character, may be asked with respect to both of these types of
activities. We may then ask, first, how we actually learn to understand words and thus acquire concepts that correspond to them, and
how we actually arrive at acceptance of certain theorems. By asking
these questions we enter the fields of the psychology of speech and the
psychology of convictions. Secondly, we may, however, ask how our
language should be built to make a scientific language, and how to arrive at theorems which are scientific theorems.

It seems to me that traditional methodology deals with both, descriptive and normative, problems. For instance, methodologists enumerate various ways by which we reach our convictions. They speak of experiment, of memory, of inference, of obviousness, and of other sources
of human convictions. They also speak of how we learn to understand
words, by, e.g., imitating the way they are used by other people or by
definitions providing the translation of a new word in terms of words
that we already understand. All these are the psychological problems
with which traditional methodologists are concerned.

However, apart from the descriptive problems, methodology also
deals with the above mentioned normative problems. It is not satisfied,
for instance with saying that we sometimes reach certain convictions
inferring them from other convictions: it also asks what conditions
should this inference fulfil, what this inference should be like so that
the conclusion thus reached could become a scientific theorem. This
way of formulating the question appears to be inappropriate, though,
for in different sciences different requirements are imposed upon inferences. There are, e.g., such ways of inferring that would guarantee
to the theorems obtained from them acceptance in some natural science,
but which would not satisfy a mathematician. For instance, some ways
of inductive inference are unacceptable from the point of view of a
mathematician.

Thus, parallelly to the establishment of norms for the ways of justified introduction of theorems in science, methodology must establish a certain division of the sciences by distinguishing such classes
of sciences so that each includes the sciences that adopt the same re-

quirements with respect to the ways of arriving at theorems. So methodology has to examine the historically existing sciences and carry out
their division according to the above mentioned methodological point
of view. The methodological division of sciences is their division into
a priori and empirical sciences. In each of these groups of sciences
different ways of introducing scientific theorems are obligatory as well
as different norms for construction of language and the conceptual
apparatus connected with it. These norms are not the product nor the
decree of methodology. Methodology does not establish these norms –
it discovers them in the practice of the historically existing sciences.
The norms are known to the specialists in the respective fields, who
do not, however, reflect on their function, but just know their practical
application like people with a practical knowledge of a language also
practically know its grammatical norms. This practical knowledge of the
norms consists in the ability to act according to them and in the ability to
observe violation of these norms. However, the practical knowledge
of norms is one thing and theoretical knowledge is another. The theoretical knowledge of the norms of scientific activities is achieved
by their codification, by their clear formulation. It is the methodologist's task to carry out the codification of the norms directing the practical activities of scientists, as it is the grammarian's task to codify
the norms applied in practice by people using a certain language.

9. Methodologists have been able to obtain such a codification for
the *a priori* sciences. We are able to provide both the norms defining
the way of conctructing languages of such sciences, i.e. also conceptual
apparatuses connected with them, and the norms defining the way
of deriving their theorems. Thanks to this the *a priori* sciences are given
the form of deductive systems. A given science becomes – as we have
said above – a deductive system, when its rules of formation, stating
which expressions belong to it and, in particular, which expressions are
its sentences are provided, as well as its rules of proof, stating when
a sentence is a theorem of this science. Now, the above mentioned
norms defining the way of constructing the language of science serve
the function of the rules of formation, and the norms defining the way
of deriving theorems correspond to the rules of proof. Since methodologists have been able to provide the rules of formation and the
rules of proof for the *a priori* sciences, they have obtained so much

knowledge about these sciences that it may form the starting point for metascientific investigations.

What, then, is the relation between traditional methodology and contemporary metascience? Traditional methodology deals with the historically existing sciences and its task is to analyse them in a way leading to a clear codification of the norms concerning construction of their languages and conceptual apparatuses connected with them as well as to codification of the norms specifying the requirements to be met to accept some sentence as a theorem of a given science. Metascience deals with the sciences for which these norms have already been clearly codified and it takes these norms as the starting point for its deductions, not infrequently leading to the discovery of interesting properties of the science under investigation. Thus traditional methodology is a sort of propaedeutics for metascience. Traditional methodology is an empirical science and maybe also an 'understanding' (in the sense of Spranger) science which tries to 'understand' the aim of the scientists who, though they do not realize it clearly, are led by it in the choice and evaluation of their procedures. Metascience, on the other hand, is a deductive science which takes the results achieved by traditional methodology as the starting point for its deductions. Thus, the relation between traditional methodology and metascience resembles, in a way, the relation between experimental and theoretical physics. As experimental physics guarantees to theoretical physics its connection with reality, so traditional methodology saves metascience from plunging into speculations on constructions detached from reality. Both these disciplines are necessary, they complement and fertilize each other.

10. The achievements of methodology in the field of the empirical sciences are not so promising. The methodology of empirical sciences has not gone far beyond the purely descriptive phase. We are not able to give exhaustive norms defining the way of constructing languages of these sciences. Although we say that all the sentences of the empirical sciences must possess empirical sense, this norm is formulated rather vaguely and there is a lively discussion among methodologists today on how to make it more precise. The argument is about whether the postulate that sentences of the empirical sciences should have an empirical sense requires that they should be derivable from observation sentences, or, whether it is sufficient if they are refutable on the basis

of such sentences. We are also unable to formulate norms stating conditions sufficient to raise some sentence to the status of a scientific theorem. We argue as to whether the empirical sciences have the structure of a building, whose firm foundation are certain observational sentences from which every other theorem must be derived in some way, or, whether the structure of the empirical sciences resembles a net rather, consisting of observation sentences, empirical generalizations and hypotheses, where the category of observational sentences is not privileged in the sense that in the case of conflict with the other sentences, the others must be rejected, and the observational sentences must be ratained at all costs. Furthermore, we are not able to formulate the difference between good and bad induction. It is probably in this example that the difference between the maturity of the methodology of *a priori* sciences and the immaturity of the methodology of empirical sciences is most apparent. As far as the *a priori* sciences are concerned we are able to decide which way of inferring is correct. The condition of correctness is not the same for all of them. In general, we may say, however, that in the *a priori* sciences any inference from whose premises the conclusion follows logically is acceptable. Such inference is called deductive inference. In the case of the empirical sciences, we can see that not only deductive inference, but also some other forms of inference are acceptable, e.g., some forms of inductive and deductive inference as well as some forms of inference by analogy, and others. However, we are not able to indicate precisely, e.g., which forms of inductive inferences are acceptable. The concept of inductive inference possesses only descriptive character, while the concept of deductive inference possesses not only a descriptive but a normative character as well. Deduction and sound deduction is the same. Bad deduction is not deduction at all, like a false diamond is not a diamond at all. On the other hand, induction may be sound or bad, and still be induction. For a long time now, methodologists have been looking for the characteristic distinguishing good inductions, i.e. the ones whose conclusions are accepted by science, and bad inductions. The search for this characteristic is one of the most frequently discussed methodological problems, known as the problem of induction.

The state of affairs is similar in many other fields of the methodology of empirical sciences. It is still at the level of description and only in

several points has it managed to reach the normative phase. This seems to be the reason why we do not have a developed metascience of the empirical sciences so far. For metascience can only go ahead after the rules of constructing a language and the rules of justifying theorems have been formulated.

11. As already stated, methodology codifies certain norms concerning the procedures used in cultivating the sciences. However, it codifies the norms which are practically known to a learned specialist. Would this theoretical knowledge of the norms help him in any way? I do not intend to discuss this question in detail. I shall content myself with saying that this question is analogical to the problem of whether a theoretical knowledge of grammar would be of any help to one who knows some language practically. However, apart from these norms of procedures in science, which are codified by the methodologist, the student of a given special science comes across certain methods of procedure dictated by an experienced expert in the field. The professor of history teaches the studednt how he should decipher incunabula, the professor of zoology teaches his student how to prepare samples and use a microscope, the professor of physics teaches the use of analytical scales, etc. Thus in each science we find norms concerning the procedures of practicing it which are beyond the competence of a methodologist: they are within the competence of a specialist. The discipline including such norms is sometimes (ambiguously) called the methodics of a given science. Where should we draw the border line between methodology and the methodics of a given science?

It seems that wherever we draw the line between methodology and the methodics of a given science, it will always be, in a sense, arbitrary, for the border line between these disciplines is not distinct. However, since in the introductory division between the norms of methodology and methodics we have used the criterion of who is competent to estab-lish them, the following suggestion would seem natural: we should include in the methodics of a given science those forms of investigation the justification of which must be based, among other things, on the theorems of this particular science, while in methodology we should include those norms of investigation which do not need reference to the theorems of any science. The norm dictating the avoidance of the error *quaternio terminorum* in a syllogism does not need reference to any particular

science for its justification. On the other hand, the norms concerning the procedures in establishing weight with the help of analytical scales, or the norms concerning the establishment of the percentage of sugar content with the help of a saccharimeter, are based on certain theorems from the field of physics. The norms of the methodics of empirical sciences would, in paricular, differ from methodology in that the former are always based on, among other things, empirical premises, while the latter are based only on a clear report on what is the essence of properly conducted research activity of a given type (e.g., what is the essence of good deduction or good definition), and constitute only detailed conclusions resulting from this statement.

Finally, I should like to emphasize that the above discussion is only an outline of certain ideas which require more detailed and deeper elaboration, explanation and justification. I realize that it will not be possible to maintain everything I have written here and that I shall have to give up some of the suggestions. Thus these notes are argumentative in character. Indeed, it is possible that they will provoke discussion that will help to explain these matters better. Thus may remarks are rather the opening for argument than a completed study.

REFERENCE

* First published in *Życie Nauki* (1948). Translated by E. Ronowicz.

K. AJDUKIEWICZ

THE PROBLEM OF THE RATIONALITY
OF FALLIBLE METHODS OF INFERENCE*

1. The discussion to which this article is devoted comes within the field of methodology and not of formal logic or metascience. We shall be dealing here with inference, convictions, with the degrees of certitude etc., i.e. the article will be full of psychological concepts. It will also deal with the people who think and act; the whole cognitive process will be considered in connection with the practical life of man and not as an idealized abstraction. The problem raised here is as follows: whether and when we can say that man acts rationally, if he believes conclusions drawn from true premises according to a fallible method, i.e. a method which may lead from true premises to false conclusions. So-called enumerative induction is one such fallible method of inference. Logicians and mathematicians have devoted much time to the question known as the problem of justification of induction. However, they have not been able so far to state clearly the nature of this problem: what should be done in order to justify induction? The aim of the present article is to give comprehensible meaning to the problem of justification of induction and, at the same time, to give a sample – unfortunately a sample that has no practical value – of solving this problem.

However, before we start discussing the problem itself, the terminology must be established and several auxiliary concepts should be introduced in order to avoid misunderstandings.

2. I shall start by explicating the term 'inference'. We call *inference* the activity of the mind consisting in that on the basis of accepting with some degree of certitude sentences called *premises*, the acceptance of another sentence, called *conclusion*, is reached with some, but always greater than before, degree of certitude.

The *degree of certitude* that accompanies the acceptance of some sentence is a feature of the subjective state of accepting it, i.e. of the belief in what this sentence states.

However, subjective degree of certitude may be measured objectively. Reasoning which leads to the establishment of this measure may be illustrated with the following example: let us suppose that we are confronted with the practical problem of whether we ought to start drilling an oil-well in some area, or whether we should give up the idea. Drilling will give in effect certain profit Z, if it is true that there is crude oil in this area, and certain loss S, if it is not true. The ratio of this loss and profit S/Z is the risk we undertake if we start drilling although we do not know whether there is crude oil or not in a given area. Now, the greater this risk, the greater must be our conviction that the state of affairs is such that it will bring profit, if we are to undertake the activity with which this risk is connected. When the risk is small we need but a small degree of certitude that the state of affairs is advantageous for us; when the risk is great we shall undertake the activity only, if the degree of certitude is high. Such and similar considerations would make us adopt as measure of the degree of certitude with which we accept sentence A the greatest risk S/Z, which we are ready to take by carrying out the activity which, in the case of sentence A being true, will bring profit Z and, in the case of sentence A being false, loss S.

Let us suppose that we are ready to bet that the weather will be good tomorrow and the terms of the bet are that, if the weather is good tomorrow, we get £2 from our partner in the bet and, if the weather is bad, we pay him £1, but that we refuse to bet under worse terms. The risk of this bet is 1/2. According to the way of measuring the degree of certitude established above we assume in this case that the degree of our certitude that the weather will be good tomorrow is 1/2.

By adopting as measurement of the degree of certitude the level of risk, i.e. the ratio of loss and profit we would obtain a scale of degrees of certitude oscillating between zero and infinity.

Readiness to act without any risk, i.e. readiness to start activity that will, if sentence A is true, bring profit other than zero and, if sentence A is false, loss equal to zero, would correspond to zero certitude about sentence A. Infinite certitude about sentence A would correspond to activity connected with an infinitely great risk, i.e. to activity which, if sentence A is true, does not bring any profit at all and, if it is false, causes some loss.

For certain reasons, however, which will become clear later, we prefer to assign to the degrees of certitude their numerical values in such a way that the scale of the degrees of certitude does not oscillate between zero and infinity, but between zero and one. We can obtain this by adopting as measure of the degree of certitude the ratio of loss and the sum of profit and loss instead of the ratio of loss and profit.

, Since,

if

$$\frac{S_1}{Z_1} > \frac{S_2}{Z_2},$$

then

$$\frac{S_1}{Z_1+S_1} > \frac{S_2}{Z_2+S_2} \quad \text{and vice versa,}$$

therefore higher degrees of certitude obtained in one of these measurement conventions will correspond to higher degrees of certitude obtained in the other convention. So in passing from the first measurement convention to the second we do not change anything in the ordering of the degrees of certitude with respect to their intensity.

Thus, we adopt the following definition of the measure of the degree of certitude with which a person X accepts a sentence A:

If a person X is ready to undertake some activity D if and only if this activity brings him at least a profit Z in the case of a sentence A being true, and when the sentence A is false at most a loss S, then we assign the measure

$$P_X(A) = \frac{S}{Z+S}$$

to the degree of certitude with which this person accepts sentence A. According to the above definition, the degree of certitude that makes us ready to undertake only the activities that are not connected with any risk, i.e. those where the expected profit $Z>0$, and the expected loss $S=0$, will be

$$\frac{0}{Z+0} = 0.$$

Respectively, the degree of certitude that makes us ready to undertake activities connected with an infinitely high risk, i.e. those where

the expected profit $Z=0$ and the expected loss $S>0$, will be

$$\frac{S}{0+S}=1.$$

If the activity D brings, in the case of the sentence A being true, the profit Z, and, in the case of its falsity, the loss S, then the same activity, will give, in the case of the truth of the sentence *non A*, the same effect as in the case of the sentence A being false, i.e. the loss S; and in the case of falsity of the sentence *non A* this activity will give the same effect as in the case of the sentence A being true, i.e. the profit Z. Since we can treat every loss as negative profit and every profit as negative loss, we may also say that, if the activity D brings, in the case of the sentence A being true, the profit Z, and, in the case of its falsity, the loss S, then the same activity will bring the profit $(-S)$ in the case of the truth of the sentence *non A*, and the loss $(-Z)$ in the case of its falsity.

Let us assume that

$$P_X(A)=\frac{S}{S+Z}.$$

This means that the person X is ready to undertake any activity which, in the case of the sentence A being true, brings him at least the profit Z, and, in the case of its falsity, at most the loss S. Or, in other words: the person X is ready to undertake any activity which, in the case of the sentence *non A* being true, brings at least the profit $(-S)$, and, in the case of its falsity, at most the loss $(-Z)$. This means, however, that the degree of certitude with which the person X accepts the sentence *non A* equals

$$P_X(non\ A)=\frac{-Z}{-S-Z}=\frac{Z}{S+Z}.$$

We have thus proved that, if

$$P_X(A)=\frac{S}{Z+S},$$

then

$$P_X(non\ A)=\frac{Z}{S+Z}=1-\frac{S}{S+Z}=1-P_X(A).$$

Let us now assume that the person X is completely undecided as to whether A or *non A* is true and he believes to an equal degree in A and *non A*. We then have

$$P_X(A) = P_X(non\ A),$$

i.e.

$$P_X(A) = 1 - P_X(A),$$

$$P_X(A) = \tfrac{1}{2}.$$

Thus, to the state of complete indecision as to whether A or *non A* is true corresponds the value 1/2. It is apparent that the way of measuring the degrees of certitude we have adopted leads to a scale similar to that of the scale of degrees of mathematical probability. This was exactly the reason why we passed from the first to the second way of assigning numerical values to the degrees of certitude.[1]

Any attempt of this kind to evaluate the degree of certitude with which the person X accepts the sentence A, according to the level of risk depending on the truth of this sentence, and connected with the activity this person is ready to undertake, may give rise to the objection that if somebody decides to undertake the activity connected with a certain risk depending on the truth of the sentence A, it depends not only on the degree of certitude with which he accepts this sentence, but also on his financial standing. Having the same degree of certitude a rich person is ready to risk a larger sum of money than a poor person. This objection would be justified, if we wanted to express profits and losses in absolute sums of money. However, the same amount of money represents different utility for people of different financial standing and of different needs. Thus, the profit and loss that were mentioned in our definition of the degree of certitude should be expressed not in absolute sums of money, but in units of utility. The concept of utility is widely discussed in mathematical economics. It has been discussed exhaustively by Carnap in § 51 of his work *The Logical Foundations of Probability* (second edition, Chicago 1951).

3. Let us now pass onto the establishment of certain terms and concepts concerning inference. I shall begin with the explication of the term 'inferential utterance'. I call an inferential utterance one that is the result of joining the sentences being the premises of some inference with the sentence which is its conclusion by means of the word 'hence' or some

expression synonymous to it. For instance, the utterance 'every human being is mortal, Socrates is a human being—hence Socrates is mortal' is an example of an inferential utterance. Instead of the word 'hence' I shall be using the symbol ' $\leadsto$ '.

By substituting in an inferential utterance variable symbols in place of one or more constants we obtain expressions called *schemes of inference*. I call *formal schemes of inference* those in which only logical constants occur in the premises and the conclusion.

If two inferences are expressed in terms of two inferential utterances that fall under the same scheme of inference and, if they, additionally, have the same degree of certitude for accepting respectively the premises and the conclusion, then we say that these inferences are carried out, in the same way or that they proceed according to the same *methods of inference*.

We shall now introduce the concept of the degree of reliability of a given scheme of inference. By *degree of reliability* of a given scheme of inference I shall mean the ratio between the number of values of the variables (or of the systems of values of the variables) occurring in this scheme which satisfy both the premises and the conclusion, and the number of those which satisfy the premises.

We shall symbolize the degree of reliability of the scheme '$P(x) \leadsto C(x)$' by

$$T\left[P(x) \leadsto C(x)\right].$$

Now, denoting by the symbol $N_x[F(x)]$ the number of those values of the variable x which satisfy the propositional formula $F(x)$, we may write

$$T\left[P(x) \leadsto C(x)\right] = \frac{N_x[P(x) \cdot C(x)]}{N_x[P(x)]}.$$

When, e.g., natural numbers are the values of the variable x

$$T\left[(1 < x < 10) \leadsto (3 < x < 6)\right] = \frac{N_x[(1 < x < 10) \cdot (3 < x < 6)]}{N_x(1 < x < 10)}$$

$$= \frac{2}{8}.$$

The degrees of reliability can rarely be computed in the way this has been done in the above example. Moreover, the concept of the

degree of reliability will not always be meaningful. Namely, with schemes of the type

$$P(x) \rightsquigarrow C(x)$$

the fraction $\dfrac{N_x[P(x) \cdot C(x)]}{N_x[P(x)]}$, which we use to measure the degree of reliability, will not have any definite value, if the number of values of the variable x which satisfy both the premises and the conclusion, is infinite. In those cases

$$T[P(x) \rightsquigarrow C(x)] = \frac{\infty}{\infty}.$$

In order to guarantee that the concept of the degree of reliability of a scheme is meaningful also in these cases, we would have to use the concept of the limit of a sequence in its definition. A similar thing is done in the case of the frequency definition of probability, when one passes from finite sets to infinite sets. This similarity is nothing extraordinary. For, our concept of the degree of reliability of a scheme of inference is exactly coextensive with the frequency concept of probability of conclusions derived according to this scheme, with respect to corresponding premises.

We have said that the degree of reliability of schemes of inference often cannot be practically computed. However, it frequently happens that although we cannot compute the degrees of reliability, we are able to compare the degrees of reliability of various schemes.

Let us take as an example the scheme of inference in which the premise is the statement that at least k objects x being S are at the same time P, and the conclusion is the general sentence that every S is P. This can be formulated as follows:

$$(*) \qquad \sum_x^k (x \varepsilon S \cdot x \varepsilon P) \rightsquigarrow S \subset P.$$

The above scheme may be considered a certain form of ordinary enumerative induction. Let us assume, for simplicity's sake, that we treat P as a constant. Then the degree of reliability of this scheme will be

$$T(*) = \frac{N_s[\sum_x^k (x \varepsilon S \cdot x \varepsilon P) \cdot S \subset P]}{N_s[\sum_x^k (x \varepsilon S \cdot x \varepsilon P)]}.$$

The above formula presents the ratio of the number of species S which have at least k members belonging to P and all of whose members

belong to P, and the number of those species S which have at least k members belonging to species P. Assuming that the number of species S is finite, it is easy to demonstrate that this ratio increases simultaneously with increase of k, i.e. that the larger the number of cases confirming the general conclusion to be found in the premise, the more reliable the schemes of enumerative induction become, or, in other words, the less unreliable they are.

It is obvious that the scheme (∗) is the least reliable when $k=0$, and this means that in inferring according to the scheme of enumerative induction we act in a more reliable way than if we stated the general conclusion independently of whether it had been asserted that there was at least one confirming case.

4. We now come to the most important concept in the course of our considerations, namely, the concept of rationality of methods of inference. The concept of rationality is not a specifically logical or methodological concept, but a general praxiological concept. For, we speak not only of a rational or irrational methods of inference, but also, generally, of a rational or irrational ways of acting.

When do we call human activity rational? We call it rational when it leads to the goal. Thus, the concept of rationality of action requires relativization to some goal. Some way of acting may be rational with respect to one goal, but irrational with respect to another. And we do not have to take into consideration the goals at which the agent aims consciously. For the agent often does not realize the goal he is aiming at, or *we* may not know the goal at which the agent aims consciously. We may then ask whether his activity would be rational, if this or that was his goal.

This is exactly how we are going to put our question concerning the rationality of the methods of inference. Namely, we shall ask when a certain method of inference would be rational, if we assumed that this method of inference is applied in order to achieve this or that goal. We shall assume in particular that the goal that we have in mind is practical in character, namely, that the goal taken into consideration is to obtain profit and to avoid losses. More precisely:

We shall accept as rational from the practical point of view a certain method of inference, if the balance of profits and losses resulting from the activities based on the conclusions obtained by this method from true

premises is not negative — after this method of inference has been applied for a long time.

5. Let us point out here that by the method of inference we understand a class of concrete inference characterized not only by a certain scheme of inference, but also by the degree of certitude with which the premises and conclusions are accepted.

Let us then take into consideration the method of inference described as follows: first — by some scheme of inference $P(x) \rightsquigarrow C(x)$, whose degree of certainty is k/l, second — by the fact that conclusions obtained according to this method of inference from true and absolutely certain premises have a degree of certitude not higher than $\dfrac{S}{Z+S}$.

The statement that the scheme $P(x) \rightsquigarrow C(x)$ has the degree of reliability k/l means that the ratio of the number of values of the variable x, for which both the premises and the conclusions are true, and the number of *all* the values, for which the premises are true, is k/l.

On the other hand, the statement that the degree of certitude of the conclusions derived according to this method from true premises does not exceed the value $\dfrac{S}{Z+S}$ means that, on the basis of accepting these conclusions, we are ready to undertake only such activities that will, in the case of our conclusions being true, give at least the profit Z, and, in the case of their falsity, at most the loss S.

The above characteristics of our method of inference will enable the computation of the lower limit of the balance of profits and losses resulting from the activities based on the acceptance of conclusions derived according to this method from true premises. Since these activities will, in the case of false conclusions, bring each time at the most the loss S, and, on the other hand, our conclusions will be k times true, and $l-k$ times false, then

$$\text{total profit} \geqslant k \cdot Z,$$

$$\text{total loss} \leqslant (l-k) \cdot S$$

and hence

$$\text{the balance of profits and losses} \geqslant k \cdot Z - (l-k) \cdot S.$$

The sufficient condition for this balance to be non-negative is that

$$k \cdot Z - (l-k) \cdot S \geqslant 0,$$

i.e. that $k \cdot Z \geqslant (l-k) \cdot S$.

Simple transformations of this inequality give

$$k \cdot Z \geqslant l \cdot S - k \cdot S,$$

$$k(Z+S) \geqslant l \cdot S,$$

$$\frac{k}{l} \geqslant \frac{S}{Z+S}.$$

This is a sufficient condition for the non-negativity of the balance of profits and losses resulting from the activities based on conclusions derived according to the above characterized method of inference from true premises, after it has been applied for a long time; thus it is a sufficient condition to accept this method of inference as rational from the practical point of view.

It is easy to see that the left side of the inequality (1) presents the degree of reliability of our method of inference, and the right side presents the upper limit of the degree of certitude for the conclusions derived according to this method of inference from true premises. Thus, we shall be able to formulate the result of our considerations in the following way: *A sufficient condition for some method of inference to be rational from the practical point of view is that the degree of certitude of conclusions derived according to this method from true premises does not exceed the degree of reliability of this method of inference.*

This is our answer to the question as to when a method of inference may be accepted as rational from the practical point of view. The above answer does not make the rationality of a certain method of inference dependent exclusively on the degree of its reliability, but on whether the degree of certitude with which the conclusions are accepted is defined with respect to the degree of reliability of this method of inference. And since we understand the degree of certitude as readiness to undertake more or less risky activities, our answer makes the rationality of a method of inference dependent on whether it leads to readiness to

undertake activities whose degree of risk corresponds to the degree of reliability of the method of inference.[2]

6. We have been dealing so far with the problem of the rationality of the methods of inference from the practical point of view. Naturally, rationality of methods of inference may be considered with respect to other goals than the practical goal. We may, e.g., having in mind a purely theoretical goal, the so-called recognition of truth and avoidance of falsity, accept as rational such, and only such methods of inference which from true premises will always lead us to true conclusions, and never to false ones. From this point of view we might accept as rational only the reliable methods of inference. We might weaken this theoretical goal and demand of the methods of inference that they lead more often from truth to truth than from truth to falsity. Then we would consider rational such, and only such methods to inference whose degree of reliability is more than half. The question as to what conditions the method of inference should fulfil will be − as can be seen from the above − different, depending on the goal we have in mind when considering its rationality, i.e. its usefullness.

7. I should also like to add that we have spoken only of the rationality of the ways of inferring and not of the rationality of concrete inferences. This concept may be defined in two ways. Namely, a concrete inference may be considered rational from the point of view of a given goal, if it realizes the goal. Then all and only those inferences which lead to true conclusions would be rational from the practical point of view. We then would have to accept as rational, e.g., the inference in which from the fact that 2 times 2 is 4 we would, by chance, reach a correct conclusion regarding the number of the ticket that would win the main prize in some lottery. Thus it would be a paradoxical concept. One might, however, consider rational, e.g., from the practical point of view, the concrete inferences which are made in the rational way from this point of view, i.e. in the way that guarantees a non-negative balance of profits and losses over a long period of time. Since by inferring according to a method that is rational from the practical point of view one might sometimes reach conclusions that make one ready to carry out harmful activity, if we accept this concept of the rationality of concrete inferences, we shall have to regard as rational certain inferences that make the achievement of the practical goal more difficult, and not vice versa

I must confess that this consequence does not seem paradoxical to me; that I do not consider the statement that somebody, although he acted rationally, did not reach his goal, to be in disagreement with the ordinary sense of the adjective 'rational'.

*

8. Finally, a few remarks devoted to the value of our considerations. Our considerations might rightly seem extremely primitive and elementary to anyone well acquainted with the theory of probability and literature devoted to the problem of probabilistic justification of induction. Our point consists in treating ways of inferring as certain systems of gambling, where one may win or lose. And the development of this idea is simply the repetition of what was said of the conditions of fairness of the game systems long ago. It has been known for a long time that a game system is fair, if the mathematical expectation of winning equals the mathematical expectation of losing. And a system is advantageous for the player and disadvantageous for the banker, if the mathematical expectation of winning is not less than the mathematical expectation of losing. The condition for the rationality of a method of inference we have given

$$k \cdot Z \geqslant (l-k) \cdot S$$

is only a different formulation of this condition. For if we divide both sides of the inequality by l, we shall obtain

$$\frac{k}{l} \cdot Z \geqslant \frac{l-k}{l} \cdot S .$$

But k/l is the ratio of the number of correct conclusions and the number of all conclusions, and hence the probability of the conclusion being correct, i.e. the probability of winning. Hence $k/l \cdot Z$ is the same as the mathematical expectation of winning. Similarly $((l-k)/l) \cdot S$ is the mathematical expectation of losing. Thus our condition for the rationality of some way of inferring is coextensive with the well known condition of the advantageousness of a game system.

Thus there are no revelations in what we have said here — it is simply the application of what has long been known about game systems to

ways of inferring, which are treated here in exactly the same way as gambling systems.

The arguments presented here may also be given a low evaluation because the criterion of rationality we have obtained cannot, in practice, be applied to such methods of inference which, like enumerative induction, are a matter of central interest to methodologists. And our criterion of rationality of the methods of inference cannot be applied, because the degree of reliability of these methods of inference cannot be practically established. This negative evaluation may be weakened, though, by the remark that in many cases where the degree of reliability cannot be computed one may compare the degrees of reliability of the different methods of inference which, as a result, facilitates the evaluation of which of two methods of inference is more rational. Nevertheless this remark indicates the weak side of our results.

In spite of this, however, I feel I may venture to say that at least three results worthy of attention have been obtained here.

The first is the precisation of the sense of the problem of the legitimacy of induction and other fallible methods of inference. This problem has worried logicians and methodologists at least since Hume's time; and I must admit that I was never quite sure what the problem actually was. Max Black spoke about this vagueness at one of the recent international philosophical congresses, but he did not clarify the problem either.

It seems that the problem of the legitimacy, i.e. rationality of the fallible methods of inference has been expressed clearly in my considerations. And it has been expressed clearly by subsuming methods of inference under the general praxiological concept of methods of acting as well as by taking advantage of the idea that a course of action is good from some point of view, if it is useful in the long run for the goal that determines our evaluation of this activity. This general praxiological concept of sound reasoning should — its seems to me — be the compass for all considerations on the logical correctness of any act of reasoning. This concerns not only methods of inference, but also such things as constructing concepts, defining, classifying, etc.

According to this approach, both the concept of rationality of some method of inference, and, generally, any concept of correctness of this or that cognitive activity requires relativization to the goal that is taken into consideration. If no such relativization is carried out, then

this goal is tacitly assumed. It is one of the most important tasks of methodology to find the aims which determine the actual evaluation of the various mental operations and their products in the sciences, if methodology aims at understanding the sciences as cultural phenomena in the humanistic sense.

The second point is the protest against the influence of the term 'probability' found in the literature devoted to the question of induction. Primarily a 'probable sentence' was understood as a sentence which it was rational to believe, though the meaning of the adjective 'rational' was only vaguely understood. This primitive content of the concept of probability has been entirely changed in the course of development of what is known as the mathematical theory of probability. Neither in the classical, nor in the frequency, nor, finally, in the measure-theoretic concept of probability, could we find any indication in its content suggesting clearly that what is highly probable deserves rational belief. In spite of this many logicians believed that the problem of rationality of the inductive or some other fallible method of inference would be solved if we could demonstrate that conclusions inductively obtained have a high degree of probability in the mathematical sense. This attitude is quite wrong, until the question has been answered why what is probable in the mathematical sense deserves rational belief. In order to answer this question, one must make precise the concept of sentence that deserves rational belief. This, however, is not done in most cases; under the influence of the term 'probability' it is believed that it suffices to demonstrate that some sentence is probable in the mathematical sense of the word, in order to demonstrate that it is probable in the primary sense of this word, i.e. that it deserves rational belief.

Drawing attention to the fact that the mathematical concept of probability is, as far as its content is concerned, different from the logical concept of probability is not our own discovery. Many authors have already drawn attention to the fact that it is not obvious at all that a sentence highly probable in the mathematical sense is, at the same time, a sentence that has a high degree of rational certitude. The point has been emphasized very strongly by Carnap who opposes the concept of logical probability (probability$_1$) to the concept of mathematical probability (probability$_2$). In his work *The Logical Foundations of Probability* quoted above Carnap has constructed a definition of the concept

of confirmation, treating the concept of confirmation as an explication of the intuitive concept of logical probability. He developed a theory of confirmation on the basis of this definition, which we call the logic of induction. But Carnap's definition of the degree of confirmation does not include anything that would allow the transition from a sentence belonging to metalanguage and attributing to hypothesis h under an assumption e a high degree of confirmation, to the adoption of any assertive attitude towards the sentence h belonging to the object language, when we accept the sentence e. Consequently, the same question arises with respect to the concept of confirmation, that arose with respect to the concept of probability: why we are allowed to and should accept with a high degree of rational certitude sentences which have a high degree of confirmation. Until we are given an answer to this question, the whole theory of confirmation will not deserve the name of 'logic of induction'. Carnap draws a parallel between the theory of logical consequence constructed in metalanguage and his theory of confirmation, which he treats as logic of induction. But the theory of logical consequence deserves the name of logic of deduction, since metalinguistic theorems stating that a sentence b is a logical consequence of a sentence a enable us to evaluate as a rational procedure inference in which, on the basis of the accepted sentence a we reach the acceptance of the sentence b. In other words, the metalinguistic theory of logical consequence deserves the name of logic of deduction, since, on the basis of its theorems, one may justify the rationality of the rules of deductive inference. In an analogous way, metalinguistic theory of confirmation would deserve to be called logic of induction only if its theorems would make it possible to show the rationality of the methods of inductive inference. But in order to do this, we must span the gap between the sentences of metalanguage assigning to the sentence h under the assumption e some degree of confirmation, and the rationality of some rule of induction that, on the basis of accepting the sentence e, allows acceptance with some degree of certitude of the sentence h in the object language.

I also venture to say that a result of this article consists in drawing attention to the fact that neither the mathematical theory of probability, nor the theory of confirmation are logic of induction until the gap has been spanned allowing, on the basis of the theorems of one or the other of these theories, to show the rationality of the rules of inductive

inference. But, in order to span this gap an analysis of the concept of rationality of a rule of induction must be carried out. Such an analysis has been carried out here and it was only this analysis that made it possible to bridge the gap between the frequency theory of probability and the rational rules, or methods, of inductive inference, though it was a primitive analysis, and applicable only to the simplest situations. The bridge is the theorem that a method of inference that allows, on the basis of premises P, to accept a conclusion C with a degree of certitude not higher than the degree of reliability of this method — is a rational method. Since the degree of reliability of a method is the relative frequency of true conclusions, derived according to this method from true premises, among all the conclusions derived according to this method from true premises, this theorem links the theorems concerning frequency probability with the rationality of certain inductive methods. If the way by which the gap between the mathematical theory of probability and the rationality of the rules of inductive inference has been bridged is, in principle, correct, then the frequency theory of probability should be considered logic of induction. The construction of some other theory of confirmation would then not be necessary.

The third idea worth the reader's attention is — I believe — as follows. It is generally believed that a fallible method of inference may be considered correct only if it very often leads to true conclusions, or at least if it leads more often to truth than to falsity. According to our approach to the problem, in which we have distinguished different degrees of certitude, also a method of inference which very rarely leads to truth may be considered rational, provided that the degree of certitude with which the conclusion is accepted is adequately low. I believe this particular detail to be important and corresponding to actually used methods of inference.

Contemporary statisticians refuse to acknowledge the existence of any fallible methods of inference. They state that in cases in which the logicians speak of fallible methods of inference, we do not deal with the acceptance of certain conclusions on the basis of certain premises, but with the decision to act. Now, the way in which we have defined here the degree of certitude of accepting sentences seems to erase to some extent the difference between accepting a sentence and the decision to act

in a way that will bring profit or loss depending on the truth of this sentence.

These are the three ideas contained in the article to which I wanted to draw the reader's attention.

REFERENCES

* First published in *Studia Filozoficzne* **4** (7) (1958). Translated by E. Ronowicz.

[1] A. Shimony in his article entitled 'Coherence and the Axioms of Confirmation' (*The Journal of Symbolic Logic* **20** (1955), 7) gives the definition of partial relative belief, which does not differ essentially from our definition of the degree of certitude with which a person X accepts a sentence h. Namely, Shimony states that x believes h on the basis of e in the degree r (symbolically: $B(h/e)=r$) "if and only if X would accept a bet on h on the following terms, or on terms more favorable to himself than these:

a) To pay rS and collect nothing for a net gain of $-rS$ in case e is true and h is false;

b) To pay rS and collect S, for a net gain of $(1-r)S$, in case e and h are both true;

c) To annul the bet, i.e., to have a net gain of 0, in case e is false."

It is easy to notice that

$$r = \frac{r \cdot S}{r \cdot S + (1-r)S} = \frac{\text{loss}}{\text{loss} + \text{profit}}$$

since negative profit is the same as loss.

The difference between the way of measuring the degree of certitude we have adopted here and Shimony's way consists only in that Shimony gives the way of measuring the degree of certitude of the sentence h given the assumption e, and we give the way of measuring the absolute degree of certitude.

[2] It seems natural that the highest degree of certitude with which we may, acting rationally, accept a sentence h as the conclusion derived from premises e according to the scheme of inference S should be called the logical probability of the sentence h with respect to the sentence e and the scheme S (symbolically $P_S(h/e)$).

Using this terminology one might derive the following consequence from our theorem formulating the condition of rationality of a method of inference: the logical probability of the sentence h with respect to the sentence e and the scheme S equals the degree of reliability of the scheme S:

$$P_S(h/e) = T(S).$$

The theorem puts the sign of equality between logical probability and frequency probability, since the degree of reliability of the scheme of inference is — as we have said above — frequency probability.

Carnap uses this or a very similar concept of logical probability in one of the unofficial explications of this concept, which he gives in a paragraph of the work quoted above. The reasoning of this paragraph is very similar to our considerations aimed at establishing the condition of the rationality of fallible methods of inference.

K. AJDUKIEWICZ

THE PROBLEM OF JUSTIFYING ANALYTIC
SENTENCES*

1. The aim of the present article is to answer the question as to whether terminological conventions are a sufficient justification for analytic sentences. We shall start by explaining the meaning of the terms used in formulating this problem.

2. We shall begin by making precise the meaning of the term 'analytic sentence'. The view has been expressed in the works in the field of logic that have appeared during the last few years that attempts to define the term present considerable difficulties; it requires reference to such concepts as the meaning of terms, or synonymy of terms — concepts, which cannot be explicated precisely. We believe that the difficulties encountered in attempts to define the analytic sentence are connected with the tendency to treat the concept of analytic sentence as an absolute concept; however, this concept — like most of the concepts of semiotics — should be understood as a relative concept. No sentence is simply analytic or non-analytic, just—as no man is simply a friend, father or son, though he may be one in relation to a certain person.

There are many ways of relativization of the concept of analytic sentence. One can say that a certain sentence is analytic for a given person, or in a given language, if only the concept of language is properly understood. Nevertheless, there is always actually one and the same relativization at the basis of all these relativizations, namely, relativization to certain terminological conventions.

By a terminological convention we understand here a statement concerning the way we are going to use certain terms. Terminological conventions may appear in both semantic or syntactic formulations. For instance, the statement: I have decided to use the word 'centimetre' as the name for the length equal to '1/100th part of a metre' is a semantic convention, for it concerns the semantic relation between the name 'centimetre' and the object that this name is to symbolize. On the other

31

hand, the convention: I have decided to use the word 'centimetre' in the same sense as the expression '1/100th part of a metre' — is a syntactic convention, for it concerns the relation between two expressions.

Let us now adopt the following convention concerning the way we are going to use the term 'language': if two persons use the same body of simple terms, observe the same rules for constructing compound expressions of them, and they accept the same terminological conventions, then we shall say that these persons speak the same language.

3. The concept of analytic sentence may be defined in two ways, depending on whether the convention adopted is semantic or syntactic in character.

In order to construct the first of these definitions we shall need the concept of postulate, which we shall explicate at the beginning. We shall say that a sentence Z is a postulate in a language J, if there is in the language J such a terminological convention which establishes that a certain term λ occurring in the sentence Z is to denote the object which, in place of the term λ, satisfies the sentence Z.

If, for instance, we agree that in the English language we call 'centimetre' the length satisfying the condition: 'centimetre is the length of 1/100th part of a metre', then this last sentence is a postulate of the English language.

The definition of a postulate may be generalized to make it apply not only to one term (the one denoting the object satisfying a certain condition), but to several terms at once. If, e.g., there is an agreement according to which the terms 'point', 'straight line', 'passes through' possess in the language of geometry denotations, which in adequate places satisfy the condition 'only one straight line passes through two points' this sentence will also be a postulate.

4. We have used the term 'denotation' in formulating the definition of a postulate, which also requires explication. Now, we shall call the denotation of a proper name (i.e. of an individual name, which is not a predicate) an object whose name this proper name is. For instance the denotation of the proper name 'Socrates' is Socrates, the philosopher. We shall call the denotation of a predicate (in the noun form, like 'alive', or in the verb form, like 'lives') the class of objects of which this predicate is true, in other words, the extension of this predicate. As a result, the

denotation of the predicate 'lives' is the class of objects of which this predicate is true, i.e. the class of living things. The denotation of a two-place predicate, such as 'loves', is the class of ordered pairs of objects to which this predicate is applicable, i.e. the class of such pairs $\langle x, y \rangle$ that x loves y. Each proper name and each predicate has one and only one denotation. The denotation of a proper name is always an individual object; the denotation of a predicate is always a class. Instead of saying that an object A is the denotation of an expression B we shall say that the expression B denotes, names, or symbolizes the object A.

5. We are now prepared to formulate the first definition of analytic sentence. We shall say that a sentence Z in a language J is *analytic* in the semantic sense, if it is a postulate of this language, or the logical consequence of one or several such postulates. The concept of postulate appearing in this definition is a semantic concept, since its definition refers to terminological conventions of a semantic character. We shall thus call the above definition of analytic sentence the *semantic concept of analytic sentence*.

6. We shall now attempt another definition of analytic sentence. This definition will not refer to terminological conventions of a semantic character, but to conventions of a syntactic character, and so we shall call the concept thus defined the *syntactic concept of analytic sentence*. Before we give the actual definition we shall introduce the definition of the auxiliary notion of logical truth. We shall call a sentence Z in a language J a *logical truth* if Z is a logical tautology, or if it is obtained from a logical tautology by substituting in it some constant descriptive term of the language J in the place of a variable. Thus the logical truths of the English language will be sentences 'Every A, which is not B, is not B', and 'Every man, who is not married, is not married'. Using such a concept of logical truth we shall define the syntactic concept of analytic sentence in the following way:

A sentence Z in a language J is analytic in the syntactic sense if it is a logical truth in this language, or it can be reduced to a logical truth by means of the syntactic terminological conventions adopted in this language.

According to this definition not only the above examples of logical truths will be analytic sentences in the English language, but also, e.g.,

the sentence: 'Every bachelor is unmarried'; for in the English language there is a syntactic terminological convention which allows for inter-changeable use of the expressions 'bachelor' and 'unmarried man', and hence the sentence 'Every bachelor is unmarried' may be brought down to the logical truth 'Every unmarried man is unmarried'.

7. Following the above terminological explications we shall now attempt to solve our main problem: are analytic sentences of a certain language sufficiently justified in terms of terminological conventions adopted in this language? In other words, we are interested in whether a sentence of a certain language must necessarily be true if there are terminological conventions in this language that guarantee analytic character to this sentence.

The problem has to be considered separately for the cases of analytic sentences in the semantic and the syntactic sense.

8. We shall start with sentences that are analytic in the semantic sense. Here is the definition again. A sentence $F(\lambda)$ in a language J is analytic if it is a postulate, or a logical consequence of postulates of the language J. In turn, the sentence $F(\lambda)$ in the language J is a postulate, if there is a convention in this language establishing that the term λ is to denote the object satisfying the condition $F(\lambda)$. It seems quite easy to demonstrate with the help of this definition that the postulates of a certain language must be true in this language. For, the following reasoning seems to be convincing at first glance:

I. If the sentence $F(\lambda)$ in the language J is a postulate, then this language has a convention establishing that the term λ is to denote an object satisfying the condition $F(\lambda)$.

II. If there is a convention in the language J establishing that the term λ is to denote an object satisfying the condition $F(\lambda)$, then the term λ in the language J denotes an object satisfying the condition $F(\lambda)$.

III. If the term λ in the language J denotes an object satisfying the condition $F(\lambda)$, then the sentence $F(\lambda)$ is true in the language J.

Ergo: If the sentence $F(\lambda)$ in the language J is a postulate, then the sentence $F(\lambda)$ in the language J is true.

How should we evaluate this reasoning which leads to a positive answer to our basic question? It is undoubtedly correct from the formal point of view, for it is an example of a correct hypothetical syllogism.

The first premise is the consequence of the definition of postulate and does not give rise to any objections. Also the third premise does not give rise to any objections: stating that the term λ denotes an object satisfying the condition $F(\lambda)$ is the same as stating that $F(\lambda)$ is true. But what is the case with the second premise? At first glance it also seems to be convincing. Indeed, when we construct some language, we are bound to believe that we are absolutely free to choose the objects which are to be the denotations of the particular words of this language. Hence, if I decide that in my language – which is a language at least partially constructed by myself – the term λ is to denote this or that object, then this term will denote this object.

But is my freedom really so very unrestricted? Let us take into consideration two different situations, in which we establish the connections between the expressions of a certain language and their corresponding objects. The first situation consists in that I have a given object and I want to assign some word to it, which is to symbolize it. In this case I am absolutely free to give to this object any name I like, and if I do not like any name, I may devise a name for it. The situation is different, however, if I have a given word and I want to assign to it some object as its denotation. Now I cannot choose any object, but only an object that exists. If an object satisfying the conditions I have devised does not exist, then my decision to assign to a given word this object as its denotation is not enough to create this object. Hence the decision that the term λ will denote the object satisfying the condition $F(\lambda)$ is not sufficient for it to be the case, if no object satisfies the condition $F(\lambda)$.

It is clear, then, that the second premise of the reasoning that has lead us to the positive solution of our problem is false. It is not true that, if only the language J has a convention establishing that the term λ is to denote the object satisfying the condition $F(\lambda)$, then this term denotes this object. Instead of the above premise we should adopt the following: if the language J has a convention establishing that the term λ is to denote the object satisfying the condition $F(\lambda)$, and, moreover, if there exists an object satisfying this condition, then the term λ in language J denotes the object satisfying the condition $F(\lambda)$.

As a result of the above considerations we may say that the terminological convention itself is not sufficient to guarantee truth to a postulate based on it. It gives such guarantee only if it is accompanied by an

existential sentence stating that there exists an object satisfying a given postulate.

9. In order to support the result of our considerations we may demonstrate the paradoxical consequences of the view, according to which the terminological convention itself gives sufficient guarantee of the truth of a postulate based on it. If this were so, one might prove the existence of objects satisfying any condition $F(x)$.

For it would suffice to decide that a certain term λ should denote an object satisfying a given condition in order to state without risking a mistake that $F(\lambda)$. Then, using the well known law of logic

$$F(\lambda) \rightarrow \underset{x}{\Sigma} F(x)$$

we would prove that

$$\underset{x}{\Sigma} F(x).$$

We might thus prove the existence of a human being 100 metres tall. It would suffice to decide, with this aim in view, that the term 'Polyphemus' should denote an object satisfying the condition:

Polyphemus is a human being 100 metres tall

and, given the above convention, state that

Polyphemus is a human being 100 metres tall.

Then the above mentioned law of logic will enable us to derive the conclusion that

there exists a human being 100 metres tall.

The result we have obtained so far reduces to the thesis stating that the terminological convention itself establishing that a certain term λ is to denote an object satisfying the condition $F(\lambda)$, is not sufficient to guarantee the truth of a sentence $F(\lambda)$ in the language where this convention has been adopted; it gives this guarantee only after it has been joined with the existential thesis $\underset{x}{\Sigma} F(x)$. The above result is not a revelation, however. It is only a generalization of the well-known theorem from the theory of definitions in mathematics. As is known, definitions in mathematics are nominal definitions, which are adopted on the

basis of terminological conventions, or which are terminological conventions themselves. These definitions may occur both in object and metalinguistic stylization. We call definitions in object stylization such definitions where the term being defined occurs itself and not its name; for instance, the sentence 'a square is an equilateral rectangle' is the definition of the term 'square' in object stylization. We call definitions in metalinguistic stylization such definitions where the term being defined does not occur, only its name. Such definition is the sentence: 'the term »square« means the same as the expression »equilateral rectangle«'.

The definitions in object stylization are adopted on the basis of semantic conventions. Thus the definition: 'a square is an equilateral rectangle' is adopted on the basis of the terminological convention establishing that the term 'square' is to symbolize the figure satisfying this condition. Generally speaking the definition $D(\lambda)$ of the term λ is adopted on the basis of a convention establishing that the term λ is to symbolize the object satisfying the condition $D(\lambda)$. It appears from this that mathematical definitions in object stylization are kinds of postulates. Now, the theory of definitions teaches us that the convention itself establishing that the term λ is to symbolize an object satisfying the condition $D(\lambda)$ does not yet enable us to accept a sentence $D(\lambda)$ as the definition of the term λ. We have the right to accept such a definition only after it has been proved that it satisfies definite conditions, called the condition of existence:

$$\sum_x D(x)$$

and the condition of uniqueness:

$$\prod_{x,y} [D(x) \cdot D(y) \rightarrow x = y].$$

The satisfaction of the condition of uniqueness is required considering the fact that sentence $D(\lambda)$ is to be the definition of the term λ, i.e. to be a unique characteristic of its denotation. By requiring the demonstration of existence before the definition is accepted, the theory of definitions puts upon real definitions (which, as we have noted, are postulates of a kind) the same qualifications we have put here upon all postulates.

It is clear, therefore, that the result of our considerations consists in extending to all postulates what is known about postulates being unique characteristics of the denotation of some term, i.e. about postulates which are definitions in object stylization. Nevertheless the above considerations are not completely devoid of value. For they allow a better understanding of why definitions or postulates cannot be accepted on the basis of convention alone, unless the demonstration of existence has been carried out. Moreover, the extension of the condition of existence on all postulates, not only definitions, carries important epistemological consequences, which I intend to deal with in another work.

10. We have demonstrated that terminological conventions alone are not sufficient to justify postulates. Hence, they are not sufficient to justify their logical consequences. But analytic sentences in the semantic sense have been defined as postulates or logical consequences of postulates. Thus our considerations lead to the conclusion that semantic conventions alone are not sufficient to justify analytic sentences in the semantic sense of the word; their justification requires the addition of corresponding existential sentences. We have thus solved the basic problem of our paper with respect to analytic sentences in the semantic sense. We shall not attempt to consider the problem with respect to analytic sentences defined syntactically.

11. Note that we have called analytic sentences in the syntactic sense of the word sentences that are logical truths, or sentences which can be derived from such truths by means of terminological conventions syntactic in character. By a 'terminological convention syntactic in character' we understand here any assertion stating that expressions A and B will be used interchangeably, i.e. in such a way that, if we accept a sentence where one of these expressions occurs, then we should also accept it after one of these expression has been substituted for the other in this sentence. Thus understood syntactic conventions are simply rules used to derive some sentences from others. Now, it is obvious that these rules themselves never suffice to justify any sentence; apart from these rules we always need premises from which a given sentence can be derived according to these rules. Hence, the terminological conventions of syntactic character alone are not sufficient to justify any sentence, and consequently, any analytic sentence. In order to justify

analytic sentences we need, apart from these conventions, the laws of logic and their substitutions, i.e. logical truths.

Thus, apart from the negative answer concerning our problem with respect to analytic sentences in the semantic sense, we have a negative answer to the problem with respect to analytic sentences in the syntactic sense: the terminological conventions alone are not sufficient to justify analytic sentences. The justification of any analytic sentence requires not only terminological conventions, but also certain premises. However, apart from this similarity there seems also to be a certain important difference between the analytic sentences of the above two types, to which I should like to draw the reader's special attention. We have demonstrated that in order to justify an analytic sentence in the semantic sense we need, apart from the terminological convention and the laws of logic, also a certain existential premise. Now, in order to justify an analytic sentence in the syntactic sense we need, apart from the terminological convention, only the laws of logic; no existential premise seems to be necessary.

12. The above mentioned difference has consequences which seem paradoxical, and which throw a certain light on the problem of definition in object stylization. We have said that real definitions in mathematics, i.e. conventional definitions, in which the term being defined occurs itself, and not its name, are postulates of a kind based on the semantic terminological conventions. As a result, their justification requires a certain existential premise. We have given as an example the sentence:

'a square is an equilateral rectangle'.

Now, it is easy to demonstrate that one may obtain real definitions with the help of syntactic conventions, justifying them with the help of the laws of logic themselves; hence real definitions may also be considered to be analytic sentences in the syntactic sense. An example will make it clear.

We adopt the terminological convention of syntactic character, establishing that the expression 'equilateral rectangle' may be substituted by the expression 'square' and vice versa in any arbitrarily chosen accepted sentence. By taking advantage of this convention we shall derive from the laws of logic alone the real definition

'a square is an equilateral rectangle'.

Here is the law of identity:

(1) $\underset{x}{\Pi}\,(x=x)$.

Hence we obtain by substitution:

(2)   'equilateral rectangle = equilateral rectangle'.

Now, taking advantage of the convention which allows the substitution of the expression 'equilateral rectangle' by the term 'square' we obtain from (2):

(3)   'square = equilateral rectangle'.

We have thus come to a sentence, which, as a postulate, seemed to require an existential premise. But we have obtained this sentence in such a way that an existential premise seems unnecessary.

This paradoxical conclusion makes us suspect that we have either made a mistake in the proof of the thesis stating that a sentence which is a postulate requires an existential premise, or that we have used implicitly an existential premise in our last demonstration.

13. This last supposition appears to be true. The way in which we have derived the real definition of the term 'square' from the terminological convention and the laws of logic, presupposes a certain existential premise. Such a premise is indispensable to pass correctly from the law of identity $\underset{x}{\Pi}\,(x=x)$ to its substitution instance: 'equilateral rectangle = equilateral rectangle'. We must not pass from a general sentence $\underset{x}{\Pi}\,f(x)$ to its substitution instance $f(a)$ unless we know that a exists.

This is connected with the meaning of the quantifier 'Π'. The general sentence '$\underset{x}{\Pi}\,f(x)$' means the same as 'for every object $x, f(x)$'. Hence we cannot derive from such a general sentence the singular conclusion '$f(a)$', unless we know that a is an object, i.e. that a exists and is unique:

$$\underset{x}{\Sigma}\,(x=a),$$

$$\underset{x,\,y}{\Pi}\,(x=a \cdot y=a \rightarrow x=y).$$

The first of the above premises is an existential one.

If the above demonstrated way of deriving real definitions from the laws of logic alone with the help of some terminological convention did not require an existential premise, it would be possible to prove the existence of any objects in this way.

We adopt the following syntactic convention:

> instead of writing 'the tallest of men over 100 metres tall' one is allowed to write 'Polyphemus', and vice versa.

Taking as the starting point the law of identity $\Pi_x (x=x)$ we obtain by substitution:

> 'the tallest of the people over 100 metres tall = the tallest of the people over 100 metres tall',

and hence, using the adopted terminological convention, we obtain:

> 'Polyphemus = the tallest of men over 100 metres tall'.

This sentence, in turn, due to the law of logic

$$F(a) \rightarrow \sum_x F(x)$$

leads to the existential sentence '$\sum_x (x =$ the tallest of men over 100 metres tall)'.

This obviously false conclusion indicates that we have made a mistake somewhere in the process of obtaining it. We have made this mistake by passing from the law of identity $\Pi_x (x=x)$ to one of its substitution instances: namely, we have substituted the name 'the tallest of men over 100 metres tall' for the variable x — a name that does not have a denotation. We have substituted this name without making sure before whether there exists an object that this name refers to, i.e. not knowing whether the existential premise has been satisfied. It is not satisfied in the case of Polyphemus and this is the way the above reasoning has lead us to an obviously false conclusion.

Everything indicates that an existential premise is equally necessary for the derivation of a real definition from the terminological convention of a syntactic character, and for the derivation of such a definition from a convention of a semantic character.

14. What we have said above about the real definitions derived from terminological conventions of a syntactic character concerns also the

overwhelming majority of analytic sentences in the syntactic sense. Note that they are either logical truths, i.e. the laws of logic and their substitution instances, or sentences which may be derived from the logical truths with the help of terminological conventions syntactic in character. We have to use an existential premise in order to justify the substitution for the law of logic; otherwise this step in the proof will not be correct. Hence such a premise is indispensable to justify any sentence we might want to derive from such a substitution, i.e., among others, also in cases when in further discussion we use terminological conventions of a syntactic character.

Thus, the difference between the justification of analytic sentences in the semantic sense and in the syntactic sense that we first looked for has proved illusory. The view that an existential premise is indispensable only in order to justify analytic sentences in the semantic sense, and unnecessary to justify analytic sentences in the syntactic sense has proved wrong. On the contrary: an existential premise is indispensable to justify any analytic sentence in the syntactic sense, unless this sentence is the law of logic; because it is indispensable to justify any substitution instance of a law of logic and any sentence that is reduced to such a substitution instance in terms of a syntactic convention; only such a premise enables us to apply the laws of logic to particular cases.

15. The fact that the derivation of a singular conclusion from a general sentence by substituting a constant expression for a variable requires an existential premise is generally tacitly assumed. The rule of substitution is always formulated in exclusively syntactic terms, without reference to the concepts of semantics, i.e. without the requirement that the substituted term should have a denotation. However, this purely syntactic rule of substitution is always formulated in a way that guarantees the satisfaction of this semantic condition.

What constant terms may be substituted for the variable according to the rule of substitution?

For sentence variables we are only allowed to substitute sentences, i.e. such constants, whose denotation is truth or falsity. For individual variables one is allowed to substitute, first, the constants which are proper names. Proper names are names which have been given to the individuals directly, i.e. they have been given to certain objects known through direct observation, and not by an indirect verbal description.

The evidence of experience, in which we are given objects referred to by proper names, is the guarantee of the existence of the denotations of these names. The rule of substituting individual descriptions for individual variables is formulated in various ways, but always so that it allows only for the substitution of non-empty descriptions, i.e. those having a denotation.

We may substitute for variables of the type of predicates general descriptions of the type '$(\hat{x})$ $(...x...)$' or '$(...x...)$', i.e. so-called propositional functions. However, we adopt at the same time a definitional axiom stating that for every propositional function there èxists a class, to which all and only those objects belong, which satisfy this function. This axiom guarantees denotation to every general description constructed of a propositional function.

We may also, under certain conditions, substitute for the variables of the type of predicates individual descriptions including variables of the type of predicates; there are also certain conditions which restrict the possibilities of substituting descriptions with variables of higher types; these conditions may be formulated in different ways but always in such a way that they do not allow the substitution of expressions devoid of denotation.

Thus the rules of substitution, despite the purely syntactic way of formulation, always guarantee denotation to those constant symbols that they allow to be substituted for the variables.

16. We have stated that terminological conventions and the laws of logic are not sufficient to justify analytic sentences either in the semantic or the syntactic sense; the only exception here are the laws of logic themselves. In order to justify any analytic sentence an existential premise is indispensable. Such a premise sometimes states the existence of an individual object satisfying a certain condition, and hence it has the form:

$$\sum_{x} f(x),$$

where the variable 'x' is of the type of individuals.

In other cases the existential premise states the existence of a class or relation determined by a certain propositional function which imposes a certain condition on its members or arguments. Then such

premise has the form:

$$\sum_f [f(x) \equiv_x (...x...)]$$

or

$$\sum_f [f(x_1 ... x_n ...) \equiv_{x_1 ... x_n} (... x_1 ... x_n ...)],$$

where the variable 'f' is of the type of predicates.

It may also happen that the existential premise states the existence of certain classes or relations, which themselves satisfy a certain condition. In this case the existential premise has the form

$$\sum_f F(f),$$

where the variable 'f' is of the type of predicates.

In the first case, i.e. when the existential premise has the form

$$\sum_x f(x)$$

and when the variable 'x' is of the type of individuals belonging to the real world, such premise must, it seems, always be justified empirically, directly or indirectly. We have here in mind a premise such as

$$\sum_x (x \text{ is red}),$$

$$\sum_x (x \text{ is a man at least 2 m tall}),$$

etc.

In the second case, i.e. when the existential premise has the form:

$$\sum_f [f(x) \equiv_x (... x ...)]$$

it is enough to refer to the laws of logic, and in particular the definitional axiom, which states that for every propositional function there exists a class determined by it. Thus we have to correct our last theorem, according to which the justification of an analytic sentence in the syntactic sense required, apart from terminological conventions and the laws of logic, an existential premise. Since the laws of logic provide the existential premise themselves, then these laws together with the conventions yield a sufficient justification.

In the third case, i.e. when the existential thesis has the form:

$$\Sigma F(f)$$
$$f$$

where 'f' is of the type of predicates, the problems of justification of such a thesis may be diversified. We believe, however, that there are cases, in which the justification of such a thesis requires reference to experience. I must admit that this last statement sounds rather unusual. For it states that, in some cases, it is indispensable to refer to experience to prove the existence of a class that has certain properties – the existence of the class itself, and not the fact that this class is non-empty.

We shall try to justify this statement by analysing the proof of existence necessary to adopt a conventional definition of the term 'gram'. We shall symbolize the definition by (D):

> 'a gram is the mass of one cubic centimetre of water in a temperature of 4°C'.

17. Let us assume that in order to introduce the definition of the term 'gram' the following convention (C) has been adopted:

> 'we decide that the denotation of the term 'gram' is to satisfy the following postulate: a gram is the mass of one cubic centimetre of water in a temperature of 4°C'.

According to this convention the denotation of the term 'gram' must satisfy the following conditions:

(1a) α is a mass,

(2a)   if x is one cubic centimetre of water in the temperature of 4°C, then x has the property α.

Introducing the abbreviations: 'M' for 'mass' and 'ccw4', for 'cubic centimetre of water in the temperature of 4°C' – the conditions (1a) and (2b) can be written in the following way:

(1b) $\alpha \in M$,

(2b) $\Pi(x \in \text{ccw4} \to x \in \alpha)$.
         x

The mass is a class of abstraction with respect to the relation of the equality of mass. This relation may be defined as follows: bodies A and

B are of equal mass, if A is identical with B, or if A and B, when acting upon each other, give to each other equal accelerations in opposite directions. The class of abstraction with respect to the relation to the equality of mass (in short 'RwM') is every such a class, to which all and only those bodies belong, that are equal with respect to mass. Thus, our terminological convention would assign to the term 'gram' as its denotation the class, which satisfies the following conditions:

(1c) $\alpha \in \mathrm{RwM}$,

(2c) $\prod\limits_{x}(x \in \mathrm{ccw4} \to x \in \alpha)$.

It should be noted that (1c) and (2c) characterize the class α by the conditions imposed on the class itself, and not on its members. Thus the theorem about the existence of such a class will not have the form:

$$\Sigma_{\alpha} \, [\alpha(x) \equiv_{x} (\ldots x \ldots)] \, ,$$

but the form

$$\Sigma_{\alpha} F(\alpha) \; .$$

Thus the existence of a class satisfying these two conditions can be proved by constructing as an example a certain class, whose existence may be demonstrated, and which satisfies these two conditions. In order to construct a class like this we take a certain existing cubic centimetre of water in a temperature of 4°C and call it w. We shall now formulate the propositional function

x equals w with respect to mass.

This function defines the class

$M_w = (\hat{x}) \; (x$ equals w with respect to mass$)$

whose existence is guaranteed by the definitional axiom. It can be seen immediately that M_w satisfies the condition (1c), i.e. that

(a) $M_w \in \mathrm{RwM}$.

In order to state that M_w satisfies also the condition (2c), i.e. that

(b) $\prod\limits_{x}(x \in \mathrm{ccw4} \to x \in M_w)$,

one must refer to experience, for only experience can decide whether all cubic centimetres of water in a temperature of 4°C belong to the class $M_w=(\hat{x})$ (x equals w with respect to mass), i.e. decide whether all cubic centimetres of water in the temperature of 4°C are equal, with respect to mass, to one such cubic centimetre, namely we.

Experience indicates that cubic centimetres of water in arbitrary temperatures are not all equal, with respect to mass, to one chosen from among them. However, if we take cubic centimetres of water in the same temperature, e.g., in the temperature of 4°C, then they are all equal with respect to mass, i.e. equal to one cubic centimetre chosen from among them, among others to w.

Thus we may state that M_w satisfies also the condition (2c), i.e. that

$$\text{(b)} \qquad \prod_x (x \in \text{ccw4} \rightarrow x \in M_w)$$

only with the help of experience. We may, on the basis of premises (a) and (b), stating that the class M_w satisfies the conditions (1c) and (2c), and according to the laws of logic, reach the conclusion that an object satisfying these conditions exists:

$$\text{(1)} \qquad \sum_\alpha \{\alpha \in \text{RwM} \wedge \prod_x (x \in \text{ccw4} \rightarrow x \in \alpha)\},$$

i.e., in other words, there exists a mass of any arbitrarily chosen cubic centimetre of water in the temperature 4°C. This is the existential premise that should be proved before adopting on the basis of the convention (C) the definition (D): 'a gram is the mass of any arbitrarily chosen cubic centimetre of water in the temperature of 4°C'.

We have seen that the premise concerns the existence of a class characterized by a condition imposed not on its members but on the class itself; hence the truth of this premise is not guaranteed by the definitional axiom and it requires proof. The example we have given here demonstrates that in order to prove existential theorems of this type (i.e. theorems of the form '$\sum_f F(f)$') one must, at least in some cases, refer to the results of experience.

18. The definition of a gram is an example of a postulate, i.e. an analytic sentence in the semantic sense. Since we have demonstrated

that the justification of this definition requires reference to experience, we must come to the conclusion that, in some cases, reference to the results of experience is indispensable to justify analytic sentences in the semantic sense and with this rather interesting conclusion, we close this paper.

REFERENCE

* First published in *Studia Logica* **VIII** (1958). Translated by E. Ronowicz.

K. AJDUKIEWICZ

AXIOMATIC SYSTEMS FROM THE
METHODOLOGICAL POINT OF VIEW*

1. METHODOLOGY AND METASCIENCE

There are two disciplines which make the axiomatic systems the object
of their research. One of them is the branch of mathematical logic
named metamathematics, or metalogic, or more generally metascience.
The other is the traditional methodology of sciences, which is treated
in separate chapters in the textbooks of logic and deals with problems
bordering upon the philosophical theory of knowledge. The achievements of metascience are so numerous, its notions so precise, and its
proofs so exact, that they have dissuaded many from examining deductive
systems from any point of view but that adopted by metascience; they
have produced an impression as if metascience could be identified with
the theory of deductive systems, and as if no other problems concerning
those systems could be stated but those treated by metascience.

In the present paper I wish, firstly, to give an account of the differences
between the point of view of metascience and that of traditional methodology; secondly, to try to find what is essential to methodological
problems; finally, to examine one of the problems of methodology,
which concerns axiomatic systems but essentially belongs to the domain
of the theory of knowledge.

The difference between the points of view of metascience and methodology consists chiefly in the fact that metascience does not take into
account the part played by man in the construction of science and considers science as a system of sentences, hereby meaning inscriptions
of definite forms, whose correspondents are the so-called models which
fulfil the system. Methodology, by contrast, investigates also the part
played by the human being that makes use of these sentences to acquire
knowledge of reality. From this difference it follows that of all the notions of semiotic or the logic of language (which includes semantics,

syntax, and pragmatics) in metascience there appear only the notions of syntax and semantics, while methodology has also recourse to the notions of pragmatics.

In metascience, of which metamathematics and metalogic are so far the only representatives, there appears the term 'sentence' but it is defined there in a purely morphological way, without any reference to what a sentence expresses, namely as an inscription of a specific form, determined by the so-called rules of the formation of expressions. In addition, metascience has recourse to the notion of 'derivability under the rules of transformation', a notion which is also defined in a strictly morphological way, viz. as a realtion holding between two sentences and with respect to their external form alone. Upon these purely morphological notions, that of a sentence and that of derivability in virtue of certain rules of transformation, several others are based. Such is the syntactic notion of a deductive system constructed according to certain rules of transformation, defined in metascience as the totality of sentences which are consequences, by virtue of these rules, of an empty class of sentences; the notion of axiom, which is defined as being a direct consequence of an empty class of sentences; the notion of proof, defined as a sequence of sentences which are consequences of axioms, etc. Besides these syntactic notions, which can be reduced to morphological ones, certain semantic notions are used in metascience, as e.g., the notion of an axiomatic system being satisfied by a model. On the basis of these notions a branch of science is constructed whose importance, theoretical at least, is considerable.

As we have said, metascience uses no pragmatic notions, i.e. no notions concerning the attitude of a thinking and acting human being towards linguistic expressions or towards what these expressions symbolize. Now, these notions play an essential part in the traditional methodology of science. Thus, for example, the notion of a sentence asserted by someone, completely alien to metatheory, is enormously important to methodology. It is true that sometimes the term 'asserted sentence' is used in metatheory too. So, for example, the *modus ponens* rule is sometimes formulated as follows: If one asserts a conditional sentence and its antecedent, one is also allowed to assert its consequent. But here the term 'asserted sentence' has quite a different meaning from that in methodology. There it denotes a sentence towards which

someone takes an assertive attitude, or — to explain in what this attitude consists – a sentence that someone is willing to make a basis of his practical activity. In metascience, by contrast, the term 'asserted sentence' means the same as 'thesis', or, more precisely, the same as 'sentence which is a direct or indirect consequence of an empty class of sentences'. It can be seen that this metascientific notion of an asserted sentence has no reference to any human attitude. Consequently, the expression 'one is allowed to assert', which appears in the above formulation of the *modus ponens* and which, if taken literally, has a normative character, lacks this character in metascience.

In metascience the statement 'if the sentence x is asserted, one is allowed to assert the sentence y' only declares that if the sentence x is a consequence of an empty class of sentences, so is the sentence y. Methodology, by contrast, deals with man's various attitudes towards sentences; consequently, it deals with a certain behaviour of man. It evaluates this behaviour and formulates norms concerning it. In particular, it evaluates this behaviour as to its efficiency in attaining given aims. In methodology the rule permitting the assertion of a consequent on the basis of the assertion of a conditional sentence and its antecedent is taken in its literal sense. By stating this rule, the methodologist evaluates the behaviour described in it as being efficient in attaining the cognitive aim and expresses his own approval of such behaviour.

The central notion of methodology, and one that does not appear at all in metascience, is the notion of the validation of asserted sentences or of the justification of their acceptance. This is a notion of an obviously normative and evaluative character. For to validate a sentence means to justify its acceptance by a method that guarantees the attainment of the aim pursued, e.g. the true knowledge of reality. So by calling a sentence validated, one states that one has justified its acceptance by a procedure which is efficient from the point of view of the aim pursued, that, therefore, this procedure deserves to be evaluated positively and that, consequently, from the point of view of the given aim, it is a good behaviour. Upon this evaluation a norm may be based which will allow this procedure to be applied whenever the same aim is pursued.

We see that methodology, unlike metascience, is a humanistic science. This is so for the following reasons: firstly, because it deals with human

behaviour, which consists in taking certain attitudes towards certain linguistic expressions; secondly, because it evaluates this behaviour and establishes the norms of good, that is to say, efficient behaviour. It evaluates not only the ways of validating sentences, but also those of forming notions, i.e. the definitions and classifications. It does so not only to know whether they satisfy so-called formal criteria, but also non-formal ones. If, for instance, methodology evaluates the logical 'division of a concept', it does so not only to know whether this division fulfils the condition of exhaustiveness (which requires that the extension of the divided concept be identical with the sum of the extensions of the members of the division) and that of exclusiveness (which demands that the members have no element in common). For these two conditions form a part of the definition of a logical division itself, and to examine whether a mental procedure fulfils them or not would be tantamount to examining whether this procedure is a division or not. Methodology also evaluates a logical division from another point of view: it asks whether the members of a division are useful notions, conducive to achieving the aim of the given research.

The aim, involved in methodological evaluations is not being understood here anthropomorphically, i.e. as an objective consciously pursued. Scientists whose behaviour it guides are often unable to define it. The teleologic terminology in this case is only metaphorical and serves to describe the actions of the scientists in question. When studying, for instance, the history of zoology or of botany, one sees that at various periods various classifications of animals or plants were suggested, but did not stand up to criticisms and were replaced by others. The old ones were evaluated as bad, the new ones as good, or at least as better. So the question arises, which was the quality that the old classification lacked and the new one possessed? If one succeeds in finding an answer to such a question, one may formulate it metaphorically in a sentence of the following form: zoologists 'seek' to build up a classification of animals that would have such-and-such a quality. Taken literally, this answer need not be true, for zoologists themselves may not even be aware of the quality they seek. To avoid the metaphor, one can say either that zoologists proceed *as if* they sought to build up a classification possessing the quality in question, or else that such is the not necessarily conscious aim, or the *quasi*-aim, of their behaviour.

Now, the study of these quasi-aims of scientific behaviour constitutes a kind of synthesis of the procedure of science, an attempt at presenting it as a system organized in the same way as are actions consciously directed by an aim to be attained. In this way the reconstruction of the actual course of scientific procedure belongs to those results which Spranger had in mind when he spoke of the 'humanistic understanding' (*das geisteswissenschaftliche Verstehen*) of cultural phenomena. Humanistic understanding is a form of synthesis peculiar to humanistic sciences, just as the construction of an explanatory theory (which allows all the laws of the given domain to be deduced from certain basic principles) is peculiar to natural science. The humanistic understanding of the development of various sciences and their branches is one of the chief tasks of the methodology of science; for it is a necessary condition of any categorical evaluation. This is so because an evaluation consists in stating that something is useful in achieving a certain aim. As long as the aim has not been fixed, nothing but conditional value-statements can be made, which affirm that such a thing is useful for such an aim. But as soon as one has revealed either a real or a quasi-aim, one can make categorical value-statements, evaluating as good whatever is conducive to the aim which directs the process under consideration. If methodology is to evaluate certain ways of reasoning and certain mental constructions, including definitions and classifications, as good and others as bad, it must attain the knowledge of the aims or quasi-aims, pursued by the sciences, the knowledge which is synonymous with the humanistic understanding of the sciences.

This is the third and most important reason why methodology ought to be regarded as a humanistic science. Unlike methodology, metascience is not at all a humanistic science. It deals with certain constructs which are called sentences, expressions, proofs, deductive systems, definitions, and theorems; the axioms of the metascience attribute to those constructs certain properties from which then further properties are derived. Consequently, metascience is a science of the mathematical type, very remote in its nature from humanistic disciplines.

Having pointed out the difference between metascience and methodology, let us proceed to our proper subject, namely to the consideration of the axiomatic systems from the methodological point of view.

2. THE SYNTACTIC AND THE PRAGMATIC NOTIONS
OF AN AXIOMATIC SYSTEM

The term 'axiomatic system' is used both in metascience and in methodology but in a different sense in each of them. In metascience it is defined in terms of syntactic notions alone, while in methodology pragmatic notions are also employed. In metascience a purely syntactic notion of an axiomatic system is used along with one defined in semantic terms. We shall from now on confine ourselves to considering the purely syntactic notion.

The syntactic notion of an axiomatic system in metascience is relativized to the rules of transformation of sentences, rules which in turn serve to define the relation of consequence. This relation, however, is understood in metascience in a peculiar way, quite different from those of ordinary language and traditional logic. In the latter, if one says that sentence B is a consequence of the sentence A, one asserts that the truth of A implies the truth of B. In metascience, firstly, one does not say that a sentence is a consequence of another one, but that a sentence is a consequence of a class of sentences; such a class may either consist of several elements, or be a unit class or else an empty class, i.e. have no elements at all. Secondly, one does not speak of a consequence in an absolute sense, but only with respect to certain rules of transformation. By saying that a sentence A is a consequence of a class of sentences α in virtue of the rules of transformation I, one means that by applying these rules of transformation to the class α a finite number of times one arrives at the sentence A. It is not required that these transformations should always lead from a class of true sentences to a true sentence, but that the relation of consequence defined by these rules should fulfil certain formal conditions, e.g. be reflexive, transitive, etc.

With the help of the notion of consequence in virtue of the rules of transformation I, the notion of an axiomatic system relativized to those rules is defined in metascience, namely as the totality of sentences which are consequences, in virtue of the rules I, of an empty class of sentences. We shall call this notion of an axiomatic system the syntactic notion. An axiom of a system based on the rules I is defined as a sentence at which one arrives from an empty class of sentences by once applying one of the rules I, i.e. as an immediate consequence of the empty class.

Sentences which are consequences of axioms, but are not themselves axioms, are called theorems.

By contrast, the notion of an axiomatic system in methodology is understood in such a way that it requires relativization not only to the rules of inference, but also to a person that constructs and develops the system. We shall refer to this notion of an axiomatic system as pragmatic. In methodology this noton can be defined as follows:

A set of sentences Z is an axiomatic system in the pragmatic sense for the person X if, firstly, Z is an axiomatic system in the syntactic sense relative to the rules of transformation I; secondly, the rules I are infallible, i.e. always lead from true sentences to true sentences; thirdly, the rules I are entirely convincing to the person X, i.e. whenever X is to a certain extent convinced of the truth of a given sentence he is to the same extent convinced of the truth of the sentence resulting from it by transformation under the rules; and, finally, the person X constructs and develops the system in question by formulating the axioms and deductively deriving their consequences, i.e. theorems.

The notion of the deductive derivation of one sentence from another used in the above definition requires some comments. It is a wider notion than that of deductive inference. Deductive inference consists in asserting the conclusion on the ground of asserting the premisses, the conclusion being accepted with the same degree of certainty as the premisses. The verbal expression of an inference takes the form of a so-called inferential statement: p, therefore q. This statement expresses, firstly, the acceptance of the motivating sentence p, i.e. of the premiss; secondly, the acceptance of the motivated sentence q, i.e. of the conclusion; finally, the fact that the acceptance of p is, for the person making the inference a motive to accept q. Deductive inference is but a special kind of deduction in general; another kind of deduction is the mental process expressed by statements of the type 'assume p – then q would also be the case'. Such a statement, unlike the inferential one 'p, therefore q', expresses neither the acceptance of the premiss p, nor of the conclusion q, but perely one's willingness to assert q if one asserts p. Mental processes expressed by such statements, and consisting in one's willingness to assert a certain sentence if one asserts another one can be called potential deductions to distinguish them from the ordinary deductive inference, which can be called actual deduction. We use potential deduction for

example in indirect proof, when we derive the consequences of the negation of the thesis to be proved, without thereby asserting – by contrast
to what is done in actual inference – either the negation of that thesis
or its consequences. Potential deduction occurs also in the procedure
of verifying hypotheses: in this case from a not yet accepted hypothesis,
we derive consequences which are not accepted either.

The general notion of deduction, which we have used to define the
pragmatic notion of a deductive system, is accordingly divided into two
kinds: actual deduction, or ordinary deductive inference, and potential
deduction.

3. CLASSIFICATION OF AXIOMATIC SYSTEMS
IN THE PRAGMATIC SENSE

With respect to the attitude of the person X that constructs or develops
an axiomatic system in the pragmatic sense, such a system may be either
assertive or neutral. It is assertive for the person X if X accepts both
its axioms and the theorems deduced from them; it is neutral for the
person that constructs it if that person has no assertive attitude either
towards the sentences of the system or towards their negations.

Assertive axiomatic systems can, in turn, be subdivided into deductive and reductive ones.

The definitions of these variants of assertive systems must be preceded
by an explanation of the difference between deductive and reductive
inference.

Deductive inference consists in accepting the consequent of a logically valid implication on the ground of having accepted the antecedent.
Reductive inference, by contrast, consists in asserting the antecedent
on the ground of having accepted the consequent of a logically valid
implication. The latter type of inference is used, e.g., in the natural
sciences when, in verifying a hypothesis, one is induced to accept it
on the basis of having ascertained that the consequences of that hypothesis are true.

Accordingly, an axiomatic system is assertive-deductive for the
person X if X accepts both its axioms and its theorems, but first accepts
the axioms and then derives the theorems from them and accepts the
latter also.

On the other hand, an axiomatic system is assertive-reductive for the person X if X first accepts at least some of its theorems and is only afterwards induced to accept its axioms by reduction on the ground of those theorems. This does not preclude, moreover, that from the axioms grounded by reduction on certain theorems, the person X may later on derive some other theorems of the system.

According to the most common view the axiomatic systems of mathematics are assertive-deductive. The methodological structure of these systems is claimed to be the following: first of all, and independently of the acceptance of the theorems, the axioms are accepted; then, by way of deduction, one is brought to accept the theorems on the ground of having accepted the axioms. It must be stressed that it is the methodological, and not the metascientific, structure of the axiomatic systems that is being discussed here. If one examines their methodological structure, the problem is to discover the *proteron pros hêmas*, i.e. to discover which sentences are accepted first thus providing the motive for accepting others. On the other hand, if the metascientific structure of the systems is examined, the question is to find the *proteron tê physei*, i.e. to ascertain the basis from which the theorems are derivable in virtue of the rules of transformation or, in other words, the assumptions from which, by these rules, the consequence follows.

Apart from the view according to which the axiomatic systems of mathematics are assertive-deductive, there is another view, which considers them to be neutral-deductive. According to this latter view, a mathematician constructing his axiomatic systems accepts neither the axioms nor the theorems, but takes a neutral attitude towards both; he only considers them; what they express are not his convictions, but simply such states of mind which the Austrian philosopher Meinong used to call '*Annahmen*'. In deducing theorems from axioms, the mathematician – according to this view – effects not an actual, but only a potential deduction. His reasoning may be expressed as follows: if it were true that A, it would be true that B.

The assertive-reductive methodological structure is characteristic of those axiomatic systems which are used to formulate theories of natural science, such as mechanics, thermodynamics, electrodynamics, etc. The methodological structure of the system of electrodynamics, for example, is the following. Concrete experiments provide the premisses; starting

from these, we arrive, by induction, at universal empirical laws; from these laws we obtain theoretical principles, which are by no means the consequences of the former; on the contrary, empirical laws are consequences of the principles. Proceeding in this way we accept, by reduction, some fundamental principles from which all the subordinate theoretical principles and all the empirical laws are derivable, so that these fundamental principles constitute the axioms of the system, while the subordinate principles and the empirical laws are the theorems.

In philosophical literature one can meet with protests against attempts to give the form of axiomatic systems to theories of natural sciences. These protests are based on the following argument: in axiomatic systems one begins by accepting the axioms from which then the theorems are deduced. By giving a theory of nature the form of such a system, one takes the basic principles of that theory a⁀ axioms and deduces its subordinate principles and empirical laws from these. Now, such procedure reverses the order of accepting theses of any scientific theory of nature, which always begins by asserting the most subordinate empirical laws, from which later on, by induction or by reduction, it draws all the principles, including the basic ones. It can be seen that this argumentation mistakes the notion of a deductive system for that of a reductive one. All it shows is that a theory of nature must not appear under the form of a deductive system; but it does not at all show that it must not take the form of any axiomatic system whatever. In reality, a scientific theory of nature can have, and sometimes has, the form of an axiomatic system in the syntactic sense of the word, which means that some of its sentences have consequences that, taken as a whole, are identical with the entire theory. It can also have, and sometimes has, the form of an axiomatic system in the pragmatic sense of the word, namely that of an assertive-reductive system.

The problem of validation or jsstification arises in methodology not only with respect to the theorems but also with respect to the axioms of assertive-deductive systems. For in those systems one is brought to assert the theorems by deriving them, in virtue of certain rules, from the axioms. If this procedure is to secure the truth of the theorems to which it leads, it is first of all necessary that the rules of inference applied should be truth-preserving, i.e. such that they never lead from true premisses to false conclusions; secondly, it is necessary that the

axioms or premises from which the theorems are derived should be accepted not in an arbitrary fashion but by applying a method that also ensures their truth; this, in turn, means that they must be adequately founded or justified.

As we have already said, the fact that the theorems are derived from the axioms in virtue of truth-preserving rules confers on the theorems validation only if the axioms are not merely true but had also been validated. However, is it indeed necessary to validate axioms? Does it not follow from their very definition – as the sentences which are not derived from anything in the given system, but from which, on the contrary, all other sentences of the system are derived – that they need no validation? To dissipate this doubt, two kinds of validation must be distinguished: direct and an indirect. Indirect validation consists in accepting a sentence on the basis of another one. In case of direct validation the ground for asserring a sentence is not the assertion of another sentence. If, for instance, I am stating at this moment that the sheet of paper in front of me is white, what convinces me of this fact is my perception of that sheet, and not the assertion of another sentence. Now the axioms of an assertive-deductive system are not validated indirectly by any other sentences of the same system. They can be validated indirectly only as theorems of another system, from whose axioms they are derivable. But even if they are derived from the axioms of another system, they are validated only provided the latter axioms are. As we see, the basis of any assertive-deductive system must, in ultimate analysis, be provided by the axioms which are no longer validated indirectly, i.e. are no longer inferred from other sentences, but whose validation is a direct one. Otherwise, we would either be involved in *regressus ad infinitum*, or base all our affirmations, ultimately, upon unfounded premises, thus committing the fallacy of *petitio principii*. Consequently, the possibility of constructing an assertive-deductive system, whose theorems would be justified, unavoidably depends on the existence of a direct method of foundation.

In philosophical literature several methods of justification are presented as direct. One of these is associated with statements immediately based on perception and called observational sentences. The perception accompanying them is believed not only to motivate their assertion subjectively, but also to secure their truth. Recently the question has

been raised whether there exist any observational sentences reporting 'pure facts' alone or whether every observational sentence is not already an interpretation of these facts, based not only upon the data of perception, but also upon previous knowledge, consequently upon certain asserted sentences. During the interwar period, the problem of observational sentences, or 'protocol sentences' as they were then called, was the object of a particularly animated discussion, in which their existence was generally doubted. It can be seen, therefore, that the problem of founding statements directly on the testimony of experience is highly controversial.

Philosophers who favour apriorism regard intuition as a method capable of directly justifying statements. This is questioned, in turn, by the adherents of empiricism. According to apriorists, besides sense experience in which particular objects and events alone are given to us, there is also experience through which we have knowledge of objects of higher order, such as universals, the so-called essences of things, etc. This other kind of experience was termed *'reine Anschauung'* by Kant, *'Wesensschau'* by Husserl. Others, who did not go so deeply into its analysis, called it intuition or simply self-evidence. As a method of direct foundation, it is called in question even more than sense experience.

Of all the methods of direct justification of statements, the least doubtful seems to be the third method, which consists in founding them upon terminological conventions. It can be described as follows: we decide that in the language we intend to use, a given term should denote an object, satisfying a condition, formulated in a sentence which involves the same term. For instance, we decide that in the language of geometry, the terms 'point', 'straight line', 'contains' should denote elements satisfying the following condition: 'If p and q are distinct points then there exists one and only one line containing p and q'. On the basis of this decision, called terminological convention, the sentence in question is asserted. This method can be more generally represented in the form of the following schema: we decide that in the language which is being constructed, the term 'λ' should denote an object that satisfies the condition '$F(\lambda)$'. On the basis of this decision, called terminological convention, one arrives at asserting the sentence '$F(\lambda)$' in the language that one is constructing. A sentence asserted on the ground of such a convention is called a postulate.

Now, it is commonly believed that this method of asserting sentences on the ground of terminological conventions secures their truth in the language in which the conventions hold. For, since I have decided that the term 'λ', in the language that I am myself constructing, should denote a certain object (viz. one that satisfies the condition '$F(\lambda)$'), it is generally believed that this will be so, i.e. that the term will really denote such an object and that the sentence '$F(\lambda)$' will be true in the given language.

Thus the method of asserting sentences on the ground of terminological conventions seems to ensure the truth of these sentences. On the other hand, it seems to be a method of direct justification: To assert a sentence, no recourse is made to any other sentence as a premiss but only to a terminological convention; the latter is a decision, an act of will, the verbal statement of which is not an indicative sentence, affirming or denying something, but rather the following imperative sentence: 'Let the term 'λ' denote an object that satisfies the condition '$F(\lambda)$'.'

Consequently, since the method of arriving at a theorem on the ground of a terminological convention secures the truth of that theorem and, moreover, founds it upon no sentence as a premiss, one would be willing to consider it a method of direct validation.

It seems that many authors who reject the method of intuition believe that an assertive-deductive system can be constructed using nothing more than the method of terminological conventions (for directly founding the basic premisses which are not further proved) and that of deduction (for founding the theorems derived from these premisses).

We intend now to subject this view to criticism. We wish to demonstrate namely that the method of terminological conventions and of deduction do not suffice, by themselves, to validate any theorem; this means that by confining ourselves to these methods, we shall not be able to construct any assertive-deductive system consisting of validated sentences.

This follows from the fact that a terminological convention, stating that a given term 'λ' shall, in our language, denote an object that fulfils the condition formulated in the sentence '$F(\lambda)$', never suffices by itself as a validation for the truth of this sentence in our language; in other words, such a terminological convention may be valid in our language

and nevertheless the relevant sentence may be false. This is so because
our decision to make a term denote an object characterized in a specified
way is not enough to make that term denote such an object. For if
there exists no object to fulfil the condition '$F(\lambda)$' – in case, for instance,
this condition is self-contradictory – then our decision does not suffice
to make such an object the denotation of the term 'λ'. In fact, although
we can decide to make every term denote any object we please, we
cannot, for all that, create such objects by those same decisions. Hence
it follows that to validate the postulate '$F(\lambda)$', one must use not only
the convention by which λ shall satisfy '$F(\lambda)$', but also an existential
premiss, stating that there is an object which satisfies this condition.

Consequently, the method of founding theorems on a terminological
convention does not secure the truth of those theorems, unless they are
also founded on a suitable existential premiss. Accordingly, if one were
to confine oneself to the method of conventions and to that of deduction,
one would possess no method of direct validation and would be committed either to the *regressus ad infinitum*, or to the use of unfounded
premisses (i.e. the fallacy of *petitio principii*).

These considerations lead to the following conclusion: In order to
be able to make justified assertions, consequently also to be able to
construct assertive-deductive systems of validated sentences, one must
possess a richer arsenal of methods than those of terminological conventions and of deduction alone. One must enrich that arsenal by
adding a method of direct foundation to it, either that of exprerience
or that of intuition, provided that these really are methods of direct
foundation. Now, the method of experience is not well suited to validate
the kind of axioms that usually appear in axiomatic systems. For, in
general, such axioms affirm nothing that could be seen or heard. By
the method of experience, one can at most validate the most remote
consequences of the axioms. If, starting from these consequences, one
were to arrive, by reduction, at asserting the axioms, one would construct a reductive but not a deductive system. Consequently, the method
of direct experience offers no prospect of constructing assertive-deductive
systems of validated sentences.

There remains the method of intuition. It seems that to save the
assertive-deductive systems, which means saving the whole of our
alleged knowledge *a priori*, we are constrained to rely upon this method.

However, with many persons it provokes very far-going reservations, especially because of its lack of precision, because of the difficulty of controlling it, of the impossibility of settling the disputes between those who appeal to its testimony. If, under the influence of these reservations, one gives up this method, one ought also to renounce the possibility of constructing assertive-deductive systems, as well as all the knowledge called *a priori*.

Would such a renouncement amount to a catastrophe? It seems that the axiomatic systems of mathematics would lose nothing if they were constructed as neutral-deductive by the mathematicians and treated as assertive-reductive by the scientists who use them.

A certain doubt arises here, however. To construct neutral-deductive systems, sentences must be derived one from another by rules of deduction which, in turn, are based on the laws of logic. Consequently, is it not necessary to presuppose a system of logic, i.e. an assertive system (since it consists of asserted sentences) in order to construct any neutral-deductive system? We wish but to sketch this doubt and our intended answer. It is true that the theorems of logic are necessary to demonstrate that rules of deduction are truth-preserving. But to deduce sentences from one another, one need not prove that one is proceeding in accordance with truth-preserving rules. It is enough simply to proceed in conformity with such rules. Therefore, to construct neutral-deductive systems it is unnecessary to presuppose the theorems of logic. These are needed only for reflecting upon such systems from the methodological point of view, for instance for evaluating the correctness of their structure.

REFERENCE

* First published in *Studia Logica* **IX** (1960). Revised translation by J. Giedymin.

ZBIGNIEW CZERWIŃSKI

THE PROBLEM OF PROBABILISTIC JUSTIFICATION OF ENUMERATIVE INDUCTION*

1. For a long time logicians have had various doubts concerning the justification of inductive inference. Is inductive inference justifiable, and when? Is inductive inference a correct one? Do the premises of inductive inference justify the acceptance of a conclusion, and when? Such are the questions which logicians have to answer, concerning the justification of induction.[1]

The problem of justification of induction arises from the fact that inductive inference is not infallible, so that it may lead from true premises to false conclusions. Inference is a methodological procedure, the aim of which is to discover new truths on the base of truths previously discovered. But in the case of inductive inference – contrarily to deduction – there is always the danger of adding falsehood to already known truths, although we should like to avoid it.

On the other hand, as it is known, we are practically inclined to consider some inductively derived statements as sufficiently justified, i.e. that at least in some cases we are willing to assign to induction ability to justify its conclusions, although it is not an infallible inference. That is why the question arises whether and when inductive inference is justified.

In order to be able to try to solve the problem of justification of induction at all, the problem itself should be formulated more precisely. With this aim in view we have to:

a) establish what is understood by 'induction' resp. 'inductive inference',

b) establish the criterion according to which justification of inductive inference will be evaluated.[2]

Ad a). As I intend in this paper to deal with only one kind of induction – namely, enumerative induction – I do not have to look for a

general definition of induction here. I shall only add that, in my opinion,
the search of such a general definition is unnecessary even if one wants
to examine the justification of all types of induction. It is enough
to characterize the respective types by giving the appropriate inferential
schemes; the question concerning justification should always be asked
separately with respect to each type of induction. The discussion of the
problem of justification of induction on the basis of some most general
scheme of inductive inference does not seem to lead to success, the more
so that the scheme is not known which could embrace all the types of
inferences called inductive.[3] As for enumerative induction, the scheme
of this type of inference is well known. Enumerative induction is an
inference in which, from the premises $a_1 \in B$, $a_2 \in B$, ..., $a_n \in B$, where
$a_1, a_2, ..., a_n$ belong to a set A, we conclude (not knowing of any such
A which is not B) that every A is B. More exactly, the scheme

$$a_1 \in B, a_2 \in B, ..., a_n \in B, \text{ so every } A \text{ is } B$$

determines not one, but a whole class of schemes depending on n.
Practically, we are inclined to value enumerative induction more as n
is larger, and to consider it as justified inference only if n is 'large'.
Therefore, when examining the justification of enumerative induction,
logicians usually restrict themselves to attempts at answering the question whether inference by enumerative induction is indeed the more
justified as n is larger and whether it is justified when n is 'large'.

Ad b). It is a much more difficult task to establish the criterion according to which inductive methods of inference could be evaluated. It
is clear that infallibility cannot be taken as such a criterion. We cannot
agree that inductive inference be justified only if it is infallible. The
acceptance of such a criterion would force us to consider unjustified
all the fallible methods of inference. Thereby we would have either to
reject — as unjustified — the inductive methods of inference, which are,
however, indispensable to the empirical sciences, or to accept the use of
unjustified methods of inference in the empirical sciences, which would
oppose the generally and reasonably accepted opinion that at least some
methods of inductive inference justify their conclusions. As Black has
rightly remarked, various objections raised in the past against induction
were based on that the critics of induction tacitly accepted infallibility
as the criterion of justification of all inferences — and obviously induc-

tion could not satisfy such a criterion.[4] It is a completely useless task to criticize induction because of its being a fallible method of inference.

What then should be taken as the criterion of justification of inductive inference?

Induction — as I understand it here — embraces various methods of inference and serves various purposes. In this situation it would be difficult to formulate one condition according to which, we could all agree, *any* inductive inference in *any situation* would be justified if and only if this condition was satisfied. In my opinion induction *can* and *should* be considered from various points of view, and only on the grounds of such a comprehensive evaluation a solution can be given of the difficult problem of justification of induction.

In this paper I shall consider enumerative induction from only one — probabilistic — point of view. What does this point of view consist in?

In the traditional approach, the probabilistic view of the problem of justification of induction consists in accepting inductive inference as justified if the premises of this inference guarantee a high probability of the conclusion;[6] that of two inferences is more justified which guarantees higher probability to its conclusions. Thus the question whether and when enumerative induction is justified is the question whether and when the premises of enumerative induction guarantee high probability to its conclusion, and possibly also the question what is the relation of this probability to the probability of other conclusions which can be drawn from the premises of enumerative induction.

This traditional approach is not the only possible one with regard to the problem of justification of induction from the probabilistic point of view. Using the concepts and theorems of the theory of probability we can demonstrate that some methods of inductive inference have several properties which can justify their application. A high degree of probability of conclusion with respect to the premises is only one of such properties. I am consciously restricting myself to such a general characterization of the latter — non-traditional — approach to the probabilistic point of view on the problem of justification of induction. A closer description is not possible before the argument that follows.

I shall first discuss here in short some traditional attempts at demonstrating the justification of enumerative induction from the probabilistic point of view, i.e. attempts tending to demonstrate that at least in some

cases the premises of enumerative induction guarantee high probability of inductive generalization. As these attempts have not given satisfactory results, I shall further deal with some other attempts to use the theory of probability in demonstrating the justification of enumerative induction.

I would like to emphasize that I do not think that the investigations of justification of induction should be limited to the probabilistic point of view. It may be doubted whether any inference justified from the probabilistic point of view should automatically be considered as worth applying in all situations. All the more we should not consider an inference as unjustified only on the grounds that its justification cannot be demonstrated in terms of the theory of probability. Nevertheless, I consider the probabilistic approach to the problem of justification of induction, and particularly of enumerative induction, as a useful contribution to the complex problem of justification of induction.

2. The attempts to use the theory of probability in solving the problem of justification of enumerative induction are not new. The question about the probability of the conclusion in enumerative induction was raised by several logicians: Keynes,[7] Hosiassonówna,[8] Wright,[9] Carnap[10] and others. Some of them linked the attempts to solve this problem with the construction of a special theory of logical – as opposed to mathematical – probability. For reasons that are too complex to be discussed here I do not think it necessary to construct such theories[11] and the (elementary) concepts of the theory of probability underlying the following discussion belong to the 'ordinary' (mathematical) theory of probability.

The best known achievement concerning our problem are two theorems proved by Keynes.[12] In formulating them I shall adopt the following terminology. An event consisting in that the n-th element of a given set A has the property (belongs to the set) B, I shall call the n-th verification of the inductive generalization 'every A is B', in symbolic notation: v_n. The symbol 'V_n' will mean the same as '$v_1, v_2, ..., v_n$'. I shall symbolize the inductive generalization 'every A is B' shortly by 'g'.

The above mentioned theorems state that:

(1) Additional verification of an inductive generalization increases its probability (symbolically: $P(g, V_n) > P(g, V_{n-1})$).

(2) The probability of an inductive generalization increases to 1 if the number of verifications increases infinitely (symbolically: $\lim\limits_{n \to \infty} p(g, V_n) = 1$).

(3) Under the condition that:

(a) the *a priori* probability of an inductive generalization is greater than zero (symbolically: $P(g) > 0$),

(b) the probability of the n-th verification, with the assumption of $n-1$ verifications, is smaller than one (symbolically: $P(v_n, V_{n-1}) < 1$),

(c) the probability of occurrence of n successive verifications, with the assumption of negation of inductive generalization, tends to zero if n increases unlimitedly (symbolically: $\lim\limits_{n \to \infty} P(V_n, \bar{g}) = 0$).[13]

If we agree — as it was accepted in the traditional, probabilistic approach to the problem of justification of induction — to consider the value of the probability of a conclusion with respect to its premises as the measure of justification of a given inference, then in virtue of (1) and (2) we can say that:

(1′) enumerative induction is the more justified inference the larger is the number of verifications of inductive generalization,

(2′) there exists such a number of verifications that enumerative induction is a justified inference.

It should not be forgotten, however, that the statements (1) and (2) are justified only to the extent to which the assumptions (a), (b) and (c) are true. Keynes tried to justify these assumptions on the basis of the so-called principle of finite independent variety.[14] However, both the formulation of this principle and the argumentation for its acceptance, as well as the way in which Keynes derives assumptions (a), (b) and (c) from this principle, have numerous shortcomings, which have already been pointed out by his critics.[15]

Without using philosophical hypotheses of the type of the principle of finite independent variety we can easily justify the adoption of the assumptions (b) and (c) assuming that the set A which the inductive generalization concerns is finite. The adoption of this assumption seems to be justified as the sets dealt with in the practice of inductive investigations are rather finite, although very numerous.[16]

On the other hand — even if finiteness is assumed — the justification of the assumption (a) is more difficult. The difficulty lies in finding an

interpretation for the *a priori* probability of the inductive generalization. Although there are some interpretations for which it can be demonstrated that this probability exceeds zero, it is difficult to dispense here with the principle of indifference. And it is widely known that the use of this principle involves serious dangers, as it sometimes leads to contradictions. Therefore the attempts to justify assumption (a) are not free of important reservations, even if it is assumed that the examined set *A* is finite.

Similar difficulties arise if we do not want to restrict ourselves to the theorem that the probability of inductive generalization increases with the number of verifications and that it tends to one when this number increases unlimitedly, and try to find the function which shows dependence of the probability of inductive generalization on the number of verifications. For we cannot do without certain assumptions concerning the *a priori* probability of inductive generalization while trying to find such a function.

Accepting various assumptions of this kind we get various functions showing the dependence of the probability of inductive generalization on the number of verifications.[17] It is a matter of discussion which of these functions is the best in the sense of being based on assumptions which correspond best to the practice of inductive investigations. But all these functions have the property of assigning a small value to the probability of inductive generalizations in the situations occurring — it seems — in the practice of inductive investigations (where the number of verifications is small compared to the number of elements of the set considered). This is the reason for the important difficulties encountered by the traditional attempts at a probabilistic justification of enumerative induction, which tend to demonstrate that the probability of inductive generalization is high when the number of verifications is large. It is doubtful whether these attempts could ever give any satisfactory results.

3. We have shortly presented above the attempts to prove that enumerative induction is an inference justified from the probabilistic point of view, in the sense that at least in some cases the premises of enumerative induction guarantee a high probability to inductive generalization. These attempts do not take into consideration the relation between the probability of inductive generalization and that of other conclusions which may be drawn from the premises on which the inductive general-

ization is based. When considering the probability of inductive conclusion we have up to now left out the question whether inference by enumerative induction — even if it does not guarantee high probability to the conclusion — is or is not more justified, from the probabilistic point of view, than some other inferences, based on the same premises and leading to different conclusions. We shall now try to answer this question.

Assuming that the examined set A is a finite one, let us denote by 'g_c' the hypothesis that the frequency of elements of A which have the property (belong to the set) B equals c. The variable c can assume values being rational numbers of the interval $[0, 1]$ and having the form i/N ($i=0, 1, ..., N$), where N stands for the number of elements of the set A. The hypothesis 'g_1' is, naturally, equivalent to the inductive generalization 'every A is B' which — as before — will be symbolized by the letter 'g'. The hypotheses g_c are mutually exclusive and complementary. From the elementary theorem of the calculus of probability

$$(1) \qquad P(g_c, V_n) = \frac{P(g_c)P(V_n, g_c)}{P(V_n)},$$

it immediately follows [18] that

$$(2) \qquad \frac{P(g, V_n)}{P(g_c, V_n)} = \frac{P(g)}{P(g_c)P(V_n, g_c)}.$$

By virtue of the assumption of the finiteness of the set A, $P(V_n, g_c)$ for $c \neq 1$ tends to zero when the number of verifications n increases infinitely.[19] We can draw the conclusion that: if the *a priori* probability of inductive generalization does not equal zero, then there is such a number of verifications n for which the probability of inductive generalization is larger than the probability of any other alternative hypothesis g_c. This conclusion, however, is a weak argument in favour of enumerative induction. First, in concrete cases we do not know how large the number of verifications guarantees that the probability of inductive generalization exceeds that of the alternative hypotheses g_c ($c \neq 1$), and second, the conclusion we have obtained is based not only on the assumption of finiteness of the set A, but also on the assumption that the *a priori* probability of inductive generalization does not equal

zero which—as I have already said in the second part of this paper—is difficult to justify in a doubtless way.

There is still another way of trying to justify the fact that it is better from the probabilistic point of view to accept the inductive generalization rather than any other alternative hypothesis g_c.

Let us consider a certain hypothesis h and a certain evidence e. Stating e will have a positive effect on justifiying h if the probability of h is greater when e is assumed than without this assumprion (symbolically: if $P(h, e) > P(h)$), independently of the absolute value of $P(h, e)$.

If $P(h, e) > P(h)$ we shall say that e *probabilistically confirms* h. This formulation should not raise any objection from the point of view of our linguistic intuitions. For to say that evidence confirms the hypothesis h is, to our understanding, equivalent to saying that if we state e the probability of h increases. Hence we may take as measure of the degree of probabilistic confirmation of h by e: $m(h, e) = \dfrac{P(h, e)}{P(h)}$, [20]

and say that e confirms the hypothesis h_1 more than the hypothesis h_2 when $m(h_1, e) > m(h_2, e)$.

After these terminological conventions have been introduced it is easy to demonstrate that the n-time verification of inductive generalization confirms inductive generalization more powerfully than any other alternative hypothesis g_c $(c \neq 1)$.

This directly results from:

$$m(h, e) = \frac{P(e, h)}{P(e)}.$$

We, thus, have:

$$m(g, V_n) = \frac{1}{P(V_n)},$$

$$m(g_c, V_n) = \frac{P(V_n, g_c)}{P(V_n)}.$$

As, for $c \neq 1$, $P(V_n, g_c) < 1$, then for $c \neq 1$ $m(g, V_n) > m(g_c, V_n)$. Moreover, it should be pointed out that when n increases $m(g_c, V_n)$ decreases for any $c \neq 1$. Therefore, n-time verification is the weaker probabilistic confirmation of any alternative hypothesis g_c the larger n is.

This result is intuitively fully comprehensible. It is clear that if an inductive generalization 'every A is B' is false, i.e. if one of the hypotheses g_c $(c \neq 1)$ is true, then as the number of verifications increases it becomes less and less probable that no A will be found that is not B. In other words, if we accepted any hypothesis g_c alternative to inductive generalization and if we did not find in the course of verification procedure any A not being B, we should have to accept that less and less probable events were occurring. The inductive generalization 'every A is B' is the only hypothesis the acceptance of which does not force us to assume that less and less probable events occur if the number of verifications increases.[21]

4. In the previous part of the paper I have presented attempts to demonstrate the justification of enumerative induction which are somewhat different from the traditional view of this problem on the grounds of probabilities.

I did not try to demonstrate there that inductive generalization is highly probable after repeated verifications; the argument consisted in showing that accepting inductive generalization was superior to accepting any alternative hypothesis g_c in the sense that the latter would force us to assume, after a sufficiently large number of verifications, the occurrence of improbable events, the less probable the larger the number of verifications.

Anybody acquainted with contemporary theory of statistical inference must have noticed that this argumentation — independently of whether it seems convincing or not — resembles the type of argumentation used in testing statistical hypotheses. The possibility therefore arises to interpret inference by enumerative induction as a statistical test. It will enable us to apply to the evaluation of enumerative induction the evaluation criteria used by contemporary mathematical statistics in the evaluation of statistical tests. In order to give such an interpretation of enumerative induction it is necessary to introduce some concepts from the theory of statistical inference.

Generally speaking, statistical inference consists in concluding from a sample about a population. The premise of statistical inference is always a sentence stating a certain property of the sample drawn from a given population, and its conclusion — a sentence stating a property of the population from which the sample had been drawn. In a specific case the

premise may be a sentence stating that a certain parameter m of the distribution of some property C in the sample S drawn from the population P is included in the interval U; its conclusion may be a sentence stating that a certain parameter μ of the distribution of the property C in the population P has the value μ_0. In the situation described above the statistical test[22] will be any rule R demanding rejection of the hypothesis h_0 saying that $\mu = \mu_0$, if m belongs to an *a priori* given set K, and allowing to accept the hypothesis h_0 if m does not belong to K.[23] The set K — called in statistics the critical region — is constructed so that, assuming h_0, the probability of m belonging to K equals a certain small number α called level of significance.

Statisticians (Neyman and Pearson) have elaborated several criteria for evaluating statistical tests. The basic idea underlying these criteria is the following:

When using any statistical test we can make two types of errors:

a) reject a true hypothesis,

b) accept a false hypothesis.

A statistical test will be the better the smaller the probability of errors of both types. The probability of errors of the first type equals the adopted level of significance α. The best of all the tests corresponding to the level of significance α will be the test giving the smallest probability of errors of the second type β. It is well known, however, that the latter probability may be univocally determined only in certain simple situations, i.e. if except the verified hypothesis h_0 there is only one possible alternative simple hypothesis h_1, saying that $\mu = \mu_1$ ($\mu_1 \neq \mu_0$). In the opposite case we can only construct a function assigning the probability of adopting the hypothesis h_0 in the case of every acceptable simple hypothesis h_1 saying that $\mu = \mu_i$. The values of this function for $i \neq 0$ give the probabilities of errors of the second type for various simple hypotheses alternative to h_0. Such a function is called the operating characteristic function of the test procedure (in short: o.c. function).

When evaluating a given statistical test statisticians take into consideration:

a) the probability of errors of the first type,

b) the shape of the o.c. function.

A good statistical test should have the following properties:

1) the probability of errors of the first type should be small,

2) the o.c. function of this test should assume its maximum value for the tested hypothesis h_0.

A test having the property 2) makes the rejection of the hypothesis h_0 more probable when any of the alternative hypotheses is true than in case when the hypothesis h_0 is true (and the acceptance of the hypothesis h_0 more probable when it is true than if any of the alternative hypotheses is true), and is called a unbiassed test.

The ideal statistical test would be one whose o.c. function:

(1) assumed the value 0 for any hypothesis h_i $(i \neq 0)$ alternative to the tested hypothesis h_0 (zero probability of errors of the second type),

(2) assumed the value 1 for the hypothesis h_0 (zero probability of errors of the first type).

After this short description of statistical tests and of methods of their evaluation I shall now present the interpretation of enumerative induction as a statistical test. In this interpretation enumerative induction will be represented as a rule of inference allowing the acceptance of the inductive generalization 'every A is B' when all the examined A are B, and demanding the rejection of this generalization when at least one of the examined is not B.

The set A is the population examined. We are interested in the distribution in the set A of the property (of belonging to) B. The frequency γ of B's among A's is the parameter of the distribution of the property B in the population A we are interested in. The tested hypothesis states that 'every A is B' what — under the assumption of finiteness of the set A — is equivalent to the hypothesis '$\gamma = 1$'. The alternative hypotheses are the hypotheses g_c of the form '$\gamma = c$' $(c \neq 1)$. The parameter of the sample, the value of which we are examining in order to verify the inductive generalization, is the relative frequency c_n of B's among n examined elements of the set A. c_n may assume the values

$$0, \frac{1}{n}, \frac{2}{n}, \dots, 1 .$$

Since in the course of inference by enumerative induction we reject the inductive generalization 'every A is B' if at least one examined a_i being A is not B, and since we accept this generalization if all the known

a_i being A are B, then the critical region is the set of numbers

$$\left\{ 0, \frac{1}{n}, \frac{2}{n}, \ldots, \frac{n-1}{n} \right\}.$$

In this interpretation enumerative induction has the following properties:

1) the probability of errors of the first type equals zero. It is not possible to reject the inductive generalization if it is true; in order to do this it would be necessary for at least one element of the examined sample not to belong to the set B, which is impossible if the inductive generalization is true;

2) the o.c. function for enumerative induction understood as a statistical test is the function $y=c^n$.[24] The variable c may assume the values $\frac{i}{N}$ $(i=0, 1, \ldots, N)$, where N is the number of elements of the set A. It can be seen immediately that the value of c^n is maximal when $c=1$, i.e. for the value of c determined by the tested hypothesis. Enumerative induction is then an unbiassed test. Moreover, when $n \to \infty$, the o.c. function for enumerative induction tends to a function which (a) assumes the value 0 for any $c \neq 1$, (b) assumes the value 1 for $c=1$, i.e. assumes the value 0 for any hypothesis alternative to inductive generalization and the value 1 for this very generalization.

Therefore, the o.c. function for enumerative induction tends to the function that the ideal statistical test would have. Comparing the shape of o.c. function for enumerative induction with that of an o.c. function for statistical tests used in practice and recommended by statisticians we must say that for n's as large as those we deal with in the practice of inductive scientific investigations, it has an advantageous shape as compared with the latter functions.

The interpretation of enumerative induction as a statistical test thus leads to the conclusion that enumerative induction is quite a reasonable inference in the light of the criteria used in the evaluation of statistical inference, and it even is an 'asymptotically ideal' statistical test.

For those who consider convincing the criteria used in mathematical statistics to evaluate statistical inference the argumentation presented in

this part of the paper may serve as a proof of the justification of enumerative induction.

5. Independently of whether the considerations presented in Part 4 can be treated as justification of enumerative induction, some additional remarks should be made.

Logicians have been inclined to assign to enumerative induction little power of justifiying its conclusions for a long time,[25] and sometimes (for example the authors of *Logique du Port Royal*[26]) even treated it as a completely fallacious inference. Traditional attempts to demonstrate the justification of induction from the probabilistic point of view, accounted for in the second part of this paper, do not give satisfactory results and seem to support this distrustful position towards enumerative induction. On the other hand, from the point of view of criteria employed to evaluate inference by mathematical statistics, that is by a science highly appreciated nowadays and constantly enlarging its field of application, enumerative induction appears to be sound inference.

The question then arises: should the logicians revise their up to now generally critical attitude towards enumerative induction, or should they extend this attitude to statistical methods of inference, warning against the dangers of their application as ardently as they have since now been warning against the use of enumerative induction?

This question seems to be important if we want logic to give the complete list of methods of inference which are applied and approved in the sciences, and to be able to evaluate their validity.

REFERENCES

* First published in *Studia Logica* V (1957). Translated by S. Wojnicki.
[1] This paper is a somewhat modified fragment of the doctoral dissertation prepared by the author.
[2] Cf. M. Black 'On the Justification of Induction', *Language and Philosophy*, pp. 63–64.
[3] The terms 'induction' resp. 'inductive inference' are used in so many different meanings that I think it an unfeasible task to find their analytical definition. In contemporary literature of logic the tendency may be observed to such understanding of these terms in which their range is very wide and embraces all the fallible inference methods which are usually approved in the empirical sciences. It is in this, very imprecise sense that I use the term 'induction' resp. 'inductive inference' in the present paper.

[4] M. Black, *op. cit.*, pp. 66–68.

[6] It cannot be generally established what probability should be called high probability. There is always some convention in choosing the lower limit of high probabilities, as for example in adopting the level of significance in statistical tests. Undoubtedly our inclination to consider a given probability high depends in practice to a large extent to the loss connected with the acceptance of a false conclusion.

[7] J. M. Keynes, *A Treatise on Probability*, 1929.

[8] J. Hosiasson, 'O prawomocności indukcji hipotetycznej' ('The Problem of Justification of Hypothetical Induction'), *Fragmenty Filozoficzne*, Warszawa 1934.

[9] G. H. von Wright, *The Logical Problem of Induction*, 1941.

[10] R. Carnap, *Logical Foundations of Probability*, 1951.

[11] Constructing such theories is nowadays fashionable among western logicians (cf. e.g. Carnap's *Logical Foundations of Probability* and the lively discussion following it which has been going on in logical and philosophical periodicals). Instead of the term 'logical probability' the term 'degree of confirmation' is used, or shortly 'confirmation'. I think construction of such theories is unnecessary for the solution of the problem of justification of induction for the following reasons: the existing axioms for the concept of confirmation usually repeat the axioms given for the mathematical concept of probability. (Cf. the axioms given by Keynes and Hosiasson in the works mentioned or the axioms given by A. Shimony in his paper 'Coherence and the Axioms of Confirmation', *Journal of Symbolic Logic* **20**.) On the other hand, attempts to give an explicit definition of the degree of confirmation usually reduce it to classical probability, defined not on events but on sentences (cf. Carnap, *op cit.*). I think this difference is not substantial. In my opinion it is obvious that classical probability may equally well be considered a function where arguments are events or sentences (cf. here i.a. K. R. Popper, *Logik der Forschung*, p. 188 etc., W. Glivenko, *Kurs teorii veroyatnosti* (*Calculus of probability*), Moscow–Leningrad 1939, pp. 231–231; R. Carnap, *op. cit.*, pp. 29–30). Hence it seems that construction of special theories of confirmation as theories of classical probability defined on sentences is unnecessary. The confirmation theories which I know do not bring anything new, neither from the formal, nor from the substantial points of view, in comparison with mathematical theories of probability, and they only repeat what is already known from the mathematical theories. The construction of a special theory of logical probability becomes reasonable only when one holds the view that logical and mathematical probability differ not only in that the former is defined on sentences and the latter on events, but when there is a difference which makes impossible the adoption of the axioms of mathematical probability as ready made axioms for logical probability. This is, for instance, Popper's opinion: according to him, logical probability (*Hypothesenwahrscheinlichkeit*), understood as the degree of justification of a hypothesis by facts, has some properties totally different from mathematical probability (*Ereigniswahrscheinlichkeit*), even if the latter were to be interpreted as function propositional arguments (cf. *Logik der Forschung*, pp. 188 ff, and Pop-

per's articles in *British Journal for the Philosophy of Science* **5,** 6). There is no place here to deal further with this interesting question.

[12] *Op. cit.*, pp. 235–236.

[13] This condition can be replaced with the following one: there is an $\varepsilon > 0$ such that $P(v_n, V_{n-1} \cdot \bar{g}) \leqslant 1 - \varepsilon$. Keynes proved his theorems on the grounds of his own axiomatic theory of probability. They can however be as easily demonstrated on the grounds of the mathematical theory of probability.

[14] *Op. cit.*, pp. 251 ff.

[15] For instance, J. Nicod, *Le problème logique de l'induction*, pp. 76 ff.

[16] If they are not finite, from the practical point of view it is important that the inductive generalization would hold for the subset of the examined set the elements of which we encounter in practical activities and this subset is always finite.

[17] Cf. i.a. B. Russel, *Human Knowledge*, G. Allen, London 1948, p. 424.

[18] As $P(V_n, g) = 1$.

[19] The assumption of finiteness of the set A does not exclude the possibility of an unlimited increase of the number of verifications, if only it is assumed that the examined elements of the set A are — statistically speaking — sampled with replacement. If we assumed that we are dealing with sampling without replacement, we would have to substitute the assumption that n tends to infinity by the assumption that n tends to N. In both cases the conclusions concerning the probabilistic justification of enumerative induction will be the same.

[20] This measure of the degree of confirmation of hypothesis h by empirical data e corresponds to what Carnap calls 'relevance quotient' (*Logical Foundations of Probability*, p. 567). We might just as well take as measure of the degree of confirmation of h by e any function strictly increasing with $m(h, e)$. This has been done by I. J. Good, who takes as measure of the degree of confirmation of h by e (Good uses the term 'weight of evidence for h, given e') the value of $\log P(h, e) + \log P(e) + \text{const.}$ Cf. *Probability and the Weighing of Evidence*, p. 63.

[21] Professor M. Kokoszyńska, in her contribution presented at the same conference and published in the same volume of *Studia Logica*, introduces among other things the non-relativized and the relativized concept of inductive fallacy. It is easy to see that my expression 'e probabilistically confirms h' has the same meaning as prof. Kokoszyńska's expression 'h is correctly inferred from e (in the absolute sense)', as well as the sentence 'e probabilistically confirms h_1 more powerfully than h_2' means the same as prof. Kokoszyńska's: 'the inference of h_1 from e is less fallacious (more correct) than the inference of h_2 from e'.

[22] Statisticians call such statistical tests parametric tests.

[23] Cf. the definition of statistical test given by J. Neyman in *First Course in Probability and Statistics*, p. 258. The description of the procedure of verifying statistical parametric hypotheses has been taken from M. Fisz's *Calculus of Probability and Mathematical Statistics*, in particular Chap. 13.

[24] I assume here that elements of the set A examined with respect to the property B are sampled with replacement. The whole reasoning might be based on the assumption of sampling without replacements and lead to a similar result.

[25] Cf. Bacon's well known opinions: *"Inductio enim quae procedit per enumerationem simplicem res peurilis est; precario concludit et periculo exponitur ab instantia contradictoria"...* And: *"Inductio mala est quae per enumerationem simplicem Principia concludit Scientiarum..."*(*Novum Organum,* Amsterdam 1964, I, 105, and I, 69). Among the contemporary logicians, Professor Kotarbiński says of enumerative induction: "A non-exhaustive induction, which proceeds by simple enumeration... is a very weak method of justifying generalization", *Kurs logiki* (*A Course in Logic*), 2nd ed., p. 135.

[26] I owe this information to Professor S. Romahnowa.

ZBIGNIEW CZERWIŃSKI

ENUMERATIVE INDUCTION AND THE THEORY
OF GAMES*

I

In may earlier paper [1] I presented an analysis of enumerative induction and showed that in spite of the severe philosophical criticism of this type of inference enumerative induction proves to be a pretty good inference rule compared with various generally accepted types of statistical inference. I also pointed out that various criteria may be applied when we ask whether it is reasonable to follow this or that rule of inductive inference. This time I will try to consider once more the problem of justification of enumerative induction applying some other criteria taken over from the game theory, and thus to make a contribution to the old philosophical discussion on the problem of induction.

It is my opinion that the problem of justification of any type of inductive inference should be stated in terms of choice of one from among several alternative inference rules. If h is not a logical consequence of e, it seems to be futile to ask whether it is reasonable to accept the hypothesis h on the evidence e (as suggested by a given rule R) unless we add which other hypotheses alternative to h could eventually be accepted on the same evidence and comparing in a way the relation of e to those hypotheses with that of e to h. Taking this for granted I shall interpret the rule of enumerative induction as a rule alternative to several other rules, i.e. as a rule of choice of one from among several alternative hypotheses, some evidence being given. The problem of choice of the best inference rule will be reduced to the problem of finding the solution of a game. The analysis of the game corresponding to the situation in which the enumerative induction rule may be applied will enable us to evaluate this controversial rule of inference.

II

It is well known that the problem of choice of a hypothesis (or decision) from among a set of some alternative hypotheses (decisions), this choice being dependent on the outcome of some experiment, may be reduced to the problem of solving a certain game, and we shall give a short outline of such a reduction.[2]

Let $\{h_i\}$ $(i=1, \ldots, m)$ be a set of mutually exclusive and exhaustive hypotheses, and let E be an experiment which may give n various results $e_j (j=1, \ldots, n)$. There are $r=m^n$ ways of assigning h_i's to e_j's, and hence $r=m^n$ logically possible rules of inference may be given, each one telling us which hypothesis is to be accepted if e_j is the outcome of the experiment. Every such rule R_l $(l=1, \ldots, r)$ is simply a function $t_l(j)$ defined on the set $\{e_j\}$ with the values out of the set $\{h_i\}$. No misunderstanding will arise if we identify the function $t_l(j)$ with the corresponding function defined on the indices of the possible outcomes of the experiment and taking the indices of hypotheses as its values. Suppose that there is some loss s_{ik} $(i, k=1, \ldots, m)$ associated with the acceptance of the hypothesis h_i, if it is the hypothesis h_k which is actually true, and suppose that there are known probabilities p_{jk} of e_j, given h_k $(j=1, \ldots, n, k=1, \ldots, m)$.

In such a situation we are able to determine the loss $z_{lk}(j)$ we incur if we follow the rule R_l, the experiment results in e_j, and the hypothesis h_k is actually true.

The loss $z_{lk}(j)=s_{qk}$, where $q=t_l(j)$ is the index of the hypothesis which is to be accepted according to the rule R_l if the experiment results in e_j.

The losses $z_{lk}(j)$ are functions of three variables: the outcome of the experiment, the actual state of nature and the rule we have chosen. We may ask what is the expected loss associated with the rule R_l, the hypothesis h_k being true. The expected loss associated with any rule R_l no longer depends on the outcome of the experiment. It depends only on the rule we have chosen and the hypothesis which is actually true, and is given by the formula

$$(2.1) \qquad u_{lk} = \sum_{j=1}^{n} z_{lk}(j) \, p_{jk}.$$

Having determined the (expected) losses u_{lk} $(l=1, \ldots, r, \; k=1, \ldots, m)$,

we may ask which of the rules $R_1, ..., R_r$ is the best in the situation we have just described. But there is no reason why should we *a priori* assume that following always one and the same rule will give us the best possible result. Perhaps we may be better off if we follow various rules $R_1, ..., R_r$ with certain frequencies $x_1, ..., x_r$ adding up to 1. Following only one rule R_l once and for all is only a special case of this policy, namely the case of all $x_1, ..., x_r$ being zero, except for x_l being equal to one. So the problem arises of finding out the frequencies with which the rules $R_1, ..., R_r$ should be followed, and this is the problem quite analogous to that of solving the game with the pay-off matrix $U = [u_{lk}]$. The solution must be based on a criterion of choice of the best rule. It is disputable whether the usual minimax rule should be applied when we are dealing with the choice of a hypothesis and not with the game in the strict sense of the term. I will come back to this question in due time. Now I shall show how the question whether the rule of enumerative induction is a reasonable rule can be dealt with within the conceptual framework which has been outlined in the last section.

III

The rule of enumerative induction or, for short, the rule of induction (no other rule of inductive inference will be discussed in the paper) is a rule which advises us to accept the hypothesis that all A's are B's if a number of A's has been examined all of which proved to be B's, and none to be non-B. It is not generally decided how large must be the number of A's having been examined. It is tacitly assumed that it is not a quite small number, but that is all we may find in traditional formulations of the rule. It is intuitively clear that the value of the rule of induction depends somehow on the number of A's known to be B's, and it may be expected that the decision whether the rule is reasonable or not will depend on that number. Let us suppose that the rule of induction advises us to accept the hypothesis that all A's are B's if all n examined A's have proved to be B's, and no A has been found not being B, without specifying the number n, assuming only that $n > 0$.

We must now take into account some set of hypotheses, alternative to the hypothesis suggested by the induction rule.

We shall assume that the set A is finite and contains $m > 1$ elements. The simplest set of mutually exclusive and exhaustive hypotheses about the relation of the property of being A to that of being B, is the set $\{h_i\}$, where h_i is the hypothesis that exactly i A's are B's. Prior to the experiment no hypothesis h_i $(i = 0, 1, ..., m)$ is excluded. The experiment consisting in the examination of n A's can give us $n + 1$ various outcomes $e_0, ..., e_n$, if by the outcome of the experiment we mean the frequency of B's among the observed A's. Hence there are $(m+1)^{n+1}$ *a priori* admissible different rules R_l assigning to any possible outcome of the experiment a hypothesis h_i.[3]

Neither of the rules R_l is, of course, the rule of induction. Any such rule is more specific than the induction rule. It tells us not only what is to be done if all examined A's turn out to be B's; it also recommends us to accept a hypothesis if only some A's (or no A at all) prove to be B's.

To make some use of the game theory in our analysis of the rule of induction we should specify the probabilities p_{jk} of e_j, given h_k $(j = 0, 1, ..., n; k = 0, 1, ..., m)$ and the losses s_{ik} $(i, k = 0, 1, ..., m)$ we incur accepting the hypothesis h_i when the hypothesis h_k is actually true.

As to the specification of the probabilities p_{jk} let us assume that the A's have been drawn in a series of independent drawings with replacement. This assumption gives us for the p_{jk} the binomial formula

$$(3.1) \qquad p_{jk} = \binom{n}{j} \left(\frac{k}{m}\right)^j \left(\frac{m-k}{m}\right)^{n-j}.$$

As we need to take into account also the cases $k = 0$, and $k = m$, we will assume (in perfect agreement with the intuition and the general rules of the probability theory)

$$p_{nm} = 1 = p_{00}, \qquad p_{jm} = 0 \quad \text{for } j = 0, 1, ..., n-1,$$

$$p_{j0} = 0 \quad \text{for } j = 1, 2, ..., n.$$

The question whether the formula (3.1) is consistent with the practice of induction is, of course, disputable. Other assumptions about the origin of the sample can also be taken into account, e.g., the assumption that the drawings without replacement took place which assumption would result in a hypergoemetric formula for p_{jk}. I would not like to

analyse the merits of various possible assumptions before tracing out the consequences of at least one assumption which is plausible and, let us add, rather easy to handle.

We have much less reason to assume any specific shape of the loss function s_{ik}, and in the present, very general, analysis we will only assume that $s_{ik}=0$ if $i=k$ and that $\infty>s_{ik}>0$ if $i\neq k$.[4] This assumption means simply that we do not incur any loss if we accept the true hypothesis, and some positive loss if we make a mistake accepting a false hypothesis, the value of the loss possibly depending on both the hypothesis accepted and the hypothesis being true.

IV

So far we have reduced the problem of choice of the best of the rules R_l (resp. of the best frequencies with which various rules R_l are to be followed) to the problem of solving the game with the pay-off matrix $U=[u_{lk}]$, where

$$(4.1) \qquad u_{lk}= \sum_{j=0}^{n} \binom{n}{j}\left(\frac{k}{m}\right)^{j}\left(\frac{m-k}{m}\right)^{n-j} z_{lk}(j).$$

Now we will pay attention to some properties of the matrix U which we will make use of later on. In the matrix U there are r rows corresponding to the rules $R_1, \ldots, R_r$ and $m+1$ columns corresponding to the hypotheses $h_0, \ldots, h_m$. The losses are not specified except for that

$$(4.2) \qquad z_{lk}(j)=0, \text{ if } t_l(j)=k \quad \text{and} \quad z_{lk}(j)>0, \text{ if } t_l(j)\neq k,$$

the value of $z_{lk}(j)$ depending in the last case in some way on k and $t_l(j)$.

In our analysis it is the m-th column (i. e., the column corresponding to the hypothesis h_m) which deserves particular attention. Let us call the rule R_l inductivist if it assigns the hypothesis h_m to the outcome of experiment e_n. In other words

$$(4.3) \qquad R_l \text{ is inductivist if and only if } t_l(n)=m.$$

The term 'inductivist' can be justified by the fact that the inductivist rules prescribe us in the case of all examined A's being B's exactly the same behaviour the induction rule does. It is easy to notice that the

inductivist rules and only such rules give us zeros in the m-th column of the matrix U, i.e.,

(4.4) $u_{lm}=0$ if and only if R_l is an inductivist rule.

Otherwise u_{lm} equals some number $z_{lm}(n)>0$.

This is because in the case the hypothesis h_m is true the probabilities of all e_j's for $j=1, ..., n-1$ are zeros, and that of e_n equals 1.

Let us notice that to any non inductivist rule R_μ there corresponds some inductivist rule R_λ assigning the same hypotheses to all e_j's $(j=1, ..., n-1)$ as the rule R_μ does, so that:

(4.5) $t_\mu(n)=i\neq m=t_\lambda(n)$.

(4.6) $t_\mu(j)=t_\lambda(j)$ for $j=0, 1, ..., n-1$.

(4.7) $u_{\lambda m}=0<z_{\mu m}(n)=u_{\mu m}$.

For any $k=0, 1, ..., m$ the first n compounds (for $j=0, 1, ..., n-1$) of the sums $u_{\mu k}$ and $u_{\lambda k}$ are identical, and so that the difference

(4.8) $u_{\mu k}-u_{\lambda k}=\left(z_{\mu k}(n)-s_{mk}\right)\left(\dfrac{k}{m}\right)^n$.

Let us denote

(4.9) $\varepsilon_k(\mu, \lambda)=z_{\mu k}(n)-s_{mk}$.

For $k=0, 1, ..., m-1$ the values $\varepsilon_k(\mu, \lambda)$ are not specified on the basis of our assumptions. They may be positive, negative or zero. But, because of the assumption $s_{ii}=0$,

(4.10) $\varepsilon_m(\mu, \lambda)=z_{\mu m}(n)>0$.

V

Before we make use of the above properties of the matrix U when solving our problem of choice of the best rule R_l (resp. of the best frequencies which those rules should be followed with) we must consider the question which criterion should be accepted to choose the best rule R_l or the frequencies with which those rules are to be applied (some of them being possibly zeros). It does not seem to make much sense to follow exactly the game theorist and to accept the minimax criterion. The

reason is that accepting the minimax rule of solution of a game implies that there are two opponents every one being sufficiently clever in adjusting his behaviour to that of the other and taking advantage of any of his partner's miscalculations.

It is hardly believable that the Nature who is our 'opponent' in the procedure of choice of a hypothesis reacts in a way on our choice, making true this or that hypothesis in such a way as to expose us to the greatest loss possible. We will assume instead that the strategy of Nature is given which means that there are some probabilities $P_0, P_1, ..., P_m$ which can be assigned to the hypothesis $h_0, h_1, ..., h_m$. We will not postulate any particular distribution of P_i's assuming only that $P_m > 0$; this assumption will be discussed later.

Though we do not assign any particular values to P_i's we may legitimately be asked about the very meaning of those probabilities. We would suggest the following interpretation:

In the history of inductive inquiries Man has examined various sets with respect to some properties (the set of metals with respect to the electric conductivity, the set of swans with respect to colour, etc). So we have the whole sequence $S = \langle A_t, B_t \rangle$ of pairs set — property with respect to which the set has been examined by Man.[5] To any pair $\langle A_t, B_t \rangle$ of the sequence S there corresponds some frequency f_t of B_t's among A_t's, and we may legitimately speak about the frequency of pairs of the sequence S with frequencies f_t falling within a given interval. To the hypothesis h_i, i.e., to the hypothesis that the frequency of B's among A's is i/m we may assign as its probability P_i the frequency of pairs of S with f_t, satisfying the condition:

$$\frac{i-\frac{1}{2}}{m} \leqslant f_t < \frac{i+\frac{1}{2}}{m} .^5$$

I would not like to go very far into the investigation of the structure of Nature or the history of inductive method. The above interpretation of the probabilities P_i seems to be sufficiently legitimate to accept it in order to enable us to find some reasonable ground for the choice of the best rule R_l. Certainly, the assumption $P_m > 0$ does not follow from our interpretation of P_i's. This assumption, however, is the necessary condition of any justification of enumerative induction. It goes without saying that if it were not satisfied any attempt to justify the enumerative

induction would be doomed to failure. I do not mean to say that enumerative induction is to be justified. Yet, I think, it is worth-while to ask whether it is in a sense reasonable rule, if the very necessary condition of its justification is satisfied. That is the reason why the assumption $P_m > 0$ is accepted in our argument.

Having assumed some given probabilities of the hypotheses h_i, i.e., some given 'strategy' of Nature we can no longer make use of the usual minimax approach to the solution of games. We must apply another (and much simpler) approach. If the 'strategy' of Nature $P_0, P_1, ..., P_m$ is given, we should choose the best rule R_l with respect to this strategy. I do not see any better criterion of choice than the expected value E_l of loss associated with the rule R_l given by the definition:

$$(5.1) \qquad E_l = \sum_{k=0}^{m} P_k u_{lk}.$$

The smaller the value E_l, the better the rule R_l. The best is the rule R_l giving the minimum value of E_l. If there are several rules giving the minimum value E_l we may apply some or all of them with any frequencies without, however, being worse or better off in the light of the accepted criterion. So in the case we face some fixed 'strategy' of Nature we cannot take advantage of 'mixing' our rules once we have chosen a rule R_l giving the mimimum value E_l.

VI

We do not know too much about the values E_l. Yet what we know is sufficient to answer the main problem we are interested in.

Let us consider the expected losses E_μ and E_λ associated with an non inductivist rule R_μ and the inductivist rule R_λ corresponding to R_μ. By (5.1) and (4.8)–(4.10)

$$(6.1) \qquad E_\mu - E_\lambda = \sum_{k=0}^{m} P_k (u_{\mu k} - u_{\lambda k}) = \sum_{k=0}^{m-1} P_k \varepsilon_k (\mu, \lambda) \left(\frac{k}{m} \right)^n +$$

$$+ P_m \varepsilon_m (\mu, \lambda) = \sum_{k=0}^{m-1} P_k \varepsilon_k (\mu, \lambda) \left(\frac{k}{m} \right)^n + P_m z_{\mu m} (n).$$

In view of our assumptions about the loss function, all the values $\varepsilon_k (\mu, \lambda)$

are finite (positive, negative or zero). Hence, by (4.10) and the assumption about P_i's we see from (6.1) that

(6.2)   for any noninductivist rule R_μ such n can be chosen that the expected loss E_λ associated with the inductivist rule R_λ corresponding to R_μ is smaller than the expected loss E_μ associated with R_μ.

What the proposition (6.2) amounts to is that if only the number of A's examined is sufficiently large we should not apply any noninductivist rule R_μ, some inductivist rule R_λ being better than R_μ, i.e., giving us the smaller expected loss. The proposition does not give us any criterion of choice between the inductivist rules. But we do not need such an information being interested only in the question whether the rule of induction should be applied if all examined A's turn out to be B's.[6]

VII

Let us summarize briefly the main results of the paper. Looking for the answer to the question whether the induction rule is reasonable we introduced first of all the general scheme of finding out the optimum rule of inference from among a set of rules, every one telling us which hypothesis (out of a given set of hypotheses) should be accepted in a given experimental situation. We decided to consider the best the rule (or rules) giving us the smallest expected loss. This has allowed us to prove that:

If: (1) the losses we incur accepting a false hypothesis are positive but finite and no loss is associated with the acceptance of the true hypothesis;

(2) there exist some probabilities of the hypothesis h_i stating that i A's are B's, the probability P_m that all A's are B's being positive;

then:

there exists such a number n that if the number of A's examined is not smaller than n and all of them turned out to be B (and none to be non-B) we should accept the hypothesis that all A's are B's.[7]

To say it quite shortly: under the assumptions (1) and (2) there exists a number of •instances confirming the inductive generalization which justifies the application of the rule of induction.

Obviously, we cannot determine the number n justifying the rule of induction as long as we do not assume some specific form of the loss function and some distribution of probabilities P_i's. In the present general logical analysis of the rule of induction it would not make much sense to specify those data of the problem. What seems to be important is that any loss function satisfying the conditions mentioned in the point (1) justifies the rule of induction the number of confirming instances being sufficiently large.

Is our conclusion a justification of induction the logicians and philosophers have been searching for a long time? Let me repeat once more what has been said at the beginning: various criteria may be applied when we ask whether this or that kind of inductive inference is reasonable. The aim of this paper was nothing else but to show that on some general conditions some plausible (and fashionable) criterion justifies at least the searching for increasing number of confirming instances, because this procedure must finally result in justifying the inductive generalization. Which number of instances justifies a particular inductive inference — this is a question which cannot be answered unless more data of the problem are specified.

REFERENCES

* First published in *Studia Logica* **X** (1960).
[1] '*Zagadnienie probablistycznego uzasadnienia indukcji enumeracyjnej*' ('The Problem of Probabilistic Justification of Enumerative Induction'), *Studia Logica* **V**, p. 91–108, see also this volume, pp. 65–80).
[2] There are many books yielding necessary information on the application of the game theory to the problem of choice of the best hypothesis (or decision) which choice depends on the outcome of some experiment. See e.g. J. C. C. McKinsey, *Introduction to the Theory of Games*, N. York 1952. The controversial problem what is the realtion between the accepting of hypotheses and undertaking some actions will not be discussed here.
[3] Certainly, we do not need to consider just the hypotheses h_i as defined above. We could instead consider a set of hypothesis H_v, the hypothesis H_v telling us that the frequency of B's among A's is v, the index v running through a given set of frequencies, say, 0, $1/w$, $2/w$, ..., 1, with $w \neq m$. This would not change the basic conclusion of the paper.

[4] The basic result of the paper would not be changed if we assumed that $s_{ik} \leqslant 0$ in case $i = k$, i.e., that there is some nonnegative profit in case we accept the true hypothesis.

[5] The sequence S includes — under the suggested interpretation — only such pairs that: (a) the sets A_i are sufficiently numerous to be subject of inductive, not exhaustive, examination, and (b) it does not follow logically from the definition of any set A_t that its elements have the property B_t. Otherwise it would be no reason to apply to the pair $\langle A_t, B_t \rangle$ inductive inference. Let us add that if the condition (b) were not satisfied it would be trivially true that $P_m > 0$.

[6] It may be added that in view of the formulae (4.8)–(4.10) if $n \to \infty$, any non-inductivist rule R_μ tends to be 'dominated' by the inductivist rule R_λ corresponding to it, as when n increases indefinitely the losses $u_{\lambda k}$ and $u_{\mu k}$ approach the same limiting value for $k = 0, 1, \ldots, m-1$, whereas $u_{\lambda m} = 0 < u_{\mu m}$. In such a situation no reasonable criterion would suggest us to use any noninductivist rule. This situation, however, is only approached and cannot be reached for any finite n, so that above mentioned property of inductivist rules is of no great practical importance.

[7] It should perhaps be noticed that we did assume throughout the finiteness of the set A. Yet, as we did not put any limit on the number m of the elements of A, this assumption is of no practical importance.

T. CZEŻOWSKI

ON TESTABILITY IN EMPIRICAL SCIENCES*

1. PROBABILISTIC JUSTIFICATION

Statements of empirical sciences — sentences stating results of observations, hypotheses and laws — can be justified by testing them. Testing such statements is a form of probabilistic reasoning, which has been given a modern shape in the investigations of J. M. Keynes, J. Nicod and others. It utilizes the concept of empirical probability, and distinguishes the probability of events and the probability of sentences. In order to determine the probability of an event one must include it into the set of events in terms of which its probability is to be determined. In order to determine the probability of a sentence, by means of which a statement is made about a probable event, one must generalize the logical relation of implication and define the *probabilistic implication.*

Implication is the logical relation which a sentence p bears to a sentence q if and only if q is not false while p is true. We denote by Cpq the implication relating the sentences p and q, and we read it 'if p, then q'. The sentences represented by the symbols p and q must therefore be chosen so that the truth of the antecedent of the implication excludes the falsehood of its consequent; such is the situation, for example, when sentences represented by p have the form 'the number x is divisible by 6', and sentences represented by q — 'the number x is divisible by 3'. We denote probabilistic implication by the symbol $C_u pq$, read as 'if p, then probably q to the degree u', where u is a real fraction $u = \dfrac{n}{m}$, n, m are whole positive numbers and n is inferior to m. The above expression is understood as follows: sentences represented by p and sentences represented by q are coordinated in such a way that n sentences true in the set of sentences q correspond to m sentences true in the set of sentences p — the set of true sentences q is the n/m-th part of all the sentences

q corresponding to the set of true sentences p. In other words: the frequency of cases when q is true, in the set of cases when p is true, equals the fraction n/m. Let the sentences p have the form 'the number x is divisible by 3', and the sentences q—the form 'the number x is divisible by 6', we can say 'if the number x is divisible by 3 then probably the number x is divisible by 6 to the degree $u=\frac{1}{2}$'—as every second number divisible by 3 is also divisible by 6.

The implication Cpq can be considered as an extreme case probabilistic implication $C_u pq$, in which $n=m$, i.e. $u=1$, as in every case in which p implies q and p represents a true sentence, also q represents a true sentence. It is worth pointing out, however, that according to the definition of the implication Cpq it is true also if p represents only false sentences. In such a case, if for example the sentences represented by p have the form 'the whole positive number x is even and inferior to 2', let the sentences represented by q be of the form 'the number x is divisible by 3', as among the whole positive numbers there is no such a number which would be even and inferior to 2; therefore, $m=n=0$ and the degree of probability $u=0/0$ is undertermined. We can, however, securely exclude such cases from further considerations.

The logic of probability determines relations between probabilities. We introduce, as assumptions of the theory of verification, three theorems of logic of probability which develop the sense of the probabilistic implication and formulate such relations. First, we shall define the expression $P(pq)$, read as: 'degree of probability from p to q',

(I)   '$P(pq)=u$' $=$ '$C_u pq$'.

It reads: 'The degree of probability from p to q equals u' means the same as 'if p, then probably q to the degree u'. In the definiendum '$=$' is the sign of arithmetical equality, and between the definiendum and the definiens the sign '$=$' is that of equivalence.

The starting point of the theory of forms of reasoning we are interested in here is the problem of inversion of implication, which can be formulated as follows: What is the relation between the sentences q an p, if the sentence p implies the sentence q? Let us suppose that there are m true sentences among those represented by p; the assumption Cpq guarantees that among the sentences represented by q there are

at least also m true sentences: they are all the sentences which correspond – as values of the variable q – to m true sentences p. But in general it cannot be excluded that among the sentences q there are even more true sentences corresponding to false values of the variable p. Again, for example, let sentences of the form 'the number x is divisible by 6' be values of the variable p, and sentences of the form 'the number x is divisible by 3' – values of the variable q; for $2n$ true sentences among the values of the variable q there are (for the same successive numbers of the natural sequence as values of the variable x) only n corresponding true sentences among the values of the variable p. Such considerations lead to the conclusion which answers the question of inversion: If Cpq and among the sentences represented by p in this implication not all are false, then $C_u \, qp$, and the value of the fraction u can be determined in terms of more detailed conditions concerning p and q – in the example just considered $u = \frac{1}{2}$. We shall write the outlined solution in the form of the theorem

(II) $\qquad CKN(p) Np (0 < u \leqslant 1)(pq) CCpq (Eu) C_u qp$.

It is the *law of inversion*: If not every sentence p is false, and u is a real fraction or equals one, then for every p and every q, if p implies q, there is an u such that q makes p probable to the degree u, or – in other words – partly justifies it to this degree. We call the sentence p the reason for the sentence q as consequence; the law of inversion determines the probabilistic relation between a consequence and its reason.

Let us now go to another theorem; let us assume:

(1) $\qquad Cpq_1$,

(2) $\qquad Cpq_2$,

(3) $\qquad Cq_1q_2$,

while accepting again that the sentences represented by p are not all false. We then formulate the probabilistic implications $C_{u_1}q_1p$ and $C_{u_2}q_2p$; the degrees of probability $u_1 = \dfrac{m}{n_1}$ and $u_2 = \dfrac{m}{n_2}$ are calculated as the quotient of m, i.e. the number of true sentences in the set of sentences represented by p, by n_1, i.e. the number of true sentences in the set of sentences represented by q_1, or by n_2, i.e. the number of true

sentences in the set of sentences represented by q_2. The conditions (1) and (2) guarantee that a true sentence corresponds in the sets of sentences q_1 and q_2 to every true sentence of the set of sentences p. The condition (3) entails that n_1 is non-superior to n_2, so that u_2 is non-superior to u_1. We thus obtain the theorem

$$(\text{III}) \qquad CCpq_1CCq_1q_2\left[P(q_2p)\leqslant P(q_1p)\right]$$

which we call the *first law of increase of degrees of probability* (for various consequences of the same reason, where q_1 is called direct consequence, and q_2 — indirect consequence): the degree of probability from the consequence to the reason is the bigger, the more direct is this consequence.

Let us now assume that:

(1) $\qquad Cp_1q$,

(2) $\qquad Cp_2q$,

(3) $\qquad Cp_1p_2$,

that is, p_1 and p_2 are reasons of q, where p_1 is called indirect reason and p_2 — the direct one. Let m_1, m_2, n be numbers of true sentences, respectively in the sets of sentences p_1, p_2, q. According to the assumptions (1), (2), (3), m_1 is non-superior to m_2, and m_2 — non-superior to n, and as $P(qp_1)=\dfrac{m_1}{n}$, $P(qp_2)=\dfrac{m_2}{n}$, we draw the conclusion that $P(qp_1)$ is non-superior to $P(qp_2)$. This reasoning can be presented in the theorem:

$$(\text{IV}) \qquad CCp_2qCCp_1p_2\left[P(qp_1)\leqslant P(qp_2)\right]$$

which is the *second law of increase of degrees of probability* (for various reasons of the same consequence): the degree of probability from the consequence to the reason is the bigger the more direct is the latter with respect to this consequence. The direct reason is a sentence less general compared to a more general one (for example, the generalization that no wild violets smell, based on the observation of some wild violets, is more probable than the supposition that wild flowers do not smell, based on the same observation), or a sentence giving a general description of an event compared to a sentence specifying exactly its time, place

and other circumstances (for example, the statement that the site of a burned settlement has been found, based on the observation of the ruins, is more probable than the statement, based on the same observation, of the date of the conflagration and of its course — because the second statement entails the first, but not conversely).

2. TESTING HYPOTHESES

Let the sentence p be a singular sentence of the form 'x is S', and the sentence q — also a singular sentence of the form 'x is P'); let us also suppose that we state about an individual a: If a is S, then a is P. The relation of implication between these sentences depends on the meaning of the terms S and P — and it is either connected with the definition of these terms, or it is based on evidence; for instance, sentences of the form 'x is divisible by 6' and 'x is divisible by 3' are connected by implication according to the definitions of divisibility by 6 and by 3, while sentences of the form 'x has graduated from a secondary school' and 'x knows Latin' are connected by implication on empirical grounds (in the European scholar system), if Peter has graduated from a secondary school then he knows Latin — conversely, if Peter knows Latin, then it is justified to a certain degree to assume that he has graduated from a secondary school, as among the people who know Latin there are some (even most of them) who have graduated from secondary schools.

A singular sentence of the form 'x is S', which, according to the considerations just presented, has a certain degree of probability with respect to the sentence of the form 'x is P' which it implies on empirical grounds, is called historical hypothesis. For such sentences are, as a rule, formulated within historical sciences, when these reconstruct facts from the past on the basis of preserved traces. A historical hypothesis is a singular sentence and concerns things or events not accessible to observation, whose existence can be determined only indirectly, by means of probabilistic reasoning. A historical hypothesis explains the facts stated in sentences implied by it, and conversely, these sentences, by giving it a degree of probability, justify it to this degree.

If a sentence of the form 'x is S' implies a sentence of the form 'x is P', then every individual being S is simultaneously P, although it usually is not the case that, at the same time, every individual being P

is S. Let m be the number of individuals which are P, and n — the number of those which are $S - n$ is not bigger than m, and the degree of probability or of justification of the hypothesis that an individual a is S if it is P, equals $u = \dfrac{n}{m}$. In general, there are many hypotheses explaining a sentence of the form 'x is P', and they have various degrees of justification. Let us suppose, for instance, that the event stated in the sentence 'a is P' is finding a human corpse with traces of violent death. The examining magistrate seeking for the explanation of this fact must chose between different hypotheses: the deceased has been murdered or he has been a victim of an accident, or else he has commited suicide. These hypotheses form an alternative: a is S_1 or S_2 or S_3, and its degree of probability depends on the degrees of probability of the hypotheses which it consists of. If (according to sufficiently exact statistical data) for m cases of violent death there are n_1 murders, n_2 accidents and n_3 suicides, then the total probability of the occurrence of the first, the second or the third possibility equals at most the fraction $\dfrac{n_1 + n_2 + n_3}{m}$ (equals if the terms of the alternative exclude each other, and is smaller if they partly coincide, for example, if death resulted jointly from murder and accident).

In order to decide between alternative hypotheses it is necessary to add some further premises, namely, singular sentences of the form 'x is R', 'x is T', etc. Let the sentence 'a is R' (for example, the statement that the fatal wound is situated in the back of the corpse) be such a premise, and let us suppose that it is implied by the hypothesis 'a is S_1', while it contradicts a consequence of the hypothesis 'a is S_2'. The premise 'a is R' therefore refutes the hypothesis 'a is S_2'; the reasoning involved here is destructive syllogism (negative testing): if p, then q — but non-q, then non-p. At the same time the degree of probability of the hypothesis 'a is S_1' increases — and in two ways: (1) the probability of the alternative, initially divided between its three terms, after one of the terms has been eliminated can be ascribed to others, so that each of the two remaining terms inherit part of the probability of the eliminated one, (2) if there are m cases x which are P, then among them (if 'x is P' does not imply 'x is R') only some are also R; let m_1 be the number of cases which are P and R — if 'x is S_1' implies both 'x is P'

and 'x is R', then each of the n_1 cases which are S_1 are also P and R; jointly then the conjunction of the two premises 'a is P' and 'a is R' makes the hypothesis 'a is S_1' probable to the degree $u_1 = \dfrac{n_1}{m_1}$, which is greater than the degree of probability $u = \dfrac{n_1}{m}$ of this same hypothesis with respect to the unique premise 'a is P', as the denominator m_1 is smaller than m. We thus have utilized the first law of increase of degrees of probability (III), because the conjunction of two consequences of a hypothesis is its more direct consequence than each of them separately.

The multiplication of premises by more and more precise examination of the elements of the event a successively eliminates some alternative hypotheses and increases the degree of probability of the others, and in the end there remains only one from among the alternative hypotheses. It may happen, however, during such examination, that all the hypotheses we took as starting point must be rejected – then a new examination should begin by finding another alternative of hypotheses. The examination ends when for an initial hypothesis a degree of probability equals or closely approximates one. Let us suppose, for example, that a is a number divisible by 3, which supports the hypothesis that a is a number divisible by 12 with the probability of the degree $u = \frac{1}{4}$. If we say, moreover, that a is a number divisible by 4, then the degree of probability of the hypothesis increases to one and the hypothesis becomes completely justified.

In our previous considerations we assumed that the sentence of the form 'x is P' and the hypothesis of the form 'x is S' explaining it refer to the same individual event a. The same reasoning can be carried out if these sentences refer to different events, if only one of them implies another. This is the case, for instance, if a set of events is understood as the parallel effects of a cause inaccessible to observation, and if a hypothesis states the existence of this presumed cause, for a sentence stating the existence of a cause implies on empirical grounds sentences about its effects.

A hypothesis is the better justified the more facts it explains and the more precise is the explanation, or else the more facts it allows to predict and the more precisely it allows to predict them. (A hypothesis explains a fact if and only if it allows at the same time to predict it; both cases

occur only when the sentence about the fact is implied by the hypothesis. Thus the hypothesis 'it looks like rain' allows to predict the fact that I shall take my umbrella when going out—and it explains this fact at the same time, if that was the case, but it is so only if the implication is assumed: if it looks like rain I take my umbrella when going out.) These dependencies are the direct consequence of the first law of increase of degrees of probability. The hypothesis p, which explains the facts a_1 and a_2 by implying the sentences q_1 and q_2 about these facts, is better justified than in case it implied only the sentence q_1, because the conjuction of sentences q_1 and q_2 is its more direct consequence than any of these sentences separately. And also the hypothesis p implying the sentence q_1, which states precisely the time, place, particular elements of the fact etc., is better justified than in case it implied instead of the sentence q_1, the sentence q_2 which determines the fact only vaguely, because the sentence q_1 implies the sentence q_2, and is therefore a more direct consequence of the hypothesis than q_2.

We have been considering the problem of justifying a historical hypothesis, and we have left out the question how it is obtained. It has however been said that the relation of implication between a historical hypothesis and its empirical premise either depends on the contents of both sentences, or it is an empirical relation. In seeking the hypothesis it is chosen for a given premise according to one of the two relations, by trials based on presumptions or chance. The actual research does not indicate any heuristic rules in these matters.

3. TESTING LAWS

Let us assume that observing an event a we have found that

(1) a is P and R and S and T.

P, R, S, T are terms referring to various properties of the event a; let for example a refer to the behaviour of a gas sample—it is described by measuring its volume, elasticity, temperature, by determining its chemical composition, colour, smell etc. To simplify let us assume that the sentence (1) exhausts our present knowledge about the event a. From the ˉentence (1) we deduce (by the *Darapti* syllogism) particular

categorical sentences:

(2)   some P is R, some P is S, etc.

These sentences serve as empirical premises to find and justify *scientific* laws of the form

(3)   every P is R, every P is S, etc.

The reasoning where sentences (2) serve as premises and one of the sentences (3) as conclusion is called inductive reasoning. There are several kinds of it; first of all simple induction and induction by elimination are distinguished. As we shall see, induction by elimination is a complex procedure, the elements of which reduce to other forms of reasoning. Simple induction has two applications: in explanations, i.e. discovery reasoning, in which having the premises (2) we try to find the law (3) explaining them, and as positive testing, i.e. justifying reasoning, in which for a given law of the form (3) we try to find the premises (2) justifying it. Induction by elimination, in its classical application, is a discovery (heuristic) reasoning, but to a degree it simultaneously justifies the law to which it leads.

Getting to a more detailed analysis of inductive reasoning let us remind that we begin from the observation of the event a, stated in the sentence (1). From this sentence we obtain a set of sentences (2) as premises of the reasoning. Let the variable q represent any of these sentences — each of them can be considered the starting point of a generalization by simple induction, the result of which is one of the sentences (3); let the sentences (3) be values of the variable p. For each value of p, for example, for the sentence 'every P is R', there is a value of q (in our example 'some P is R') such that the implication Cpq holds. According to the law of inversion (II) we conclude from this implication that $C_u pq$, but the degree of probability u for each generalization thus obtained is very low. Eliminative induction is used in order to select among them the generalizations having the greatest degree of probability.

In view of the low degree of probability of the generalizations p obtained by simple induction some of them are supposedly false; it can be expected, however, that may be not all of them are false, and that the alternative obtained by joining them all is true. We eliminate from that alternative false generalizations by means of alternative syllogism

(*modus tollendo ponens*). Two methods are distinguished in this respect, that of agreement and that of difference, depending on whether different predicates are chosen for the same subject in the generalizations *p*, or different subjects for the same predicate. The distinction between the two methods is connected with the non-symmetry of the roles of the both terms of the general sentence (3), contrarily to their symmetrical position in the particular sentence (2) — particular sentences are directly inversible, while general sentences are not. Therefore in actual research problems of the general law may be set forth either as the question what properties the objects *P* have (the method of agreement is then used), or as the question to what objects the property *R* belongs (the method of difference is then used).

Let us consider generalizations having the term *P* in the subject and other terms, i.e. *R*, *S* or *T* in the predicate. In order to eliminate false generalizations about the subject *P* we refer to further observations stated in sentences of the form (1). In these sentences we predicate about individual events *b*, *c*, *d* and we choose them so that the term *P* appears in each of them, while other terms appear or not. Let us suppose that we have stated new observations in the sentences

(1a)   *b* is *P* and non-*R* and *S* and *T*,

(1b)   *c* is *P* and *R* and non-*S* and *T*.

From them we obtain premises of the form (2): some *P* is non-*R*, some *P* is non-*S*, etc. Together with the alternative of all the generalizations (3) obtained before we use these premises to construct an alternative polisyllogism:

Every *P* is *R* or every *P* is *S*, or every *P* is *T*, etc.

some *P* is non-*R*,

some *P* is non-*S*,

which gives as conclusion the elimination of the two first terms of the alternative premise. By choosing successive premises *q* in an analogical way false generalizations about the subjects *R*, *S*, *T* are successively eliminated.

In order to eliminate false generalizations about the term *T* in the predicate alternative syllogisms are formulated by choosing empirical

premises not containing T, while other terms appear or not. For instance, the following observations are given:

(1c) d is non-P, R, non-S, non-T,

(1d) e is non-P, non-R, S, non-T.

From these data we obtain the premises: some R is non-T, some S is non-T, etc. We now order the generalizations which the alternative premise consists of so that the sentences about the predicate T are at the beginning:

Every P is T or every R is T, or every S is T, etc.

some R is non-T,

some S is non-T.

This syllogism gives as conclusion the elimination of the second and the third term of the alternative premise, the false generalizations about the terms P, R, S in the predicate are successively eliminated analogically.

The methods of reasoning described above can be considered from various points of view. With respect to the generalizations that are eliminated they constitute negative testing, i.e. demonstrations of falsity. With respect to the terms of the alternative which have not been eliminated our reasoning becomes eliminative induction, and it gives to the remaining terms of the alternative all the probability that the latter had at the beginning. As some of its terms have been eliminated, the others gain a higher degree of probability than they had before. Eliminative induction does not, however, totally discount the justificative power of the premises. Let us return to the observations (1a) and (1b); they supply not only premises serving to eliminate false generalizations, but also such which confirm generalizations that are not subject to elimination. For, these observations are instances of the generalizations which they supposed to justify (the situation is different when the method of difference is used, as the observations (1c) and (1d) do not contain the term T, which appears in the generalization being justified). The conjunction of the premise obtained from the observation (1) and similar premises obtained from the observations (1a) and (1b) implies each of these premises and is simultaneously implied by the law being justified,

applying to all these premises; such a conjunction is therefore a more direct consequence of the law than each of the premises separately. This allows to apply the first law of increase of degrees of probability. Every new observation, which tests the law applying to it, increases the degree of probability of this law, independently of its role in eliminative induction. The reasoning carried out here is a positive test; it conforms the scheme of simple induction in which, for a given law of the form (3), premises of the form (2) justifying it are searched for.

The positive test is carried out further after of the procedure of eliminative induction is completed, by multiplying confirming observations and thus increasing the degree of probability of the tested law. Moreover, the first law of increase of degrees of probability — here as well as with respect to the hypotheses — makes it that a law is the better justified the more facts it explains and the more precisely it explains them. The theoretical end of justification is the so-called complete induction, i.e. the exhaustion of the range of cases explained by the law — if it is attainable.

4. TESTING OBSERVATIONS

The question about what is the justification of elementary singular sentences about facts, for example 'it is raining here now', may usually be answered in two ways. First, one can refer to a perceptive idea: I maintain that it is raining because I can see it, hear it, feel it with my skin. Such a justification of a statement is called the direct justification. It is not a justification in the logical sense, but it is meant by epistemologists when they speak of experience as basis of knowledge. But even within epistemological considerations such a justification is blamed for being subjective, as a sensible idea is a subjective psychological phenomenon. Therefore yet another justification is required for statements of that sort, namely their testability. And testing is a kind of justificative reasoning, i.e. a reasoning in which justifying premises are searched for a tested sentence. It is then not wholly correct to maintain, in the methodology of empirical sciences, that singular sentences about facts are ultimate premises of those sciences. A sentence which requires testing is not an ultimate premise of the theory to which it belongs; this role is assumed by the sentence which has served to test the former. A sin-

gular sentence about a fact is testable by another such singular sentence —
and to that extent the statement just mentioned is true. But as, at the
same time, the process of testing in empirical sciences is never complete
from the theoretical point of view, then we can say that in these sciences
no singular sentence is an ultimate premise. The analysis of the justification of sentences in terms of their testability is a matter of logic, not of the
theory of knowledge. The present considerations aim at an analysis
of the former sort, with regard to the testability of observational sentences.

Singular sentences about facts are formulated in one way in historical
sciences and in another — in natural sciences. We therefore distinguish
historical sentences about facts from sentences of natural sciences about
facts. Here we have an example of a historical sentence about a fact
which assumes the already known form:

(1) a is P and R and S and T,

where a is the name of an individual event or object, P, R, S, T are terms
by which this individual event is described, — for example 'the session
of the Roman Senate on March 15th, 44 B.C., proceeded as follows:
...'' A fact stated in such a sentence becomes inaccessible to observation
in the moment when it is finished, and the sentence stating this fact enters
the category of sentences about events or objects inaccessible to observation. Such sentences are historical hypotheses. They are testable not
by confrontation with a new observation of the same event, but only
indirectly by referring to observations of other facts, which supplies
premises implied by the tested hypotheses (as it has been presented
in the chapter on testing hypotheses). In the actual historical investigation the starting point of a testing is usually a historic note as
'source'; the readings of such notes are the facts confronted with
hypotheses. The statement, on the grounds of reading, that the source
describes a given event may be explained either by that the chronicler
had observed and exactly described the event or by that he had invented
or changed it for some reasons. One or the other hypothesis explains
the contents of a note the better the more detailed it can be, i.e. the more
exactly it is conform to various elements of the contents of the note.
On the other hand, one or the other hypothesis explains more if it is
applicable directly or indirectly to various notes concerning the given

fact, accounts of which agree or disagree in various reports, both contemporary and elaborated later. We thus see that testing historical sentences about facts does not constitute a separate form, but it is reducible to the case of testing a hypothesis presented before.

In natural sciences testing does not begin with an observation which would provide premises implied by the tested sentence, as it was the case in testing a historical sentence about a fact, but from the observation of facts of the same kind as that referred to in the tested sentence. Therefore a sentence of natural sciences about a fact describes the fact otherwise than a historical sentence (1) — it does so by specifying an individual a, about which it predicates (1) in a general way in terms of a certain set of its properties, for instance, P. We replace the sentence (1) by (1_1) which is deducible from it (on the grounds of the *Darapti* syllogism):

(1_1)   some (first) P is R and S and T

and we can consider it tested if it appears that also some second, third etc. — i.e. every P is R and S and T. Therefore the observation (1_1) is tested in the same way as the general sentence:

(2)   every P is R and S and T,

in which P (as it is often said) is a sufficient condition for R, S and T. Let us suppose for example that the observation to be tested is the experimental stating of the phenomenon of diffraction of a light beam passing through the aperture of a screen; in determining P the experimental conditions of the occurrence of that phenomenon are given, i.e. a description of the equipment used and an enumeration of the steps which are to be undertaken — R, S, T represent the results of the experiment.

The sentence (2) is testable by simple induction, as scientific laws. We observe a new fact belonging to the set of objects P and we assume that the primary observation will be confirmed if R, S and T are observed, and it will not be confirmed in the opposite case. Let us suppose the first possibility occurs; the new observation

(1′) b is P and R and S and T

yields the premise

(1$'_1$)   some (second) P is R and S and T,

which together with the sentence (2) belongs to the scheme of simple induction. The multiplication of premises of that sort increases the degree of probability of the sentence (2) and confers the same degree of probability to the primary observation (1). If, on the other hand, the new observation leads to a negative result, the law (2) is refuted and it appears at the same time that our interpretation of the observation (1) by including the observed fact in the set of objects P is not utilizable in the testing. This does not, however, prove the falsehood of the historical sentence (1), in which the fact was stated. We only must correct the primary interpretation by restricting the range of the set to which the observed fact was included. We shall therefore now assume, for example

(2′)   every P and R is S and T

and the testing of this sentence proceeds in exactly the same way as that of the sentence (2).

The sentence (2′) is implied by (2) (as, if every P is R and S and T, then – as every P and R is P – every P and R is R and S and T; and as every R and S and T is S and T, then – as every P and R is R and S and T – every P and R is S and T) – let us then assume that the next observation tests (2′), then according to the second law of increase of degrees of probability (IV) the degree of probability of (2′) with regard to the result of the new observation of the type (1) is greater than the degree of probability of (2) would be with regard to the same result; the same degree of probability may be assigned to the result of the primary observation. Thus replacing the sentence (2) by (2′), which is less general and therefore describes more precisely the object of the primary observation, entails that every successive case of positive testing increases the degree of probability of the primary observation more than it would if (2) had remained. The more elements of the observed individual a are transferred to the domain of the conditions determining it, the less undetermined elements are left, that is, elements for which the result of testing is not determined. In actual investigations this is achieved by more and more precise knowledge of the individuals

examined and by excluding so-called disturbing circumstances. On the other hand, the economy of investigation requires that the description of the object or phenomenon examined should not be charged with elements which are indifferent to the investigation and can therefore be left out. Many indications about the procedure to be taken in order to satisfy the requirements of exactness and economy are provided by experiment and experience.

We shall formulate the result of the present discussion in a statement which we shall call the *principle of testability of observation*: The more precisely determined is the object of examination, the more the probability of the observational sentence about that object increases in every case of testing, which entails that, at the same time, it becomes the more probable that a new observation will confirm the primary one. This conclusion is based exclusively on the logical analysis of probabilistic reasoning, and we have referred particularly to the second law of increase of degrees of probability. The law does not require the assumption of the principle of causal determinism, nor that of uniformity of nature, which were considered previously by scientists as necessary assumptions in such reasonings.

The reasoning we speak of as of the testing of observations is a sort of analogical inference, which is accounted for in handbooks of traditional logic. For analogical inference is one from which — starting from a statement that a particular of a sort *P* is *R*, as premise — we conclude that also another particular of the kind *P* is *R*. The general law 'every *P* is *R*' applies to both cases, which implies both the premise of the reasoning and its conclusion.

This law determines the analogy between the objects *P*; the less general it is, the closer is the analogy. Our principle of testability of an observation finds its counterpart in the analogical inference, namely in the rule saying that the more the second case resembles the first in properties known to us (i.e. the closer is the analogy and, at the same time, the richer is the content of *P*), the more probable is the assumption that it does not differ from the first in respect of further properties *R*; if we have established that two similar apples have been picked up from the same tree, then there is a greater probability of the conclusion that they will taste similarly, than the probability in case of the same conclusion based only on the similarity of their appearances.

However, analogical inference is usually considered heuristic, i.e. it proceeds by taking the primary observation as the starting point to predict an analogical result of observation in further instances. Thus, for example, when we employ the analogical inference in its heuristic form, we predict that a medicine which has proved efficient in a case (a) of disease will also prove efficient in another sufficiently similar case (b). In using the analogical inference as method of testing the efficiency of the medicine in the case (b) serves as premise testing the statement that it was efficient in the case (a).

When testing an observational sentence in the way indicated above it is not possible to demonstrate its truth by the exhaustion of all the possible instances (which takes place when a law is tested by complete induction), as no set of testing premises is equivalent to the sentence under test, but it is also impossible to demonstrate the falsehood of an observational sentence, as the supposition cannot be excluded that the observation is not tested because a substantial factor in the determination of the examined object has been omitted. This is an important point as in the discussion concerning so-called metaphysical sentences attempts have been made to define them as sentences having this characteristic property. As we see, also ordinary observational sentences have this property, sharing it with all the existential sentences. Hume's statement about the uncertainty of observational sentences should therefore be extended also to their negations.

REFERENCE

* First published in *Revue Internationale de Philosophie* **17–18** (1951). Translated by S. Wojnicki.

JERZY GIEDYMIN

ON THE THEORETICAL SENSE OF
SO-CALLED OBSERVATIONAL TERMS AND
SENTENCES*

1. PRESENTATION OF THE PROBLEM

In contemporary methodological literature concerned with the structure
of scientific theories, the empirical sense of expressions and testing
hypotheses, the view prevails that it is possible to distinguish observational and theoretical terms among the descriptive terms of empirical
sciences, and that among the sentences of these sciences it is possible
to distinguish observational (perceptual, basic, elementary, evidential)
and theoretical sentences.[1] Names of colours of things (macroscopic
objects) are given as examples of observational terms: red, green, blue,
etc.; names for their shapes: round, triangle, with a sharply outlined
contour, etc.; relational names: left from ..., congruent with ..., longer
than ..., etc. Simple sentences (without quantifiers), containing only
observational terms among their descriptive terms and negations of
these sentences, are given as examples of observational sentences: 'This
lemon is yellow now', 'The Eiffel Tower is higher than the building
next to it', 'The pointer of this instrument is now indicating zero', etc.
Observational expressions are characterized according to the following
definitions: A term M belongs to the observational vocabulary if, under
suitable conditions, it is possible to decide (definitely), on the basis of
(only) direct experience, whether the term M applies to a given situation
or not.[2]

A sentence Z is an observational sentence if it is possible to decide
directly, on the basis of observation, without reference to any other
sentences, whether what this sentence states (i.e., if it is possible to
justify sentence Z or its negation) is the case.[3]

The class of observational sentences of a language constitutes so-
called empirical basis (empirical foundation) which provides with

111

empirical sense expressions logically related to its elements, more-over—if we consider the subclass of observational sentences accepted at a given moment—this basis is used to decide or partially test (to confirm, to question) non-observational sentences. The contents of observations wholly determines the sense, and thereby the conditions of truth of observational sentences, or—in a different terminology— every observational sentence of a language belongs to the domain of an empirical meaning rule, which assigns it univocally either to sensory experiences or to an observable situation. The rejection of a given sentence in the presence of sensory experiences or a definite observable situation, which are assigned to this sentence by an empirical rule, violates the sense of the sentence.[4] In this respect all observational sentences are non-hypothetical, ultimate. Observational terms have meanings independent of the sense of theoretical terms which occur in theories systematizing results of observations; observational sentences and empirical laws, i.e. general sentences containing, apart from logical expressions, only observational terms, have a sense independent of the theories which systematize them.[5] In this case the independence consists in that (a) observational sentences and empirical laws are justified independently of the theory, (b) observational sentences and empirical laws may prove to be true in spite of the falsity of the systematizing theory.

The empirical sense and the method of deciding observational sentences seem unproblematic, and therefore contemporary literature dealing with the empirical sense of expressions deals mainly with the conditions, sufficient or necessary, of the empirical meaningfulness of non-observational expressions, theoretical terms and sentences, as all the criteria of meaningfulness hitherto proposed by advocates of empiricism have proved inadequate. Although it is admitted that the distinction between observational and theoretical terms in the natural language of a given empirical science meets numerous difficulties because of the vagueness of the distinguishing criteria, it is nevertheless said that discussions concerning the criteria of meaningfulness of theoretical terms are independent of the actual boundary between the observational and theoretical vocabularies.[6] For these considerations it suffices to assume a given language J, the vocabulary of which contains descriptive as well as logical terms, and that there is at least one finite model M,

in which at least some descriptive terms of the language *J* have a direct interpretation, i.e. the objects they denote are assigned to them by indication, that is ostensively. In the sections that follow I shall attempt to give arguments which, in my opinion, seem to support the view that:

(1.1)   The account outlined above is not a satisfactory reconstruction of the language of empirical sciences.

(1.2)   Examples of observational expressions found in contemporary methodological literature, as well as other relatively simple sentences of empirical sciences, on the basis of which theories and hypotheses are tested, do not have the properties that are assigned to them according to the above-presented definitions of observational term and sentence; from the point of view of the criteria presented in these definitions all the terms and sentences of the natural language of empirical sciences have a non-observational, theoretical character, i.e. they are revokable, and their rejection need not involve breaking linguistic rules.

(1.3)   All the elementary sentences of empirical sciences, on the basis of which hypotheses and theories are tested, are sentences about objective facts, and not sentences about the observer's experiences, and they have the character of refutable hypotheses.

(1.4)   The nominalistic and constructivistic view of the empirical sciences, utilizing the concept of direct interpretation of elementary expressions from which significant compound expressions are then constructed according to definite syntactic rules, is not a satisfactory reconstruction of the procedures of empirical sciences (especially physics); nor is the conception according to which an empirical theory is a formal calculus, enriched by so-called correspondence rules, which assign to it at least one domain of observable objects as a model (in the semantic sense). The shortcoming of both these conceptions seems to be the fact that they both assume definite observational interpretations of the elementary expressions of the language of empirical

sciences, they assume the existence of a constant 'empirical basis' and they do not take into consideration the mutual dependency between the elementary expressions and those of the 'higher levels' of the theory.

(1.5)   It is erroneous to assume that from the point of view of the theory of knowledge and methodology only so-called theoretical terms require analysis, while the status of so-called observational terms and observational sentences is non-controversial.

(1.6)   Revisions of the negative results of various experiments in view of later results, obtained by using more sensitive instruments seem to show that the impressions received by the observer, or even by many 'experimentally independent' observers, do not constitute criteria of sense nor criteria of truth of the observational expressions of the language of empirical sciences, while the sentences about the observer's reactions, about the observer's reliability or about his systematic errors are elements — either as explicitly formulated premisses or as tacitly accepted assumptions — of the procedure of testing hypotheses and theories in terms of observational sentences.

In search of arguments supporting this view, I shall present in Section 2 a text-book version of the so-called Aristotelian experiment and of its revision; in Section 3 a certain analysis of the Aristotelian experiment is given; in Section 4 we shall find some other examples of revisions of observational sentences, Section 5 contains general conclusions of the discussion.

2. A TEXT-BOOK VERSION OF THE SO-CALLED ARISTOTELIAN EXPERIMENT AND OF ITS REVISION

In order to find arguments supporting the theses (1.1)–(1.6), let us first consider the reasoning (known from elementary text-books) the author of which is said to be Aristotle (or Archimedes) and which is therefore called the Aristotelian experiment.[7] We shall also present the revision of this argument resulting from the development of methods

of observation. We are interested in the following question:

(H_0?)   Does the Earth go round the Sun?

and whether the positive answer to that question is to be rejected. The positive answer in this case is the heliocentric theory, which concerns a non-observable phenomenon, or at least one that is not accessible to direct observation for an observer on the Earth. In order to decide the question on the grounds of experience, we must find sentences describing possible observations [8] and connected by logical relations (we restrict ourselves to such relations in the considered example) with direct answers to the question (H_0?), that is to the heliocentric hypothesis and its negation. As Aristotle tried to reject the heliocentric hypothesis, we shall call it — according to a common convention — the zero hypothesis. The relation between the zero hypothesis and possible observations is stated in the following sentence:

(2.1)   If the Earth goes round the Sun then stellar parallax occurs, i.e. an apparent shift of a near star with regard to a distant star.

Sentence (2.1) in conjunction with (H_0) logically entails, according to *modus ponendo ponens*, the sentence:

(2.2)   The phenomenon of parallax occurs,

which is a partial answer to the question (H_0?),[9] and its negation is a potential falsifier of the zero hypothesis, and therefore a full indirect answer to the question (H_0?).[10] If the phenomenon of parallax is an observable one, and such was the opinion of the author of the analyzed experiment, sentence (2.1) may be considered the assumption of the empirical decidability of the question (H_0?),[11] shifting the problem from the domain of non-observable phenomena to that of observable ones and supplying the observer with information about what phenomenon he is to look for.

In text-book descriptions of the Aristotelian experiment we usually find an explanation saying that the phenomenon of stellar parallax, referred to in statement (2.1) belongs to the class of phenomena we know well from everyday experience. This category of phenomena also includes,

e.g., the effect which can be observed if we look at a pencil held in an immobile outstretched hand, first only with the left eye (the right one being closed), then only with the right eye (the left being closed), and if we determine the position of the pencil with respect to some object which is more distant from us than the pencil, for example with respect to the corner of the room; the pencil is then subject to an apparent shift towards the left with respect to the corner. Looking in turn only with the right or left eye at the pencil held in an immobile outstretched hand we observe the pencil from two different points. Similarly, if the Earth together with an observer situated on it goes round the Sun, then at two different moments, say, December 15th and June 15th, the observer determines the position of a nearer star with respect to a farther star [12] from two different points in space, and we can assume that a difference will be observed between their mutual positions. An analogous phenomenon is instrumental parallax which appears when we look at the pointer of a measuring instrument not perpendicularly but at an angle less than 90°, which — as we know — may result in an erroneous reading. All these examples have in common the phenomenon of an apparent shift of an object x with respect to an object y, when the observer determines the position of x with respect to y from two different points in space, while x is actually immobile with respect to y. Speaking only of the occurrence or not occurrence of parallax we use the term 'parallax' as a classificatory term; naturally, there also exists a metric concept of parallax. In astronomy we mean by parallax in the metric sense a half of the angle, the apex of which is a given nearer star and which is formed by the lines connecting that star in its two extreme positions, for instance on June 15th and December 15th, with the observer on the Earth.

Text-books also tell us that Aristotle carried out observations in order to find the phenomenon of stellar parallax, which would have to appear under the assumption of the zero hypothesis, and that, having obtained a negative result, i.e. having observed no parallax, he concluded the sentence (2.2) to be false and on these grounds he rejected the heliocentric hypothesis. This information is usually completed in text-books by a comment saying that:

(2.3)   Aristotle was right in looking for stellar parallax under the assumption of the heliocentric hypothesis.

(2.4)   The phenomenon of stellar parallax does occur, although it was observed and measured for the first time only in the years 1837–39.

(2.5)   Aristotle did not realize the enormous distances between the stars and the Earth, and probably for this reason he did not take into consideration the possibility that the parallax may be too small and, therefore, unobservable; he thus did not take into consideration that his observations might have been unreliable, and from their negative result he drew the conclusion that the phenomenon he looked for did not occur.

3. THE ANALYSIS OF THE ARISTOTELIAN EXPERIMENT

In this section we shall analyze the premises of Aristotle's reasoning and the theorems (2.3)–(2.5), in order to consider, on the basis of the results of this analysis, two antagonistic views on empirical sciences presented in Section 1.

The Aristotelian experiment was – or at least can be – designed on the basis of models known from common experience. The experiment with a pencil held against the background of a room corner provides such a model, as well as the situations in which we observe instrumental parallax. Thanks to these models the phrase: 'goes round the Sun' has a certain figurative or visual sense which enables us to have some expectations, and at least some of these expectations may prove observable.

It is worth noting on the ground of the later history of the heliocentric hypothesis that cardinal Bellarmino proposed to Galileo a merely instrumental interpretation of the heliocentric hypothesis, on which it was to be only a convenient mathematical instrument for making predictions and possess semantic function only in an abstract geometric model.[13] It should be added here that in the previous sentence we have used the term 'model' in the semantic sense, while speaking of models known from common experience on which the Aristotelian experiment was based we mean models in the heuristic sense, which provide hypotheses under the assumption of an analogy with the examined systems inaccessible to direct observation.

The sense of the term 'parallax' is also understood by analogy owing to the knowledge of the above mentioned models. Let us now try and define the sense of this term more precisely, as well as the sense of sentences (2.1) and (2.2), formulated in the previous section in an incomplete way, as we did not, e.g., provide the time co-ordinates.

We shall give an explication following which the occurrence of parallax is understood as an objective fact, independent of the observer, i.e. as non-identity of two relations between objects, or otherwise as a change in the spatial location of three objects:

(3.1)   The relative positions of the Near Star, the Distant Star and the constant point O_z on the Earth are different in the moments t_1 and t_2, where t_1 and t_2 are two definite, non-identical moments of one, although arbitrary, year.

For short we shall write the sentence (3.1) in symbols:

$$e_{t_1 t_2}.$$

The negation of the sentence (3.1) naturally states the identity of the relative positions of the Near Star, the Distant Star and the point O_z on the Earth in the moments t_1, t_2 and thus it states the non-occurrence of parallax in the time interval $t_1 - t_2$.

In describing in the preceding section the text-book version of the Aristotelian experiment we have defined parallax as an 'apparent shift'. The objects of whose relative position sentence (3.1) speaks have been choosen so that we know about two of them, i.e. about the Near and Distant Stars, on the grounds of prior astronomic knowledge, that they are immobile in relation to each other. We shall now write explicitely this information as an assumption.

(3.2)   The Near Star and the Distant Star are immobile with respect to each other.

It is only due to the assumption (3.2) that the sentence (3.1) states the occurrence of parallax in the sense accepted in natural sciences, i.e. in the meaning of an 'apparent shift'. Owing to the assumption (3.2) we interpret the movement of the Near Star with respect to the Distant Star observed by an observer as illusory, i.e. as the projection of the

actual movement of the point O_z, and knowing about the latter that it is a fixed point on the Earth we interpret it as a projection of the motion (the change of position) of the Earth. Assumption (3.2), as well as the information concerning the stars nearer and farther from the Earth, belongs in Aristotle's reasoning to unquestioned assumptions, and has its counterparts also in the reasonings concerning instrumental parallax and the experiment with the pencil—from Section 2. Their role consists in eliminating an alternative hypothesis with respect to (H_0) (or to the hypothesis under test in other reasonings), without which the result of experiment would be ambiguous.

It is easy to guess that sentence (3.1) is obtained in Aristotle's reasoning from a general sentence stating the occurrence of parallax for any two moments i, j of any year from the history of the Earth:

(3.3)   For any two moments i, j (from any year of the history of the Earth), if $i \neq j$, the mutual position of the Near Star, the Distant Star and the constant point O_z on the Earth is different.

Let the following notation be the abbreviation of (3.3):

$$\bigwedge_i \bigwedge_j e_{ij}.$$

The premise (2.1) of Aristotle's reasoning will be symbolized now as follows:

(3.4) $H_0 \rightarrow \bigwedge_i \bigwedge_j e_{ij}.$

The change of the relative positions of three objects in space has many degrees of freedom. If these are limited, sentences (3.1) and (3.3) can be simplified. Let us say at the same time that we are above all interested in cases where there is an observer observing the position of the Near Star and the Distant Star from a fixed point O_z on the Earth. We are considering a system where the position of the observer with respect to the other elements of the system plays an important role, i.e. we are considering the relation of the spatial location of the Near Star and the Distant Star and of the observer situated in a fixed point O_z on the Earth. From this system two elements are accessible

to observation from the Earth: the positions of the Near and the Distant
Star. We obtain sentences simpler than (3.1) by taking the position
of the Distant Star as the beginning of the system and by determining
the position of the Near Star for two different moments t_1 and t_2 of
an arbitrary year in terms of 'left from the beginning of the system for
an observer from the Earth' and the distance. The position of the Near
Star in the moments t_1 and t_2 is symbolized as follows:

$$\overleftarrow{(p_{t_1}, 0)}, \quad \overleftarrow{(p_{t_2}, 0)},$$

which reads: 'the situation when in a moment t_1 (respectively t_2) the
Near Star is in a point p to the left from the Distant Star for an observer from the Earth'. With these simplifications we shall write the
explication of the sentence stating the occurrence of parallax as follows:

(3.5)   the parallax of the Near Star and the Distant Star occurs
          between the moments t_1 and t_2 if and only if:

$$\overleftarrow{(p_{t_1}, 0)} \neq \overleftarrow{(p_{t_2}, 0)}.$$

The non-occurrence of the parallax will, under these simplified assumptions, be the identity of positions, i.e. of the
direction and the distance of the Near and the Distant Stars.

A difference in the positions of two objects, for example their distance,
may assume various values, starting from zero. Only for some values
will this difference be noticeable for the observer using a given observational method. Every measuring instrument, as well as the observer's
eye, has a limit of sensitivity, beyond which it does not react to impulses
and therefore does not differentiate weak impulses, weaker than the
minimum value distinguishable by the eye or a given instrument, from
the lack of impulses. In taking a negative result of an observation or of
a measurement, i.e. the absence of reaction of the instrument or of the
observer, as an indication of the absence of impulses rather than of too
weak impulses, we are following direct experience but some explicitly
or tacitly assumed theory or hypothesis. Let us call the failure to observe
of any difference between the positions of the Near and the Distant
Stars in two different moments t_1 and t_2 the negative result of the search

for stellar parallax. In order to be able to justify the acceptance in the Aristotelian reasoning of the negative result of the search for parallax as an indication of the non-occurrence of the parallax, the following assumption should be made:

(3.6)   If the parallax of the Near Star with respect to the Distant Star occurs in the period $t_1 - t_2$, it will be observable for every properly situated observer on the Earth:

Let $e_{t_1 t_2} \rightarrow \bigwedge_x \mathrm{Obs}(x, e_{t_1 t_2})$ be the abbreviation of the sentence (3.6).

An appropriate size of parallax is a necessary condition for its observability. Sentence (3.6) is then justified by the following assumption:

(3.7)   If we assume (H_0) the relation of the distance between the two extreme positions of the Earth on the solar orbit and the distance between the Near Star and the Earth in these extreme positions is sufficiently large for the parallax to be observable with an unaided eye.

We shall note the negative result of the search for parrallax as follows:

(3.8)   In the period $t_1 - t_2$, where t_1 and t_2 belong to one year, no difference was observed in the relative positions of the Near and the Distant Star. Sentence (3.8) can be abbreviated as:

$$\sim V(a, e_{t_1 t_2}).$$

If the assumption (3.2) is implicitly contained in the sentence (3.1), we can now formulate Aristotle's reasoning as follows:

$$(H_0 \rightarrow e_{t_1 t_2}) \wedge \bigwedge_x \mathrm{Obs}(x, e_{t_1 t_2})$$

$$\sim V[a, e_{t_1 t_2})$$

$$\sim e_{t_1 t_2}$$

$$\overline{}$$

$$\sim H_0$$

What is the value of Aristotle's argument from today's point of view? Sentences (2.3)–(2.5) of the preceding section are relevant here. On their ground the premise (3.4) is now accepted, while the premise (3.6) and the conclusion $(\sim H_0)$ are rejected. Also the report (3.8) is accepted. The error of Aristotle and his followers consisted in accepting the false assumption (3.6) on the ground of the false claim (3.7). Assumptions (3.6) and (3.7) were the source of a systematic error which occurred in every negative result of the search for stellar parallax based on methods of observation similar to those utilized by Aristotle (this covers observation methods used until 19th century when — as it has been said — stellar parallax was measured for the first time). These results, although obtained in a way commonly called 'experimentally independent' (i.e. observers do not utilize each other's results not even knowing of each other's research), should be considered dependent, namely, dependent on the assumptions (3.6), (3.7) and on the measurement method they shared.

Let us also stress that, if we now reject the premise $(\sim e_{t_1 t_2})$ of Aristotle's reasoning and if we assume the occurrence of parallax also for the moments $t_1 - t_2$ in which Aristotle or anybody following the so-called Aristotelian experiment carried out observations, we do not do so on the basis of direct experience, but on the basis of $\bigwedge_i \bigwedge_j e_{ij}$ and of the negation of (3.6).

We can see that the negative result of the search for parallax noted in the observational report (3.8) did not uniqually determine the decision to reject or accept hypothesis (H_0). The negative result could be explained either by assuming that the parallax had not taken place or by assuming that this phenomenon had not been observable in all conditions or for any observer appropriately placed on the Earth. These two explanations of the negative result have their counterparts in two possible systems of sentences:

(1) $(H_0 \to e_{t_1 t_2}) \wedge \bigwedge_x \mathrm{Obs}(x, e_{t_1 t_2})$,

(2) $(H_0 \to e_{t_1 t_2}) \wedge \sim \bigwedge_x \mathrm{Obs}(x, e_{t_1 t_2})$.

The choice between (1) and (2) depends on the acceptance of hypothesis (3.7) or of its negation, i.e. of the hypothesis concerning the

distances of the stars and the Earth, and the extreme points of the terrestial orbit. The hypothesis (3.7) and its negation naturally have theoretical character and owe their meaning to the heliocentric model.

4. SOME ANALOGOUS EXAMPLES

I shall quote in this paragraph some facts from the history of natural sciences (physics) which seem to me analogical from the point of view of methodology to the case of the Aristotelian experiment. These examples which are by no means exceptional, show that the revision of Aristotle's reasoning was not an isolated event and that negative results of experiments are often burdened with systematic error.

We shall take the first example from the history of the controversy between the advocates of the corpuscular and the undulatory theories of light. As we know, I. Newton and later other advocates of the corpuscular theory rejected the undulatory theory before the Young and Fresnel experiment on the ground of the failure to observe diffraction of light: for the prediction of the diffraction effect is a consequence of the undulatory theory. The controversy might be formulated in the form of a decision-question: [14]

$(H_f, H_k?)$ Is light of undulatory or of corpuscular nature?

The reasoning leading to the solution of this question will be presented as follows:

(4.1) (1) If light is of undulatory nature then diffraction of light occurs.

(2) If light is of corpuscular nature (and propagates from the source in straight lines) then diffraction does not occur.

(3) If a shield (e.g. a sheet of paper) is put between the source of light and the screen, then diffraction occurs if and only if the shadow cast on the screen by the shield is blurred.

(4) In experiments carried out according to rule (3) we always obtain a clear-cut shadow on the screen.

(5) The phenomenon of diffraction did not occur in the experiments carried out, and hence:

(6) Light is not of undulatory but of corpuscular nature.

If we add to the premises of the reasoning (4.1), as a positive (non-tautological) assumption, a disjunction of the two direct answers:

(4.1)   (7) Light is either of undulatory or of corpuscular nature,

then the reasoning (4.1) has the form of *experimentum crucis*.

We can, moreover, guess there is a premiss analogous to the assumption (3.6) from the preceding paragraph, i.e. the sentence:

(4.1)   (8) If diffraction occurred it would be observable (in the experiments defined in rule (3)).

Premise (3) is the assumption of empirical decidability, premise (4) is an observational report, the counterpart of the sentence (3.7) in Aristotle's reasoning; sentence (5) is a complete indirect answer to the question (H_f, H_k?) with respect to sentences (1) and (2).

Rule (3) relates the term: 'diffraction' to the term: 'sharp shadow cast by the shield on the screen', which refers not to the observer's impressions but to an objective phenomenon.

The revision of this reasoning, as a result of the development of the techniques of research, has been described as follows:

"... Experiments performed many years after Newton revealed that the bending of waves into the region behind an obstacle is only noticeable when the obstacle is small compared with the length of the waves ... if the wave length is small compared to the object casting the shadow ... only the most delicate tests will help clarify the issue between the corpuscular and the wave theory of light. When experimental techniques and instruments had been sufficiently refined ... physicists were able to show that light really does bend somewhat into the shadowed region ...".[15]

Accordingly we can say that the revision of the argument (4.1) in the light of the results of Young's and Fresnel's experiment consisted in rejecting the premises (4.1), (5) and (8) as well as the conclusion (6).

Further development of the controversy, leading to the recognition of the dual nature of light, may be understood as a rejection of the negative assumption of the question (H_f, H_k?), i.e. of the disjunction of the negations of the two direct answers:

(4.1) (9) Either light is not of undulatory nature or it is not of corpuscular nature.

As we know, justified acceptance of (5) required not only a true observational report (4) but also a non-observational premise (8). We can see, moreover, that premise (5) was rejected retrospectively, obviously not on the grounds of observation, but, among others, on the basis of the accepted undulatory theory, that is (H_f).

Another example of a revision of observational sentences of an empirical science is connected with the discovery of the fine structure of the spectrum. N. Bohr when constructing his model of the structure of the atom, thought — according to experimental results of the time — that the lines of the spectrum of hydrogen (produced by the various atoms dropping from highest to lower energy levels) were single lines.[16] The splitting of spectral lines was first observed by F. Paschen in 1916, in his experiments with the spectrum of helium. One month later A. Sommerfeld's theory was published which, on the basis of the special theory of relativity, predicted the splitting of the spectral lines.

"Refined measurements, considerable increase of exactitude in measuring wave lengths (which correspond to respective lines of hydrogen) enable us to discover a new fact. A sensitive spectroscope reveals the fine structure of the hydrogen lines. They do not usually consist of a single line, but of a series of closely grouped lines of various degree of brightness. We thus discover the fine structure of spectra. In order to discover this fine structure of spectral lines, we must use spectroscopes showing lines of wave lengths differing only by tenths of Ångström. If less sensitive methods are used, the lines with a fine structure will become blurred, and appear like a single hydrogen line ...".[17]

Spectroscopy is an experimental science. Sentences describing the spectra of substances are relatively simple sentences, in terms of which we note the results of experiments carried out in order to test hypotheses and theories — they are, therefore, elementary observational sentences. We shall now present, in a simplified, popularized version, argument

based on Sommerfeld's theory and leading, on the basis of the results
of spectroscopy, to the solution of the question:

(H_e?)   Does the mass of an electron (in an atom of hydrogen)
          change (as a result of the change in orbital velocity, accord-
          ing to the special theory of relativity)?

We shall write this argument as follows:

(4.2)   (1) If the mass of an electron changes (as a result of a
              change in orbital velocity) then the path of the electron
              is a rosette.
          (2) If the mass of the electron does not change then the
              path of the electron is a single ellipse.
          (3) If the trajectory of the electron is a rosette then, if we
              produce a spectrum (of hydrogen) using a spectroscope,
              the spectral lines (e.g. of the Balmer's series) are dif-
              fracted.
          (4) If the path of the electron is a single ellipse then, if we
              produce a spectrum using a spectroscope, the spectral
              lines are not diffracted.
          (5) Using sufficiently sensitive spectroscopes we observe
              the diffraction of the spectral lines (e.g. of the Balmer's
              series) of the given portion of substance.
          (6) The spectral lines are diffracted, and hence:
          (7) The mass of the electron changes.

Rule (3) in the argument (4.2), which like argument (4.1) has a structure characteristic for *experimentum crucis*, expresses only a qualitative
prediction of the diffraction of spectral lines. As we know, Sommerfeld's
theory and additional premises [18] entail that each line of Balmer's
series consists of three lines; initial observations seemed to confirm this
consequence, but later spectroscopic research proved the existence
of five lines and attempts have been made to explain this divergence
between the theory and the results of observations, by the spin hypothesis. [19] In connection with our question of the dependence of observational sentences on theory let us note that in spite of the discovery
of the fine structure of spectrum we consider as reliable the observations
of single lines of Lyman's series, as such a result of observation agrees

with the theory which predicts the diffraction of the lines of Balmer's, Paschen's, Brackett's series and, at the same time, predicts that the lines of Lyman's series will remain single even if the most sensitive spectroscopes are used. Without the theory we would not know whether the observed structure of Lyman's series can be considered an objective fact or whether it is a consequence of too small precision of our measuring and observational methods.

5. CONCLUSIONS

Finally, I shall propose several interpretations of the facts of the history of science presented in the Sections 2–4, and connecting them directly with the theses (1.1)–(1.7) from the first section.

The sentences:

'The relative position of G_b and G_d in the moments t_1 and t_2 is different',

'In the moment t_1 G_b is situated to the left of G_d with respect to the observer on the Earth and in the moment t_2 G_b is situated to the right of G_d with respect to the observer on the Earth',

'The spectral lines of this substance are now split (consist of several lines)',

'The shadow cast now by this shield on the screen is sharp',

and their negations and conjunctions are examples of observational sentences of the empirical sciences, i.e. — they are relatively simple sentences (not containing quantifiers) which describe objective facts (and not experiences of the observer), usually the relative positions of macroscopic objects and their observable properties. In the empirical sciences such sentences are used to test hypotheses and theories.

The above quoted sentences are considered observational in the empirical sciences, and the phenomena they speak about — observable.

In Sections 2–4 some cases have been shown in which observational sentences, used to test hypotheses and theories, had been accepted during some period and then retrospectively rejected on the

grounds of laws or theories and premises concerning the unobservability of the examined phenomena or the unreliability of the observers. The use of better methods and instruments of observation led, as these examples show, to the revision of the verdicts about theories or hypotheses tested previously by means of more primitive methods and instruments, while in turn the rehabilitation of a theory or hypothesis hitherto rejected led to the retrospective rejection (acceptance) of the hitherto accepted (rejected) observational sentences. Thus examples discussed in Sections 2–4 seem to indicate the hypothetical (tentative) character of observational sentences of empirical science, in accordance with thesis (1.2) of the first section. They seem to indicate that the language of the empirical sciences does not always change with the changes of observation and of measurements methods. This applies not only to 'higher level' theoretical statements removed from experimental basis, but also to observational sentences which report results of experiments. This is so, anyway, if we accept the view that retrospective changes in the status of observational sentences discussed in Sections 2–4 did not entail any change of the meaning of terms contained in them, that they did not lead to a change of language. Otherwise, i.e., if we assume that in the examples discussed changes of meaning always occurred under the influence of the changes of the methods of observation or of measurement, then the concept of parallax in the language of antiquity would be different from that of the language of 19th and 20th century astronomy, and the same would be true of the concept of 'diffracted (single) spectral lines', the concept of 'sharp shadow cast on the screen by the shield'. An astronomer saying today that stellar parallax does occur and an ancient astronomer saying that parallax does not occur would be speaking about different things. An experiment carried out by using better instruments or methods would never be a crucial experiment (even provisionally) for alternative theories, formulated on the grounds of a less perfect experimental technique, ergo Young and Fresnel experiment could not even provisionally be the *experimentum crucis* for the corpuscular theory in Newton's formulation or for the undulatory theory of light in Huygen's conception. Moreover, a historian of the natural sciences could not claim, for example, that the premise 'there is no stellar parallax in the period of time $t_1 - t_2$', of Aristotle's reasoning, and the premise (4.1), (5), of the reconstruction of

Newton's argument against the undulatory theory, have proved false, nor could he claim that they were rejected later in the light of later results, because they would not be the same statements. Although the development of science sometimes consists in changing the problems, it does not seem to be true that it consists exclusively in changing them. From the point of view of the history of science and methodology concerned with the development of the empirical sciences the view claiming that every change of methods of observation or measurement leads to a corresponding change in the language of science is as unacceptable as the thesis of operationism, claiming that there are, for instance, as many different concepts of temperature as methods of measurement.

The acceptance of R. Carnap's criterion of the distinction between observational and theoretical terms, would lead to the same consequence as does the relativization of the sense of observational expressions to the methods of observation and measurement, i.e. to splitting of the language of science into individual and group languages. Carnap writes: "Thus, if a scientist has decided to use a certain term 'M' in such a way, that for certain sentences about M any possible observational can never be absolutely conclusive evidence lent at best evidence yielding a high probability, then the appropriate place for 'M' in a dual-language system (containing the observational vocabulary L_0 and the theoretical vocabulary L_T), like our system $L_0 - L_T$, is in L_T rather than L_0..."[20] It follows that, if Aristotle employed the term parallax as a non-theoretical, observational term in Carnap's sense, then he regarded the sentence: 'There is no parallax in the period $t_1 - t_2$' as consciousively justified (for himself?) by observation. Let us suppose that this was indeed the case. Since a contemporary astronomer rejects an equiform sentence, then the term 'parallax' is not for him an observational term in Carnap's sense; thus we would have two different concepts of parallax and two non-equivalent though equiform sentences: 'parallax occurs in the period $t_1 - t_2$'. Consequently, there would be no crucial experiments to decide, however provisionally, between two alternative theories or hypotheses contrary to claims often made by scientists and historians of the natural sciences. It would turned out, moreover, that all the experimenters we have thought that by devising new experiments they were solving problems unsuccessfully attempted by their predecessors were only the victims of linguistic misunderstandings; similarly historians of science

claiming for example that "... experience at last pronounced its verdict a hundred years after the controversy. In the first half of the 19th century Young and Fresnel carried out the experiment on the diffraction of light".[21]

The observational sentences in the examples of reasonings considered in Sections 2–4, as well as the 'observational' terms contained in them have a theoretical sense in the light of Carnap's criterion. As argued in Section 3, the following are indications that the expression 'parallax occurs' has a theoretical sense: (a) the assumption (3.2), in the light of which we interpret the observed shift as apparent, (b) the assumption (3.6), in the light of which we interpret the negative outcome of the experiment not as a result of too small a sensitivity of the measuring instruments, but as an indication of the objective real situation, (c) the retrospective rejection of the premise: 'parallax does not occur in the period $t_1 - t_2$', among others on the basis of the acceptance of the universal sentence $\bigwedge_i \bigwedge_j e_{ij}$.

It seems, moreover, that all the sentences of empirical sciences used in contemporary methodological literature as examples of observational sentences also have a theoretical content, because:

(1) If they have been taken simply from everyday language, they are understood under the assumption of realism characteristic of everyday language; if their sense has been modified by theories then *eo ipso* they have a theoretical sense; such sentences may be accepted and rejected on the basis of direct experience,[22] but only if it is considered that there are no reasonable grounds for questioning the associated non-observational assumptions (such for example as the assumptions (3.2) or (3.6)).

(2) In the case of every negative result of observation or measurement it is necessary to make the decision which is not uniquely determined by that observation or that measurement, whether to accept the negative result as an indication of the objective situation, and thereby the method of observation (measurement) as reliable, or whether to refute the result on the grounds of the hypothesis of too small sensitivity of the observation and measuring instruments.

(3) Every observational sentence accepted in a given moment may be *ex post* rejected on the basis of a theory or of a law.

The retrospective rejection of the sentences about parallax, about the sharp contour of the shadow made by the shield, etc. on the grounds of theories or laws seems to be incompatible with Campbell's thesis of the independence of the sense of observational terms and sentences from the sense of theoretical terms and sentences. If the sentence 'parallax does not occur in the period $t_1 - t_2$', at first accepted on the basis of observation and of the theoretical assumptions (3.2) and (3.6), which raised no objections at the time, was afterwards rejected together with the assumption (3.6) on the grounds of theoretical premises, including the heliocentric hypothesis, then this sentence did not have a justification completely independent of the justification of the hypotheses (H_0) and ($\sim H_0$), between which it was to decide, i.e. the first of Campbell's conditions of independence would not be fullfilled.[23]

In my opinion examples analysed in Sections 2–4 show the role of systematic errors in determining results of observations. I have tried to show that Aristotle's error was on the one hand connected with the insufficiently sensitive method of observation and measurement, and, on the other, with the acceptance of an erroneous hypothesis as to the distance between the stars to the Earth; this error was later repeated many times when Aristotle's reasoning was used and the search for parallax based on analogous methods. When he identified the observer's competence with the observance of the empirical rules of meaning in accepting observational sentences,[24] did he take into consideration such systematic errors, and expect a competent observer to be free from any systematic error? If so, then there is no effective method to decide whether in accepting a sentence, one is guided by empirical rules of meaning or not; we can at most put forth hypotheses. It is then impossible to demonstrate the non-revokable (non-hypothetical) character of the basic (elementary, observational) sentences of empirical sciences by showing that they are subject to empirical meaning rules.[25] Similarly, because of the role of systematic errors in determining results of observation it is not possible in my opinion, to maintain D. G. Ellson's claim that the statistical dependece of the observational reports given by 'experimentally independent observers' constitutes simultaneously the criterion of meaning and of truth of observational sentences. Experimental independence mentioned by Ellson does not at all guarantee the elimination of all systematic errors, which may result, for instance,

from the use of the same method of observation or measurement, from carrying out observations in the same unfavourable conditions or from the ignorance of other possible situations, which would lead to different results of observations.

REFERENCES

* Paper read at a conference of the Institute of Philosophy and Sociology of the Polish Academy of Sciences in April 1964. First published in: *Teoria i doświadczenie* (edited by H. Eilstein), Warszawa 1966. Translated into English for the present anthology by Stanisław Wojnicki.
[1] R. Carnap, 'The Methodological Character of Theoretical Concepts', *Minnesota Studies in the Philosophy of Science* 1 (1956); C. G. Hempel, 'The Theoretician's Dilemma', *Minnesota Studies in the Philosophy of Science* 2 (1958); C. G. Hempel, 'Problems and Changes in the Empiricist Criterion of Meaning', *Revue Internationale de Philosophie* 2 (1950), reprinted in: *Semantics and the Philosophy of Language* (ed. by L. Linsky), Urbana 1952.
[2] C. G. Hempel, 'Problems...', pp. 164–5; R. Carnap, 'The methodological character...', p. 69.
[3] The definition of observational sentence given here is based on the reportive definition of observational sentence which can be found in M. Przełęcki's paper 'W sprawie uzasadniania zadań spostrzeżeniowych' ('On Justyfying Observational Sentences'), *Studia Logica* 13, p. 213.

I believe that the authors utilizing various metalinguistic terms, as observational, perceptional, basic, phenomenal, evidential, elementary etc. sentences, all mean the same sentences of empirical sentences, although they characterize them differently. Cf. also K. Ajdukiewicz's paper 'Empiryczny fundament poznania' ('The Empirical Foundation of Knowledge'), *Prace PTPN* (1936), and his 'Subiektywność i niepowtarzalność metody bezpośredniego doświadczenia' ('The Subjectivity and Non-Repeatability of the Method of Direct Experience'), *Studia Logica XIII* (1962).

R. Carnap (*op. cit.*) includes so-called dispositional terms in an enlarged observational vocabulary. I do not consider dispositional terms separately as all the observational terms seem to have a dispositional character, as it is shown by the expression 'under the given conditions', in the definition of observational term presented in the text. There is certainly a difference between the 'suitable conditions', in which the dispositional property and the so-called non-dispositional one are observable, so that we can say that terms are dispositional in different degrees. This is discussed by K. R. Popper in *The Logic of Scientific Discovery*, London 1959, Appendix **.
[4] K. Ajdukiewicz, 'Język i znaczenie' ('Language and Meaning') in: *Język i poznanie (Language and Knowledge)*, Warszawa 1960.

⁵ N. R. Campbell, *Physics, The Elements*, Cambridge 1920; E. Nagel, *The Structure of Science*, London 1961, pp. 86–7.

⁶ C. G. Hempel, *The Theoretician's Dilemma*, pp. 41–2.

⁷ D. S. Allen and R. J. Ordway, *Physical Science*, Princeton 1960, pp. 318–20, 349–351.

⁸ Sentences describing possible observations and those with the help of which theories and hypotheses are tested are called basic sentences. They usually have the following form: 'In a place m and at a moment t there is a (macroscopic) object S having an (observable) property W'.

Finite conjunctions of basic sentences are also considered basic sentences. The position of an object S_1 at the moment t is determined relatively to another macroscopic object S_2, and hence we say that basic sentences state the mutual position of macroscopic objects.

⁹ Sentences 'S' and 'non-S' are called direct answers to the question $(S?)-$'$S?$'. If a sentence Z is a direct answer to a given question, then every logical consequence of the sentence Z is a partial answer to this question. If Z is a direct answer, and a sentence Z' logically follows from the conjunction of T and Z, then Z' is a partial answer to the question with respect to the theorem T.

¹⁰ If Z is a direct answer to a decision-question then every sentence W, from which Z logically results (possibly with respect to T) is called a complete indirect answer to this question (possibly with respect to T).

¹¹ The assumption of empirical decidability of a question says that among the given observational sentences there is a true indirect answer (partial or complete indirect) to this question.

¹² In Aristotle's reasoning the brighter stars are considered nearer, which – as we now know – is not a reliable criterion.

¹³ A similar suggestion had already been put forth in the preface to Copernicus' *De revolutionibus* by Andreas Osiander.

¹⁴ L. Infeld, *Nowe drogi nauki* (*New Ways in Science*), Warszawa 1957, pp. 26–7; E. Condon, 'Physics', Ex. *What is Science*, New York 1955, pp. 115–120.

¹⁵ E. U. Condon, *op. cit.*, pp. 116–117.

¹⁶ A. d'Abro, *The Rise of the New Physics*, vol. 2, 2nd ed., New York 1951, p. 513.

¹⁷ L. Infeld, *op. cit.*, pp. 92–93.

¹⁸ A. d'Abro, *op. cit.*, pp. 510–13; L. Infeld, *op. cit.*, pp. 96–7.

¹⁹ A. d'Abro, *op. cit.*, p. 514.

²⁰ R. Carnap, *The Methodological Character*, p. 69.

²¹ L. Infeld, *Nowe drogi nauki* (*New Ways in Science*), p. 27.

²² In 'Subiektywność i niepowtarzalność metody bezpośredniego doświadczenia' ('The Subjectivity and Non-Repeatability of the Method of Direct Experience), *Studia Logica* **13**, p. 209, K. Ajdukiewicz calls an observational sentence for a person x a sentence about a unique, non-repeatable fact, accepted by x on the grounds of observation. Somebody may for example consider as such the sentence: 'a thunder has just stroken a near tree'. The question arises whether in order that a given sentence be an observational one for somebody it is necessary to accept

it only on the grounds of observation, it seems that, although accepting sentences as a rule on the grounds of observation, we at the same time tacitly make various assumptions not justified by that observation, but raising no objections in the given moment; if any of these assumptions proved false, we would revoke the acceptance of that observational sentence, and if it raised objections at the moment of making the observation, we would not accept that observational sentence, which shows that its only justification is observation. For example, I accept on the grounds of observation the sentence: 'There is a glass standing on the table in front of me now', but only as long as the object on the table behaves 'properly', just as I expect a glass to behave on the grounds of my general knowledge. Let us also remember of the plastic 'crushed foot', showed to the passengers by F. Morton in J. Steinbeck's book *The Wayward Bus*.

[23] Campbell's thesis about the independence of observational terms and sentences from theoretical terms and sentences is also opposed, although by different arguments, by B. Hesse in *Forces and Fields*, London 1961, pp. 13–21. The author rejects the distinction between observational and theoretical terms as well as the corresponding distinction of sentences.

[24] K. Ajdukiewicz 'Subiektywność i niepowtarzalność...', p. 210.

[25] K. Ajdukiewicz's pre-war paper 'Empiryczny fundament poznania' (The Empirical Foundation of Knowledge), *Prace PTPN* (1936), was devoted to the justification of the thesis that if in the empirical sciences there are empirical meaning rules then basic sentences (in the terminology of K. R. Popper whose hypotheticism Ajdukiewicz criticized) of these sciences are not hypotheses.

[26] D. G. Ellson, 'The Scientist's Criterion of True Observation', *Philosophy of Science*, **30**, No. 1.

ANDRZEJ GRZEGORCZYK

THE PRAGMATIC FOUNDATIONS OF SEMANTICS*

1. INTRODUCTION

The present work is an attempt at defining certain semantic concepts which properly belong to the so-called pragmatics of language, that is science which deals with the relations existing between the language itself and the people who use it. For that purpose five primitive expressions of essentially pragmatic character have been chosen, by means of which it is proposed to define the relation of naming, the concept of the true sentence and other semantic notions. It should be made clear that none of these concepts strictly refer to any natural language, though some sections of the latter may satisfy certain conditions of the definitions accepted. Such a situation may arise in many other branches of science, especially in those of the humanist branch. Here we are often unable to describe concrete phenomena by means of a definition which operates a small number of precise terms, as the real phenomena are of much more complex character. Consequently, after we have made the decision to accept an exact definition, it may appear as often as not that there are many borderline cases which satisfy that definition, even if we have not intended so, or do not satisfy it, even if we have intended that they should do it.

Thus, the semantic relations which can be found in concrete social encounters present a field full of anomalies and various complexities. However, it is not with such concrete relations that we are concerned, but with the study of a certain language which we will call 'descriptive language of science'. This language, in the form attributed by us, does not *de facto* exist in any science, although certain fragments of any concrete descriptive languages may freely come very near to it. It is to that fictitious, descriptive language of science that all our semantic concepts and definitions will refer.

In order to justify such a procedure we must turn to the general tendency in grasping the foundations of all empirical sciences. In any analysis of them we can either investigate the existing real state of the science, scrutinizing the actual behaviour of the investigators, that is, to pursue the science of real science, defined as a set of human actions and their products placed at a given time and space; or, glancing at a science which exists at present or which existed in the past, we can try to describe a certain state of science which so far was perhaps never fully realized, and which in actual fact perhaps will never be completely realized. Nonetheless, such a procedure may be of some interest as its realization is not impossible and, moreover, by being a methodological ideal, it can give us a general idea of the 'structure' of science.

As will be seen later, our main argument will be mostly devoted to the study of the latter type. And it may not be irrelevant to mention in this connection that a similar attitude is usually taken in the study of the foundations of deductive sciences, For both in the methodology of deductive systems and in the constructing of individual deductive sciences it is customary to regard them as sets of written statements of a shape strictly fulfilling certain accepted conditions. However, only very seldom and only a few fragments of deductive sciences are built in such a way as to conform strictly to the conditions accepted. In mathematics complete proofs are not given, as this would take too much time and space, and may thus result in a complete stalemate in deductive sciences. Although such proofs would, methodologically, represent an ideal state, yet this ideal would bring little benefit to the further development of science. In point of fact, every mathematician uses many short cuts, simplifications, and employs several didactic devices which would not be permissible in any exact deductive system. Despite this, it have often been said that mathematics constitutes a deductive system or a set of such systems. This, however, is obviously not true, if by deductive systems we mean sets of written statements of certain shape strictly defined in a recursive way, and if we understand mathematics as those statements which have actually come from the pens or chalk of the mathematicians. Therefore, the expression 'mathematics is a deductive system' has a different meaning; it expresses the conviction that all akcnowledged mathematical reasoning may be shaped into exact theorems belonging to formalized deductive systems.

This conviction is based on experience and on manifold attempts at the formalization of fundamental fragments of reasoning, and appears to be true in most cases. Nevertheless, such a state of affairs may cause various doubts, especially among non-mathematicians. Hence in meta-mathematics, which studies formalized deductive systems, we do not deal with those statements which the mathematicians actually write, but we make the study of certain relations between those statements which the mathematicians could have written, had they wished to formalize strictly their reasoning; furthermore, we deal here with such sets of statements which certainly will never all be written down because they contain an infinite number of statements. Yet it is possible to write any individual sentence of such a set, though it may be superfluous to the study of the given system. It is therefore customary in metamathe-matical investigations not to write any sentences belonging to objective language, but only to define recursively their shape and to study their interrelations resulting from the given definitions. In this way their existence is assumed, though actually these sentences might never have been written down. Thus, if we accept that deductive systems represent sets of statements conceived in one way or another, it follows that in the methodology of deductive sciences it is enough to investigate those statements all of which could potentially be written down, but which most likely, at least the majority of them, will have never been written down.

One can, therefore, say that the study of deductive sciences extends over two distinct fields: (1) the study of the real work of the scientists and of its products which belongs to the domain of history, sociology, or the psychology of science, and (2) the study of certain 'ideal' [1] products of that science, that is, products which are, in principle, possible, but in fact not realized by anybody. The latter is the chief subject-matter of the methodology of a given deductive science.

In the type of study which concerns the foundations of empirical sciences the line of division between the two corresponding domains of investigations is often blurred. Yet it would be difficult to understand much of the reasoning of the methodologists, for example, those of the Vienna Circle, in any other way than by assuming that it concerns only a certain possible state of a given science which it achieves by pointing to an 'ideal' form of that science. This ideal defines both the

way in which certain sentences which belong to that science should be built, and how they should be joinded with the results of experiments.

Although intellectual satisfaction and practical applicability are the final objectives of any scientific work, yet the most tangible products are language-products. Most methodological requirements made of science first and foremost determine a certain ideal state of these products.

In empirical sciences, just as in deductive, it is necessary to use abbreviations as well as to apply various didactic devices, both in language itself and in its linking with experience. Hence concrete linguistic products of these sciences are apt to deviate from their methodological ideal, even to a greater extent than in purely deductive sciences. For that reason, if we want to understand the structure of empirical sciences, it is necsssary to divide with equal precision as in the case of deductive sciences our study of science into two distinct fields: (1) the science of the real science which is concerned with the behaviour of the investigator and all its concrete manifestations, such as his linguistic activity, didactic tricks, and (2) the methodology of a science which is concerned with its potential ideal state devoid of any facilities or practical devices, and reflecting solely its methodological structure.

In order to describe this ideal state of an empirical science it is, of course, not enough to put down in a purely graphic form the shape of its linguistic products as was sufficient in the methodology of deductive sciences. Here it is equally necessary to describe the ideal behaviour of a well-trained and thorough investigator who by means of his own behaviour in the concrete situations gives a precise empirical meaning to certain linguistic expressions of his science. In this way we make one step further in the direction of the method of deductive sciences. For there we speak about the structure of a deductive science, describing its partially fictitious linguistic shape, whereas in the methodology of empirical sciences, in order to describe the methodological structure of science, we must describe not only its partially fictitious linguistic shape, but also the partially fictitious and accurate (in the ideal sense) behaviour of the investigators. For the semantic relations of the descriptive language are produced through the behaviour of the investigators in certain situations. This is so because a thorough and methodical investigator will make a certain statement when, and only when during

an experiment he will receive certain strictly defined products of his action.

It should be remarked that the methodological description of a structure of science is not, of course, completely fictitious, as it originates with the study of the real state of science and is based on certain concrete observations. The following observations may become relevant for the distinction of the descriptive language in experimental sciences.

In an actual scientific work we can distinguish: (1) the experimental observational work which results ultimately in the description of an experiment performed, a document discovered, a field scrutinized and the like, (2) the deductive work which results in new theorems, deduced from the old ones in an accepted manner, and (3) the systematizing work in the course of which we generalize our previous observations, make hypotheses, discover types, formulate or change scientific rules or laws, or even the entire deductive theories. To these correspond the final linguistic products of the scientific work: the description of a certain event directly observed and the deductive theory presented in a more or less formalized and axiomatized form. The descriptions of direct observations in physics and kindred sciences are quite distinct from the theory. In humanistic sciences the difference between the description and the theory seems to be less marked.[2] The distinctions between descriptions of things directly observed by the investigator and the statements concerning non-observable phenomena (i.e. hypotheses and generalizations) is, of course, always empirically legitimate. Hence the division of the scientific language into descriptive and theoretical is not 'unnatural' — a distinction, in fact, made a long time ago by the Vienna Circle by the discrimination of the protocol sentences. This distinction seems to bring certain advantages in so far as it accounts for the division of methodological problems. Thus, the problem of the empirical foundation of a scientific theory entails two problems: (1) the problem of the empirical foundation of the descriptive language, and (2) the problem of the reduction of the sentences of a theory to those of the accepted descriptive language. This methodological distinction is useful, as it can help in the solving of certain controversial problems. For instance, the essential conditions of Bridgman's operationism appear to be justified in their entirety in relation to the descriptive language; on the other hand, it can be assumed that it would be

sufficient to reduce [3] the language of theory to the descriptive language. In the course of this argument the requirements placed upon the descriptive language should be treated as a variety of the requirements placed upon the descriptive language by operationism.

In contrast to the deductive theory whose methodological ideal, consisting of complete formalization and axiomatization, is, one can assume, generally accepted, the ideal of a scientific description is not clearly defined and is subject to discussion, as, e.g. the discussions on the shape of protocol sentences in the Vienna Circle. The descriptions of experiments, observations, or documents are often built by means of occasional phrases in a way which is schematic and the briefest possible. Sometimes, when the facts of everyday experience make the starting-point of scientific analysis, their description is often even entirely omitted. Also, often in scientific descriptions, phrases of everyday language are used which have not been made precise enough. The desciption is regarded as good when by virtue of linguistic habits it causes corresponding 'adequate' associations among the readers, irrespective of the type of expressions by means of which it had been written down.

In order to discover the methodological structure of the descriptive language, it will be necessary to define its ideal shape. Before, however, we pass to its precise definitions, we will first describe certain general conditions which, it would seem, the descriptive language of science could fulfil. For instance, the description of a physicist's experiment may be limited to an account of the arrangement of physical instruments and their indications. Such a description need not necessarily include any terms referring to nuclear physics. It is enough, if it has such a shape as might be given to it by any intelligent observer without even the knowledge of physics, and only observing carefully the movements of the investigator and his apparatus, and knowing the names of the basic parts of the instruments as well as the principal substances involved. The description of the experiment, therefore, may be free from the whole theoretical background of the working of instruments, though it is indispensable for the explanation of the experiment. Similarly, the description of an astronomical observation, as well as of a microscopic one, may become reduced to the description of apparatus or that of certain photographs. Generally speaking, the scientific description need not be solely macroscopic, but it may also contain only the names of

such objects which are visible and tangible. The latter can be called the objects of everyday experience. Scientific description, thus, may be limited solely to the description of the things of everyday experience of the investigator of a given period. Hence the descriptive language of empirical sciences may be limited to a descriptive language of everyday experience.[4]

This language can subsequently be made more precise syntactically. Thus, its grammatical forms can be shaped like logical systems. In this we must distinguish among the expressions of the descriptive language semantic categories (logical types), accepted in logic, and by giving, for example, to the linguistic proposition a shape which is composed of a certain functor and certain brackets, containing arguments of required type, while observing in it the ordinary logical principle of not mixing the types. We can further assume that the lowest semantic category in the hierarchy of logical types is the semantic category of the names of things of everyday experience and in this way to turn the descriptive language into reistic.[5]

The acceptance of such a structure of the descriptive language shifts the whole difficulty — that is that of giving a strictly empirical sense to the expressions of language — to the conferring of a strict extension upon the names of various sets of things of everyday experience. Such a conferring may take place either by verbal designation, or by a certain action which distinguishes the given thing from others and which points to that thing as a designate of the name. The stock of names of objects of everyday experience, acquired in childhood, is usually so large that it is seldom necessary to resort to the actual pointing. Were we not able to trust the names acquired in childhood, verbal designations would not suffice, and it would be necessary for us to introduce certain names by other means, not exclusively verbal. It would appear that all non-verbal methods of connecting certain names with certain things, as their designates, could be always reduced to a certain pointing to a thing called by a given name. In any case, by giving a somewhat broad interpretation to the form of 'pointing', one could include here also the origin of the names in childhood, and to say that all situations in which the people around him use a certain name, point to the child its designate.

Assuming that the meaning of the name is ultimately fixed in situa-

tions involving pointing, we must make the condition that the ideal descriptive language must contain the names of things of everyday experience introduced by means of certain ideal action of pointing.[6] Thus, speaking about ideal descriptive language, we must *de facto* speak about a certain ideal setting of investigators who at a certain time had decided to form in an ideal manner a descriptive language by means of certain actions of pointing, in doing which, however, they broke away from nearly all accepted semantic habits. Thus, while considering the descriptive language we shall deal with all these actions which no investigator as a rule performs, though, in principle, he can perform them—just as in the methodology of deductive sciences when we speak about languages of deductive sciences we consider such statements which neither a logician nor a mathematician is ever likely to write.

In order to present our argument in a somewhat general form, we will not confine ourselves to any one action of pointing, and by using the expression:

(1)   *A man A points to the object B*

we shall understand by it a certain action whose specification depends upon the restriction to the descriptive language of any single science. Nevertheless, the very indication must consist of certain separation of a given thing from others. This can be achieved by e.g. lifting the given thing or by touching its entire surface. Naturally, not all things which are visible or tangible are pointable in this sense. A mountain peak is also visible and tangible, but it cannot be separated in this way from other mountains. It would appear, however, that if we restrict the descriptive language to the language of things pointable in everyday experience, then this language will still be sufficiently rich in order to formulate in it the interpretations of all scientific descriptions.

Apart from the expression (1) we shall further use other expressions of praxeological character which resemble those analysed by Professor T. Kotarbiński in his numerous papers on praxeology.[7] We shall discuss them, though briefly. They are as follows:

(2)   *A man A performs the action f upon the thing B.*

The substitutions for the expression (2) are, for instance, 'a man *A*

measures the thing B', 'a man A dissolves a particle of the thing B in the hydrochloric acid', 'a man A looks at the object B'.

(3)   *A man A obtains the product X of his action f performed upon the thing B.*

Here the following expressions are substitutions: e.g. 'a man A obtains the score X which is the result of his measurement of the objects B'; 'a man A obtains green transparent solution by dissolving a little of the thing B in hydrochloric acid'; 'a man A has a sensation of redness when looking at the object B'; for visual sensation is the product of the action of looking. (Speaking more precisely in the reistic manner it should be said that man himself as looker-on is the product of the action of looking at the time when he looks — that is to say, as 'seeing red' in the above case.) The expression (3) can be regarded as equivalent to the following expression: (3) *A man A, by performing the action f upon the object B, became the producer of the object X.*

(4)   *A man A can perform the action f upon anything of the set b.*

Here the substitutions are: 'a man A can measure (by means of a certain method) any thing of the set b', 'a man A can drive a car of a certain make'). The meaning of the expression (4) may cause the greatest reservations, yet in our methodological analysis it would be difficult to do without it. One can suppose that the definitions of certain contractions of the expression (4), which are necessary in order to grasp the methodological problems, can be stated operationally and yet correctly in the field of experimental psychology.

(5)   *The man A says 'a'.*

Here the substitutions are e.g.: 'Peter says, "the sun shines" '. Here it might be observed that though in any exact formulation we ought not to have used expressions in inverted commas ('...'), we shall continue to do so for the sake of convenience. But these expressions can always be replaced by correct structural and descriptive names of given expressions. Also, in order not to complicate unnecessarily these primitive expressions, we will not add to them temporal determinations, as this should not be very difficult for anyone who wished to do so. Without such precise formulation it is obvious enough that when we use an

expression saying, e.g., that A performs some action upon the thing B, we always mean a certain action performed in a certain time. The most frequent logical symbol which we shall apply is the symbol '$\equiv$', in place of the expression 'if and only if then'.

We have given above a list of expressions which will be used in the building of the following methodological and semantic definitions. These expressions, according to the terminology of Morris and Carnap, may be called pragmatic, and the analyses which follow can be included into the pragmatic study of the language [8]. Hence it can also be said that the following analysis attempts to prove the thesis that the semantics of the descriptive language is in a certain sense a part of pragmatics.

II. SEMANTIC DEFINITIONS

These definitions refer to individual elements of the descriptive languages. They lead to the definitions of two kinds of descriptive languages which can be called the *narrow* descriptive language and the *broader* descriptive language. The first group contains only atomic sentences of the shape 'α is β', and the second may include the atomic sentences of the following shape:

$$\varphi\{\alpha\}, \psi\{\alpha, \beta\}, \chi\{\alpha, \beta, \gamma\}, \ldots, \text{etc.}$$

which can be extended to the sentences of higher types. Neither of both kinds of language contains compound sentences but only atomic ones, none contains variables. All names 'α', 'β', 'γ', ... and the functors 'φ', 'ψ', 'χ', ... and other of both languages are understood here as constant names and functors. The ordinary descriptive language of science also seems to contain no variables. Moreover, there is always in these languages a finite number of names and functors, though it is always possible to extend the language to new names and functors necessary for the description. These languages do not contain definitions. All their names and functors are, therefore, primitive, though from the empirical point of view they can be divided into empirically primitive and empirically introduced. In our language, however, empirically primitive is only one functor, that is, 'I am now pointing to...'; all other names and functors are empirically introduced by such a type of behaviour of those who use that language which is described in the following definitions.

D.1.   The name 'α' is spread among the people a by means of reconnoitring action f, applied to the objects b, and by means of the indices c. $\equiv$.

(1)   Every mean a knows how to perform the action f upon any object of the set b, and

(2)   if a man A being an a performs the action f upon a certain object B of the set b and will obtain the product X belonging to the set c, then A pointing to the object X will say: 'now I am pointing to 'a' ', and

(3)   if an A who belongs to the group of people a performs the action f upon a certain object B of the set b, only then ponting to the object B will say: 'now I am pointing to 'a' ' when A obtained a certain product X which belongs to the set c, performing the action f upon the object B.

For example, the name 'iron ball' has become spread among the physicists by means of actions of measuring and analysis applied to solid objects. In this way, this name is spread by means of indices which consist of the results of measurements and the indices of iron in chemical analysis, since (1) every physicist knows how to measure and analyse any solid object; (2) if a physicist performs these actions and if he receives the expected results of measurements as well as the indices of the analysis, then the mentioned physicist, when pointing to the object upon which he performed the above-mentioned actions, will say: 'Now I am pointing to the iron ball'; and (3) only then the physicist will say: 'Now I am pointing to the iron ball', pointing to a certain object when, after performing measuring and analytical actions upon the said object, he will obtain the expected results. Naturally in that sense the name 'iron ball' is known solely to those physicists who would agree, having obtained certain results of investigations, to perform the mentioned action of pointing together with saying: 'Now I am pointing to the iron ball', and who would agree to perform that action only after they obtain results in question. What was said above refers to all examples which will be given below. This action of pointing and speaking is, to the physicists, mostly a superfluous action, and can be substituted by a verbal indication of the object under investigation. However, had certain physicists wished to revise their descriptive language, it seems that they

would be able to establish certain names only by means of some pointing (specifying) actions combined with speaking. For without pointing operations they would not be able to go beyond linguistic definitions. Naturally, the action of pointing together with the speaking, 'Now I am pointing...', which is the basic action of giving a name in our world of ideal physicists, is based — as one might say colloquially — on the understanding of the role of one's own action of pointing (specifying) in establishing the name as well as on the understanding of the formula 'Now I am pointing...' simultaneously uttered.

The physicists in question understand the formula 'Now I am pointing...', and for that reason they know how to point and also for that reason they know when, according to the accepted meaning of the term 'iron ball', they should, while pointing to a certain object, say: 'Now I am pointing to the iron ball'. Although this is what we would like to say, yet it would be necessary for that purpose to investigate the empirical sense value of the expressions containing the term 'understand'. However, we can pass over this difficult question and speak only about the behaviour of people of certain classes (e.g. physicists) — that they point to something and in certain moments speak something. For, even if we could speak with the full sense value about understanding, this understanding must be manifested in observable behaviour. Hence, when looking for observable features of an empirically well-grounded language, we need not speak about anything more than about the action of an ideally conceived investigator and about the products (including the verbal ones) of this action. The expression, 'I am now pointing to...' is the only formula which the said physicists, who seek to establish their descriptive language, must take over from the colloquial language, though they must use it only in the sense which had previously been made precise. We have called this expression the primitive formula of the descriptive language, but one can regard it as not belonging to the descriptive language, though as the one which helps to establish it. Thus, the definition of the descriptive language which we adopt, will tend to eliminate this formula from the descriptive language.

In the same way the names of chemical compounds are made known to the chemists by means of the action of chemical analysis and by the indices which are, for example, the chemical sets endowed with certain qualities. In order, however, to make this example valid we

must assume similar things about the chemists as we have already assumed about the physicists, that is, that they agree to perform the action of pointing and that they carry it out scrupulously, repeating when necessary the expression 'Now I am pointing...' (In quoting further examples it will be superfluous to repeat these 'ideal' assumptions.) Apart from these rather complex actions of reconnoitring which, as already stated, consisted of measuring and chemical analysis, there exist also several simpler actions of reconnoitring, as, for example, looking at an object. Here the index, as one of the products of looking, is the observer himself in that very moment when he experiences the act of seeing, or, in other words, as one who sees in a definite way and also one who experiences other sensations. For example, the name 'hydrogen sulphide' is spread among the chemists with some sense of smell by means of an action of reconnoitring consisting among others of smelling and by means of an index − in this case the smelling person himself as experiencing that sensation in a characteristic way. Since it can be assumed that the complex reconnoitring action can always be in a certain sense reduced to a simple reconnoitring action which consists of the direct reaction to certain stimuli, then we should say that basic reconnoitring indices are the people themselves as experiencing sensations in a certain way. (Or, speaking less precisely from a reistic point of view, these indices consist of human experiences.)

Also, for example, the name 'kaberifop' can be spread among a certain group of people by means of the action of tossing a coin and by means of an index 'heads'; these people having come across an object and pointing to it will say, 'Now I am pointing to kaberifop', when and only when a coin tossed by them previously fell heads. In connection with examples taken from scientific terminology, it should be remembered that the descriptive names 'iron ball', 'hydrogen sulphide' etc. which are spread by means of the said reconnoitring activities, are not the same names as, for example, 'iron ball' or 'hydrogen sulphide' in the field of theoretical physicochemistry. In theory we can consider a small quantity of gas, consisting of one or a few molecules, which will not produce a characteristic smell or only other index, or which perhaps will be contained in a distant star, yet despite it it will still be a portion of hydrogen sulphide. Thus theoretical concepts appear to have broader extension ranges than the corresponding descriptive

concepts whose extension is restricted by the limited range of the existing reconnoitring methods.

D.2. The name 'a' is unambiguously spread among a by means of the reconnoitring action f applied to the objects b as well as by means of indices c. $\equiv$.

(1) The name 'a' is spread among a by means of the reconnoitring action f applied to the objects b as well as by means of indices c, and

(2) if X and Y are people of the class a, and if any of them will independently perform the action f upon an object B of the class b, then if X pointing to B will say, 'Now I am pointing to α', then Y too, pointing to B will say, 'Now I am pointing to a'.

For instance, the name 'metal hexagon' is unambiguously spread among the chemists who are trained in measuring, which they do by means of an accepted method and by means of an analysis as well as by means of indices which are the result of measurements and analyses. Most names of chemical compounds are spread unambiguously among the chemists by means of the action of chemical analysis and its indices. In a similar way the names of plants are unambiguously spread among the botanists by means of the action of naming and by the products of the action. A similar position is that of the names of minerals: they are unambiguously spread only among the trained mineralogists by means of the action of naming, consisting of, for example, the testing of hardness and he like, and by means of indices consisting of cuts obtained by moving one mineral over the surface of the other. In general, the names are unambiguously spread only among specialists who are properly trained. On the other hand, the name 'kaberifop' is not unambiguously spread among any group of people by means of a reconnoitring action consisting of tossing a coin, as this action performed by X in relation to the object B may result in obtaining 'heads', which will make X say that B is kaberifop, whereas applied by Y in relation to the same object B may result in falling 'tails', owing to which Y will not be able to name the object B kaberifop.

D.3.　　'α' is a name normalized among people a and referring to some objects of the class $b. \equiv$.

(1)　　The name 'α' is unambiguously spread among people a by means of a certain reconnoitring action applied to objects b and by means of certain indices, and

(2)　　if 'α' is a name unambiguously spread among people a by means of the action f, applied to objects b and by means of indices c and the same name 'α' is unambiguously spread among people a by means of the action g applied to objects b_1, and by means of indices d, then $f=g$, and $c=d$, and $b=b_1$.

For example, the names of plants are normalized among the botanists and they refer to plants because they are unambiguously spread among the botanists; and also because the indices of particular plants and the methods of naming them have been also made uniform among the botanists and are applicable to plants only. In a similar way, the names of minerals are normalized among the mineralogists and the names of chemical compounds are normalized among many groups of chemists who use the same methods and indices.

D.4　　L is a narrower descriptive language of objects b, spread among the people a, when L is a set of all expressions: 'α is β', in which both 'α' and 'β' are names normalized among people a and referring to some objects of the class b.

A certain narrower descriptive language of some botanists may, for instance, contain a sentence, 'karabakus is a cactus', in which the name 'karabakus' denotes the cactus which stands on my window. This sentence may be even true in that language. For the narrower descriptive language we can adopt a simpler definition of truth than the general definition D.12 (see below). In the first place we call a certain normalized name 'α' individual if there is only one object about which people a, having performed the corresponding reconnoitring actions, will say: 'I am now pointing to α'. Further, we call the formula 'α is β' true in the narrower descriptive language L spread among people a, when L is such a language, 'α is β' is a sentence which belongs to it, 'α' is an individual

name in the above sense, and a certain person of the class a having performed the reconnoitring actions of the names 'α' and 'β' upon a certain object X of the class b, pointing to X said: 'Now I am pointing to α' and also, 'Now I am pointing to β'. For the term 'is' used in that sense there is a system of logic constructed by S. Leśniewski and called by him 'ontology'.[9]

For the definition of narrower descriptive language the concept of the relation of naming is not necessary, though it will be necessary for the definition of the broader descriptive language. We shall pass now to the definition of the relation of naming.

D.5.   A term 'α' is a precise descriptive name of the objects d spread among people a by means of reconnoitring action f, applied to the objects b, and also by means of indices c. $\equiv$.

(1)   The name 'α' is spread among people a by means of reconnoitring action f, applied to objects b, and with the aid of indices c,

(2)   if a man A belonging to the class a performs an action f in realtion to certain object B among the class b and B is d, then A pointing to object B will say: 'now I am pointing to α' and

(3)   if a man A belonging to the class a performs an action f in relation to a certain object B among the class b, and pointing to B, will say 'now I am pointing to α' then B is of the class d.

It is not difficult to see that if 'α' is a precise descriptive name of certain objects in the sense of the definition D.5, then 'α' is a name spread unambiguously in the sense of D.2.

Having the definition of the name D.5, we may define 'α' as an empty name, spread among people a by means of... etc., when 'α' is a precise name of objects d, spread among people a by means of... etc., and that the objects d do not exist. On the other hand, when every man a knows how to point out freely many objects d, then we can call the name 'α' full name. If there exists one and only one object d, then we will call the name 'α' individual name. If 'α' is a name of objects d, spread among people a by means of... etc., then every object d is the designate of the name 'α'. From these conventions follow familiar sentences which state

that an empty name is the name devoid of designates, that the individual name is the name which has one and only one designate, and that every name is the name of its designates.

As examples we can give again the same scientific names which served as examples of unambiguous names. Thus, the name 'quartz' is the precise descriptive name of free quantity of crystallized silicon-dioxide, spread among the chemists by means of relevant reconnoitring activities. In a similar way, every name which is of the shape of 'salmiac' (sal ammoniac) is a name of some portion of salmiac, spread among the chemists by means of... etc. The name 'equus' is a precise descriptive name for horse, spread among the zoologists by means of research activities and such indices as, e.g., the type of bones.

It is well known that the definitions or the reconnoitring methods seldom eliminate completely the possibility of error or ambiguity in statements about certain objects. For each reconnoitring method it is easy to think about some borderline object, about which it is not known whether it should be still included in the designates of a given name or whether it should be already eliminated from it. Scientific names and methods satisfy the definitions which had been formulated before and which will be still formulated, in a precise way, only when they are applied to the objects which are distant from the said borderline objects. The delimitation of the field of objects b, which in our definitions fix the extent of applicability of reconnoitring methods encounters, naturally, similar difficulties. Hence in actual scientific procedure the extent of the applicability of methods and terms is, on the whole, intuitively defined within narrower bounds, by using them only in those cases where they produce clear effect.[10]

D.6.   The names d form the basis of the names of objects among the class b spread among the people $a. \equiv.$

(1)   each name within the collection d is a precise, descriptive name of certain objects of the class b, spread among the people a by means of certain reconnoitring activities, applied to objects b and by means of certain indices, and

(2)   each name of the collection d is normalized among people a and pertaining to some objects of the class b, and

(3)   if '*a*' belongs to the collection *d* and '*a*' is the name of objects *n*, spread among people *a*, and '*α*' is the name of objects *m* spread among people *a*, then the objects *m* are the same as the objects *n*.

For example, the names of minerals, plants and chemical compounds are the bases of names, known to the experts who use similar methods and indices.

We shall pass now to the analysis of further elements of the broader descriptive language. It has been already mentioned that the sentences of broader descriptive language conform to logical formulae as e.g.

$$\varphi\{\alpha\}, \psi\{\alpha, \beta\}, \chi\{\alpha, \beta, \gamma\}, \text{etc.}$$

However, the sentences of the descriptive language differ from the sentences of the majority of logical systems in this, that the names '*α*', '*β*', '*γ*', ... of the descriptive language, introduced as above, may equally be individual names as the names having a great number of designates. The names with one designate do no differ in practice from the names with greater number of designates, at least not to such a degree as it would be necessary to create for them a special semantic category (logical type). Therefore, they all belong to one logical type in our language, just as e.g. in the logical language of S. Leśniewski. Hence also to one logical type we must include, for instance, all functors of one name-argument, irrespective of the fact whether they are names of the attributes of one object or names of certain attributes of a certain set of objects. Further definitions will concern such functors of arguments which may consist both of individual names nad of the names with greater number of designates. Thus, the names of features of individual objects form here individaul cases of the functors of that type. Therefore, in their further analysis we can frequently regard the sign of inclusion '⊂' (which will be subsequently used) as equivalent to the sign '*ε*' of membership in a set. From the formal standpoint it is correct on the ground of the system of logic of S. Leśniewski[11] which contains a thesis:

$$(mb) : m\varepsilon b . \equiv . m \subset b . m\varepsilon V$$

D.7.   The functor '*φ*' of one argument is spread among people *a* by means of reconnoitring action *f* applied to objects *b*

and by means of indices c and in respect of the collection of names $d. \equiv$.

(1)   The names d form a basis of the names of objects among the class b spread among people a, and,

(2)   each person a can perform an action f in relation to any subset $m \subset b$, and

(3)   if a man A among the class a performs an action f in relation to a certain subset $m \subset b$, and 'α' is a name of objects m belonging to the basis d, then A will state a sentence '$\varphi\{\alpha\}$', if and only if A will attain as a result of his action f in relation to the class m such an object X which belongs to the class c.

The subset m may also be one object according to what was said before. In this case the formula $m \subset b$ is equivalent to the formula $m \varepsilon b$.

It would be, of course, possible to define analogically the spreading of the functors of more than one argument.

For instance, the functor 'is a wooden object' is spread among the students of the resistance of material by means of the pertaining action and indices in relation to the collection of the names of objects. Similarly, 'dissolves in water', 'is an acid solution', and the like are the functors spread among chemists by means of familiar actions and indices such as the litmus paper turned red in the case of acids.

Certain modifications of the above definitions are possible according to the fact whether the descriptive language refers only to permanent features of certain objects, as e.g. the solubility of a certain chemical compound, or whether it refers to features operating at a certain time, as e.g. joy of a man. We are concerned here only with the general form of that definition.

D.8.   The functor 'φ' of one argument is univocally spread among people a by means of reconnoitring action f applied to objects b as well as by means of indices C, and in respect of the collection of names $d. \equiv$.

(1)   The functor 'φ' is spread among people a by means of reconnoitring action f applied to the objects b, and by means of indices c, and in respect of the collection of names d, and

(2)   if two people A and B among the class perform an action f
             in relation to the same subset $m \subset b$, then if A obtains
             the product X belonging to the class c, then B will also
             obtain a certain product Y belonging to the class c.

The functors written above are those which are univocally spread,
just as are many other functors used in sciences.

D.9.   The functor 'φ' is a name of the trait g, spread among
             people a by means of reconnoitring action f, applied to
             objects b and by means of indices c and in respect of the
             collection d of names. $\equiv$.

(1)   The functor 'φ' is spread among people a by means of
             reconnoitring action f applied to objects b as well as by
             means of indices c and in respect of the collection of names d,
             and

(2)   if a man A belonging to the class a performs an action f
             in relation to the objects of the subset $m \subset b$ and will obtain
             a product belonging to indices c, then $g(m)$, and

(3)   if a man A of the class a performs an action f in relation to a
             certain subset $m \subset b$, and if $g(m)$, then he will obtain the
             product which belongs to the indices c.

Analogous definitions can be built for various other functors of the
first order. If we have the definitions of how certain functors of the
first order spread and become names, we can further build the
definitions of how functors of higher orders spread and act as names.
In this, from the functors of the first order to the functors which become
the names of traits of the designates of the above functors we can pass
in the same way as from the names of things to the functors of the
first order. In most descriptions, however, only the functors of the
first order are applicable.

The functor 'dissolves in water' is the name of the trait of solubility
in water spread to the students of chemistry, etc. The functor 'is a liquid
of a certain viscosity' is a name of the trait of a viscosity spread among...,
etc.

D.10.   The functors K constitute a basis of descriptive functors of
             one argument spread among people a and supported by the
             basis d of the names of objects among the class b. $\equiv$.

(1)   the names *d* constitute a basis of the names of objects among the class *b*, spread among people *a*,

(2)   each functor *K* is a name of a certain trait, spread among the class *a*, by means of a certain action applied to objects *b*, and certain indices in respect of the basis of names *d*,

(3)   if the functor 'φ' is a member of the collection *K* and if 'φ' is a name of the trait *g*, spread among the class *a* by means of action *f* and the indices *c*, and 'φ' is a name of the trait *h* spread among the class *a* by means of the action *j* and indices *n*, then the trait *g* is the same as the trait *h*, the action *f* is the same as the action *j*, and the indices *c* are the same as the indices *n*.

Analogically, we can define the bases of more than one argument functors. As examples of the bases of descriptive functors, we can take the collections of functors which can be found in various sciences, e.g. the physico-chemical functors.

After we have defined the bases of names and the bases of descriptive functors we can arrive at the definitions of a broader descriptive language.

D.11.   *L* is a broader descriptive language of objects *b*, based on bases: *d* (i.e. those of names), *k*, *l*, ..., etc. (the bases of various functors), spread among people *a*. $\equiv$.

(1)   each collection: *d*, *k*, *l*,... is a basis (of names, of functors of one argument, etc.), applied to objects *b* and spread among people *a*; and if *k* is the basis of functors of one argument then *k* is a basis supported by the basis *d* of names; in a similar way all other bases of the functors of higher semantic categories are supported by the bases *d* of the names, and by other bases which belong to that series of bases: *d*, *k*, *l*, ... and which are bases of functors of lower categories, and

(2)   *L* is a collection of all sentences well constructed according of the rules of sentence construction of descriptive language, and arguments and functors being in these sentences are descriptive names and functors belonging to bases *d*, *k*, *l*, ... and so on.

Having thus constructed the definition of descriptive language¡ one can easily define the concept of a true sentence in the classical sense in relation to that language.

D.12.   The sentence of the shape '$\varphi\{\alpha\}$' is empirically true in a broader descriptive language L, of the objects b and founded on the bases $d, k, l, \ldots$ and used by people $a. \equiv$.

(1) L is a broader descriptive language of the objects b based on the bases $d, k, l, \ldots$ which are spread among people a, and

(2)   'α' is a name of certain objects m spread among people a by means of certain reconnoitring actions and certain indices, and 'α' belongs to the basis d, and

(3)   'φ' functor is a name, of a certain trait g, spread among people c by means of certain actions and certain indices and 'φ' belongs to the basis k and

(4) $g(m)$.

D.13, 14, 15. Given the same conditions 1–3 and *non* $g(m)$, the sentence of the shape '$\varphi\{\alpha\}$' is empirically false in the language L. If the conditions 1–3 are fulfilled, irrespective of whether $g(m)$, *or non* $g(m)$, we call the sentence of the shape '$\varphi\{\alpha\}$' empirically meaningful in the language L. On the other hand, if any of the conditions 1–3 is not fulfilled, then we will call the same sentence empirically meaningless in the language L.

Analogically one can define empirical truth in a broader descriptive language L of the sentences of the shape '$\psi\{\alpha\beta\}$', as well as of the sentences with much more complicated shapes. A precise formulation of the general definition of an empirically true sentence in the descriptive language L does not present great difficulties.[12]

For instance, every sentence of the shape 'the diamond scratches glass' is empirically true in the precise mineralogical descriptive language, based on mineralogical bases of names and functors and which is used by mineralogists as the names 'diamond' and 'glass' are the names of these objects in that language, and the functor 'scratches' is a name of the relation consisting of the capability of one object to produce a

mark on another when they touch each other in a familiar way. Apart from that, the names and functor belong to the bases of the precise descriptive language of mineralogy; and, by moving the diamond on glass, we effect a mark on it. On the other hand the sentence: 'kitchen salt cratches glass' is empirically false in the precise descriptive language of mineralogy; while the sentence: 'the donkey scratches glass' is empirically meaningless in the precise descriptive language of mineralogy, as the name 'donkey' does not belong to the basis of the descriptive language of mineralogy.

III. SEMANTIC ANTINOMIES

An inquiry whether our semantics does not lead to contradiction would be worth our while. It would appear that the descriptive language could be purely extensional, the modal expressions being here unnecessary, while its finitistic character precludes Richard's antinomies. There remain, therefore, the antinomy of the liar and that of heterological expressions. In order, however, to formulate them it is necessary to extend the descriptive language so that it could contain variables, truth-functors, and quantifiers. It is quite enough to accept as meaningful within the descriptive language those sentences which are logically correct and which contain, apart from primitive functors of the descriptive language, also the truth-functors and quantifiers. It is also not difficult to define the concept of a true sentence for such an extended descriptive language. For, since we have a defined concept of the truth of elementary sentences, we can recursively define the general concept of truth, accepting that: general sentences 'for each x $\varphi\{x\}$' is empirically true in language L when every sentence is empirically true which is a substitution in the expression '$\varphi\{x\}$' for the variable 'x' of a certain concrete name belonging to the basis of names of a given semantic category of that language – and accepting that the sentence of the shape $\sim P$ is true if and only if the sentence P is false, and that the sentence $P \vee Q$ is true if and only if at least one of the sentences P or Q is true.

The extension of the descriptive language into expressions which contain variables makes possible the introduction of definitions into language. In connection with this the problem arises whether the names defined by means of exact descriptive names are precise descriptive

names of that language, and if so, by which reconnoitring action and indices they are made known? This is a problem which is difficult to solve. It would appear at first glance that this question can be answered positively and that the reconnoitring action for the name defined is the sum of reconnoitring actions of the defining names while the indices are the sets indices of the defining names, or their absence. This, however, does not contain a complete explanation, and on closer inquiry certain doubts may arise. One can, for example, indicate such names which, through correctly defined in the language which seems to resemble the descriptive one, yet are not precise descriptive names. There are two semantic names, namely, the name 'false sentence' defined in such a way as we have done in the definition D.13; and the name: 'a name which is not its own designate', or, in a shortened form, 'a heterological name' which can be defined for our purpose in the following way:

D.16.   The name A is heterological if, and only if, A is an expression which belongs to the basis of names of the semantic language, and A is a precise descriptive name of certain objects d in the language of semantics and A is not an object of the class d.

These names, as is well known, in various types of semantics lead of necessity to contradiction. In our semantic approach, however, no contradiction arises here, but by means of reasoning which is analogous to antinomial reasoning it can be shown that these names, defined in such a way as was just done here, are not precise descriptive names in the sense D.5. This is so despite the fact that the language of our semantics resembles the precise descriptive language, as the primitive expressions of semantics, stated in the introductory part of this essay, give the impression of being precise descriptive names of certain relations (in line with the definitions modelled on D.9). The reasoning leading up to this conclusion can be expounded as the traditional antinomial reasoning. Thus we write a sentence:

(P1)   P1 *is a false sentence in the language of semantics.*

Where the name 'P1' denotes that sentence written above and by semantics we mean the system of semantics which we are now elaborating.

It will be easily seen from the definition D.12 and D.13. that, had P1 been a true sentence, then it would have been false, and vice versa. This situation can at once lead to contradiction in some systems of semantics. In the semantics under discussion there is still the third possibility that P1 is a meaningless sentence in the language of the semantics in question. Hence it follows from D.15 that either 'P1' is not a precise descriptive name of the sentence P1, or 'is a false sentence in the language of semantics' is not the name of the trait of being a false sentence in the language of our semantics. It would appear that the name 'P1' is a precise descriptive name of the sentence P1 put above and composed of these and these only traces of printers-ink which everyone who accepts our convention will point to unambiguously and unmistakably, and which constitute on the previous page the 3rd line from the bottom. It should be rather assumed that functor 'is a false sentence in the language of semantics' is not a name of that trait, which is equivalent to the assumption that the name 'false sentence in the language of semantics' is not a precise descriptive name of false sentences in the language of our semantics. We would not have reached that conclusion had we narrowed the meaning of the term 'false sentence' by adding another condition to the definition D.13, viz. one which states that a sentence, in order to be false, must be constructed in a non-eubulidean way, that is, in such a way that it could not be similar in shape to P1–thus, it cannot be composed of its own name and the predicate 'is a false sentence'. Therefore, if we narrow the definition D.13., then the sentence P1 would appear meaningless, as it can be neither true nor false on account of its eubulidean shape. Therefore, the name 'false entence', defined as in D.13 is not a precise descriptive name of false sentences, if the language of the semantics in question is a descriptive one. On the other hand, the name 'false sentence', determined with the addition of the condition of non-eubulidean construction, is free from that qualification and may be regarded as a descriptive name. In any case we see that there are names which are defined correctly by means of precise names and descriptive functors which are not precise descriptive names. (There remains the problem: by which definition schemes certainly precise descriptive names and functors can be obtained from precise descriptive names and functors? Before this problem is solved one should not introduce definitions to the descriptive language.)

With regard to the names 'heterological' our reasoning is analogous when we write first the expression:

(P2)   *heterological*

and name this very expression, written in that place, P.2. Had P2 been a precise descriptive name of the heterological names in the language of semantics, then if P2 was heterological names in the language of semantics, then by virtue of D.16 it must have been not heterological in the language of semantics. Also, had P2 not been heterological in the language of semantics, then being a precise descriptive name of the heterological names must have been itself heterological (which follows from D.16). It follows, therefore, that P2 cannot be a precise descriptive name of the heterological names. The same refers, of course, to all written expressions which are uniform with P2. One can, therefore, state in general that no name of the shape 'heterological' is a precise descriptive name of the heterological names. The above reasoning also proves that in the discussed system of semantics traditional semantical antinomies do not take place, although we allowed an inquiry within the bounds of our semantics of any descriptive language, and therefore also an inquiry of the language of semantics itself which can be regarded as a descriptive language.

The fact that we deal with names equiform with parts of their designates and nevertheless this did not result in contradictory statements, seems to vouch for the usefulness of the above semantic definitions. The order not to mix the language under study with metalanguage, often applied in the construction of the systems of semantics, is a shockingly unnatural means of avoiding contradiction. For if we come across any written statement, whether any other individual, concrete object, it would seem obvious that we can call it as we wish, provided that we use that name consistently solely as a name of that object. The consistent use of a name is determined by the definition of an exact descriptive name which, as it has become evident, does not lead itself to contradiction. Therefore, instead of an artificial prohibition, it is enough to accept the above definition of a name and, as a result of it one can obtain a language which in a certain sense is universal as the everyday language. But it is consistent. It would appear, therefore, that it is not solely the universality of the everyday language that is a source

of antinomy, but the lack of empirical precision of semantical concepts which are contained in the normally formulated antinomies. For the semantic concepts of everydy language have also an empirical charac- ter—the meaning being given to language by people through their behaviour when using the linguistic expressions. Hence it seems that the construction of semantic concepts based on the analysis of human behaviour is a procedure which is fundamentally right. The above definitions represented an attempt at such a construction, and the descriptive language to which they refer is among the idealized scientific languages nearest to everyday language.

REFERENCES

* First published in *Przegląd Filozoficzny* **XLIV** (1948). Reprinted from *Synthese* **VIII** (1950–1951) F. G. Kroonder Bussum, Netherlands.

[1] The word 'ideal' is used throughout in a somewhat not precise sense, being equivalent in its meaning to 'non-existing but possible to realize'. Hence, strictly speaking, it is in some contexts superfluous. A different question must be asked: how can there exist sentences which have not been written by anybody and which are studied by the metalogicians? The question may be answered in different ways. We can suppose that the existing sets of elements of the material world contain in themselves any shapes which are defined as expressions of the system investigated. Hence to write a certain expression belonging to the system is tantamount to the repetition of a certain shape which already exists in nature. One can also apply certain theory of possibility. The solution of this problem is not nescesary for our argument.

[2] Nevertheless a similar distinction can also be drawn here. For instance, in histor- ical sciences all correct descriptions of objects existing at present and which can be seen by every living man (i.e. documents) may become scientific descriptions. All other sentences about events in the past belong to the realm of theory and are usually individual hypotheses. Apart from these hypotheses, which refer to past events, the historian accepts in a more or less conscious manner various laws of physics, chemistry, biology and psychology which, together with purely historical hypotheses, offer the explanation of the fact that such and such a document can be found nowadays at such and such a place. Or, speaking more precisely, a his- torian adopts such historical hypothese, from which, together with general laws, he would be able to deduct the descriptive sentences which refer to the present state of documents. Since a very great number of various assumptions enter into historical theories, their formalization may appear practically impossible and perhaps even purposeless, for reasonings are not so involved that their formali- zation would result in their simplification.

[3] For instance, in the sense of Rudolf Carnap, cf. his 'Testability and Meaning', *Philosophy of Science* **3–4** (1936–37).

⁴ The concept of a thing of everyday experience is, therefore, relative. It signifies those objects which the said people can see and touch. Of course, the word 'can', as it is used here, does not have a clearly defined meaning. Nevertheless, it is difficult to avoid that word in any methodological consideration where we speak constantly about possible forms of science. The word 'can', or its various narrowings, appear to be typically theoretical and not descriptive. It can only be reduced to certain descriptive operational terms, but cannot itself become descriptive term.

⁵ The descriptive language in this sense will not be limited to the description of someone's sensations, but may contain all descriptions of the things of everyday experience, hence also, among others, the descriptions of the experiencing people. On the other hand, by reistic language we mean such a language which can be reduced, in an easily visible fashion, to such a language which is characterized by the following conditions:

(1) Among logical types there is the lowest category and all expressions of logical types are functors of the lower ones. (The lowest type can be called the category of the names of objects.)

(2) The primitive expressions of language (except quantifiers and truth functions such as implication or negation) are functions in which only the expressions of the lowest logical type are arguments (e.g. the variables of the lowest category).

(3) All names of bodies which can be seen or touched, hence the names of animals, plants, objects and so on, or, in general, all names of bodies in the somatic sense, are regarded as constant names belonging to the lowest logical type. (If only the names of bodies are constant names of the lowest type, then such a language can be called somatistic.)

(4) The names of psychological experiences do not belong to the lowest logical type, but to higher types. Sensations, notions and ideas are not objects, but rather they constitute a way of psychological experience. (For instance, if I want to describe that A imagines a mount of gold, we say that A imagines in such a way in which he would experience remembering such a mount of gold, had he seen it previously), cf. T. Kotarbiński: 'Sur l'attitude réiste (ou concrétiste)', *Synthese* **VII** (1948/49).

⁶ The idea of making the meaning of terms relative to the situation and the emphasis upon the role of indication comes here from K. Ajdukiewicz, cf. esp. his 'Sprache und Sinn', *Erkenntnis* **IV** (1934).

⁷ Praxeology is a general science of the modes of action; it is the most general methodology of all action, cf. e.g. T. Kotarbiński 'O istocie i zadaniach metodologii ogólnej' ('On the Nature and Tasks of General Methodology'), *Przegląd Filozoficzny* **41** (1938); 'Principes de bon travail', *Studia Philosophica* **3** (1948).

⁸ C. W. Morris, 'Foundations of the Theory of Signs', *International Encyclopedia of Unified Science* **I**, Chicago 1938; also R. Carnap, *Introduction to Semantics*, Studies in Semantics, vol. 1, Cambridge, Mass. 1946.

⁹ S. Leśniewski, 'O podstawach matematyki' ('On the Foundations of Mathematics'), *Przegląd Filozoficzny* **30** (1927). Cf. also Z. Jordan, *The Development*

of Mathematical Logic and of Logical Positivism in Poland between the Two Wars, London 1945. Polish Science and Learning No. 6.

[10] K. Popper, *The Open Society and its Enemies*, vol. 2, p. 15.

[11] S. Leśniewski, *l.c.* **34** (1931).

[12] Some of the definitions given here in full have been already published in my communiqué: 'Un essai d'établir la sémantique du langage descriptif', *Proceedings of the International Congress of Philosophy* I, Amsterdam 1948, pp. 776–778.

ANDRZEJ GRZEGORCZYK

MATHEMATICAL AND EMPIRICAL VERIFIABILITY *

In the last twenty years mathematicians have elaborated a hierarchy of
mathematical concepts from the point of view of, generally speaking,
their effectiveness. This classification has an intuitive sense, suggesting
wider analogies. The hierarchy is based on generally recursive concepts
(recursive relations, functions, functionals), also called computable.
These concepts belong to the arithmetic of natural numbers and are
clearly distinct from other mathematical concepts. If $R(n, m)$ is a recursive relation between natural numbers, then for any pair of natural
numbers n, m, there is a method of verifying, in a finite number of
steps, whether the relation R holds between n and m or not. This property is not a mathematical definition but only an intuitive characterization of computable relations; it is, however, a characterization well
accounting for the sense of computability. All strictly mathematical
definitions of recursive relations are, so to say, attempts to make this
intuitive characterization more precise. After closer examination it can
be easily seen that the characterization, as well as its precise counterparts, has a clear empirical sense.

Let us analyze the sense of the sentence: 'For any pair of natural
numbers n, m there is a method of verifying, in a finite number of steps,
whether the relation R holds between n and m or not'. First, then, the
numbers n and m must be given. They are given concretely in the form
of a notation in some notational system, that is in the form of a certain
formula. This is the sense of the concept that a number is given in
strict terms of computability. The number is then given by a material
notation. Next, an effective method permitting the verification in a finite
number of steps whether a relation holds or not is given as a rule of
transformation of the notations representing numbers. For example,
the relation of minority is computable. For any two numbers written
in the decimal system, the method of stating minority may be described

as follows: if you want to verify whether $n > m$, write the number n over the number m and subtract by the usual method. If the subtraction is possible, then $n \geqslant m$, if not, then the relation $n > m$ does not hold. The complete description of this procedure would of course be much longer. Subtraction, although we are used to it, is a rather complex graphical operation. It should be precisely described that if we have, for instance, the number 4 over the number 3, then the number 1 is written below in the same column, etc. The reservation should be made that the procedure begins from the end and the complicated method of borrowing the tens should be described.

Similarly, we say that a function is computable when there is such a rule of transforming the notation of its arguments which leads to the notation of the function's value. Thus, for instance, all the commonly used arithmetic functions as addition, multiplication, division, involution, etc., are computable.

From the empirical point of view it should then be said that the objects themselves, the natural numbers, are considered here as given in form of some empirical entities; the method of dealing with them is also described as an empirical procedure. This empirical character of the concept of computability can be the most clearly seen in the theory of algorithms, which is equivalent to that of recursive functions. Algorithm theory deals with words constructed of a finite number of letters and with various simple rules of transforming the words, as for example the rule of changing places of first neighbouring letters or of moving the final letter to the beginning. It appears that all the rules of computing recursive functions may be presented by repeating some such simple algorithmic rules.

Analogical empirical rules of course exist in experimental sciences. They can be used to formulate methods of measuring length, mass, time, temperature, methods of stating numerous properties, as chemical composition, radioactivity, etc. The empirical methods of experimental sciences include a much wider range of activities than the algorithmic methods. First of all, they concern a larger domain of objects (not only notations), and also — which is more important — they usually are approximative methods. Physical magnitudes are generally characterized as continuous or as using units so small that it is practically impossible to calculate exactly the number of these units. Therefore a measurement

is never perfect and a chemical analysis never precise. It does not mean that there are no empirically decidable physical properties. Such properties do exist, but their very determination must involve physical magnitudes with an approximation. We can weigh objects of about 1 kilogram exact to 0.01 gr. The empirically decidable properties are then not identical with the theoretical ones, i.e. these which we assign to objects as their exact characterization. Physical objects are assigned as characteristic some empirical properties which are not decidable with full precision.

Empirically decidable properties also differ from theoretical ones. Numerous natural laws are not obeyed. For instance, equality is not transitive. If $A = B$ exact to 1 cm, and $B = C$ also exact to 1 cm, it does not entail that $A = C$ exact to 1 cm. The errors of measurement add up and may exceed 1 cm.

This example shows how convenient are idealizations, undeterminable magnitudes which nevertheless have numerous ideal properties that faciliate our thinking about physical objects to which we ascribe them. Kant's opinion may be reminded of, saying that the properties ascribed to things are forms of conceiving these things by the mind.

Mathematical decidability also involves a specific idealization of another kind. There are many mathematical properties potentially decidable. Practically nobody would have the time, and may be there would not be enough matter to the notation, although time and the necessary notations are finite; they only exceed the dimensions of the real world. Therefore, the notations of mathematical theory should not be conceived of as real material notations, for instance with a pen on some paper, but as geometric entities in an infinite space. The whole algorithm theory, or that of recursive functions, obtains its adequate place only in the ideal sphere of mathematical fiction.

The recursive concepts are then linked with some idealizations, but not in a greater degree than the concepts considered as empirical.

As the recursive concepts have their counterparts in the decidable empirical concepts, so the higher levels of the logical classification of predicates have theirs in some conceptual constructions based on empirical concepts. The logical classification of predicates can be divided into two branches, depending on whether the computable concept involves a general or an existential quantifier. If $R(x, y, z)$ is a computable

relation, then the higher classes include for instance the concepts A_1, A_2, B_1, B_2, characterized as follows:

$$A_1(x, y) \equiv \bigvee_z R(x, y, z), \qquad B_1(x, y) \equiv \bigwedge_z R(x, y, z),$$

$$A_2(x) \equiv \bigwedge_y \bigvee_z R(x, y, z), \qquad B_2(x) \equiv \bigvee_y \bigwedge_z R(x, y, z).$$

The concepts A_1 and B_1 result from R by means of one-quantifier operations, and the concepts A_2 and B_2 — by means of two-quantifier operations. Analogically, concepts of higher level result from the use of an increasing number of quantifiers for computable concepts. For example, the concept of limit of a sequence of fractions is a three-quantifier concept, as the definition of limit is based on the arithmetic of natural numbers; for instance in the simplest case of the approaching of the sequence of fraction $\dfrac{f(n)}{g(n)}$ to zero, it has the following form

$$\lim \frac{f(n)}{g(n)} = 0 \equiv \bigwedge_k \bigvee_m \bigwedge_n \left(n > m \rightarrow \frac{f(n)}{g(n)} < \frac{1}{k+1} \right).$$

The relation in brackets is computable. It involves three quantifiers. At the same time, it has been demonstrated (A. Mostowski) that all the three of them are necessary in the sense that the concept of limit cannot be expressed in any equivalent way in terms of a smaller number of quantifiers, or even the same in a different order, followed by a computable arithmetical relation.

In real sciences an analogical classification can be set up. It will order all concepts, as more and more detached from experience. The concept of equality of two real segments seems, for example, to be a one-quantifier concept, of the same type as B_1. The segments a and b are equal if and only if for each (optionally exact) method of measurement they are not distinguishable with the exactitude characteristic for the given method. On the other hand, the concept of difference is of the type $A_1 \cdot a \neq b$ if and only if there exists a measuring method (an exactitude) for which we obtain different results of measurement. The relation $<$ is also of the type A_1, and the relation $\leqslant$ of the type B_1. All these relations are reducible in a one-quantifier way to empirically decidable relations. They might be called one-quantifier decidable, or one-quanti-

fier detached from experience. The existence in a definition of a single quantifier which cannot be eliminated entails a fundamental undecidability of a given property in a finite time. It is impossible, in a finite time, to carry out an infinite number of measurements of an optionally great exactitude. The existence of the concept of limit in the definitions of numerous physical concepts entails a very high degree of their undecidability. The non-continuity of some physical magnitudes can, of course, considerably reduce this degree of undecidability.

The intuitive sense of the mathematical hierarchy of predicates is analogical. Concepts with more quantifiers are more distant from algorithmical decidability than concepts with fewer quantifiers.

Thus the logical classification of predicates enables us to realize, and even to measure, to what degree our thinking about physical and mathematical reality differs from what is easily noticeable, empirically determinable, decidable in a finite number of measurements or computational procedures.

REFERENCE

* First published in *Rozprawy Logiczne*, Warszawa 1964. Translated by S. Wojnicki.

JERZY KMITA

MEANING AND FUNCTIONAL REASON *

1. This paper has a double purpose. First, I wish to draw a distinction between two types of research procedures, or two types of explanation: between such explanation which answers why something occurs, or is the case, by pointing to its meaning (purpose), and such explanation which refers to its functional reason. The former type I call humanistic interpretation, while in the latter case I speak of functional explanation. The terms here employed, such as meaning, functional reason, humanistic interpretation, functional explanation will be elucidated with some more precision in the course of further discussion.

In the second place, I wish to sketch an answer to the question, what are the attitudes towards these two types of explanation of some approaches usually labelled as 'structuralism' on one hand, and Marxism on the other; such a comparison will allow to grasp and relate with each other the methodological consequences of the two outlooks.

2. At the outset it might be asked if the opposition between humanistic interpretation and functional explanation, and the parallel opposition of meaning and functional reason, is not something familiar to, and well kept in mind by all? Certainly, this opposition is related with some common intuitions, but they are dim enough to be but a poor protection against the frequently observed misunderstandings. Besides, these misunderstandings are by no means always spontaneous; more often they result from reflection neglecting the distance between meaning and functional reason. An illustration may be a statement by L. Goldmann, which I shall quote below, in which it is assumed that the 'genetic' or 'dynamic' structuralism, propagated and applied by J. Piaget, constitutes a peculiar synthesis of 'understanding', i.e. of what I prefer to call explanation by reference to meaning — humanistic interpretation — with naturalistic explanation. I wish to bring to attention the first component of the alleged synthesis, i.e., the 'understanding'. According

171

to Goldmann, 'genetic structuralism' which can be described as adher-
ing to a methodological postulate advicing a certain variety of explanation referring to functional reason, at the same time assumes an 'understanding' or establishment of meaning.

Here are the essential fragments of Goldmann's statement: "...un-
genetic structuralism — the names designated by this term are familiar:
from Husserl to the Gestalt theorists, Lévi-Strauss, and the recent
works by Roland Barthes in France — opposes against atomism an existence of structures which provide the only frames for accounting for a
significance and meaning of an element. However, inasmuch as such
structures are conceived as stable and universal, any idea of explanation
becomes meaningless on their level".[1] On the other hand, 'genetic
structuralism' "...introduces wholly new perspectives inasmuch as it assumes that understanding and explanation are not only interconnected
intellectual processes, but one and the same process, referring to two
different levels of an object's cross-section".[2] Thus "... genetic structuralism assumes a synthesis of fact and value judgments, of understanding and explanation, of determinism and finalism".[3]

3. I think the opinion that Piaget's functional explanation is at the
same time an 'understanding' (establishing of meaning) to be a misunderstanding, but there is another mistaken notion in the quoted
fragments: it consists in combining the structuralism of Lévi-Strauss with
classical concepts of 'understanding' of the German antinaturalists. In
fact, neither in C. Lévi-Strauss, nor in J. Piaget's conceptions it is 'understanding' or meaning, but functional reason which is involved. The
differences between these two approaches will be discussed later, now
we shall draw a rough explication of the classical concept of 'understanding'.

The concept has its origin in W. Dilthey. He conceived it as a peculiar
variety of experience, different from both extraspection and introspection; it was said to consist in experiencing by an 'understanding' subject
the same experiential whole which had found its expression in a definite
manifestation of 'spiritual life' of some other individual. Such a manifestation could be a simple 'expression of spiritual life' like a single
statement, notion, gesture, some simple action; it could also consist in
more complex behaviours of products, as a work of art, a work of

scholarship, political or legislative activity, etc. In the latter case 'understanding' is more complex too, and it is defined as 'higher understanding'.

This is not a proper place to analyse later modifications of the notion of 'understanding' suggested by W. Dilthey. Sometimes they were rather essential, as in the dispute, whether 'understanding' was aintellectual, according to Dilthey, or intellectual, or whether it made another's experiential whole directly accessible, or opened up only what was common in it for different subjects and defined by the supra-individual sphere of values and ideas. However, in spite of all these disputes, two basic moments are always there: 1) 'Understanding' is a kind of intuition allowing for direct cognition; 2) such cognition always makes some whole accessible, organized by some central value. More recent adherents of 'understanding' and of the related conceptions have been putting strong emphasis on the proposition that a whole made accessible by 'understanding' is not psychological in character, that it is not shaped by any psychological mechanisms such as associations, that it is not a collection of elements gathered in any mechanistically conceived manner. Also Dilthey shared this attitude, though he defended it by other arguments than the later antinaturalists; according to him, 'understanding' is able to grasp wholes because of its aintellectual, nonconceptual character.

4. The vogue still enjoyed by antinaturalistic ideas among human scholars cannot be accounted for — as champions of neopositivism would like to believe — by the fact alone that the advocates of these ideas have been using persuasive language; in the debate of antinaturalists with positivists the former reveal a much better orientation in research practice of the humane disciplines, and this fact determines their having the upper hand. It will be convenient to use W. V. Quine's statement to define the scope within which, I believe, the antinaturalists' right should be admitted. This statement, quite recent as is it, witnesses the fact that the debate between antinaturalism and positivism about the methodological basis of the humane disciplines is still going on. Quine speaks here about the study of language, and the process of learning it, but it can be understood more generally, as concerning all the studies of cultural phenomena. It may be added that Quine's term 'naturalism'

roughly corresponds with my usage of 'positivism'. Here is Quine's statement:

"What the naturalist insists on is that, even in the complex and obscure parts of language learning, the learner has no data to work with but the behaviour of other speakers. When with Dewey we turn thus toward a naturalistic view of language and behavioural view of meaning, what we give up is not just the museum figure of speech. We give up an assurance of determinacy. Seen arccording to the museum myth, the words and sentences of a language have their determinate meanings. To discover the meanings of the native's words we may have to observe his behaviour, but still the meanings of the words are supposed to be determinate in the native's mind, his mental museum, even in cases where behavioural criteria are powerless to discover them for us. When on the other hand we recognize with Dewey that 'meaning [...] is primarily a property of behaviour', we recognize that there are no meanings, nor likenesses nor distinctions of meaning, beyond what are implicit in people's speech dispositions to overt behaviour".[4]

The position of Quine's 'naturalist', or of a positivist in my terminology,[5] can be interpreted in two ways: 1) as a normative position, questioning the cognitive validity of a 'mentalist's' research procedure; and 2) as a descriptive methodological position, reporting a linguist's, or more generally, a human scholar's, research procedure. Now, this position interpretetd in the second of the two ways is certainly inconsistent with actual facts. Actually, a linguist's or any other human scholar's research experience, as well as the experience of linguistic or other human communication is 'mentalistic' in character. When we study human activity or its products, or when we communicate with each other, we assume much more about the subjects of the studied activities, or about our partners in the process of communication, than that they are apt to react in definite, observable ways in definite, observable conditions.

Quine's 'naturalist' can be reproached that he does not distinguish between two kinds of 'mentalism' which he attacks. In studying human activities or in communicating with others, we assume a number of non-behavioural propositions about the subjects of the studied activities, or about our partners in the communication process, but these propositions can be conceived in two basically opposed ways:

a. they can be considered as theoretical hypotheses, for which observable human behaviour constitutes the role of empirical evidence, or

b. it can be believed that these propositions have got a direct justification of some other kind, that they refer to some special sort of intuition, different from extraspection, or even from introspection.

An adherent of the view a. shares an antinaturalistic belief that a student of a human discipline, or a communicating individual, makes use of some knowledge about the respective subject, which cannot be expressed in behavioural terms. He agrees with a contemporary positivist, i.e., with Quine's naturalist, in that he does not admit any empirical evidence other than behavioural.

Because of this second moment, I call an adherent of the view a. a methodological naturalist, similarly to Quine's 'naturalist'; thus I give the term a wider meaning than W. V. Quine. However, to distinguish a naturalist in the sense of the view a. from a naturalist in Quine's sense, I add the qualifications 'antipositivistic' and 'positivistic'.

5. Within the antipositivistic variety of naturalism here proposed, it is assumed that the extrabehavioural knowledge, employed by a human scholar with respect to a subject of an analysed human activity or its products, consists of

a. an assumption about rationality of such subject;

b. hypothesis about a definite preferential order of the considered results of the undertaken activity and of its alternatives;

c. a hypothesis to the effect that the subject is aware of the 'casual' links between the alternative activities considered by him, and their above mentioned results. In a case when the subject is sure that the considered alternative activities would bring definite and no other results, the rationality assumption is a statement that the subject would undertake such action as would lead to the most preferred result.

Thus the fact an activity had been undertaken, or that its product has a given property, follows from the premises a., b. and c.; the premises together explain the given fact. I call this type of explanation humanistic interpretation, and an activity thus interpreted I call a rational activity. Correspondingly, we may speak about a product of a rational activity. Finally, if a rational activity has been undertaken in conditions of certainty, its most preferred result (purpose of the activity)

is called its meaning. The notion of meaning can also be defined for other conditions, but the task is rather complex and I shall not deal with it here. Suffice it to say that in conditions other than subjective certainty, the meaning of an action may not always be its most preferred result. However, humanistic interpretation always aims to establish a meaning, defined in some way or another of an activity, or of a property of its product. Thus it is an explanation by pointing to a meaning.

A rational activity or its products are most often complex entities. If we are dealing with a complex rational activity C with the meaning S, we can single out a set of elementary activities $C_1, ..., C_n$, such that from the standpoint of knowledge of the respective subject, each of the elementary activities C_i ($i = 1, ..., n$) is a necessary element of the sufficient condition—the latter, being the performance of the whole set $C_1, ..., C_n$—for a realization of the meaning of the activity C. I shall describe this relation between C_i and a complex activity C by the phrase, 'C_i is a direct instrumental component of C'. It is obvious that direct instrumental components of C_i can in turn have their own instrumental components; in such a case a complex activity is a hierarchical whole, sometimes a very involved one. In parallel to the notion of a complex rational activity, we can introduce the notion of a product of such activity. The difference between the two is not great: the components of a complex activity are themselves activities, whereas the components of respective products are, analogously, products.

I call a complex rational activity or its product a humanistic structure. It corresponds to Dilthey's 'complex expression of spiritual life', among other things, in that its particular instrumental components are subjected to its meaning in the way indicated above. At the same time, in accordance with Dilthey, the whole which is a humanistic structure, is constituted in effect of a consideration of the relevant subject's viewpoint. It is only from this viewpoint that the relation of instrumental hierarchy, constitutive for the whole, does at all hold; it is not, to use Dilthey's language, an 'outer' relation, or an objective one, independent on anyone's consciousness.

6. Besides explanation by reference to meaning, another type is commonly used, formally similar to the former, as it operates with the concept of structure, too. A certain variety of it is offered by J. Piaget.

However, it is essentially different from explanation by reference to meaning, which fact seems to have remained unnoticed by L. Goldmann whom I have quoted above. This can be witnessed by a following statement by J. Piaget.

"Evolution may have a direction without necessarily having an end. As an example may serve the physical evolution of entropy. The increase of entropy is a model of evolution which has a direction, but not a trace of finality [...]. From my point of view, the method of reconciling of genesis with structure consists in making the structure into a form of balance, towards which genesis aims; however, this does not mean that finality is thereby called forth. Subjectively, the process can be expressed in finalistic propositions, but as soon as explanation is concerned, I do not see any advantage in reviving this old idea which doesn't explain anything; and can be reduced to a pointing of a direction [...]. The finalistic approach has its role in the sphere of consciousness, in opposition to the cause-and-effect pattern, where the tendency towards a balance interferes".[6]

It can be seen that J. Piaget is not interested in explanation by reference to meaning, and thus he is not interested in humanistic structures. Let us then try to answer, what kind of structure may be involved here. However, before we do it, we must emphasize, to avoid terminological misunderstandings, that we use the term 'structure' as equivalent to 'a relational system' whereas both J. Piaget and C. Lévi-Strauss use it as roughly equivalent to: 'a family of relations common for a given class of relational systems'.

Now, the concept of whole which interests us now, and about which J. Piaget would be apt to say that it possesses a structure, can also be encountered in biology. Every organism, sometimes together with its nearest environment, is conceived in biology as a functional structure. Such a structure can be roughly described as follows: 1) it is subdivided into a number of components, and for each component a certain repertory of its possible states always exists; 2) there is a set of all possible sequences of states of components which correspond to periods or moments in time; these sequences I call global states. On this set its proper subset is defined, such that the structure has an established property W, called its state of balance, if and only if the global state of the structure is an element of the proper subset; 3) for each subsequence of

components there are such subsequences of the states of their components that for any states of the other components, the whole structure would not have the property W; I shall call this kind of subsequences of states of components antifunctional series of the states of components; 4) a functional structure always maintains the relevant property W.[7]

If a structure is conceived as a functional one with respect to the given property W — such a relativization is necessary, as a given structure may well be functional with respect to a property W_1 but nonfunctional with respect to another property W_2 — features of its components or of subsequences of its components can be soundly explained according to the following pattern of inference:

1) The given structure S_1 belongs to the type S.
2) Every structure of the type S possesses the property W.
3) The complement of a feature C of a (simple or complex) fragment f of a structure of the type S is antifunctional (with respect to the property W).

The fragment f of the structure S_1 possesses the feature C.

An explanation along this pattern may be called a functional explanation, while what is stated by its premises taken together, I call the functional reason for the explained features of a fragment of the given structure. Functional explanation thus consists in pointing to a functional reason. Here is a typical example of a functional explanation as an answer to the question, why there is chlorophyll in the leaves of some plant.

1) This organism is a higher plant.
2) Every higher plant is capable of photosynthesis.
3) A lack of chlorophyll in the leaves of a higher plant would cause a loss of its ability of photosynthesis (it is antifunctional with respect to this ability).

There is chlorophyll in the leaves of this plant organism.

I call the premise 2) of a functional explanation a law of balance maintenance, and the premise 3) a functional law, as it states what function is performed by a definite feature C of a fragment of a given type of structure, or otherwise speaking, to what is C necessary.

7. It can be seen now in what consists the so often misleading similarity between explanation in terms of meaning, and explanation in terms of functional reason. Both the meaning and the functional reason, or more precisely, the property W, constitute one of the factors determining a humanistic and a functional structure respectively. It is still more necessary to emphasize the difference between humanistic and functional structure, and between explanation by reference to meaning and explanation by reference to functional reason. The relation between an activity, or a property of its product, and its meaning, is subjective in character; it occurs because of somebody's point of view, while the relation between a feature of some fragment of a functional structure, and a property W of this structure, is an objective one, independent on anybody's being aware of it; a meaning is realized, because a certain value order has been accepted by a given subject, while property W is possessed by a functional structure in a natural way, and it is not any purpose whatever of this structure.

A failure to perceive this difference introduces teleology into functional explanation on one hand, and leads to a 'dehumanization' of humanistic interpretation on the other. Many relevant examples could be induced from biology, the social sciences, or linguistics. In particular, constant heasitation between meaning and functional reason is typical for the European 'structural' linguistics, in opposition to the American, distributionistic version of linguistic 'structuralism'.

8. A functional structure can be sometimes direction-oriented. The relevant property W is then involved in a process of development, and the structure approaches in time, in a monotonic manner, or by oscillations, to a certain type of a global state. It is this kind of directional global state, which is called 'structure' by J. Piaget; according to him, it is a 'form of balance toward which genesis strives'.

An explanation presupposing the notion of a direction-oriented structure can be described as diachronic-functional. Thus, J. Piaget represents in his theory and research practice (studies on development of intelligence in children) a conception of diachronic explanation by pointing to functional reason. Another variety of functional explanation is the synchronic-functional one, presupposing the notion of a functional structure relative to some static, unchanging property W.

This kind of explanatory procedure is represented by C. Lévi-Strauss.

The latter statement is not as obvious as it might seem. The 'systems' studied by the author of *Pensée sauvage*, or structures in my terminology, are always some kinds of classifications, carried out by ethnic groups on a material which is given to their awareness, and drawn from their own natural or social environment. Now, classifications are certainly humanistic structures; they are products of complex rational activities, the most general meaning of which consists in dividing a certain set into a number of subsets which are mutually exclusive, and add up together to the divided set. Thus a justified doubt arises, whether the 'structuralism' of Lévi-Strauss presupposes a resource to the notion of functional reason, or to that of meaning?

A consideration of a property ascribed by Levi-Strauss to his 'systems' and called 'structure' by him, is crucial to this question. As it has been mentioned, the 'structure' is here a family of relations organizing the particular 'systems'. Those relations are always oppositions of elements of the parts of the particular systems. The fact that a relation of opposition holds between a and b can be formulated by saying that a has a feature C_1, while b has a feature C_2, and that these features are distinctive, i.e., that they are mutually exclusive and a substitution of one type of feature for the other brings about a shift of the pattern of reference of meaning of the given element. The particular systems can change, some elements can give way to others, but the relations of opposition between them remain, the 'structures' are maintained: "Synchronism resists diachronism", as the French ethnologist has put it aphoristically.

Now, considering that the stability of 'structures' is a wholly unintentional property of the 'systems', that is, not a consciously accepted purpose of the makers and perpetrators of the 'systems', that Lévi-Strauss has repeatedly emphasized the unimportance of any conscious motives that might be at work in individuals who keep and observe intra-systemic oppositions in their customs and rites,[8] considering all this, we can reiterate our initial statement: the 'structuralism' of Lévi-Strauss is a conception postulating explanation in terms of synchronically conceived functional reason.

9. A precise distinction between explanation by pointing to meaning and by pointing to functional reason allows to put the question, whether Marxist explanation of human activities is diachronic-functional. Many philosophers considered by others and considering themselves as Marxists have expressed opinions presupposing a positive answer to this question. Now, I would like to present some arguments in favour of the notion that at least the founder of Marxism, Karl Marx, had neither postulated nor employed this kind of explanation.

It will be convenient to begin with turning attention to a certain carefully elaborated conception, according to which "...human societies, or more precisely, definite social formations,"[9] are, from a Marxian point of view, a particular case of what I have called formerly a direction-oriented structure. I mean here the conception of O. Lange presented in his work *The Whole and Development in the Light of Cybernetics.*

According to Lange, the global states of a direction-oriented structure have their corresponding compound vectors of inputs, or alternatively, of outputs of the elements of the structure. The subset of global states at which the structure maintains a definite development property W, is constituted of such complex vectors of inputs, or outputs, and thus of such global states, that each state of this kind ascribed to a given period, or moment, t by a value not greater than a deviation defined in advance as an admissible deviation from the value of a so-called directional function of the system, for a period or moment t. The limit of the series of the admissible deviations for the consecutive moments is 0, and thus the consecutive global states approach their corresponding values of the directional function. The property W consists here in the fact that the direction-oriented structure develops ergodically, aiming at a definite, directional global state.

If a socio-economic formation is conceived as an ergodic structure the phenomena occurring in it must be explained in the diachronic-functional manner. In such an explanation, there are three premises: a. the given structure is of the type R, b. every structure of the type R is ergodic with respect to a certain directional function, c. if a structure of the type R had assumed a global state beyond the scope of deviation admissible for a given period or moment, it would have ceased to be ergodic. From those premises an explanandum follows, stating that at a given

time the lobal state of the structure remains within the limits determined by the value of the directional function of the structure valid for this time, and by the extent of the admissible deviation.

Evidently, events occurring within a socio-economic formation entail, among other things, human activities which are relevant with respect to their results for the formation as a directional structure. Thus these activities are subject to diachronic-functional explanation. The consequence of this statement is as follows: irrespective of their subjective determinations or motives, human activities are never dis-functional with respect to the direction and development phase of the formation; their subjective determinations must be made to 'fit into' the direction of development and its phase.

This is how J. Ładosz sees it, when he writes: "Marx explained the emergence of slavery without resorting to any marvels. The very 'idea' to make someone into a slave could have only occurred to people, when private property had already existed, and the exploitation of labour of the people associated with it had begun. And these phenomena result from a regular, and by no means accidental, development of material production, from the emergence of surplus value. It was only then that the making of someone into a slave had acquired meaning and con-tributed to economic progress. Earlier it would have been of no use."[10]

10. The view that 'it occurs' to people to do, in a domain of activities which are socially and economically relevant, only what 'contributes to economic progress' (or at least does not inhibit it essentially), is a consequence of the acceptance of the diachronic-functional conception of explanation of human activities, and on the other hand it conduces to the conclusion, formulated expressly in the quoted fragment, that activities are meaningful, i.e., are of some use for the agents. The con-clusion assumes in addition a rationality of human endeavours: if people never act in a manner inadmissible with respect to the direction and phase of development of a socio-economic structure, and if they at the same time act rationally, i.e., they aim consciously at a realization of definite purposes, then they must set forth only such purposes, the realization of which is not excluded in advance by the direction and development phase of the formation, which means that their objectives must be 'realistic'.

The rationality assumption is doubtless one of the basic theoretical premises in Marxism; however, if the several socio-economic formations are conceived as direction-oriented structures, the methodological import of this assumption becomes negligible. It justifies explanation of human activities being, as it has been said, one of the premises of such explanation, but the explanation itself becomes a procedure of little cognitive import, some kind of a literary addition to the basic type of diachronic-functional explanation. The kind of human activities to be undertaken, are determined in advance by the given direction and phase of development, and humanistic interpretation can at its best answer an extra question, why among the given type of activities, imposed by the direction and phase of development, some particular variety had been undertaken instead of another.

11. To support my contention that K. Marx did not assume the diachronic-functional conception of explanation, I would like to offer two basic arguments: a. the author of *The Capital* many times expressed opinions incompatible with that conception; b. he explained human activities employing premises other than those involved in the diachronic-functional pattern of explanation.

Here are two examples to illustrate the statement a. In each socio-economic formation in which the development of commerce, urban industry, production of commodities, and thus the circulation of money, has not attained a proper degree, the use of monetary rent instead of natural rent is disfunctional. Thus, according with the diachronic-functional conception of explanation, it should not 'occur' to anyone to substitute monetary rent for natural rent. But according to Marx, we record as early as "...the Roman Empire, several failed attempts at a similar change, and returns to natural rent, when at least the part of natural rent charged as state tax was intended to become monetary rent."[11]

Another example. In the capitalist conditions monetary rent has already become functional, while natural rent is disfunctional. It should not thus occur to anyone to keep the natural rent. However, Marx states that "[...] natural form of rent paid in goods, adequate for the natural economy of the Middle Ages, and wholly incompatible with the capitalist way of production, has survived up to the recent times".[12]

How, then, should the economic, legislative, political, etc. activities be explained according to Marx? Similarly, as any other rational activity: by knowledge and norms (purposes) of a rationally behaving agent. What, then, is the import of describing their activities in terms of their objective effects, or in terms of their functions in the context of a socio-economic formation? It is quite essential. The objective effects of an action, and their relation to its subjectively defined meaning, determine that the motivating beliefs will either be reinforced, and thus the action will be repeated, sometimes on mass scale because of communication processes, or else it will remain a marginal phenomenon, like monetary rent in antiquity, or natural rent in modern times.

In maintain, then, that according to Marx, a character of a given social and economic formation allows to explain only facts of spreading, persistence, and waning of the types of rational activities, motivated by subjective beliefs, and thereby facts of spreading, persistence and waning of these beliefs.

12. Explanation of this kind of facts is of course not diachronic-functional. They do not occur because the relevant beliefs lead to actions which are functional or disfunctional with respect to a given direction-oriented socio-economic structure, about which it is assumed that in its automatic course of development outside of human awareness, it selects some of their beliefs and actions. Conversely, they do occur because the respective beliefs and activities lead to such objective consequences within the pattern of a developing socio-economic structure that individuals, behaving as rational subjects, are either reinforced in their initial beliefs, and they continue their respective actions, or else they give them up. Original beliefs, and traditionally inherited ones in particular, are objectively conditioned at all only because individuals acting upon these beliefs are rational beings. There are always two factors. explaining the spreading, persistence and disappearance of definite forms of consciousness: 1) the subjective factor, i.e., the original state of human consciousness inspiring particular rational activities; 2) the objective factor, i.e., an actual phase of development of a socio-economic structure, within which these activities result in definite objective consequences.

The kind of explanation described above belongs to the more general

type of functional-genetic explanation. I shall define it as the Marxian functional-genetic explanation.

Many exemplary illustrations of the use of the Marxian directive of functional-genetic explanation can be found in the *The Eighteenth Brumaire of Louis Bonaparte*. Perhaps the most frequent in that work by Karl Marx is the explanation of facts of giving up subjective beliefs in face of such objective effects of actions undertaken upon those beliefs, which are inconsistent with their purposes.

Thus, e.g., the basic inconsistency between aims and objective effects of political activity of the two main fractions of the party of order: the Legitimists and the Orleanists, caused that traditional political views have been gradually undermined in favour of new ones, based on a more adequate recognition of the real situation: "They do their real business as the party of Order, that is, under a social, not under a political title, as representatives of the bourgeois world-order, not as errant knights of princesses; as they bourgeois class against other classes, not as royalistic against the republicans. And as the party of Order they exercised more unrestricted and sterner domination over the other classes of society than ever previously under the Restoration or under the July Monarchy, a domination which, in general, was only possible under the form of the parliamentary republic, for only under this form could the two great divisions of the French bourgeoisie unite, and thus put the rule of their class instead of the regime of a priviliged fraction of it on the order of the day".[13] This is a description of political activity of the party of Order in terms of its real consequences, in the context of the objective socio-economic situation in France in the half of the 19th century. Hence, "... the leaders of this party, Thiers and Berryer, the Orleanist and the Legitimist, were compelled openly to proclaim themselves republicans, to confess that their hearts were royalist but their heads republican, that the parliamentary republic was the sole possible form for the rule of the bourgeoisie as a whole. Thus they were compelled, before the eyes of the bourgeois class itself, to stigmatise the Restoration plans, which they continued indefatigably to pursue behind the parliament's back as an intrigue as dangerous as it was brainless."[14]

But not only facts of the waning of delusive beliefs are explained along the functional-genetic pattern in *The Eighteenth Brumaire*. Per-

sistence of delusions also yields to this kind of explanatory procedure. The views of the peasants, the social force which installed Louis Bonaparte in power, originated from the period of Napoleon's rule; "Napoleon confirmed and regulated the conditions on which they could exploit undisturbed the soil of France which had only just fallen to their lot and slake their youthful passion for property."[15] Thus it was an inherited historical tradition of the French peasants which "gave rise to the belief [...] in the miracle that a man named Napoleon would bring all the glory back to them. And an individual turns up who gives himself out as the man because he bears the name of Napoleon [...]. After a vagabondage of twenty years and after a series of grotesque adventures, the legend finds fulfilment and the man becomes Emperor of the French. The fixed idea of the Nephew was realized, because it coincided with the fixed idea of the most numerous class of the French people."[16]

However, what had the objective circumstances been which caused that the traditional Bonapartism, i.e., a certain subjective factor, prevailed among a major part of French peasants in the half of the 19th century? Otherwise speaking, what objective factors had enabled this subjective factor to have played its political role? They had been nothing else but the objective consequences of political activities of the bourgeoisie. "The three years' rigorous rule of the parliamentary republic had freed a part of the French peasants from the Napoleonic illusion and had revolutionized them, even if only superficially; but the bourgeoisie violently represssed them, as often as they set themselves in motion. Under the parliamentary republic, the modern and the traditional consciousness of the French peasant contended for mastery. This progress took the form of an incessant struggle between the schoolmasters and the priests. The bourgeoisie struck down the schoolmasters [...]. It has itself forcibly strengthened the empire sentiments of the peasant class, it conserved the conditions that formed the birthplace of this peasant religion."[17]

The Marxist functional-genetic explanation, as it has been partly made clear by the quoted fragments, concerns not only facts of spreading, persistence, or waning of certain sets of beliefs, but it deals as well with appearance of definite real states of socio-economic structures. In the latter case, too, the subjective factor constitutes, alongside with the objective one, represented by a former state of a structure, a necessary

element of the sufficient condition. However, I shall not enter into details of this problem, as what has been said seems to prove clearly enough that the Marxist functional-genetic explanation is closely related to humanistic interpretation. Indeed, it presupposes such interpretation with respect to the explained course of human activity. Without consideration of subjective factors, human endeavours cannot be explained, as the adherents of the diachronic-functional explanation are apt to believe for whom the knowledge of subjective factors constitutes, at its best, a purely literary addition to the strictly scientific knowledge.

REFERENCES

* First published in *Teoria i doświadczenie*, Warszawa 1966. Reprinted from *Quality and Quantity* V (2), 1971, Società Editrice il Mulino, Bologna.

[1] *Entretiens sur les notions de 'genèse' et de 'structure'*, sous la direction de M. de Gandillac, L. Goldmann, J. Piaget, Paris, 1965, pp. 8–9.

[2] *Ibid.*, p. 10.

[3] *Ibid.*, p. 16.

[4] W. V. Quine, *Ontological Relativity and Other Essays*, New York 1969, p. 28.

[5] It may be noted, however, that Quine's 'naturalist' who expresses the philosopher's own opinion in some other respects, not mentioned in the quoted fragment, steps beyond the limits of positivism even in the most extended usage of the word.

[6] *Entretiens ...*, cit. pp. 18–19.

[7] See E. Nagel, *The Structure of Science*, New York 1961, pp. 411–418.

[8] One example may be the proposition that religious beliefs do not sufficiently account for the frequent prohibition to utter names of the dead; according to C. Lévi-Strauss, such prohibition is a 'structural feature of the system of names', *Pensée sauvage*, Ch. VII.

[9] O. Lange, *Wholes and Parts. A General Theory of System Behaviour*, Oxford — London — Warszawa 1965.

[10] J. Ładosz, *Marksistowska teoria walki klas (The Marxist Theory of Class Struggle)*, Warszawa 1969, p. 61.

[11] Cf. Karl Marx, *The Capital*, vol. II, part 2.

[12] *Ibid.*

[13] Karl Marx, *The Eighteenth Brumaire of Louis Bonaparte*, Moscow 1967, pp. 38–39.

[14] *Ibid.*, p. 76.

[15] *Ibid.*, p. 108.

[16] *Ibid.*, pp. 106–107.

[17] *Ibid.*, pp. 107–108.

MARIA KOKOSZYŃSKA

ON A CERTAIN CONDITION OF SEMANTIC THEORY OF KNOWLEDGE*

1. LANGUAGES UNDER INVESTIGATION

The structure of the language in which our knowledge is expressed is more and more frequently the object of investigation of contemporary epistemological research where it is being studied from the point of view of syntax (relations between the expressions of language), semantics (relations between expressions and the objects to which these expressions refer) and pragmatics (relations between the expressions and the individuals who use them). The above studies endeavour to speak about languages which are characterized as precisely as possible by the definitions of 1) a sentence, and 2) direct consequence. In general, these languages possess the property that both of the above mentioned definitions can be formulated without reference to extralinguistic reality. They are the so-called *structural* languages. In the syntactic theory of such languages there is no place for, e.g., the concept of empirical thesis, which may be very useful in the methodology of the empirical sciences. On the other hand this concept can be obtained, like many others, in the syntax of *empirical* languages [1] for which the concept of direct consequence should be characterized with extralinguistic reality in view. Since, in our opinion, languages like these can, too, be precisely described by providing the definitions of a sentence and of direct consequence (relativized to the classes of empirical situations), and since such investigations are promising and useful for the methodology of the sciences based on experience, it seems that they should be covered equally well as the structural languages by epistemological discussions. Hence, speaking of language, I shall have in mind here a language that may, equally well, be either structural or empirical. I shall only assume that it is characterized by 1) the definition of a sentence, and 2) the definition of direct consequence with regard to the classes of empirical situations (when these classes are reduced to an empty class of empirical situations, we

189

deal with a structural language; if at least one such class is non-empty, the language is empirical). All languages characterized in this way may be considered formalized languages, since their form is precisely stated.

The concept of direct consequence relativized to the classes of empirical situations — symbolically: $Wb_E(K, z)$ (read: sentence z is entailed directly by the class of premises K with regard to the class of empirical situations E) is in need of some explication. The classes of situations ranged over by the variable 'E' for a given language always form some definite set. We assume that one class from such a set is empty. Thus the relation of direct relativized consequence includes direct consequence. When E is empty — $E = \Lambda$ — and the class of premises K is empty — $K = \Lambda$ —, the relation $Wb_E(K, z)$ defines the set of sentences which are the direct consequence of the empty class of premises, i.e. the set of axioms of the language. When $E = \Lambda$, and the class of premises $K \neq \Lambda$, such relation defines all pairs $\{K, z\}$ of the type that z results directly from class K. When E is a non-empty class of empirical situations — $E \neq \Lambda$ —, we have again to distinguish two cases: $K = \Lambda$ and $K \neq \Lambda$. In the former case, i.e. $E \neq \Lambda$ and $K = \Lambda$, the relation $Wb_E(K, z)$ determines sentences which Ajdukiewicz calls sentences acceptable directly on the basis of the empirical rules of the language;[2] here, they are sentences which are the direct consequence of an empty class of premises with regard to a non-empty class of empirical situations. The latter case — $E \neq \Lambda$, $K \neq \Lambda$ — is not realized for any sentences in normal languages (the direct consequence of sentences from their premises usually does not depend upon additional occurrence of some empirical situations). However, our concept allows also the discussion of languages in which relations of this type hold.

In using the concept of the relation $Wb_E(K, z)$ when $E \neq \Lambda$ we do not come across greater difficulties than in using the concept of non-relativized direct consequence: while, on the basis of the latter concept, we are allowed to accept z after we have established that the premises of the class K are sentences accepted in the language, we are allowed to accept z on the basis of our concept, after we have additionally established that some empirical situations of the type E have been realized as well.

The introduction of the concept of direct consequence relative to classes of empirical situations, one of which is empty, is useful for many reasons. Thus, 1) it makes possible to characterize empirical

languages in the same way as structural languages, 2) it is useful in emphasizing the common features of the axioms of the language and sentences that are accepted in this language on the basis of empirical rules, and, at the same time, in distinguishing differences existing between these classes of sentences, 3) it makes possible the construction of a theory based on further non-defined concepts of a sentence and relative direct consequence for the empirical languages, which might be called the *empirical syntax* as distinct from the syntax of the structural languages, and finally 4) it allows the presentation of the structural and empirical languages as special cases of formalized languages, and the normal and empirical syntax as cases of the *general syntactic theory* of the latter group of languages.

2. TRANSLATION PRESERVING THE MEANING OF EXPRESSIONS

According to the above discussion we can regard language as a field of the relation 1) of forming a sentence z by expressions $x_1, x_2, ..., x_k$ in this order—symbolically $Z (x_1, x_2, ..., x_k, z)$—within the scope of which also the relation 2) of direct consequence of sentences z from the class of premises K with regard to the classes of empirical situations E—symbolically $Wb_E(K, z)$—has been defined. Thus, every language will be uniquely characterized by a certain system of relations Z and Wb_E. It may happen that the relations in one of these systems will be— in case of an identity correspondence of classes E—isomorphic with the relations in another formation. We shall say about languages characterized by two such systems that they are mutually translatable in the sense I. We might, therefore, attempt the following definition:

(I)   Languages J_1 and J_2 are translatable = relations Z_1 and $Wb_E^{(1)}$ defining J_1, and relations Z_2 and $Wb_E^{(2)}$ defining J_2 are respectively isomorphic.

(The postulated isomorphism in this definition would depend on the existence of such a one-one correspondence between the members of the field of relations Z_1 and Z_2 that $Z_1 (x_1, x_2, ..., x_k, z)$ holds if and only if $Z_2(x'_1, x'_2, ..., x'_k, z')$ and—in case of every $E - Wb_E^{(1)}(K, z)$ holds if and only if $Wb_E^{(2)}(K', z')$, where the members x_1, etc., z, K respectively correspond to members x'_1, etc., z' and K' (where those classes K and K' are

considered to correspond, which consist of respectively corresponding
sentences). We shall call synonymous those members which play the
same role in the systems isomorphic with a given pair of relations Z
and Wb_E (i.e. isotopic with regard to these systems of relations). We
have thus arrived, with the help of somewhat different concepts, to the
same, it seems, definition of the sense of the terms 'translatable languages'
and 'synonymous expressions' that Ajdukiewicz endeavoured to state
precisely in his work 'Sprache und Sinn'.[3]

3. THE LANGUAGE UNDER INVESTIGATION AND THE LANGUAGE OF ITS SEMANTICS AS UNTRANSLATABLE

There exist numerous languages which are mutually untranslatable in the
sense of I. Especially languages which possess a different number of
expressions are not translatable into one another in this sense; similarly
even in numbers sets of expressions from two languages are not translatable into one another, if the relations Z and Wb_E defined for them
associate them with different number of expressions in one language
than in the other. For these reasons no language of science is translatable into the language of semantics of this science nor into a subset
of expressions of the latter language. Even if we assume, as is often
done, that a one-one correspondence can be established between the set
of expressions in the language of science and certain subset S of expres-
sions in the language of its semantics, so that whenever $Z(x_1, x_2, ..., x_k, z)$ in the language of science, then $Z(x'_1, x'_2, ..., x'_k, z')$ in the language
of semantics and—for every E—whenever $Wb_E(K, z)$ in the language
of science, then $Wb_E(K', z')$ in the language of semantics, the relations
Z and Wb_E are not isomorphic for these languages. For, even if we
disregard the fact that the language of semantics contains additionally
the names for all the terms of the science under investigation and the
structural relations between them, it must be emphasized that a wider
range of logical terms must occur in the language of the semantics
of some science than in that science itself. Thus, the language of semantics is essentially richer than the language of the science that is being
investigated by this semantics.[4] For these reasons the language of
semantics and the language of the science investigated by it cannot
contain any synonymous terms.

4. THE GENERALIZATION OF THE CONCEPT
OF TRANSLATION

The lack of translatability in the sense I of the language of science into the language of its semantics leads to the following difficulties. The semantics of some science deals, *ex definitione*, with the relations that hold among the expressions of this science and the objects of which it speaks. Hence both the terms denoting the expressions of the science under investigation and those denoting the objects to which these expressions refer, should be present in the language of semantics. Semantics should, therefore, contain terms which possess the same extension as those in the science under investigation. If such terms are absent in semantics it does not investigate the relations that it should investigate, but only relations apparently indentical with them. Thus, the possession of terms denoting the same as the terms of the science under investigation is the condition determining the possibility of the semantics of this science. We know, however, that the language of semantics does not contain terms synonymous to the terms in the science that is being investigated by this semantics. In this way the most important arguments for equal extension of the correspondent expressions has to be rejected.

One who does not want to doubt in the possibility of the semantics of a given science, the expressions of which have some meaning and refer to something, must seek for some different justification of the proposition that the expressions of science and their corresponding expressions from the language of semantics mean one and the same thing.[5] The following way suggests itself: some more general concept of translatability of languages must be formulated which — although not securing synonymity — would secure the identity of the denoted objects to the expressions corresponding respectively and demonstiate that the language of science and the language of its semantics may, in some sense, be translatable into one another. We shall now try to find this notion.

It seems that the simplest step leading to this goal would be to give up the isomorphism of the whole relations Z and Wb_E defining these languages (with an identity correspondence of classes E) as the necessary condition for the translatability of languages, and require (with the same correspondence of these classes) only the isomorphism of certain subrelations Z and Wb_E, defined for the respective languages. (By a *sub-*

relation S of some relation R we understand such a relation that if S holds between some elements we can always infer that R holds between these elements; symbolically $S \subset R$. When the conclusion in the other direction is also justified, $S = R$ and S is the improper subrelation for R.) We might, therefore, take into consideration the following concept of translatability of two languages:

(II)   Languages J_1 and J_2 are translatable = there exist sub-relations with regard to Z and Wb_E, defined for J_1 and J_2 and respectively isomorphic.

(By providing an appropriate modification of the explication of iso-morphism given for case I we shall obtain the explication of this re-lation in the present context.) The concept of translatability in this second sense is the generalization of the concept of translatability in the sense of I: foɪ, if the subrelations under discussion are all improper, then the translatability in the sense of II passes into translatability in the sense of I. However, the concept that has presently been defined is too general for our purposes. Any two languages will be translatable in the sense of II with regard to the isomorphism of empty subrelations. Even the qualification that none of the subrelations be empty will not change much. We shall, therefore, endeavour to narrow the notion.

In the first place we might—having in mind the languages of a science and its semantics—require that the subrelations for one of the trans-latable languages correspondingly isomorphic to the subrelations of another language be identical with the complete relations Z and Wb_E, i.e. that they be improper subrelations. There remains to be stated precisely the requirement that must be set up as regards the subrelations that correspond to those relations in the other (richer) language. Let us, with this aim in view, turn our attention separately to the descriptive terms and the logical and mathematical terms of languages. The de-scriptive terms of a large group of languages are characterized by the fact that, if they are mentioned by the rules of the language (as constituents of appropriate sentences) at all, they are mentioned by its empirical rules. On the other hand, the logical and mathematical terms are involved in the axiomatic or deductive rules (or terms defined by them). When shall we say that two descriptive terms 'A' and 'B' denote the same? It seems that if and only if we are inclined to accept the sentences 'A ex-

ists' and '*B exists*' in the same empirical situations. As for the logical and mathematical terms, it is difficult to speak of them as denoting anything in general. On the other hand, it is possible to state about complexes of these terms that they refer to the same, when, with their help, we can formulate 'the same' true statements. And we can say that two complexes of expressions allow the formulation of 'the same' true statements when there is a one-one correspondence among these expressions assigned so that the relevant sentences are simultaneously. true. It seems, therefore, that we shall be able to state about sets of terms containing both descriptive and logical and mathematical terms that they 'refer to the same', if we can establish such a correspondence among them that whenever we deal with a sentence formed exclusively of terms from one set, which should be accepted in certain situations, then the sentence from the second set that corresponds to it will also have to be accepted in the same situations, and vice versa. This leads to the following concept of translatability of two languages, which we shall call translatability in the sense of III:

(III)   Languages J_1 and J_2 are translatable = there exist subrelations of Z and Wb_E defined for languages J_1 and J_2 and correspondingly isomorphic, while, for one of the languages, these subrelations are improper, and for the other, the subrelation of the relation Wb_E contains the whole relation Wb_E holding between sentences constructed exclusively of terms that belong to the field of the mentioned subrelation of Z from this language.

(The isomorphism postulated here should — like before — occur when there is an identity correspondence of classes E.)

The translatability in the sense of III secures for the corresponding terms reference to the same objects with a high degree of probability and is a generalization of translatability in the sense of I. We may, therefore, assume that we have found the concept we had intended. It remains to be mentioned that the occurrence of translatability in the sense of III between two languages seems to be a sufficient, but not necessary condition for the corresponding expressions to denote the same. Thus, it seems that we shall be able to obtain translatability securing equal extension to the terms as well if in defining it in statement (III) we use,

instead of the relation of relative direct consequence, the relation of
relative consequence (the definition of this concept is based on the
previous one and does not present special difficulties being, basically,
analogous to the definition of the concept of the non-relative consequence
which, in turn, is based on such a concept of direct consequence). For,
the order in which we arrive at the statements is irrelevant to what
we are talking about—it is only their acceptability that matters. We shall,
in this way, obtain the concept of translatability in the sense of III′
which, it seems, also fulfils the required conditions. (By putting the
concept of relative consequence—symbolically W_E—in place of Wb_E
in the notion of translatability I and II we obtain correspondingly new
generalizations of the concept of translatability: I′ and II′—both useless
for our purposes, however.)

Note that it is usually assumed in mathematics that the theory of
integral numbers speaks about the same as a part of the theory of rational
numbers, the theory of rational numbers about the same as a certain
part of the theory of real numbers, etc. The isomorphism of the set
of all integral numbers with a certain subset of rational numbers and of
the set of all the rational numbers with a certain subset of real numbers
etc. is quoted as justification here. This isomorphism occurs with respect
to the operations of addition, multiplication and other arithmetical
operations, every time determined in the narrower of the sets. It may
be reduced to the isomorphism of the languages of respective theories
or of their parts with regard to the relations Z and Wb_E (or Z and W_E)
or their subrelations. It seems that the translatability of languages in the
sense of III (or III′) is, therefore, actually considered to be the sufficient
condition for the corresponding terms to speak of the same things in the
mathematical sciences. (The sets of classes E are the same in those
languages, i.e. they reduce themselves to an empty class.)

5. THE APPLICATION TO THE LANGUAGE OF SCIENCE
AND ITS SEMANTICS

We shall presently demonstrate that the languages of science and its
semantics are translatable in the sense of III, if only we make some additional qualifications and that, therefore, we have grounds to accept
that the terms of science and corresponding terms in the language of

semantics refer to the same. In this way we shall demonstrate at the same time that the requirement of there being a possibility of a semantics of some a priori given, autonomically functioning science can be fulfilled. Since, in the language of semantics there usually exists a class S of expressions for which the relations Z and Wb_E hold whenever the same relations hold for the corresponding expressions of science, the existence of subrelations with regard to Z and Wb_E in the language of semantics isomorphic with the corresponding relations in the language of science with regard to the identity correspondence of classes E is secured. If we assume additionally that there are no rules of direct acceptance for sentences formed exclusively of terms from the set S in the language of semantics which would not have corresponding rules in the language of science, the translatability in the sense of III will be secured.

According to these additional assumptions there cannot exist in semantics, among others, any new axioms or statements acceptable directly in the context of some empirical situations and formulated exclusively in terms which correspond to the language of science, if there is nothing similar corresponding to it in the science itself. However, this is only the condition for the translatability of these languages in the sense of III and we cannot exclude the possibility that the corresponding expressions refer to the same, even if this condition is not fulfilled. Its formulation, however, draws our attention to the necessity of being cautious in building the semantics at a point usually not considered. As is known, the language of semantics must possess assumptions not less powerful than those of the theory under investigation.[6] However, the reinforcement of the assumptions cannot go too far, since it could turn the semantics of a given science into something that is not the semantics of this science.

At the same time, a possibility of constructing the language of semantics of some theory in another way reveals itself. Namely, we might require that a set of expressions in which, for the same E, not the relations Z and Wb_E, but Z and W_E from the language of science are reproduced, be present in the language of semantics. With the fulfilment by this language of some appropriate additional conditions we might state that the language of science and the language of semantics that has been constructed in this way are translatable in the sense of III′

which also, according to what has been said, secures the same reference to the corresponding terms. What is an axiom in science might then pass into a distant theorem of semantics, but every thesis of semantics that could be formulated in terms assigned to the expressions of science would find then counterpart in the science under investigation.

6. RELATED PROBLEMS

Finally, the question should be answered, why we have been looking for the conditions for the same reference of the terms of semantics and those of the science investigated under the conditions of syntactic character and not under conditions objective or semantic in nature. After all, if we want to demonstrate that 'A' means the same as 'B', it suffices to indicate that A is B, or that 'A' denotes B. We must notice in this connection that the terms 'A' and 'B' belong to two separate languages which were not translatable into one another in the sense of I. Thus, the sentence 'A is B' would have to belong to some third language that would be translatable in the sense I neither into the first nor into the second language. Thus, the terms 'A' and 'B' from this language would not be synonymous with the analogous terms from those languages so that the determination of the sentence 'A is B' would not contribute to the solution of our problem. Note that, using sentences like: 'A' denotes B, we usually determine them by indicating that B is C, where 'C' means the same as 'A'. The term 'C' is then, however, the term of semantics, while 'A' is the term under investigation, and we are just looking for the conditions determining that the term of semantics denotes the same as the term corresponding to it of the language under investigation. Hence solutions either objective, or semantic in nature will lead us nowhere in our discussion.

They lead nowhere also in the case of a wider problem: what are the conditions necessary for synonymy of two terms 'A' and 'B' from any untranslatable languages? Nevertheless, it seems, that the establishment of criteria for determining such a question is necessary. For, in many cases — apart from the languages of semantics and the science investigated by it — we deal with languages that are not translatable in the sense of I, about which we should like to state that they refer — at least with the help of some terms — to the same objects. Thus, we must, it

seems, e.g. recognize the everyday language of things and the language of atomic-electronic physics as untranslatable languages, whose expressions are not synonymous. However, we should like to state that, e.g. the phrase 'this stone' from the first language refers to the same thing as 'a cloud of protons and electrons that fills this spatial area'. Similarly, it would seem that descriptive terms from the language of any empirical science based on two-valued logic refer to the same objects as analogous terms from the language of the same science that uses three — or more — valued logic: for we speak about the same, even though our reasoning proceeds differently with the help of respective sentences. And, finally, also in the language of physics based on Euclidean geometry and in the language of physics using Riemann's geometry we could distinguish parts that speak about the same, although these languages are not translatable either (the isomorphism of relations Wb_E for $E \neq \Lambda$, requires different correspondence of terms here than the isomorphism of relations Wb_E for $E = \Lambda$). The above mentioned intuitions might be made precise on the basis of some new notions of translatability of two languages, namely, such notions which, although looser than the notions III and III' — would secure to the corresponding terms the same extension. More detailed discussion of these matters goes beyond the framework of the present paper.

REFERENCES

* First published in *Przegląd Filozoficzny* **XLIV** (1948). Translated by E. Rono-wicz.

[1] The concept of empirical languages that we have in mind here was introduced by K. Ajdukiewicz in his work 'Sprache und Sinn', *Erkenntnis* **IV**, 100–138. Cf. especially § 5.

[2] See Ajdukiewicz, *op. cit.*

[3] *Ibid.*, especially §§ 9–10.

[4] Cf. A. Tarski, 'Der Wahrheitsbegriff in den formalisierten Sprachen', *Studia Philosophica* **I** (1935) 261–405, § 4 and 'Nachwort', especially p. 398 ff.

[5] R. Carnap in *Introduction to Semantics*, Cambridge, Mass., 1942, begins the construction of semantics with the definition of the relation of designation for the expressions in a given syntactic system, therefore the way in which he solves our problem is conventional. As far as syntactic systems are concerned, if they possess — as in Carnap — the character of certain games with senseless signs, a different solution is not, however, necessary nor possible.

[6] See Tarski, *op. cit.*, p. 330.

MARIA KOKOSZYŃSKA

ON THE DIFFERENCE BETWEEN
DEDUCTIVE AND NON-DEDUCTIVE SCIENCES *

In his already classic article 'On A Priori and A Posteriori Sciences' K. Twardowski[1] distinguishes deductive sciences, i.e. those which refer to axioms, definitions and postulates in finally determining their theorems, and non-deductive sciences, which in case of doubt refer to observational sentences, i.e., to experience.[2] In Twardowski's opinion, however, the difference between these sciences does not consist only in the ultimate foundations of the justification of theorems. It also consists in the method of justifying the theorems of these sciences. While namely the deductive sciences utilize only deduction to that effect, the non-deductive sciences make use of various methods (e.g. induction) but never of deduction.[3]

Although presented very clearly, Twardowski's argument raises some interpretational difficulties. We shall present an interpretation of it resulting from a not very profound study. We shall make use of the distinction between direct and indirect justification. Generally speaking, by 'indirect justification' in one step of a sentence I understand a justification in which we refer to some sentences, while by one step 'direct justification' — one in which we refer to something which itself is not a sentence. I shall, moreover, call the justification of a sentence absolute if it terminates — in a finite number of steps — by referring to something which itself is not a sentence, and relative — if in the same number of steps it terminates by referring again (if not exclusively, then also) to some sentences. The latter sentences consitute the class of final premises of justification, and a given sentence is then justified only with regard to this class of premises while in the first case it is — simply — justified. As we can see, the expresion 'is justified' belongs in these cases to different syntactic categories. It also has different semantic referents: while in the first case this expression denotes a certain property of a sentence,

in the second it stands for a relation, at least of two arguments, between the given sentence and some sentences, namely, its final premises. It is obvious that every justification in the absolute sense — absolute justification, as we have called it — requires at a certain moment that we make use of the method of direct justification. If one wants to obtain the (absolute) justification of a sentence without using direct justification, by gathering still new processes of relative justification, which is always indirect justification — he commits the known error called *regressus in infinitum*. Indeed, indirect justification cannot (simply) justify anything.

And here is the interpretation of Twardowski's argument, in terms of the distinctions presented above:

Deductive sciences, as referring finally in justifying their theorems to axioms, definitions and postulates, i.e. to some sentences, justify their theorems only in the relative sense, with respect to their final premises. Deduction, the only method they make use of in this justification, is a method of indirect justification: sentences are always deduced from others. Deductive sciences do not, therefore, utilize any method of direct justification; none of their theorems is justified in the absolute sense. Deductive sciences then do not (simply) justify anything, they do not assert anything. They have a character — as K. Ajdukiewicz says — of only hypothetical-deductive systems.[4] The case of non-deductive sciences is different. Although Twardowski says that they finally also refer to some sentences, namely to observational sentences, these sentences are not, however, understood here as arbitrary sentences suitable for expressing what would be observed if corresponding sensible ideas appeared. These sentences account for our actual perceptions, they are based on experiences of sensible ideas which have actually occurred. We are inclined to such an interpretation of Twardowski's observational sentences because of his identifying them with experience. In Twardowski's view, then, non-deductive sciences do not finally refer to any sentences, but to what these observational sentences are based on — i.e. to the sensible ideas one is actually experiencing. These sciences then make use of a method of direct justification, i.e. of a method consisting in referring in the course of sentence justification to sensible ideas, which sometimes are included in 'experience' and consist — in successful cases — in a specific contact of the feeling being with the objective reality. In Twar-

dowski's view, then, non-deductive sciences finally justify their theorems directly, by referring to something which is not a sentence. These sciences then justify at least some of their theorems simply, in the absolute sense. They are sciences which really assert something. While nothing is justified in deductive sciences, here some sentences are (simply) justified. It is then clear that deduction is not suitable for this sort of justification in non-deductive sciences. As a method of only relative, indirect justification, it cannot itself justify anything (simply, in the absolute sense).

I do not intend to stick to such an interpretation of Twardowski's text. One might remark that Twardowski does not stress the word 'itself' when speaking of deduction as a possible method of justification of theorems in non-deductive sciences, and that it is therefore simpler to assume that he simply has made a mistake in treating *a posteriori* (non-deductive) sciences as never utilizing deduction in justifying theorems. One could quote as an obvious example of justifying sentences by deduction in natural or social sciences the rejection of hypothesis, i.e. demonstrating for a hypothesis '*h*' that non-*h*, which is generally carried out by using a reasoning based on the *modus tollendo tollens* (if *h* then *f*, and non-*f*, then non-*h*). There are, however, still other reasons why we could hardly stick to the above-presented interpretation of Twardowski's text. Indeed, final remarks of the article clearly oppose such an interpretation. Twardowski says there that deductive sciences "can boast of undeniable theorems" and "never concern facts".[5] In his opinion, then, something is asserted in these sciences (and therefore also justified in the absolute sense), and this assertion refers to something that is not given in perception.

Are we then confronted here with a situation in which the author contradicts himself, and is it not possible to present his views as a coherent whole? On the contrary. I believe that the author has extremely deeply penetrated the nature of deductive and non-deductive sciences, but in order to understand him correctly we must take into consideration the recent findings concerning the nature of deductive sciences. These can be seen in the argument presented by a group of mathematicians (publishing under the name Nicolas Bourbaki), and concerning the nature of mathematics, as well as in the contemporary trend to identify mathematics with all deductive sciences. This trend is accounted

for by Tarski in his *Wstęp do logiki matematycznej i metody dedukcyj-nej*[6] (*Introduction to Logic and to the Methodology of Deductive Sciences*). On the basis of this argument we shall try to describe some characteristics differentiating deductive from non-deductive sciences, and we shall see that in the light of these analyses Twardowski's enunciations will not only turn into a coherent whole but will also obtain an unexpected confirmation. We shall have to modify some of his formulations, however.

In his article 'L'Architecture des mathématiques' N. Bourbaki[7] presents the thesis that mathematical sciences have a specific object, different from that of empirical sciences.[8] This object consists of structures of various sorts.[9] Let us examine what Bourbaki understands by 'structures'. It will be best to start with an example of structure characterized by himself, i.e., the structure of a group. Here is the relevant fragment of the article:

"Let us consider the following three operations: 1) adding of real numbers, in the course of which the sum of two real numbers (...) is determined in an ordinary way; 2) multiplying whole numbers following a simple module r, when the considered elements appear to be the numbers $1, 2, 3, ..., r-1$, and, by definition, the product of such numbers is the rest remaining after the dividing by r of their product in the ordinary sense of this word; 3) the 'composition' of translocations in Euclidean, three-dimensional space, where, by definition, the result of such a composition (i.e. the product) of two translocations T and S (in that order) is the translocation obtained by performing, first, the translocation T and then, S. In each of these three theories, for two members x and y taken in that order from the set considered (in the first case from the set of all real numbers, in the second—from the set of numbers $1, 2, 3, ..., r-1$, in the third—from 'the set of all translocations) the corresponding, uniquely determined third element of this set is obtained (by using a procedure different for each set). We shall denote this third member, by $x\tau y$ in all the three cases (it will be a sum if f and y are real numbers; their product following the module r, if they are natural numbers $\leqslant r-1$; the result of the composition if they are translocations). If we now consider the properties of the 'operation' in each of the three theories, a parallelism worth noting will occur; within each of these theories the properties depend on each other, and

the analysis of the logical relations among them leads to the distinction of a small number of properties which are independent of each other. For example, the following three properties can be considered, which we shall express by using our symbolic notation, easily translatable into the language of each of these theories:

a) Whatever the members x, y, z, we have $(x\tau y)\tau z = x\tau(y\tau z)$ (the operation $x\tau y$ is associative);

b) there is a member e such that for each member x, $e\tau x = x\tau e = x$ (for adding real numbers it is the number 0, for the multiplication following the module r — the number 1, for the composition of translocations — the 'identical translocation', which leaves every point of space in its place);

c) for each member x there is an x' such that $x'\tau x = x\tau x' = e$ (for adding real numbers it is the opposite number $-x$, for the composition of translocations — the opposite translocation, i.e. such that it takes back to the original position every point translocated as result of a change of x; for the multiplication following the module r the existence of x' results from very simple artihmetic reasoning)."[10]

We leave out here the reasoning, which was described in the footnote in the article. Further on the author says that the properties which can be expressed within the range of each of the considered theories in the same notation appear to be consequences of the three properties enumerated above. Therefore, in order to avoid unnecessary repetitions, we have decided to develop at one time all the logical consequences of these properties. In this connection the necessity arises of accepting an appropriate terminology. Let Bourbaki speak again: "We say that a set on which an operation $x\tau y$ is characterized by three properties a), b) and c) is equipped with *group structure* (or, more simply, is a *group*). The conditions a), b) and c) are called axioms of the group, and inferring consequences from them — constructing the axiomatic theory of the group."[11]

Further on, the author says: "we can now explain what is to be understood in the general case by a *mathematic structure*. The different concepts called by this name have the common feature that they are applied to a set of members the nature of which is not determined."[12]

The author says: "in order to determine the structure, one or more relations are given which connect its elements (in the case of a group

the relation is $x\tau y = z$ for any three members); it is then postulated that the given relation or relations satisfy some conditions (which are enumerated and which are *axioms* of the structure considered)."

The next sentence is: "To construct the axiomatic theory of a given structure—means to derive logical consequences from the axioms of the structure, *without taking into consideration any other information* concerning the considered members (particularly without taking into consideration any hypotheses concerning their 'nature')".[13]

These explanations, given by Bourbaki in connection with the concept of structure, are not clear. We shall, therefore, quote Bourbaki's remarks on an example of structure other than group structure, namely on the structure of order. The author says:

"Another important variety (sc. of structures) are the structures determined by the relation of ordering; in this case it is the relation between two members x and y which we most often express as 'x less or equal to y' and which is written in the general case as xRy." And further on: "The axioms that govern this relation are as follows: a) for every x, xRx; b) if xRy, yRx hold then $x=y$; c) if xRy, yRz hold then xRz. An obvious example of a set characterized by such a structure is the set of whole numbers (...), where the sign R is in this case modified into $\leqslant$." Bourbaki remarks that in the axioms a), b), c) the requirement that the relation R be connex has been left out. In some structures of order such a connex relation will not hold, for example "if X and Y denote subsets of a set and $X R Y$ means 'X is included in Y', or if x and y are natural numbers and $x Ry$ means 'x divides y', or else if $f(x)$ and $g(x)$ are functions of the real variable which are defined on the interval $a \leqslant x \leqslant b$, and $f(x) R g(x)$ means 'no matter what x, $f(x) \leqslant g(x)$."[14]

Let us now try to answer the question what the structure is in Bourbaki's opinion. We at once encounter an inconsistency in the use of this term by the author himself. He calls structure both the feature common to sets which—because of some definite relations in them—satisfy given axioms (i.e. a class of systems ⟨set, relation⟩), and the systems themselves. We can see the first case in his discussion of *group structure,* and the second—in his discussion of the *structure of order.* It cannot be excluded, though, that in the second case the author speaks of structures of order because the order can be determined not only

by the three axioms given by Bourbaki, but also by four axioms (the three presented and that of connexity), which, as a matter of fact, is often the case. It seems that we can, therefore, accept that in Bourbaki's opinion there always is only one structure assigned to a given system of axioms. It will be the class of systems; some set of members and the relation or relations determined on this set. Or maybe a set, some individuals from this set, the classes of these individuals and the relations determined on this set? Anyway, it seems that the class considered will contain all and only the systems which satisfy given axioms. Model theory enables us to determine this concept of structure more precisely.

In order to utilize model theory to determine Bourbaki's structure, let us first recall the concept of the model of a language. Let there be in a language in which some sentences are formulated (we shall denote their class by A), apart from logical constants (we assume for the sake of simplicity that the language contains at its most the lower functional calculus), and as specific constants, the individual constants $i_1, i_2, ..., i_k$, the class constants (one-place predicates) $K_1, K_2, ..., K_l$ and the relational constants $r_1, r_2, ..., r_m$ (n-place predicates, for a finite $n \geq 2$). By 'model' of this language I understand a $(1+k+l+m)$-member system, the first member of which is the universe of individuals, the next k members are individuals of that universe (where an individual with a given index is assigned to the individual constant having the same index), the next l members are the classes, the members of which belong to that universe (and the i-th class corresponds to the i-th one-place predicate) and, finally, the next m members are the relations determined on that universe, where the i-th relation for $1 \leq i \leq m$ has as many arguments as the i-th predicate r_i. If we denote the model of the language by M, the universe by U and the other objects by corresponding capital letters, we shall write:

$$M = \langle U, I_1, I_2, ..., I_k, K_1, K_2, ..., K_l, R_1, R_2, ..., R_m \rangle$$

or, for short, $M = \langle U, I, K, R \rangle$.

Let us now recall the concept of model of the sentences A of a language. By 'model of the sentences A of the considered language' we understand every model of that language in which the sentences A are true, i.e., loosely speaking, in which they become true when the specific

constants obtain as denotata (interpretations in the semantic sense) the objects corresponding to them in the model. Here is the formulation of the enunciation that the model M is the model of the class of sentences A: $A \subset \mathrm{Ver}\,(M)$.[15]

On the grounds of these concepts we can now attempt a precise definition of Bourbaki's structure, corresponding to a set of sentences A. It will be the class of all the models of the language to which the sentences A belong, models which are simultaneously models of the set of sentences A. Here is the formulation of this definition for a specified language:

$$M \in \mathrm{Str}\,(A) \equiv A \subset \mathrm{Ver}\,(M).$$

The sentences A are axioms of the corresponding structure. By deriving from A logical consequences we construct a deductive theory T based on the axiomatics A. We can therefore identify this theory with the set of all logical consequences of A, so:

$$T(A) = \mathrm{Cn}\,(A).$$

Because of the well-known properties of the concept of logical consequence, it is clear that the set of models of the class of sentences A and the set of models of the theory based on those sentences, i.e., of the class of sentences $\mathrm{Cn}(A)$, is the same. Then the structure determined by the axiomatics A is identical with that determined by the whole theory based on that axiomatics.

In Bourbaki's view, the world of mathematics is a world of structures. By finding the features common to all the models belonging to some different structures we obtain the axiomatics for a more general structure, embracing all the members of the previous ones. On the other hand, by supplying new axioms satisfied only by some members of a given structure we obtain more detailed structures. Through an adequate modification of axioms it is thus possible to obtain structures situated at the point of intersection (i.e. which are the profile) of several other structures.[16] The traditional branches of mathematics have turned out to be precisely theories of such structures.[17] In this way a whole developed hierarchy of structures is created. Bourbaki sees a possibility of an appropriate ordering of the whole of mathematics by referring to such a hierarchy of structures.[18]

It may be doubted whether Bourbaki would like to accept a structure determined by any system of axioms as object of the mathematical science, and the set of consequences of such a system—as such science. Most probably the structure, as he sees it, cannot be an empty class, and the theory—a contradictory one. However—according to the view under discussion—a structure being the object of mathematical science does not have to contain any model which would be a fragment of the real world (i.e. where the universe would be identical with that of real individuals or with a subset of it, and the interpretation of the class and relational terms would come from this universe). Bourbaki admits [19] that mathematical theories quite unexpectedly find a model within the range of the real world (of micro- or macro-individuals), but it seems to be for him rather a peculiar coincidence than a natural phenomenon, or, as he says, and which seems to be the same, the result of connections which are hidden more deeply than one might suppose.[20]

No matter how deep this reasoning might seem, it does not at all embrace everything Bourbaki puts into the characterization of mathematical sciences. We even could go as far as saying that speaking about structures as objects of those sciences we were dealing with the marginal part of that characterization. Indeed, Bourbaki pays incomparably more attention to the method utilized in his opinion by the mathematical sciences than to the object of these sciences. It is the *axiomatic method*. It consists not only in using a language with a defined vocabulary and a precisely determined syntax (logical formalism), but above all in including in the axioms what is common in sometimes very distant domains,[21] and in the simultaneous study of all these domains from the point of view of precisely these common features presented in adequate axiomatics.[22] It is in view of that that "*it is postulated*" (emph. mine — *M.K.*) that the individuals, the classes and the relations referred to in axioms satisfy them, and by logical inference their other properties are then derived.[23] It is very important that in the course of this procedure all other assumptions about the objects under investigation are left out, and particularly no hypotheses concerning their 'nature' are taken into consideration, as it has already been pointed out.

As we can see from these enunciations on the axiomatic method, Bourbaki not only requires that the person using it infers logical consequences of a set of sentences (axioms), but also that this person con-

siders these axioms as so-called *implicit definitions* of specific terms of a given theory. Indeed, when are the axioms of a theory considered as an implicit definition of the specific terms of that theory? The concept of implicit definition has never been adequately studied. In my opinion, it can be best explained by a generalization of the concept of arbitrary definition, presented by Ajdukiewicz in his 'Trzy pojęcia definicji' ('Three Concepts of Definition').[24] Ajdukiewicz defines there this notion as follows: "We call a sentence... *arbitrary definition*, i.e. postulate of a language, if there is in this language a terminological convention saying that the terms contained in this sentence are to symbolize the objects which satisfy this sentence if substituted for these terms".[25] We shall then adopt the following definition:

The set of sentences Z in a language J is the *implicit definition* of terms T contained in them$\equiv$there is for the language J a terminological convention saying that the terms T contained in the sentences of the set Z are to symbolize respectively the objects which satisfy these sentences if substituted for these terms.

Thus by postulating with respect to a language in which we are constructing the theory with the axioms A that the specific terms of this theory denote the objects satisfying these axioms, we are making of the set of the axioms the implicit definition of these terms. In Bourbaki's view, then, one of the basic properties of the axiomatic method is treating the axioms of the theory thus constructed as implicit definitions of the specific terms of this theory. The corresponding terminological conventions can be called denotation (or interpretation) rules for the terms they refer to; they will usually be rules not determining the denotata of these terms uniquely.[26]

We then have in Bourbaki's article a characterization of mathematics in terms of its object and its method. Is it really necessary, however, to speak separately of the object once the method has been determined? By taking the axioms of the theory as the implicit definition of the specific terms of this theory we at the same time determine to some degree the objects which will be talked about. This determination is very ambiguous, with the exactitude to the class of all models of axioms. But it is. Thus each of these models becomes the object of the given theory. But as none is distinguished, it can also be said that the theory has as object the class of all models. It is clear that the expression '*is an object*' has

in these two cases a different, although related meaning. Indeed, if we agree to say that every system of acceptable (on the grounds of the terminological conventions accepted) denotata for the specific terms of a theory, additionally enlarged by the universe from which these denotata have been taken, is an object in the 1-st sense of the theory (it will already be a model of this theory), then we can define this sense of the expression 'is an object', in which the whole class of models (i.e. the structure) is the object of the theory, in the following way:

A class α of models M is an *object in the 2-nd sense* of the theory $T \equiv$ for every M such that $M \in \alpha$, M is an object in the 1-st sense of the theory T.

The moment we have accepted Bourbaki's axiomatic method as characteristic for the mathematical sciences, it becomes a simple consequence that the structures (the classes of models) of respective theories are an object — in a natural sense of the expression 'is an object' — of these sciences. In Bourbaki's view, then — if we interpret him adequately — the axiomatic method, although understood in a rather particular way, is sufficient to characterize the mathematical science.

Bourbaki himself has put forth reservations as to whether his argument is right. He calls it schematic, idealizing, rigid.[27] Let me express the opinion that even if he did not succeed in characterizing the mathematical sciences, he succeeded in pointing out to a relevant feature of the deductive sciences, at least in their contemporary role. Maybe such was Bourbaki's intention. It was certainly so if he belongs to those who — as Tarski says — identify mathematics with deductive sciences (which justify their theorems by means of the deductive method).[28] It seems, however, that even not sharing this view we could agree without reservations with Bourbaki's characterization of deductive sciences, taking into consideration their contemporary aspect. Therefore, even without undertaking the problem whether mathematics can be identified with the deductive sciences, I would like to express the opinion that the deductive sciences are sciences using the axiomatic method in Bourbaki's sense. The only reservation is that if they are constructed without enumerating the axioms, as a deductive science based only on rules, then Bourbaki's axiomatic method should be understood as postulating interpretations for the specific terms of a given theory such that the rules be either infallible or — if the truth of the premises does not matter — that they

assure the truth of the conclusions (this should be the case for the so-called axiomatic rules). It seems that such an extended understanding of the axiomatic method would conform with Bourbaki's intention, although he does not distinguish the case of a mathematical science based exclusively on rules; he would probably not like to overlook it. For the sake of simplicity we shall refrain from insisting on this case.

Let us agree to call Bourbaki's axiomatic method the deductive method. We shall then further on use the term 'deductive method' in one of the two basic meanings [29] in which it can be used: first (I), for denoting the procedure in which, in order to justify a sentence on the grounds of a set of sentences, one starts from the convention that the terms distinguished in these sentences as specific for the given considerations will be given only such interpretations (will be allowed only such denotata) by which the sentences of this set become true and then one derives this sentence from that set of sentences by using logical rules. The set of sentences is called axioms (A), and the set $Cn(A)$—the theory based on these axioms. In its second meaning the term 'deductive method' (II) denotes a method by the help of which in order to justify a sentence no convention is accepted as to the denotata of the specific terms and the latter remain with their previous meanings (the denotation rules are unchanged), while the sentence is logically derived from primitive sentences, or else the specific terms are not interpreted at all. According to the opinion considered a characteristic feature of the deductive sciences would be using of the deductive method in the first sense for the justification of theorems.

Such a characterization of deductive sciences allows to derive several properties often assigned to them. Thus:

1) Every deductive science has as object not only a distinguished model of its axioms, but equally well any model of them. We can also say that a deductive science has as object—in a natural understanding of the expression 'is an object'—the class of all the models of its axioms. The deductive science would then always have as object the structure in Bourbaki's sense. As we know, this class is empty for inconsistent theories. It is always something different from all its members. This, of course, also occurs when this class is empty: it exists, while it has no members.

2) The theorems of a deductive science are always analytic (true,

if only the denotata of their terms required by the linguistic conventions exist). This immediately results from the definition of the denotata of specific terms: indeed, a system of such denotata consists of every and only such a system of objects $M = \langle U, J, K, R \rangle$, in which the axioms of the science (and hence also the theorems) are true. If then M is a system of denotata for terms of a deductive science, then the axioms and the theorems of that science must — *ex definitione* — be true in it. Thus the truth of every thesis of such a science logically results from the assumption that M is the system of denotata required for the terms of this science. And this is a necessary and sufficient condition of the analyticity of its theses according to the understanding of analyticity adopted here.[30]

3) The justification of a sentence by using the deductive method in the sense spoken of is a method of justification in the absolute sense in the language of the deductive science and guarantees that these sentences are true, if only there exists a model of this science. Apart from deriving some sentences from others on the ground of logical rules (the usual understanding of deduction), which is a relative justification, in the use of the deductive method understood as above also a way of direct justification is included: by referring to something that no longer is a sentence of a language (and is no sentence at all in the sense of truth or falsity), namely by referring to decisions concerning the interpretation of specific terms of the given science (decisions expressed by the terminological conventions adopted at the beginning of this science). Treating the reference to decisions as a method of direct justification should not be surprising. It is true that the decisions to denote objects of a certain sort by terms alone do not guarantee the existence of the objects (members of a model). But does the sensible experience (sensible idea) alone guarantee the existence of an object with given features? Nevertheless basing on observational ideas is one of the methods of direct justification of our enunciations. The theory of knowledge has many ways of controlling the justifying value of a given sensible idea (intra-subjective and inter-subjective control which — in both cases — may be in turn intrasensual and intersensual). Similarly, one can control (and does control) the justifying value of a terminological convention, the latter possessing this value only if the objects of the required sort actually exist. This can be made sure in particular cases by showing these objects (i.e. by referring to

sensible ideas). A model of deductive theories does not usually have as members perceptible things. They are classes of individuals or classes of systems of such individuals, and these are not always perceptible. The consistency of the theory is, at least for elementary theories, a proof of the existence of such a model. On the other hand, this consistency—if there is no separate proof for it, which is quite a common case—can be confirmed only in the course of the development of the given theory. This makes us think of a currently often used phrase: practice is the decisive way of justifying theorems in deductive sciences. I would prefer to put it differently: there are different methods of direct justification of our theorems; but practice is the only method of controlling the justifying value of these methods (practice understood widely enough as to include, among other things, control through new sensible ideas and through demonstration—or confirmation—of the consistency of the theory under discussion).

Thus I should like to call reference to terminological conventions a method of direct justification. This can be done if we reject the superstition that direct justification must be infallible. The fact that such a conviction is a superstition can be clearly seen in that reference to sensible ideas—as it already has been said—is considered as a method of direct justification, although it may lead—and it does in many cases (illusions, hallucinations)—to falsehoods.

As a method of direct justification (we have called so the reference to the understanding of words), the deductive method, as it is understood here, is a method of justification in an absolute sense (and not only relative, as it is the case of deductive method in the second sense distinguished). And there is no need to demonstrate that, if there exists a model of a given deductive science, the theses of this science must be true (in the case of the accepted interpretation of its specific terms).

We have thus characterized the deductive sciences by a specifically understood method. We have derived from this characteristics the following: 1) they concern *structures* (classes of models), and not individual models; 2) their theses are *analytic*, 3) their theses are justified in the *absolute* sense (by using—apart from ordinary deduction—also a method of direct justification, which, however, does not give any absolute guarantee of the truth of what it justifies).

It is time now to compare the deductive sciences and the non-deductive sciences which, as they utilize language, also contain some axioms. This character is common to all those sentences of the language in which the given non-deductive science is constructed which can be assigned the role of meaning-postulates.[31] They limit the ways of interpretation of the respective extralogical terms. The sentence 'a bachelor is unmarried',[32] often considered in connection with the problem of analytic sentences, has, within a language where the terms 'bachelor' and 'married' are undefined, precisely the character of such a postulate with respect to these terms: the sentence is accepted with the tacit convention that no matter how we interpret the terms considered, it always should be done so that their·ranges are mutually exclusive. It is very difficult to decide in the field of natural language whether something is a meaning-postulate or not. Nevertheless, it can be done with respect to many sentences. Thus, we certainly are decided to interpret the words 'book' and 'dog' so that the sentence 'a book is not a dog' be true, the words 'nymph' and 'satyr' — so that 'a nymph is not a satyr' be true, etc. Whether we call these sentences meaning-postulates or directly analytic sentences — will be only the question of terminology. These sentences — and their logical consequences — must be true in all the models considered in the given language. Let us recall, using an idealizing device, that there is a finite number of such sentences, and let us denote the conjunction of all such sentences of the language of a non-deductive science by $A(w_1, w_2, ..., w_n)$, where the symbols 'w_1' etc. represent the extralogical terms of the language. For the sake of simplicity we assume here that 1) there is a finite number of these terms in the language and 2) that every extralogical term of the language appears in that conjunction. All the sentences found in the conjunction $A(w_1, w_2, ..., w_n)$, together with their logical consequences, will constitute the analytic basis of the given non-deductive science formulated in that language.

The models of these sentences will be $(n+1)$-member systems and all consist of a universe as first member and of n objects (individuals, classes, relations) taken from that universe, so they will assume the form $M = \langle U, I, K, R \rangle$, where there are l_1 individuals I, l_2 classes K, l_3 relations R, and where $l_1 + l_2 + l_3 = n$.

A non-deductive science formulated in some language is not interested at all, however, in every model of the sentences constituting its analytic

basis, nor in the class (structure) of them. Contrary to a deductive science it aims at finding one distinguished model of that axiomatics, namely the universe of individuals, those of its members, those properties of these individuals and those relations between them (possibly also other objects of higher levels) which form the *real world*, resp. a fragment of it. The real world (its fragment), however, may not exist at all among these models. Indeed, our axiomatics for extralogical terms may be inconsistent. There will then be no models at all for the class of analytic sentences of our science. *A fortiori* the real world will not be such a model. The supposition that it — at least its fragment which we intend to study in our science — is none of the models of the axiomatics assumed may arise for still another reason. In order to realize it, let us consider the way in which that unique model of non-deductive science, belonging to the real world, is distinguished, or more precisely: how is the model of the meaning-postulates of that science distinguished. We shall call this model *'proper model'*.[33]

In order to speak of the real world in terms of any words, we must somehow assign the objects of that world to the words. This is done by using so-called ostensive definitions (deictic definitions), resp. the procedures discussed by Reichenbach as 'coordinate definitions' (*Zuordnungsdefinitionen*). We shall further on speak of ostensive definitions. Without attempting a deeper analysis let us remark that the only terms that can be considered as undoubtedly observational are those for which ostensive definitions have been given. As to natural language it is almost impossible to determine for which terms such definitions have been given and for which not. For the sake of simplicity, however, we shall assume in the theory of non-deductive sciences that this matter has been cleared up. We shall also remember that such definitions are given for all the individual terms, the class terms and those denoting relations. Such 'definitions' assign objects (individuals, properties, relations) of the real world to words which — as we know — is done in a way usually far from univocality.[34]

Dependingly on whether, in the language of a given non-deductive science, there are as extralogical constants only observational terms or not, and dependingly on whether the ostensive definitions assign objects of the real world to words univocally or ambiguously — the real world which we want to study in this science will be determined uniquely or

not (and even very ambiguously). This univocality resp. ambiguity is also influenced by whether the relations assumed in the analytic basis of non-deductive science between observational and non-observational (theoretical) terms are such that every interpretation of the observational terms uniquely determines the appropriate interpretation of the theoretical terms or not. In general, six cases may occur. We shall discuss them in more detail.

The case $(\alpha, 1)$

Let us assume that (α) a language of science contains only observational terms as extralogical terms, and that (1) the ostensive 'definitions' assign things to terms in a unique way. Then, for a given universe, the system consisting of that universe and of these objects—if only it is a model of the analytic basis of the given science—is a uniquely distinguished fragment of the real world, the fragment studied by this science. It is the proper model of the assumed meaning-postulates.

The case $(\alpha, 2)$

While the assumption that (α) all the extralogical terms of some science are observational terms holds, let us now assume that (2) the ostensive definitions assign denotata at least to some observational terms in an ambiguous way. Then, even for a definite universe, there can exist more than one model of the analytic sentences assumed, a model consisting of objects assigned to observational terms and of that universe. If every such system were the model of the postulates assumed, and if the interpretations of the extralogical terms were independent of each other, then the number of these models might easily be calculated—for a given universe. It would equal the cardinal number of the Cartesian product of the classes $K_1 \times K_2 \times \ldots \times K_n$, where K_i, for $i = 1, 2, \ldots, n$, denotes the class of interpretations assigned to the term w_i by ostensive definitions. However, because of the dependence between the permitted interpretations of observational terms, involved in the analytic assumptions of the science, the number of the models discussed may be—for a given universe—much smaller. The proper model would then be one of the members of this set.

The case $(\beta, 1, a)$

Let us now consider the situation in which (β) the extralogical terms of the science—apart from observational terms—also contain some theoretical terms, but (1) the ostensive definitions uniquely assign objects to terms of the first type and (a) every interpretation of the observational terms, uniquely determines the interpretation for the theoretical terms. We can easily see that—for a given universe—the system of objects composed of that universe and of interpretations assigned to extralogical terms, if only it exists among the models of the analytic postulates of the language of the given science, is uniquely determined and represents the real world studied in that science. It is the proper model of the assumed postulates.

The case $(\beta, 1, b)$

Let (β) the extralogical terms be observational and theoretical, and let (1) the 'definitions' assigning objects to observational terms do it uniquely. Let us assume, however, that (b) at least some of the meaning-postulates connecting theoretical and observational terms are such that the interpretations of the observational terms do not uniquely determine the interpretations of the theoretical terms. The real world, which is the object of interest of the given deductive science, is then ambiguously distinguished as one of the systems of a universe and of the objects of a finite sequence belonging to a subset of the Cartesian product $K_1 \times K_2 \times ... \times K_n$, where $K_1, K_2, ..., K_n$ are —as above—classes of objects of the appropriate order, assigned to the extralogical terms $w_1, w_2, ..., w_n$ by assigning 'definitions' and by meaning-postulates. Again, this system of objects—constituting the real world studied in the given science—will be the proper model of the assumed meaning-postulates, if it is at all the model of these postulates.

The case $(\beta, 2, a)$ and the case $(\beta, 2, b)$

There is no need to discuss the remaining cases separately. In both these cases (β) the language of the non-deductive science contains both observational terms and theoretical terms, but (2) the 'definitions' assigning objects to observational terms do it ambiguously. Then—no

matter whether (a) the interpretations of observational terms uniquely determine the interpretations of the theoretical terms connected with them by the meaning-postulates or (b) not—the real system of objects studied in the science will be ambiguously determined as one of the members of the appropriate subset of the Cartesian product constructed as above, preceded by a universe, and will be the proper model of the meaning-postulates of the language of the science, if it is at all the model of these postulates.

Having thus some knowledge as to the ways in which in classes of models which satisfy the meaning-postulates of the language of a non-deductive science the proper model, this science is interested in, becomes distinguished, let us now examine closer the reservation continuously made above, saying that the science may not succeed in including this system of objects within its models, even if it has models. Namely, it may come out that among sentences dictated by experience (observations) and concerning the objects assigned to terms by ostensive definitions, resp. objects connected with the former by specified relations, there will be sentences inconsistent with the meaning-postulates of the language, resp. with their consequences. This will prove that the real objects spoken of are not contained in any model of the analytical basis of this science. This, precisely, will be the other reason for which the given science will not have its proper model even if the meaning-postulates it is based on are consistent.

It is precisely because of these possible divergences between what is dictated by experience and the meaning-postulates assumed on the grounds of the accepted ostensive definitions that every non-deductive science must be able to develop; this development should, for example, consist in a modification of the meaning-postulates and of the assigning definitions as well as of the conditions of accepting the empirical theses dictated by experience. Such a modification should, as a result, harmonize the empirical theses of the science with the meaning postulates constituting its analytic basis. It would sometimes be difficult then to determine where the old science ends and the new one begins. This development does not require the change of the richness of language. Another kind of development takes place when, in course of precising the meanings of vague terms in various directions by making univocal the interpretative procedures, there appear several new terms apart from—or instead

of—the old ones (for example, apart from the term 'yellow', the terms 'yellowish-green' and 'yellowish-red'). In such cases the number of extralogical terms changes and the vocabulary of the language is modified, while new meaning-postulates and new ostensive 'definitions' appear. And the modification of the linguistic vocabulary may occur in many other ways.[35] I do not intend to characterize here these changes in more detail. I should only like to remark that even if the objects determined by ostensive procedures are not suitable to be the members of any model of the accepted postulates, we can, nevertheless, say in some natural sense that the model taken from the real world is the object of the given science: for this science tends, through its development, to a situation in which the objects indicated by the assigning definitions can be members of the model of the analytic assumptions on which the science is based in that situation.

A non-deductive science then always has as its object a proper model, either in the sense that a) the objects distinguished by the ostensive definitions together with the corresponding universe and with the interpretations for the other terms actually form a model of the analytic postulates which the given science takes with the language it utilizes, or in the sense that b) by changing the analytic postulates, the assigning definitions and possibly the conditions of acceptance of empirical theses, which is sometimes accompanied by changes in the vocabulary, the science tends to harmonize all those factors so that the real objects determined by the ostensive definitions, together with the appropriate universe and with definite interpretations of other terms, constitute a model for the analytic basis of the science and thus form some actual model of the meaning-postulates connected with its language. The 'proper model' is then either the actual member of the class of models of the meaning postulates accepted by the non-deductive science or a model intended in the process of its multidirectional development. In the former case the proper model is the object of the non-deductive science in a sense close to the first of the above-distinguished senses of the expression 'is an object'.[36] In the second case, however, we have to do with an entirely new, third sense of that expression. Here is its definition:

M is an object in the 3-rd sense of the science $N \equiv$ the development of the science N tends to harmonize the factors composing it so that M can be found among the models of its meaning-postulates.

The real world, understood as the system of the real universe of individuals, some of its members, their observable properties and the relations between them and of some unobservable objects the existence of which has been made probable experimentally — is then always the object of a non-deductive science in one of the senses presented. The fact of the uniqueness of this system of objects is not disturbed by the probability of an ambiguous way of distinguishing it. With regard to a non-deductive science there is only one possible way of learning its 'proper model': by observation, experiment, experience, empirical evidence. It is within this framework that, utilizing some language and its meaning postulates, this science gathers its empirical theses, with the set of which it might even be identified. As a matter of fact, it is difficult to say when in the actual development of a science a sentence is no longer an empirically confirmable hypothesis and becomes a meaning-postulate and vice versa; but it is generally difficult to determine the methodological status of any theorems of a science in the course of its actual development. Indeed, we do not precisely characterize the language in which a given science is developed, while the distinction between hypotheses and meaning-postulates may be made only after such a characterization has been done. It seems necessary at every stage of the development of the methodological analysis of a science to distinguish empirical hypotheses from meaning-postulates. I would even say that it is decisive for the content of the science. For, while we shall readily consider part of a non-deductive science all its empirical theses (and hypotheses), we shall treat the meaning-postulates on which it is based as related rather to the conceptual apparatus with the help of which this science studies reality, i.e. to its language, which may simultaneously be that of many other sciences. By utilizing experience to justify their theses, the non-deductive sciences inevitably become empirical sciences. All their theses then have a synthetic character (if they are decidable, it can be done only by experience).

Do non-deductive sciences justify their theses in the absolute sense, or only in a relative sense? As they are finally based on a method of direct justification, i.e. on reference to observational ideas (as empirical), we then have to do here with justification in the absolute sense. It does not interfere that often complicated control is necessary to determine whether a given idea has not been an illusion or a hallucination, that

we must refer to the efficiency of our deeds based on respective theses in order to exclude errors. This absolute character we are talking about does not deprive the justification of an element of uncertainty, but it makes it simple justification, as distinguished from ordinary derivability according to some rules from a class of premises, which is the case of justification in a relative sense. Of course, justification has in non-deductive sciences this 'absolute' character even in spite of the fact that besides infallible steps in the course of inference also fallible steps are allowed (among other things, by incomplete induction or even by reduction). We simply mean to guarantee — at least to some degree — truth to our theses, and not that they be only probably true if some others are true.

If we want to put together the main features by the use of which we have tried to characterize the non-deductive sciences as compared with the deductive sciences, we can say that 1) an object of a non-deductive science always is *one distinguished* (although not always uniquely distinguished) model of meaning-postulates of the language employed by that science, 2) proper theses of non-deductive sciences are always *synthetic* (although when justifying them we use previously adopted analytic meaning-postulates) and 3) theses of non-deductive sciences are justified in the *absolute* sense, and the method of direct justification used by these sciences consists in basing their final premises, i.e. observation judgments, on sensible ideas; as a consequence these theses are true if only the objects of those ideas exist and are such as presented by these ideas, and if the steps of the inference have led from truth to truth.

What can be said of Twardowski's article from the point of view of this confrontation between deductive and non-deductive sciences? Twardowski's characterization of non-deductive sciences, as referring to observation judgments in the course of justification of their theorems — given a conceptual apparatus — is fully confirmed. Is it so also with his theorem that deduction is never used in these sciences as a method of justification? It is enough to substitute the term 'deductive method' for 'deduction' in the understanding of Bourbaki's axiomatic method to obtain a theorem true without reservations, as no synthetic sentence can be justified in this way. The use of the axiomatic method means just leaving out any concrete model and its properties and concentration

exclusively on the class of all models of the given set of sentences. As long as we are interested in a distinguished model we do not use the axiomatic method or — in our terminology — the deductive method in the 1-st sense. Nonetheless, we use deduction in the 2-nd sense for the justification of theorems, which is not a method of absolute justification. I think that after these reformulations it is clear that Twardowski's intuition, concerning the non-applicability of deduction as method of absolute justification of theorems in the empirical sciences, proved right. An empirical science utilizing the deductive method as method of absolute justification would be no longer empirical and would become deductive. In real investigation it often happens that using empirical methods is succeeded by using the deductive method, and vice versa. This, however, does not eliminate the difference between these two sorts of investigations. The deductive method as method of absolute justification has no application in the empirical sciences — as Twardowski said.

Let us now turn towards Twardowski's characterization of the deductive sciences. Do they really refer to axioms, definitions and postulates in the justification of their theorems? We would rather say, on the grounds of the characterization presented above, that the ultimate method of justification is the reference to terminological *conventions*, due to which axioms and postulates become implicit definitions for the specific terms of these sciences. If only the model of these sciences exists, the convention of understanding by 'specific terms' of a science the members of this model does indeed guarantee the truth to the axioms and the postulates of that science. But the final step in indirect justification really terminates in these sciences on axioms and postulates. Similarly, in the empirical sciences observational judgments are the final element of indirect justification. Twardowski simply did not want to go further than sentences (judgments) in justification, and he did not use any other concept of justification. But this should be done if we want to see clearly the difference, which he himself probably suspected. What, however, about definitions? They can be understood as sentences of the language, and they do not then differ from axioms and postulates, or from the point of view of metalanguage, correspondingly to the decision of understanding a given term as denoting the same that a group of words with a previously characterized meaning does. The first case

has already been discussed. In the second case the definition only introduces an abbreviation in the notation, and has no influence on the models of the theory, which in order to be established first require getting rid of the definitions. This would be all about Twardowski's theorem saying that deductive sciences ultimately refer to their axioms, definitions and postulates.

What about the theorem saying that the results of these sciences are certain and do not concern facts? As we have seen it above, the sentences of deductive sciences in the sense considered above are analytic and if only the postulated denotata exist the theses of these sciences referring to them must be true. Therefore — with this reservation — the theses of these sciences are certain. And what do they concern? We have said, following Bourbaki, that they concern structures, and of various orders. Structures are not simple models but classes of models. All the more, then, they are not models belonging to the real world or — as we say — being the world, resp. its fragments. They are then not facts if by 'facts' we are to understand fragments of the real world. Although the concept of fact is very vague, I think that such a precising of it seems quite natural. Thus also the last of Twardowski's enunciations, which are surprising for the reader, can be maintained. Do the non-deductive sciences, however, from this point of view, concern facts, as Twardowski says? As they concern the proper model of the meaning-postulates they have assumed, which model is the real world or a fragment of it, they obviously concern facts.

There is, however, one fragment of Twardowski's paper which leads to apparent divergences between what he says and the concept of deductive sciences discussed here. Namely, Twardowski does not exclude the possibility that the axioms and the postulates of deductive sciences are generalizations of some former experiences.[37] We have here two possibilities: either this fragment leads to a contradiction with the enunciation that deductive sciences do not concern facts (they would have to concern facts as generalizations of experiences) and that their results are certain (no inductive generalization can pretend to certitude), or else the fragment should not be understood literally. In my opinion the second possibility is involved here, the more so that a non-literal interpretation of this fragment is obvious for anybody studying the development of knowledge in general. Indeed, the phenomenon can

be seen in the process of the passing of inductive sciences into deductive sciences, the phenomenon which has been noticed by different authors. Mill [38] has discussed it, and among the contemporaries Quine's opinion [39] is worth recalling. The facts are simply as follows: we gather experimentally various properties of objects characterized for example as objects having the properties a, b and c. Among these new properties we have for instance the properties d, and f. If we come to the conviction that every object P, having the properties a, b and c, also has the properties d and f, we proceed to change the concepts. We begin to connect with the same term 'P', not the first three properties, but five properties: a, b, c, d and f. After this sentences assigning the properties d and f to objects like P become analytic, while before they had a synthetic character, as they were generalizations from experience. This is how the actual process of constructing of some deductive sciences took place. Some theorems about points, straight lines and other concepts were known from geodesy and were theorems based on experience. At a certain moment they become elements of the implicit definitions for the terms 'point', 'straight line', etc. When saying that axioms could be generalizations from experience, Twardowski could have the same situation in mind, namely, that the sentences which became implicit definitions of the extralogical terms contained in them could have before expressed generalizations from experience. Only the shortened character of his presentation of the problem did not allow the author to speak in more detail.

Thus all Twardowski's enunciations — after slight reformulations — appear to be consistent and right. For this reason, I would like to stress that the confrontation of deductive and non-deductive sciences which we have done in this paper was already present *in nuce* in the discussed Twardowski's article. Bourbaki with his concept of mathematics and Tarski, who identified mathematics with sciences employing the deductive method in justifying their theorems, allow us to perceive this confrontation in Twardowski's paper.

However, are all Twardowski's theses really right? The thesis saying that the deductive sciences never concern facts seems to be obviously non-evident. Of course, structures are something which is not a single model belonging to the real world. But if the class of models of a set of sentences contains at least one model, the members of which (for

example individuals), resp. their belonging to the classes and relations of that model, are observational, then — as it has already been said — this model can also be an object of the given deductive science, although the expression 'is an object' will be understood here somewhat differently. A deductive science, although it concerns a class of models, concerns at the same time all the members of that class. And if among these members there is a fragment of the real world, then this science concerns this fragment. This use of the term 'to concern' seems to be very natural. Therefore, while agreeing with Bourbaki that the mathematical sciences concern structures, I would like to stress that — in the presented above, natural understanding of the word 'to concern' — they also concern the real world, if a fragment of that world (a system of real individuals and of their properties and relations) is member of such a structure. This is the situation for most of the mathematical disciplines. And in that sense can the deductive sciences concern 'facts', and indeed they often do concern them. This is not contradictory, however, with Twardowski's theses, as he dealt with a different concept of the object of science.

There is still another problem arising objections. In his paper Twardowski also calls deductive sciences — and he emphasizes it — *a priori* sciences, and non-deductive sciences — *a posteriori* sciences. These terms may lead to numerous misunderstandings. They can be maintained only if we understand by the expression '*a priori* science' exclusively that theorems of that science are ultimately justified on the ground of axioms treated as implicit definitions for the specific terms of the science that occur in these axioms, and if by the expression '*a posteriori* science' we mean that the science tries to guarantee the truth of its theorems not in the denoting rules for the terms contained in it, but in observations of the objects it speaks about. It has already been said that an implicit definition alone, imposing conditions on terms contained in it, cannot guarantee that these conditions will be satisfied by certain objects.[40] Such objects may not exist (for example if the conditions are contradictory). Sometimes we can learn about the existence of a model of some given conditions only from the practice of several centuries (among other things from the consistent development of a theory). This clearly shows how the above understanding differs from the traditional view of *a priori* knowledge. However, the knowledge

contained in a deductive science, if it is assumed thad its model exists, is not conditioned by any observations. And to this extent it is *a priori* independent from observation. On the other hand, the use of the expression '*a posteriori* science' to denote non-deductive sciences does not raise any objections.

When speaking about the opposition of deductive and non-deductive sciences, we have up to now meant the deductive sciences different from logic. How about logic? To end these considerations we shall devote some lines to this difficult and controversial matter. Logic is not included in the comparison presented. Knowledge, understood as the sum of deductive and non-deductive sciences here considered, does not contain logic. In Artistotle's classification of sciences logic is prior to the whole knowledge and constitutes its tool. In the view which I want to suggest here, however, logic occupies a different position. By using the deductive method it starts from the postulate of interpreting its constants (so-called logical constants) so that, under all possible interpretations of the variables in the universe of the real world (of the model distinguished by non-deductive sciences), resp. — as far as the propositional logic is concerned — in the universe of logical values, such and such of sentences be true. What these sentences will be like, depends on the regularities that have previously been repeatedly observed in experience (cf. Lenin on the axioms of logic). These sentences constructed only of logical constants and variables are the axioms of logic A. Moreover, the condition is imposed upon the logical constants that under all interpretations of variables in the domains mentioned above (the universe of real individuals and the universe of logical values) some rules of inference D (usually, among others, the rule of *modus ponens* and substitution) be infallible. Then, on the basis of these rules, theorems are derived from axioms. The deductive science $L = Cn_D(A)$ so constructed is logic. It has all the features corresponding to the features of the so far discussed deductive sciences: 1) Its object is a 'structure' (the class of acceptable 'models' of the sentences A).[41] 2) Its theses are analytic and as such they are true if only there exist the required models for the axioms assumed (and for the relations determined by the rules). 3) The theses of logic are justified in the absolute sense, namely by the ultimate reference to the decisions of this and not other understanding of the logical constants. Whether there exists the required 'model' for the

axioms (rules) assumed and whether the axioms are true (and the rules infallible) for every one-to-one transformation of the real world of individuals and of the world of logical values – is confirmed in the course of the development of our knowledge and its application.

Contradiction, no matter where encountered, may among other things be an argument also for the non-existence of one of the logical functions mentioned in the axioms, and this can lead to the change of axioms or the rules of logic.

Once the logic has been determined with relation to the real world of individuals and to a chosen world of logical values, it is possible to define the general concept of the *universe of a model* as a set of objects such that the axioms of logic and the rules – of the functional calculus namely – remain true, resp. infallible, when the variables occurring in them range over that set. Only then it is known what systems $\langle U, I, K, R \rangle$ can be models of respective languages and sentences.

The most difficult problem in the process of development of knowledge is to determine interpretations for logical constants. As this cannot be delayed until the 'invariants of the real world' are known – such knowledge is something like the limit of the development of all sciences – from the beginning some interpretations of them *are arbitrarily chosen* (interpretations suggested by the existing experiences). Thereby it is not excluded that also these interpretations may change some day – for the sake of accurate reflecting reality by our knowledge – in the process of adjusting the synthetic and analytic parts of the empirical sciences with one another.

BIBLIOGRAPHY

[1] K. Ajdukiewicz, 'The Axiomatic Systems from the Methodological Point of View', *Studia Logica* **9** (1960).

[2] K. Ajdukiewicz, 'Trzy pojęcia definicji' ('Three Concepts of Definition') in: *Język i poznanie* (*Language and Knowledge*), vol. 2, Warszawa 1965, pp. 296–307 (first printed in 1958).

[3] K. Ajdukiewicz, 'Zagadnienie uzasadniania zdań analitycznych' ('The Problem of Justifying Analytic Sentences', in: *Język i poznanie* (*Language and Knowledge*), vol. 2, pp. 308–321 (first published in French in 1958, included in the present volume).

[4] N. Bourbaki, 'L'architecture des mathématiques', in: *Les grands courants de la pensee mathematique*, 1948, pp. 35–47. Russian translation in: *Otcherki po istorii matematiki*, Moscow 1963, pp. 245–249.

[5] R. Carnap, 'Meaning-Postulates', in: *Meaning and Necessity*, 2nd ed., Chicago 1958, pp. 222–229 (first printed 1952).

[6] J. G. Kemeny, 'A New Approach to Semantics', *Journal of Symbolic Logic* 21 (1956).

[7] M. Kokoszyńska, 'O dwojakim rozumieniu uzasadniania dedukcyjnego' ('On the Double Understanding of Deductive Justification'), *Studia Logica* 13 (1962).

[8] M. Kokoszyńska, 'O stosowalności metody dedukcyjnej w naukach niededukcyjnych' ('On the Applicability of the Deductive Method in Non-Deductive Sciences'), in: Sprawozdania Wrocławskiego Towarzystwa Naukowego (Reports of the Wrocław Scientific Society) 15 A.1096.

[9] M. Kokoszyńska, 'W sprawie dyrektyw inferencji' ('On Inference Rules'), in: *Rozprawy Logiczne. Księga pamiątkowa ku czci Profesora Kazimierza Ajdukiewicza (Logical Dissertations. Papers in the Honour of Professor Kazimierz Ajdukiewicz)*, Warszawa 1964, pp. 77–90.

[10] J. Kotarbińska, 'Tak zwana definicja deiktyczna' ('On Ostensive Definitions'), in: *Fragmenty filozoficzne. Księga pamiątkowa ku uczczeniu 40-lecia pracy nauczycielskiej Tadeusza Kotarbińskkego (Philosophical Fragments. Book in the Honour of 40 Years of Tadeusz Kotarbiński's Didactic Work)*, second series, Warszawa 1959; included in the present volume.

[11] J. Łoś, 'On the Extending of Models' (1), *Fund. Math.* 52 (1955).

[12] J. St. Mill, *A System of Logic Ratiocinative and Inductive*, new impression, London–New York–Toronto 1947.

[13] M. Przełęcki, 'O definiowaniu terminów spostrzeżeniowych', ('On Defining Observational Terms') in: *Rozprawy Logiczne. Księga pamiątkowa ku czci Profesora Kazimierza Ajdukiewicza (Logical Dissertations. Papers in the Honour of Professor Kazimierz Ajdukiewicz)*, Warszawa 1964, pp. 155–183.

[14] W. V. Quine, 'Truth by Convention', in: *Readings in Philosophical Analysis*, ed. by H. Feigl and W. Sellars, New York 1949, pp. 250–273.

[15] W. V. Quine, 'Two Dogmas of Empiricism', in: *From a Logical Point of View*, Mass. 1953, pp. 20–46.

[16] R. Suszko, 'Logika formalna a niektóre zagadnienia teorii poznania' ('Formal Logic and Some Problems of Epistemology'), *Myśl Filozoficzna* 2 and 3 (1957).

[17] A. Tarski, *Wstęp do logiki i metody dedukcyjnej (Introduction to Logic and to the Deductive Method)*, Biblioteczka Matematyczna, 3-4-5, Lwów–Warszawa 1929.

[18] A. Tarski, *Introduction to Logic and to the Methodology of Deductive Sciences*, 2nd ed., New York 1946.

[19] K. Twardowski, 'O naukach apriorycznych, czyli racjonalnych (dedukcyjnych) i naukach aposteriorycznych, czyli empirycznych (indukcyjnych)', ('On A Priori, i.e. Rational (Deductive) and A Posteriori, i.e. Empirical (Inductive) Sciences', in: *Wybrane pisma filozoficzne (Selected Philosophical Writings)*, Warszawa 1965, pp. 364–372 (originally published in 1923).

REFERENCES

* First published in *Fragmenty Filozoficzne* **III**, Warszawa 1967. Translated by S. Wojnicki.

[1] Cf. [19].

[2] *Ibid.*, p. 367_7.

[3] *Ibid.*, p. 370_7.

[4] Cf. [1].

[5] Cf. [19], p. 372^{19} and p. 372^{21} ff.

[6] Cf. [17], p. 85, where we read: "The view becomes more and more common that the deductive method is the only essential feature by means of which the mathematical disciplines can be distinguished from all other sciences; not only is every mathematical discipline a deductive theory, but also conversely, every deductive theory is a mathematical discipline". Cf. also the English edition of this book [18], p. 120.

[7] Cf. [4]. I am using here the Russian translation of this article. Although the name 'Bourbaki' stands for a whole group of mathematicians, I shall continue to accept the fiction proposed by this group in order to make things simpler and I shall speak of Bourbaki as of one author.

[8] Cf. [4], p. 258. As a matter of fact Bourbaki speaks here of the problem of the relation between the 'mathematical world' and the 'experimental world'; but the fact itself of his speaking of two worlds, which sometimes appear to be unexpectedly close to each other, allows to assign him the view that they are different worlds.

[9] Cf. [4], pp. 254 ff.

[10] *Ibid.*, p. 249 ff.

[11] *Ibid.*, p. 250.

[12] *Ibid.*, p. 251.

[13] *Ibid.*

[14] *Ibid.*, p. 252.

[15] With respect to the distinction between the model of a language and the model of a class of sentences cf. [11], where the concepts of 'models in' (or 'models for') and of 'models of' are discussed. Cf. also [16], particularly the part I of this work, in which R. Suszko uses the term 'model of a language' in the sense understood in the present paper.

[16] Cf. [4], p. 255.

[17] *Ibid.*, p. 256.

[18] *Ibid.*, p. 255.

[19] *Ibid.*, p. 258.

[20] *Ibid.*

[21] *Ibid.*, p. 248.

[22] *Ibid.*, p. 250.

[23] *Ibid.*, p. 251.

[24] Cf. [2].

[25] *Ibid.*, p. 297.

[26] I have discussed such rules among other things in [8] and [7].

[27] Cf. [4], p. 256.

[28] Cf. Note 6.

[29] I have presented the distinction of these meanings in paper [7].

[30] I define here the analytic sentences of a language as Cn(BAZ) or—in another notation—as $W_{LD}(BAZ)$, where BAZ denotes the set of sentences being an implicit definition of the extralogical terms of that language, LD—the logical directives, and $W_{LD}(X)$—the set sentences which are derivable on the grounds of logical rules from the class of sentences X. I utilize then a simplified concept of analytic sentence of a language. If we wanted to use the unsimplified concept of analytic sentence, we would have to utilize the concept of analytic rule of inference (AD) and to define the set of analytic sentences of a language as the set of all the sentences derivable on the grounds of the analytical rules of that language from the empty class, i.e. as $W_{AD}(\wedge)$. Concerning this, cf. [9], esp. p. 87, Def. 6.

[31] The concept of meaning-postulates is utilized by R. Carnap in his paper [5]. It is also used by Kemeny, cf. [6].

[32] Cf. [15].

[33] The 'proper model' of a language, distinguished among the family of its models in a period of time by using extralogical means, has been discussed by R. Suszko in [16], Part 1, § 5; Suszko is right in pointing out that the language is in this case 'added' to that model (more exactly: to the system of objects which will in time become its model) rather than conversely. When speaking in the text about the assigning of objects to terms, we do not intend to mean that language is primary.

[34] As to the ambiguity in determining the denotations of terms by so-called ostensive definitions etc.—cf. [10] and [13].

[35] R. Suszko, in [16], pays much attention to the problems of the development of a science from the point of view of the relation between its language and the world this science describes, although his interests are different from those dealt with in this paper.

[36] While the definition of the expression 'is an object 'in the 1-st sense could be written:

M is an object (in the 1-st sense) of the deductive science $Cn(A) \equiv M$ is a model of axioms A,

the definition of the sense (I'), referred to in the text, which is utilized in the case of non-deductive sciences, is:

M is an object (in the sense (I')) of a non-deductive science with meaning-postulates $A \equiv M$ is a model of meaning-postulates A of that science.

[37] Cf. [19], p. 368

[38] Cf. [12], Book 3, Ch. 13, § 7.

[39] Cf. [14], esp. p. 269 where W. V. Quine discusses the change of empirical truths into conventional ones.

[40] It is the basic thesis of paper [3].

[41] The terms 'model' and 'structure', of course, have here a somewhat different meaning than before. We refrain from discussing these meanings here.

JANINA KOTARBIŃSKA

ON OSTENSIVE DEFINITIONS*

1

Language serves the purpose of communication, it is used to express someone's thoughts about this or that object and thereby to evoke in someone else thoughts of similar meaning. But it can be used so only if a semantic relationship is established between some of its expressions, at least some names, as verbal signs on the one hand, and those very objects, the thoughts about which one wants to express on the other. The common method of establishing such relationships is the usual, that is, non-ostensive method of defining: introducing a name through a definition of equivalence or a conditional definition (or reduction), we establish a relationship between that name and its extension which is fully or partly determined by the extensions of the names used as the defining terms. By its very nature, however, such a method has a limited scope. Some names can be defined by some other names, those other names in turn by still other ones, but sooner or later we reach the primitive (elementary, basic) names such as 'acid', 'hard', 'red', etc. In the case of the last group, the relationships between the names and their respective extensions are established in another way than by simply relating them to the extensions of the names that have been previously introduced. The so-called ostensive definition serves precisely, at least among other things, the purpose of establishing relationships of the last-named kind.

We say 'so-called', because the problem of its *genus proximum* is not clear. The term 'ostensive definition', or 'deictic definition' (the latter comes from Greek), commonly used in the literature of the subject, seems to point to the fact that we have to do with a certain kind of definition; yet this term is generally used so that the normal meaning of definition does not cover that of ostensive definition. This point will

233

not be discussed for the time being. Putting aside its pedigree let us first
have a look at the method itself, without taking the trouble for the
present, to decide whether this is a definitional method or whether we
have to class it as a non-definitional method. One thing, however, has
to be stated plainly: we have to do with a method which is often used
in common practice when perceptual terms are explained, and which
is almost without exception believed to be the only method suitable
for the introduction of primitive perceptual terms.

What is ostensive definition? According to the common opinion
it consists in pointing with a suitable gesture at a single designatum
of the term which is being defined and in making at the same time
a statement of the type 'This is N', where 'N' stands for the term being
defined. A typical example: we explain ostensively to someone the
meaning of the term 'yellow' if, while pointing towards a yellow object,
for instance a ripe lemon, we say 'This is yellow'.

In such an interpretation ostensive definition always is a complex
procedure consisting of two distinct actions only one of which belongs
to the sphere of verbal language. That is why it is often stressed that ostensive definition, as opposed to normal definition, is not purely verbal.
Russell even describes ostensive definition, in a not very precise way,
as a process whereby we teach someone to understand words in a different
way than by the use of words.[1]

In a broader interpretation, however, what is essential for ostensive
definition is not the pointing gesture but the result which we want to
obtain thereby. The aim is to distinguish in our surroundings a certain
definite object and to draw to it the attention of the addressee of the
definition in question.

And that can be done in various ways, not only by pointing at the
object with the finger or by lifting that object or handling it in some other
way, but also a purely verbal manner, by mentioning its individual name
in the definition statement. In the latter case it is enough to pronounce
the defining sentence to point thereby to the object to which that sentence
refers, and no additional actions are necessary. In such a case the defining procedure is purely a language (verbal) operation.

If ostensive definition is understood so — and we suggest confining
ourselves in principle to such an interpretation — then the schema of the
defining sentence quoted above requires a certain modification: instead

of the pronoun 'this' the sentence ought to include a more general symbol — we shall hereafter use the symbol 'A' — representing any individual name, and of course, among other things, also the pronoun 'this' as a special case. (More precisely: certain specimens of the pronoun 'this', namely those which are accompanied with a gesture referring to a given specimen of a certain definite object.) Such schema would then be given the form: 'A is N', where 'N' stands for the term being defined.

Such a form is usually given to the defining sentences when we use ostensive definition in practice. This is not always the final form. Let us distinguish two situations: a) the term 'N' is an individual name referring to a definite object, and b) 'N' is a general name. The form 'A is N' is adjusted to the former case, it being additionally assumed that the role of individual name is not restricted to the role of subjects in sentences.[2] In the latter case another schema seems more suitable. When introducing a name by means of ostensive definition we point to certain objects not so much as those objects which are to exhaust the extension of the name in question, but rather as those objects which are to be covered by its extension, that is, so to speak, as standard designata to which the other designata of the defined term are to be similar. If we then pronounce the sentence 'A is N', it is, strictly speaking, in view of the intention of the speaker, an abbreviation of the sentence 'N is such an object as A', or, more pedantically,

(1) $\Pi x(x$ is $N \equiv x$ is such as $A)$.

In our investigation we shall confine ourselves to that kind of ostensive definition which is used to define general names, that is, in a different terminology, names that can appear as predicates of one or many subjects. The term 'ostensive definition' will be used to denote both the defining operation and the sentence resulting therefrom, since such an ambiguity should not give birth to any misunderstandings in practice.

The question arises, how to understand the statement of the kind 'Πx (x is $N \equiv x$ is such as $A)$' in the case in which it is used as an ostensive definition of the term 'N'. A literal interpretation is, of course, out of question, because then any object having some common trait with the object A would be covered by the extension of the term 'N'; this would cover any object that would be similar to the object A in any respect,

consequently, any object in general; since, as we know, any two objects are similar to one another in some respect. This would result in the universality and equal extension of all the general names introduced by means of ostensive definition which would in turn cancel all value of such definition as a method of relating extensions to names.

Consequently, another interpretation must be adopted which assumes not just any similarity with the standard object, but similarity in some definite respect and within definite limits. If we, e.g., define the word 'yellow' by saying those and only those objects are yellow which are like this lemon, we undoubtedly mean similarity with respect to colour, and a sufficient similarity at that. Therefore, we will not be willing to consider as 'such as this lemon' those objects whose common trait with the lemon in question consists in the fact that, like the lemon concerned, they are non-red, non-violet, etc.; nor even those whose colour comes much closer to yellow, for instance orange objects, although in the latter case the similarity is much greater than it was in the former. On the other hand, we will not restrict identity to those case only in which a given object has a colour that is indistinguishable from that of our lemon.

With such an interpretation, the definition statement of the kind under discussion is not complete. It lacks a certain additional element that could relate similarity to a certain definite respect and to a certain definite degree. It also lacks such data as would permit one to guess how such an additional element would be formulated.

If we consider all this, we might suggest that it would, perhaps, be proper to formulate ostensive definition, at least for the time being, in the following way:

(2a) Πx (x is $N \equiv x$ is such as A in the respect R and in the degree D),

or, in an abridged and more symbolic notation,

(2b) $\Pi x [N(x) \equiv x \operatorname{sim}_{R,D} A]$,

or, having noticed that similarity in some respect and in some degree consists in having, in that respect, certain definite common traits, in the

following way:

(3) $\Pi x\,[N(x)\equiv Y(x)\,Y(A)].$[3]

The incompleteness of the definition statement appears much clearer then. It states only this, (and in an indirect way at that), that some similarity, defined as to respect and degree, specifically connects objects of the type N with the object A. More precisely, that

(4) $(\Sigma R)(\Sigma D)(\Pi x)\,[N(x)\equiv x\ \mathrm{sim}_{R,D}A].$[4]

But it does not state at all, in what respect and to what a degree that similarity ought to obtain. That is, it does not provide sufficient information as to how one should distinguish those objects which could legitimately be considered to be designata of the term 'N'.

Thus, ostensive definition informs about the criteria of applicability of the defined term in a partial way only. The rest must be supplied by the addressee for whom the given definition was destined, and who thus faces the following problem (which we may call the definition problem): for what R and D is it the case that $\Pi x\,[N(x)\equiv x\ \mathrm{sim}_{R,\ D}\ A]$? or alternatively for what Y is it the case that $\Pi x\,[N(x)\equiv Y(x)\cdot Y(A)]$? If he cannot solve this problem, even in part, he will not be able to decide about any object that is not identical with A, whether it is, or is not, covered by the term being defined; that is, he will not grasp the meaning of that term.

As we can see, ostensive definition presupposes co-operation between two human subjects. If we resort to such definition to explain the use of a given word, we, so to speak, require active co-operation on the part of the person who wants to avail himself of the explanation; we set him a certain task to perform. Whether that task can be performed or not, depends on the amount of data at the disposal of the person concerned. Hence the following practical requirement: if we want our ostensive definition to be efficacious, we must provide the data which will permit the other party to perform his task. What data are needed for that purpose? On what grounds can one reasonably guess the criteria of applicability of the term being defined, since those criteria are only in part described in the definition? And how can the above-formulated requirement be satisfied? Let us try to examine those problems more closely, explaining and reconstructing the various links of the process.

2

But before we do that we must stop to think about the very notion of solving the definition problem. It can be understood either in a broader or in a narrower way. If we adopt a very strict interpretation, we are inclined to consider the definition problem to be solved only when we have found a general sentence which is a true substitution for a given definition statement, in other words, when we have found such substitutions for the variables which appear in that definition statement (of course, the free variables to the right of the symbol of equivalence) which turn that statement into a true sentence.[5] This is the situation when, for instance, someone to whom we explain ostensively what a cube is, by stating that it is such an object as that one to which we point, guesses rightly on the strength of our explanation that we mean objects having strictly the same shape as the object that was pointed to, that is, objects which can be distinguished by some specific spatial structure.

These, however, are not the situations most typical of the practical use of ostensive definition. Usually, the defining operations result in the addressee's merely intuitive grasping of the meaning of the defined term. He realizes, somehow, under the influence of our operation, (if, of course, that operation has proved effective) what similarity is in point in the case concerned; and he succeeds in distinguishing that very trait which is decisive for the scope of application of the term in question. But he does not formulate this verbally, even in the so-called inner speech; and, in particular, he does not come to use sentences of the kind mentioned above. Thus the aforesaid conditions are not satisfied. However, if we adopt a broader, more liberal, interpretation we may after all consider the definition problem to be solved not only in those cases when the addressee of an ostensive definition, following the defining operation, comes to use certain definite verbal formulations. It is rather sufficient if he starts to use the defined term in a proper way, and applies it to those, and only those, objects among all the objects he meets with which are similar to the standard object in the respect concerned and in the degree concerned, that is, to those objects which have precisely that trait with respect to which the term in question has been related to its extension. The very fact of correctly using the defined term is in this case considered a sufficient proof of the fact that the addressee

of the ostensive definition in question has somehow grasped the criteria of applicability of that term; though this may have been done only in a purely intuitive way and only in part, that is, under limitation to only *some* sufficient and only *some* necessary conditions.[6]

It is worth while to note that when the term being defined is an elementary term — and such cases are most typical of the use of ostensive definitions — the intuitive method is by its very nature the only one available. All known attempts to formulate the result in words would lead either to a vicious circle or to a *regressus ad infinitum*.

It is not necessary to add that a broader interpretation of the solution of the definition problem is much more useful here, since it does not involve such consequences as that outlined above.

3

After all these explanations we can revert to the questions formulated before. As we remember, they pertain to the conditions of an efficacious use of the ostensive method in determining the meanings of expressions, in other words, of the conditions on which depends the possibility of the basic solution of the definition problem.

We know that the definition statement does not supply in a ready form the criteria of applicability of the term being defined. If we establish a relation between the term '*N*' and all those, and only those, objects which are such as a certain selected object, called '*A*', we supply practically only this information: that some of the traits characteristic of the object *A* are also characteristic of the extension of the term '*N*'. Thus we show only the direction of investigations, or rather, the territory in which these investigations should be undertaken. Now it is obvious that to be able to make use of that information its recipient must necessarily possess knowledge of that object *A*; and if the entire procedure is to yield the desired result, that knowledge of the object *A* must cover the knowledge in that respect, precisely, in which the designata of the term '*N*' are to be similar to it. Hence the obvious conclusion: if the recipient of the definition is to carry out his task, we must employ, as the standard object, an object which is known to him in the respect in question. The question can of course be raised here, how we understand that knowledge. Opinion on that point, from which opinion we shall not

deviate here, is unanimous: the characteristic trait of ostensive definition is that it appeals to direct knowledge based on the recipient's
own perceptions. If so, then the object employed as the standard must
be an object which in the given respect belongs to the sphere of perception of the recipient of the definition.

Must it necessarily belong to his sphere of perception *hic et nunc*, as
it is generally assumed? This reservation seems obvious if ostensive
definition is interpreted in the usual way described above. In that case
the gesture pointing to the demonstrated object is an indispensable
element of the defining operation, and such a gesture would contradict its purpose if the recipient of the definition did not simultaneously
perceive the object which the gesture was intended to demonstrate. Things
look different in the case of the broader interpretation of ostensive
definition, in favour of which we have declared above and which does
not require that the definition statement should be accompanied by a
special demonstrating gesture. In the case of such an interpretation it
seems that no time limits need be set. From the point of view of the
fundamental purpose which ostensive definition is to serve it is inessential
whether, in explaining to someone what a dwarf is, we say: 'A dwarf
is such a man as that whom you now see standing before you', or 'A
dwarf is such a man as that whom you saw yesterday there and there',
or else 'A dwarf is such a man as you will see this evening in such and
such circumstances'. In all the three cases, the inquirer can obtain the
information he asks for. But in the second case this depends on the
good functioning of the recipients' memory, which increases the risk of
failure, usually the more so the more remote is the fact referred to. In
the third case a certain lapse of time is necessary, and the procedure
depends on whether the expected perception actually occurs. In view of
these inconveniences it is in fact better to appeal to present time perceptions than to those in the past or in the future, but it must be borne
in mind that the most convenient solution is not always possible (e.g.,
it might be difficult to bring in a dwarf on a moment's notice) and after
all not always necessary.

Therefore we stick to the original and more general formulation of
the condition investigated, the formulation which does not include any
time limits. Practically, this is tantamount to the recommendation that
only those objects should be mentioned in the definition statement which

are known to to the given addressee (or will be known to him to all probability) from his own perceptional experience.[7]

These two, and only these two, sources of information — the definition statement and direct knowledge of the standard object — are included in the characteristics of ostensive definition which served us as the point of departure. These two sources are, however, not sufficient. If the role of ostensive definition resembles the role of indication restricting the territory to be investigated, then the acquaintance with the standard object in turn resembles a general acquaintance with the territory to be investigated. The information acquired in this way is preparatory to, and constitutes, the initial rather than the final phase of the cognitive process. It is the starting point and not the point of arrival. It permits, at the most, a determination of the alternatives of the very numerous possible cases which include the proper solution, but it does not permit ascertainment of which of these alternative and possible cases is the proper solution.

If this is so, then the problem which has led us up to this point has not lost its topical interest for us. For the question remains to be answered how, and on the strength of what, can one distinguish, among the vast number of properties which characterize a certain designatum of the term being defined, those properties which are decisive for the extension of that term. One point is clear: the problem can be solved only if one has at one's disposal some other data than those referred to so far. If one knows only that the given term covers all and only those objects which are *such as* a definite standard object known in direct experience, then one has practically no grounds for distinguishing some traits of the demonstrated object rather than some other traits as criteria of comparison. Thus it is equally clear that when we supply an ostensive definition we must also supply the addressee with some additional sources of information from which he can obtain his missing data.

4

In practice, this requirement is met in various ways; though this is not to say that one is always conscious of the fact that these ways serve precisely the purpose in question.

It happens, sometimes, that when we mention an object having the character of the standard designatum of the term being defined we must see to the fact that the addressee of the definition should notice that object in the respect in question: we must, so to speak, introduce that object into the addressee's sphere of perception. This is usually attained by means of some manipulation accompanied by a verbal instruction, or by means of such verbal instruction alone. The manipulation most often consists in moving the standard object so that the adressee of the definition should find himself in a certain perceptional situation with respect to that object; e.g., that he could see, touch, smell it, etc. The verbal instruction just tells him how to behave with regard to that object in order to obtain the required perceptual data. It generally takes the imperative form 'Touch this piece of wood, it is rough', or 'Look, the stone which you are holding just now is oval', (the defined terms being 'rough' and 'oval', respectively) etc. If the object selected as a standard belongs, without our interference, to the direct environment of the addressee, the manipulation is of course not necessary, and the verbal instruction is sufficient. Now this verbal instruction (and occasionally the manipulation as well), apart from its principal role, also plays an additional role; namely, that of supplying the indirect information that the properties which are decisive for the inclusion of the given standard object in the extension of the term being defined, must be sought exclusively among the properties which are the attributes of that standard object only in that respect (or those respects) which can be perceived if we behave in conformity with the instruction, (that is, if we look at the object, touch it, smell it, etc.). This is the second elimination step, which still more restricts the field of investigation; whereas the first step eliminated all those properties that are not attributes of the standard object, the second excludes, out of the remaining properties, all those which do not belong to certain distinguished spheres of perception.[8] This function of the verbal instruction does not, generally, receive the attention it deserves.

Attention is also usually not paid to the analogous role of certain descriptive names used in ostensive definitions to distinguish the demonstrated object as the standard. Thus, if explaining to someone the meaning of the word 'viscous', we say that viscous is such an object as that which that person actually holds in his hand — we suggest indirectly

that viscosity is a property that is perceived by touch. Similarly, if we say that 'The crucian is the fish which you now see in this aquarium', we suggest that 'crucianness' is determined by some of the properties which can be percieved by sight and which account for the appearance of the fish in question. It seems superfluous to add that the certainty of such guesses is much less in those cases where such guesses are justified only by the fact that the definition statement included a descriptive name of the kind in question, than where they are based on verbal instructions explained above. This is so because such descriptive names are sometimes used unequivocally to characterize the given standard object, regardless of whether they point to properties which make that object to be one of the designata of the term concerned. Thus, for instance, if we want to explain to a person the meaning of the word 'purple' and if we see that that person is actually smelling a purple flower, we say: 'The flower which you are smelling now is purple' — in spite of the fact that we define a term from the visual sphere of perception. The choice of the descriptive name ('the flower which you are smelling now') is in this case explained only by the fact that by its means it was very easy to distinguish the object to which we wanted to draw the addressee's attention.

It must also be noted that sometimes the definition statement supplies direct information of the respect in which the designata of the term being defined are to be specifically similar to one another. For instance, in explaining to someone the meaning of the word 'magenta' we may say: 'Magenta is the *colour* of your frock', and the like. The task facing the addressee of such a definition statement is easier, as it was in the cases analysed before, when the respect essential from the point of view of the given definition has been already correctly distinguished.

This does not mean, however, that the addressee's task is already carried out in full. To understand this it is sufficient to realize that if an object has at all a trait distinguishable in some definite respect (let it be the trait Y^1 in the respect R, or, briefly, the trait Y_R^1), then it has, in the same respect, not only that trait but also a number of other traits, (even if these be only traits which are entailed by, but not equivalent to, the trait in question — called here $Y_R^2, Y_R^3, Y_R^4, \ldots$, etc.), so that it is not clear, which of these traits is meant. A simple example: if someone

stands 1.60 m, it is equally true that he is shorter than all those who are 1.61 m, 1.62 m, ..., etc., high, and taller than all those who are 1.59 m, 1.58 m, ..., etc., high. By analogy, we can state about each red object that it is characterized by a warm colour, a light or deep colour, an intense or pale colour, etc. Moreover, to make this point plain, let us bear in mind that if one and the same object can be a designatum of many different terms which characterize it in the same respect but differ from one another in the degree of generality, then each such term can be defined in an ostensive way by pointing every time to one and the same object A as the standard object and thus by defining the designata of that term as being, in the respect concerned, the same as the object A. In all those cases the degree of similarity between the given designatum of the term being defined and the standard object, A, will, of course, be different. In all these cases a different trait, from among the traits $Y_R^1, Y_R^2, Y_R^3, ...$, etc., all of them attributes of the object A but differing in the degree of generality, will determine the extension of the term in question.

Thus, in principle, we face the same situation as before. Here again, out of many competing possibilities we must distinguish the right one, to the exclusion of all the remaining ones. But now we must distinguish not the respect in which the designata of the term being defined are to be similar to each other, but the limits within which that similarity is to be confined.

Thus, further eliminating steps are necessary. But this time the verbal text of the ostensive definition is not sufficient for such an elimination. The carrying out of that task depends then on data supplied by means of examples. The elimination of wrong possibilities is feasible if the data at one's disposal permit the application of the schemata of eliminating induction — the only methods suitable for the solution of tasks of such kind. If so, then the assumption made — and adopted in view of the common opinion that ostensive definition refers to only one standard example — is untenable. The demonstration of only one standard designatum can suffice only if it is known in advance that the applicability of the term in question is determined by similarity to that standard understood in the most literal sense, that is, complete lack of difference (indiscernibility) in the given respect. But in practice we rarely meet with such an interpretation of similarity when ostensive definition comes

in question. In all other situations acquaintance with only one standard object permits at most a grasp of the narrowest of the sufficient conditions of applicability of the term (that is, the complete lack of difference in the respect concerned) but offers no information on whether the term is applicable or not if similarity is not complete. If we consider that our task is to find those conditions of applicability of the term being defined which would determine its extension in an unequivocal way, or at least as close to unequivocality as possible, then we realize immediately that the results obtained thus far are not satisfactory.

Consequently, the demonstration of only one standard object can not suffice. There must be many examples, and among them not only suitably selected positive cases, that is objects falling within the extension of the term being defined, but also negative cases suitably selected. Thus, if we define the word 'yellow' – to stick to the example already used – it is not enough to say that something is yellow if, and only if, it is such as this lemon (pointed to by a gesture). We must further point to some objects having other shades of the yellow hue, as well as to some not-yellow objects, e.g., greenish, orange, etc. Elimination will then include not only those traits which are attributes of not all the designata of the term in question – in this case the yellow hue *identical with* the colour of the given lemon (since the possession of such a colour is not a necessary condition of falling within the extension of the term 'yellow'). Elimination will also include those traits which are attributes not only of the designata of the given term – for instance, warm colour or not-violet colour (since the possession of those traits is not a sufficient condition of falling within the extension of the given term, the degree of similarity to the given standard being insufficient).

In summing up the last part of our investigations we may say that the basic method of finding out the traits, which form the criterion of applicability of the term being defined, consists in narrowing the field of search, in gradually eliminating those properties which have proved wrong from the given point of view. Among the eliminated properties we have also noted a gradation of generality. We have seen that at the lower level, the level of determinata, as Johnson would say,[9] the elimination of wrong possibilities takes place as a rule on the basis of eliminating induction. As far as elimination at the upper level, the level of more general properties, those 'respects' in which the given determinata are

attributes of the objects in question (*determinabilia* in Johnson's terminology) is concerned, we have examined non-inductive methods only. It is worth adding that here too induction can be used as one of the alternative methods. This requires of course a proper selection of data. They must be more variegated than those of which we have spoken before. In defining the word 'yellow' we would have to demonstrate, apart from the cases specified above, some other cases for instance ones in which the yellow colour is an attribute of objects differing in shape, and ones in which it is not an attribute of objects having the same shape. Thus the inductive method can be used on a narrower or a broader scale, and for the implementation of more or fewer partial tasks. But its application at least on a narrow scale seems an indispensable condition of solving some of these tasks in a justified way.[10]

At this point many readers would probably be inclined to raise the objection, whether any substantiating value may be ascribed to eliminating induction. This question is being more and more often answered in the negative, all the critical reservations being formulated in a very general way. There is a suggestion that these reservations pertain to those cases also in which induction has definitional tasks to perform. Accordingly we must push our analysis further on and ask in turn whether, and to what extent, the conclusions obtained on the strength of eliminating induction can be relied on in the cases under investigation.

Not to complicate matters too much let us make a certain simplifying assumption. Let us assume, first, that the term being defined is an elementary term, and second, that the respect in which the designata of that term are to be similar, or dissimilar, to the standard objects specified in the definition, is already known to the addressee (as we have seen, there are usually no serious difficulties in finding out that respect). Now it seems that under these assumptions ostensive definition makes it possible to substantiate for sure, or at least practically for sure, some partial solutions of the definition problem, solutions which come more or less close (this depends on the number and selection of the standards referred to) to the complete solution. Let us try to reconstruct, step by step, the various links of reasoning so as to bear out the role of those standards referred to, both positive and negative.

We already know that a definition which refers to only one standard —let us continue to call it the object A—justifies the conclusion that

any object, which in the respect R is not distinguishable from the object A, falls within the extension of the term 'N'. This conclusion follows from the definition statement (we mean here a statement of type (2) on the strength of the very meanings of the terms which appear in that statement. They are so understood that if, it is true that similarity in the respect R to the object A within some limits D is a sufficient condition of any object being N, regardless of what is the lower limit of that similarity, (that is, how small a similarity is sufficient) then the upper limit is always constant: maximum similarity, or indistinguishability from A, is anyhow sufficient. Thus the following general theorem

Πx (if x is indistinguishable from A in the respect R, then x is N)

is a logical consequence of the definition statement, i.e., a fully substantiated theorem. Knowledge of the object A in the respect R thus supplies a reliable criterion of partial applicability of the term 'N'; but on a very small scale: it does not show whether that term is applicable or not to objects which are similar to the object A in the respect R, but are not indistinguishable from it.

Let us see in turn what conclusions can be drawn from knowledge of further positive examples. Let it be assumed that, apart from the object A, some other object, e.g., the object B, has been demonstrated as a positive example. The theorem

Πx (if x is industinguishable from B in the respect R, then x is N)

is, of course, a consequence of the definition statement. Does that consequence broaden our knowledge of the scope of applicability of the term 'N'? Yes or no, depending on whether the object B is, or is not, distinguishable from the object A (this is stated empirically, on the strength of perceptual acquaintance with the objects A and B). If it is not distinguishable, the situation does not change at all. If it is, the definition supplies an additional partial criterion consisting, as we know, in the indistinguishability in the respect R from the object B. Moreover, in the latter case ostensive definition leads to far-reaching conclusions. Namely, it makes it possible to state — also on the strength of the mean -

ing of the terms which appear in it, and in particular on the strength
of the meaning of the phrase 'such as... in the respect R' – that any
object which is similar in the given respect both to A and B not less
than A and B are similar to one another, is N (because if lesser similarity
is sufficient, then greater similarity is sufficient too).

Of course, things look analogous if, apart from the objects A and B,
the ostensive definition points to some other objects, e.g., the objects
C, D and E, as positive examples: addition of a new positive example
broadens the partial criterion if such an example consists in an object
which goes beyond the criterion binding so far; that is, it is an object
which is less similar to previously specified standard objects than the
criterion, binding so far, had determined. This example makes it also
possible to state that satisfaction of the criterion binding so far is not
necessary for the given object to fall within the extension of the term
being defined. Thus, the positive examples play a double role. On the
one hand, they serve the purpose of determining the sufficient conditions satisfied by the ever growing number of objects falling within the
scope of applicability of the term concerned, and on the other hand,
they eliminate those of sufficient conditions which are not necessary conditions.

As to the negative examples, they supply the information that such
and such degrees of similarity to the standard objects are not sufficient
to qualify a certain object as falling within the extension of the term
'N'.[11] These results, like the results referred to above, can be considered
practically certain since they are fully substantiated on the strength of
the definition statement and of the experimental data supplied by ostensive definition.

For any selection of examples, such general theorems can be found
as partial solutions of the given definition problem and fully substantiated
by the whole of the data supplied by ostensive definition. The recipe
for such theorems is simple. The theorems determining the sufficient
conditions of applicability of the term 'N' should specify the narrowest
condition satisfied jointly by all the positive examples. Generalizations
which do not go beyond these limits are certain; broader generalizations
are merely hypothetical. If the examples are well chosen the limits within
which the given term may be used can be determined in a way which
is nearly unequivocal. This explains the considerable uniformity in the

use of elementary terms introduced exclusively by means of the ostensive method.

To realize well the informative role of positive and negative examples it is advisable to form in one's mind a series consisting of objects similar to the object A in the given respect and ordered according to the degree of that similarity in decreasing direction. Now an increase in the number of properly chosen positive examples makes it possible to shift the lower level of similarity from left to right; and it does so in two ways: by extending the scope of sufficient conditions, and by eliminating those sufficient conditions which are not at the same time necessary conditions. This permits a reduction in the sphere of uncertainty. The negative examples eliminate in turn those similarities which are insufficient to permit the application of the term 'N'. They reduce the sphere of uncertainty from right to left. The smaller the sphere of uncertainty, the closer the partial solutions of the definition problem come to a complete solution. It can also be said that the greater is the degree of effectiveness of the given ostensive definition, that is, the degree in which that definition makes it possible to attain the goal of a correct and substantiated solution of the problem in question.

Let us avail ourselves of this opportunity to make some observations about the conditions on which that degree of effectiveness depends. As follows from the above, the lesser the sphere of indeterminateness left by the choice of examples, (or, the lesser is the sphere not covered by the partial criteria of applicability of the given term, determined by the given choice of examples) the greater the effectiveness of the definition. If this is so, then from the point of view of the principle of effectiveness it is, of course, better to demonstrate as positive examples (standards) those of the designata of the given term which are, in the given respect, less similar to each other than those which are more similar to each other. It is also obvious that addition, to already chosen positive examp'es, of new examples satisfying the criterion determined by examples specified before, does not increase the effectiveness of the entire operation and consequently is useless. It is also not economical; since the same result is being obtained with the help of more examples which could be obtained by means of fewer examples.[12] As far as negative examples (standards) are concerned, it is better to specify as such examples objects which are not designata of the term 'N', but which, in the given respect, are more

similar to the standard designata than those which are less similar to them; since the former have greater force of elimination than the latter and consequently more closely restrict the sphere of indeterminateness. If some negative examples have already been specified then addition of new negative examples is useful only if they are not covered by the partial criteria determined by those specified previously.

Thus it is obvious that the selection of examples can be better or worse not only from the point of view of effectiveness, but also from the point of view of economy. Generally speaking, from the latter point of view a choice of examples is better which makes it possible to perform mass eliminations, (that is, to exclude many alternative possibilities at the same time) than a choice which requires the performance of the same task by means of a number of different operations. One further illustration may be useful. Now, elimination 'by installments', at first on the level of more general properties, (let us call them the properties, P_1, P_2, ..., P_n, they are the 'respects', that is, such properties as colour, tune, smell, etc.) and only then on the level of more specific properties, is more economical than, e.g., the elimination exclusively on the level of more specific properties. This is so because by eliminating, for instance, the property P_2, we eliminate all those properties which are subordinate to it, instead of eliminating those subordinate properties one by one. If mass elimination is needed, the economy of the operation is a necessary condition of its practicability.

The question arises whether a selection of examples is possible which would unequivocally determine the complete solution of the definition problem, that is, a solution which would — as we have already mentioned several times — consist in finding such a degree of similarity to the standard object as to be not only the sufficient but also the necessary condition of applicability of the term 'N'. The answer does not require any long investigations. It is enough to realize that similarity comes in grades in a continuous way to understand that it is not possible to determine any sharp limits by means of ostensive definition. Hence the inavoidable vagueness of the terms introduced exclusively by means of ostensive definition; hence, further, the inavoidable vagueness of the terms defined, not ostensively, but by means of terms previously defined in an ostensive way and thus necessarily vague.

5

Having come to know the peculiarities of ostensive definition we can now consider the problem, only mentioned at the very beginning, to which we were to revert at a convenient opportunity. It consists in the question, whether ostnesive definition is a definition *sensu stricto*. It is obvious that the answer to this question will depend on the criteria of the limits within which the use of the term 'definition' is permissible. That is why attention must be paid to that problem.

In the literature of the subject definition is understood in many differing ways. Let us recall the controversies around the problem of whether both so-called real and so-called nominal definitions are definitions, or whether this term may be applied only to real definitions or only to nominal definitions. Let us also recall that the status of definition is controversial in the case of the so-called analytic definition. Finally, let us recall that there is a clear difference of opinion concerning the classification of ostensive definitions.

This is not the place to analyse in detail all those controversial matters.[13] Our task consists rather in pointing to certain general principles which would permit us to decide such questions in this or that doubtful case, and which, let it be noted, would not deviate too much from the actual usage in the sphere of the specialized disciplines. In view of the above, we suggest that the extension of the term 'definition' be restricted in conformity with the following principles.

First, a definition must satisfy certain specified structural conditions which warrant its consistency and translatability, e.g., equational form or a form that can be reduced to equation. If the term being defined is a predicate with one argument — and only such cases are discussed here — then each definition should have the form of: '$\Pi x[N(x) \equiv \ldots \ldots x \ldots]$',[14] where '$N$' is the term being defined and where in place of '$\ldots x \ldots$' there is a sentential function having 'x' as the only free variable. It seems unnecessary to add that a definition of such a type determines the conditions of applicability of the term concerned, those conditions being understood as both sufficient and necessary.

Second, definition must have certain specified properties characterizing it not from the point of view of syntax, but from the point of view

of the pragmatic aspect. Only those statements can function as definitions which are accepted as axioms, statements whose truth is warranted by the very meaning of the terms included in those statements: in adopting a definition we thereby so give a meaning to the term being defined, that the definition is determined as a true sentence. Therefore it can be said that definitions are analytic sentences. Let it be added that such an interpretation of definition is strictly connected with conceiving its tasks as means serving to determine the meanings of expressions.

The consequences of the above assumptions are obvious. It can easily be seen that for certain interpretations of analytic definition on the one hand and real definition on the other (namely, for those interpretations which are most common in the controversies mentioned above) the problem of the definitional character of those so-called definitions is predetermined in the negative. An analytic definition is usually understood as a theorem of specific form, e.g., as a theorem of the type 'the expression E_1 is used in the language L in the same sense as the expression E_2', describing the actual use of an expression in this or that language; real definition is understood as an unequivocal description of an extension distinguished beforehand, in other words, as a 'definition of a thing', as opposed to nominal definition understood as a 'definition of words'. In both cases these so-called definitions are statements which are substantiated empirically, in most cases by the inductive method; that is, they are statements which do not satisfy the second condition adopted above and restricting the extension of the notion of definition exclusively to non-empirical statements. Consequently, these so-called definitions do not fall within the scope of applicability of the term 'definition'. On the other hand, both the so-called real definitions and the so-called analytic definitions (the latter after being given a certain specified form) can be transformed into definitions proper if we decide to use just the terms concerned in such meanings for which those statements would be true sentences not subject to control by experience.

Let it be noted in this connection that the second condition imparts to the notion of definition a relative character. This becomes clear if we take into account that statements having identical form can, or cannot, be definitions; for this depends on the way in which the given statement is adopted — as an axiom or following some sort of substantiation.

Therefore the notion of definition must be related either to some definite discipline, or even to some phase of development of that discipline, or to some person. This holds true, of course, also for the notion of analytic sentence, synthetic sentence, empirical sentence, etc., as well as for all the derivative notions. It holds true, too, for the notion of ostensive definition, as will be demonstrated shortly.

We have not yet answered the question, whether ostensive definition is a definition *sensu stricto*. Our answer to this is neither a simple *yes* nor a simple *no*, becuase it depends on whether we have to do with an ostensive definition 'without gaps', supplying clear information about the criteria of applicability of the term concerned, or with an ostensive definition 'with gaps'. In the former case, ostensive definition satisfies all the conditions required of definition in the proper sense of the word, both as to its external form and as to its pragmatic character, and is just a variation of definition in the narrower sense of the word. As we have seen, this occurs when the same quality in a certain respect is understood as equality in that respect and when that respect is defined without ambiguity. The latter case, however, is much more typical. Strictly speaking, ostensive definition in such cases is not formulated as a sentence but as a propositional function including variables and changing into a sentence only when the various values are substituted for those variables.[15] In such cases ostensive definition is not, of course, a definition *sensu stricto*. But it can be a step leading to the formulation and adoption of such a definition. It happens so when the recipient of the definition succeeds in finding out those properties of positive standards, because of which the said standards fall within the scope of applicability of the term concerned, that is, it happens when he succeeds in finding the proper substitutions for the free variables in the right-hand part of the definition statement. An example of such a situation was adduced on p. 238. This, as we can see, is a different road to definition *sensu stricto* from that along which we proceed when we take the so-called analytic definition or the so-called real definition as our starting point. It is only the last step which is common to them: the axiomatic adoption of the sentence first assumed to be an empirical generalization.

We shall probably not deviate from truth in stating that the so-called inductive method of building definitions *sensu stricto* is often nothing

else than reaching a definition *sensu stricto* through the preparatory
stage of ostensive definition; since the inductive method consists in selecting certain applications of a given term as standard cases and then
deciding in a general way that all those, and only those, objects fall
within the extension of the term concerned which are such as those
standard objects and, finally, in finding out, by means already described,
the characteristics of that extension. If those standard objects are given
to us in perceptual experience, we have to do with an ostensive definition.[16] The peculiarity of such a situation usually consists in the fact
that the author of the given ostensive definition is identical with its
recipient. It must also be added that ostensive definition much more
often leads to so-called conditional definition than to equivalence
definition. For, as was already mentioned several times, the most common are those cases in which we have to do not with a complete but
with a partial solution of the definition problem. Such a partial solution
consists in grasping the criteria which, only with regard to some objects,
make it possible to decide whether such objects fall within the extension
of the term concerned. Should we formulate these results verbally, we
would receive a conditional definition (also called a partial definition or
reduction) composed of two formal implications of the following type:

$$\Pi x \left[\ldots x \ldots \to N(x) \right],$$

$$\Pi x \left[\ldots x \ldots \to \sim N(x) \right].^{17}$$

The first of these two implications determines the sufficient condition
of the applicability of the term 'N', the second does the same for the term
'$\sim N$'. Neither of these two implications is a converse of the other (at
least in the cases discussed here) so that their conjunction may not be
transformed into an equivalence definition. The terms introduced in this
way have, of course, no strictly limited extensions, there is always
a sphere of uncertainty for which the criteria of applicability of a given
term are not determined. Hence the already known vagueness of ostensive terms.

6

Let us now put together all that has been said above on the properties
which contribute to the specific distinctiveness of ostensive definition.

We have drawn the readers' attention to the three essential questions:

the external form of ostensive definition, the kind of terms which appear in such definition, and the pragmatic character of such definition.

As far as the external form is concerned, we have provisionally assumed that in its final structure ostensive definition is an expression of the type '$\Pi x[N(x)=...x...]$',[18] where in place of '$...x...$' there is a propositional function which includes individual names, and usually also some free variable other than 'x'. We have also concluded that the constant terms which appear in that expression — that is, the names to the right of the equivalence symbol — must be individual names of objects known to the recipient of the definition in the respect concerned and from his own perceptual experience. We have also assumed that if an ostensive definition is a sentence, then — like all definitions — it is an analytic sentence (let us add that the definitional assumption on which ostensive definition is based, always is an analytic sentence).

Let us begin with the first question. The proposed schema of ostensive definition was suggested to us, as we remember, by the interpretation of those statements to which we resort in practice when explaining a term by means of ostensive method. Generally these statements are sentences of the type 'This is N' or 'This is not N' where in place of 'N' the term being defined appears. It is of course a controversial point whether it would not be more natural to consider ostensive definition as a sort of definition by postulates, consisting of sentences of the above type as individual definitional postulates. Should we decide to formulate the problem in that way, our analysis would not be essentially affected since in that case, too, it would be correct to conceive ostensive definition as a means of only partially determining the meanings of expressions, and consequently as a means of defining which leaves for the recipient certain tasks still to perform. Alternatively it would be correct to think that our task consists in solving the problem: for what Y is it the case that $\Pi x[N(x) \equiv Y(A) \cdot Y(B)... \sim Y(K) \cdot \sim Y(L)]$? — or, in other words, for which R and D is it the case that $\Pi x[N(x) \equiv x$ is in the respect R and in the degree D such as $A, B...$ and is not such as $K, L...]$? — since this task would consist, as did the former one, in finding those properties which are attributes of $A, B...$, but not of $K, L...$, and which determine the range of applicability of the term 'N'. Yet we prefer the previous formulations (2), in spite of their being more complicated, because we think that they more adequately cor-

respond to that intention which marks the ostensive defining of general terms; namely, the intention of using a given term not only in reference to those objects which are pointed to in the definition as standard designata, but in reference to the entire class of objects which, in a certain definite respect and within certain definite limits, are the same as those standard designata.

Two other facts also are in favour of such a formulation: on the one hand, it stresses the incompleteness of the data included in ostensive definition, and on the other, it bears out the relationship between ostensive definition and definition proper, the relationship which is already known to us. Finally, it seems that the fact of being of the same quality as the standard object is precisely that criterion on which we rely in practice if we are unable to decide whether a certain object falls, or does not fall, within the extension of an ostensive term.

As far as the second question is concerned, it is worth while realizing that, in consequence of that assumption, ostensive definition must be treated as a relative notion since the analytic character of a sentence is, as we know, a relative property. Strictly speaking, we should say that a certain statement is an ostensive definition with respect to a certain person, e.g. the author of the statement or its addressee. Thus far we have left out such an additional provision, but we tacitly assumed that the statement must be referred to the author of the definition concerned. And, when using hereafter the term 'ostensive definition' without such addition, we shall mean its reference with respect to the author. But it must be noted that in general a sentence which is an ostensive definition with respect to its author is also so with respect to the addressee.

In connection with the third question we must push relativization still further and speak of ostensive definition not only with respect to its author, but also with respect to its addressee (or, if already previously it was related to the addressee and not to the author, it must be related to the addressee twice, that is, in two different respects). This results from the fact, as can easily be guessed, that the same standard object can be perceived by somebody and not be perceived by somebody else; consequently, from the point of view of one of these two persons a statement may be an ostensive definition, but from the point of view of the other it may not be so. Should we like, therefore, to make our formulation still more exact, we would have to accompany the term 'ostensive

definition' with the following double addition: 'with respect to this author and from the point of view of that addressee'. Since, however, it is always the same addition which is implied, we leave it out as being tacitly assumed.

It is obvious that as a result of the above condition only perceptual terms, that is terms relating to observable objects with respect to their perceivable properties, can be defined by ostensive method. Does this cover only extraspective terms, as all our examples might have suggested, or introspective terms as well? There is no foregone conclusion as to that point. It seems, however, that if we assume that all elementary empirical terms are introduced by ostensive method (and this assumption is practically universal) and if we at the same time assume that the meanings of elementary introspective terms cannot be reduced to extraspective terms, then we must also assume that introspective terms can be defined by ostensive method. Robinson, e.g. seems to understand the problem just in this way since he gives [19] the following example of ostensive definition: somebody explains to somebody else the meaning of the word 'pleasure' by giving him a rose to smell and a chocolate to eat, and saying each time: 'Now you experience pleasure'.

So much for the conditions binding ostensive definition and the problems connected with them.

We are coming towards the end of our investigations. They were concentrated chiefly on problems connected with the specific, distinct, characteristics of ostensive definition, on the mechanism of its functioning, and on the properties of the terms introduced by ostensive method. These problems have so far been very little studied, probably because ostensive definition is usually considered to be a makeshift of little use for scientific purposes. In fact, ostensive definition has little application in science. But it must be borne in mind that, as already mentioned, all non-ostensive empirical terms (we mean here predicative terms) are finally reduced, through equivalence definition or conditional definition, to some ostensive terms, usually borrowed from colloquial language, and that it is to this fact that they owe their empirical character. We may say safely that a language which would include no terms introduced by ostensive method would be useless for the implementation of those tasks which face the disciplines that have to investigate reality, because it would include no terms by means of which that reality could

be described. If this is taken into consideration, it is difficult not to realize that in every descriptive language ostensive definition, directly or indirectly, plays an essential role. It is also difficult not to realize that problems connected with ostensive definition deserve more time and place than they have received so far.

REFERENCES

* First published in *Fragmenty Filozoficzne* **II**, Warszawa 1959. Reprinted from *Philosophy and Science* **27 (I)** (1960).

[1] B. Russell, *Human Knowledge, Its Scope and Limits*, London 1948, p. 78.

[2] If the other view were assumed concerning individual names, the definition sentence would have to run '$N \equiv A$', e.g., 'John Smith is identical with this man' (and not, as before, 'This man is John Smith'). But it must be borne in mind that the problem of ostensive definability of individual names is a controversial issue. Cf., for instance J. Xenakis, 'Function and Meaning of Names', *Theoria* **XXII** (1956).

[3] Of these two schemata, (2) and (3), type (2) is considered to be the basic one. Let it also be noted that the symbol 'sim' is an abbreviation of 'similar'.

[4] This statement might be called a definitional assumption since ostensive definition performs its function, i.e., establishes a relation between the term defined and the scope of its applicability, on the condition only that statement (4) is true. Usually, ostensive definition is resorted to only by those who adopt the definitional assumption on which a given definition is based.

[5] Cf. K. Ajdukiewicz, *Zarys logiki* (*An Outline of Logic*), Warszawa 1955, p. 192 ff. It must be noted that this book discusses a solution of the problem in general, without paying any special attention to problems of defining.

[6] A complete solution of the definition problem would consist in finding a condition that would be sufficient and at the same time necessary. A complete and a partial solution of the definition problem could be distinguished from the point of view of the first interpretation, too. The characteristic given on p. 238 refers to a complete solution. A partial solution would consist in finding correct substitutions for variables not in the definition statement itself, but in some of its consequences, namely those which differ from the definition statement only in this, that the symbol of equivalence is replaced by the symbol of implication in either direction. This means that it would consist in finding sentences determining partial criteria.

[7] This condition is treated, as the usage is, not only as a condition of efficacy of the defining operation, but also as a condition without which ostensive definition would not be 'what it is', i.e., would not be an ostensive definition at all.

[8] It is perhaps worth mentioning that the lesser the scope, so to speak, of the operation recommended by the given instruction, the greater the eliminating force of that step. For instance, the instruction 'Smell it' eliminates all sense qualities except those which belong to the scope of the olfactory, since such an operation

usually serves only the purpose of perceiving the various smells. On the other hand, the instruction 'Touch it' has a much lesser eliminating value, since by touch we perceive temperature of objects, their consistence, shape, type of surface, etc. Further elimination is in such cases usually carried out by an inductive method, to be discussed below.

⁹ W. E. Johnson, *Logic*, Part I, Cambridge 1921, p. 4.

¹⁰ The notion of induction requires here comments analogous to those given before in connection with the notion of a solution of the definition problem (pp. 238–239). To avoid misunderstandings we must explain that induction must here be interpreted broadly enough so as to cover those cases, too, in which reasoning is based not on verbal premises but on preception, and the conclusion need not necessarily be a general sentence. As an example of induction interpreted so broadly we could give a mental process which leads the recipient of the definition to a correct solution of the definition problem even if neither clearly formulated premises nor a clearly formulated conclusion come in question. The very fact of a correct use of the term defined seems to prove that we have to do with some generalization: the recipient of the definition behaves as if he based himself on a general statement and applied it to particular cases, as if he understood clearly the criteria of applicability of the term in question. In view of that similarity to induction in the narrower sense it seems legitimate to extend the notion of induction so as to cover the cases mentioned here; the more so since such an interpretation of induction is in conformity with the common practice of using that term. It can easily be seen that the terms 'reasoning', 'premises' and 'conclusion' have been used here in sense broader that those to be found usually in school textbooks of logic. This broadening of meaning follows the same direction as in the case of induction and solution of problems.

We, of course, fully realize the vagueness of notions characterized as they have here been. But it is difficult to abstain, when discussing the problems under consideration, from referring to the pragmatic aspect of language, and, as is known, that pragmatic aspect defies attempts to be formulated clearly and with precision. We prefer, however, rather to risk being blamed for vagueness than to pass over in silence those matters which seem essential.

¹¹ Since the statement that a certain degree of similarity is not sufficient for an object to be N results from the statement that the given degree of similarity is sufficient for an object not to be N, that is, to be non-N, therefore the reasoning here is analogous to the former case. The assumption on which that reasoning is based, and which is a consequence of the definition statement on the strength of meaning of the terms appearing in that statement, can briefly be formulated as follows: if a greater similarity is not sufficient, a lesser one is also not sufficient.

¹² The general concepts of the efficacy and economical character of action are taken from *Traktat o dobrej robocie* (Treatise on Good Work) by T. Kotarbiński (Łódź 1957).

¹³ These problems have been dealt with in our paper 'Definicja' ('Definition'), *Studia Logica* **II** (1955).

[14] Or a similar form, corresponding to a formulation in a metalanguage, in which definition directly refers to the term being defined, names it, and ascribes it a definite semantic function. These differences in formulating definitions do not seem essential at all.

[15] If, despite this fact, it is difficult to resist the impression that by formulating an ostensive definition we after all state something by means of such a definition, then the reason must be sought in our taking into consideration not only the definition statement, but also the definitional assumption on which it is based and which, therefore, is associated with it. Let us recall that the definitional assumption differs from the definition statement only in this, that the variables which are free in the definition statement are in the definitional assumption bound by the existential quantifier (see p. 237). E.g., if the definition statement is of the type '$\Pi x [N(x) \equiv x \, \text{sim}_{R,D} A]$' then the statement which it assumes is a sentence of the type '$\Sigma R \Sigma D \Pi x [N(x) \equiv \text{sim}_{R,D} A]$'.

[16] If this condition is not satisfied, we can speak, following Johnson's example, of denotative definition. This would be broader than ostensive definition, regardless of whether they refer to standards that are known to the recipient from his experience or not.

[17] The symbol '...x...' can, of course, stand in each of these implications for a different propositional function including 'x' as the only free variable.

The notion of conditional definition was introduced by R. Carnap in' Testability and Meaning', *Philosophy of Science* **III–IV** (1936–7).

[18] This schema is a generalization of schemata (2) and (3).

[19] R. Robinson, *Definition*, Oxford 1950.

JANINA KOTARBIŃSKA

THE CONTROVERSY: DEDUCTIVISM
VERSUS INDUCTIVISM*

1

Inductivism is the view, universally known and almost universally adopted, which recognizes the inductive method to be the basic method in the empirical sciences. Deductivism is upheld by Professor K. Popper, whose standpoint originated in his criticism of inductivism.[1] His criticism is quite revolutionary, since it calls for the elimination of induction from the so-called inductive sciences and for its replacement by the deductive method as the only one which does not sin against the requirements of logical correctness and which makes it possible, in general, to avoid the difficulties encountered by induction. It is Professor Popper's conviction that the deductivist theory has succeeded in working out a detailed program of such a method. As indicated by the title, the purpose of this paper is to examine the issues emerging from the controversy between deductivism and inductivism. According to K. Popper, the critical analysis of the two rival theories, which was carried out in *The Logic of Scientific Discovery*,[2] unequivocally resolved the problem in favour of deductivism. Namely, it was to demonstrate that deductivism enables to avoid those difficulties, which are present in the other theory, and that it does not lead at the same time to new difficulties. I should like to analyse critically this particular result. The matter is worth considering if only, because Popper sees the main, and may be the only, reason for the existence of deductivism in this putative advantage.

2

Without entering into details, it can be said that both deductivism and inductivism endeavor to solve the same major problems. In both cases the point is to indicate a method which would make it possible to justify general statements pertaining to an indefinite number of cases — in other

words, statements of universal generality – on the strength of individual statements of a certain distinguished type (basic statements, as Popper calls them) and at the same time to establish for the justified statements both an empirical character and a sufficient cognitive value. Discussion is focused, above all, on the issues pertaining to the specific distinct properties of empirical statements, i.e., the criterion of empiricism, and to the conditions on which the cognitive value of such statements depends, i.e., the criterion of their acceptability. The first issue has been widely examined in the literature of the subject, so I confine myself to discussing more significant divergences, devoting attention mainly to the second issue.

When it comes to empiricism, the two theories have a common starting point, which consists in the assumption that certain statements, namely basic statements, are empirical in nature regardless of their relation to other statements, and that precisely those (basic) statements determine the empirical character of all other statements. The empirical statements include, on the one hand, the basic statements, and on the other, all and only such statements which bear a certain definite logical relation to the basic statements. But both the basic statements and the logical relations involved are interpreted in different ways.

In the case of inductivism, basic statements are usually characterized as individual statements of the type 'A is B' which describe observable phenomena. In the case of deductivism they are characterized as individual existential statements, that is, statements of the type 'There is such and such an object in such and such a place and at such and such time', which describe phenomena that are observable intersubjectively.[3]

More important is another difference, which is revealed in the different characterizations of the empirical non-basic statements. Although the inductivists see, according to Popper, the empirical character of the non-basic statements in the fact that they are subject both to verification and to falsification with the aid of finite and non-contradictory sets of basic statements, they have to, consequently, deny empirical character to all statements of universal generality, i.e. among others, to all laws of nature, since such statements *a priori* exclude the possibility of a full empirical justification. In order to avoid this consequence, which obliterates the difference between the empirical sciences on the one hand, and metaphysics on the other, Professor Popper drops the conditions of veri-

fiability and makes the empirical character of statements depend solely on whether they admit the possibility of being falsified by means of basic statements.[4]

Indeed, according to this view, the problem of the laws of nature no longer exists. There is, however, another problem, that has been pointed out by Popper himself. Namely, similarly to the previous standpoint, existential universal statements, i.e. those which do not contain a covert or overt reference to some defined place or time, have to be considered non-empirical statements — for the negations of such sentences do not result from any finite and non-contradictory set of basic statements. Thus, among the metaphysical statements, i.e. among statements which, according to Popper, have no right to appear in the field of science, there will be statements such as 'There are people', 'There are black ravens', etc., despite the fact that they are not only fully justifiable, but even fully justified on the basis of experiment.[5] May be this consequence is not as bad as that excluding from science all the theorems of universal generality, but it demonstrates a considerable deficiency on the part of deductivism, especially as universal existential theses play a considerable role in the empirical sciences.

On the other hand, one may doubt if the difficulty connected with the verificational criterion demonstrates a deficiency of the standpoint of inductivism. For this criterion does not seem to be characteristic of inductivism as such, but only of one of its neopositivistic varieties with very few proponents. Inductivism usually contents itself with a less strict criterion, consisting in full testability — either positive verifiability), or negative (falsifiability). As is known, this widened criterion raises the objection that it is too narrow with some respects, and too wide with others. Too narrow — for it does not include universal statements containing both the universal and the existential quantifier, too wide — for it includes, similarly to the criteria discussed previously, certain types of complex statements which contain as constituents meta-physical statements.[6] However, the range of statements determined by this widened criterion is closer to the range that was intended than that distinguished with the help of Popper's criterions. And, one should add, according to Popper's view the comparison of the standpoints of deductivism and inductivism in this respect was to provide one of the most important arguments in favour of deductivism.[7]

3

The problem of the criterion of acceptability of empirical statements is usually examined separately with reference to the basic statements and the remaining empirical statements.

If the only requirement with respect to the basic statements is that they must have a certain definite outward form and describe observable or intersubjectively observable phenomena—and only such a requirement has been taken into consideration in the case of the interpretations discussed so far—then one obtains an infinite number of statements satisfying these conditions, including contradictory statements. Among those statements which Prof. Popper sometimes calls 'possible basic statements' (this terminology will be used hereafter) one only has to choose the basic statements in the proper sense of the word; that is, choose those which in the empirical sciences would decide the acceptance or rejection of all the remaining statements without being themselves justified by any other statements. What would be the criterion of choice of such basic statements in the proper sense of the word?

The inductivists usually ascribe this role to perceptual statements as statements which are justified directly by perceptual experiences, and as such require no further justification. To this Professor Popper raises the objection of psychologism which, in his opinion, is not admissible in the field of methodology. He compares it with his own view, which is a different one and which he himself calls conventionalism with reference to the basic statements. According to his standpoint the basic statements, like all statements in general, may be justified exclusively by means of other statements. And since such a process of justification cannot be continued idenfinitely, one has to stop sooner or later at some statements as the basic statements (in the proper sense of the term), in spite of the fact that they in principle require a justification and may be falsified as a result of further investigations. Now the selection of such basic statements is always arbitrary and in that sense conventional, since we just accept them by virtue of our decision, as there are no logical reasons which would force us to stop at these statements and not at others.

The intentions behind conventionalism conceived in this way are not quite clear. It is opposed to psychologism, but in what interpretation of that term? Does the psychologism stigmatized here consist in the fact that

the choice of the proper basic statements depends on their perceptual nature, in the requirement that such basic statements should state phenomena which not only are observable but have actually been observed by someone? Or does it consist only in the fact that the perceptual nature of the proper basic statements is considered to be the criterion of their complete validity, the reliable guarantee of their being true statements? The critical argument used by Prof. Popper is aimed almost entirely at the second version of psychologism, which is much more far-reaching and, let it be added here, not accepted by all inductivists. But his own suggestions and conclusions, reached as a conclusion of his considerations, seem to testify to much more radical tendencies, which oppose any demands to make the choice of the basic statements depend on psychological criteria. And that probably is Prof. Popper's true intention.

If so, then the methodological program of conventionalism, and consequently of deductivism, of which conventionalism is one of the component parts, leads to extremely paradoxical results: it is programmatically not required when laying the foundations of the empirical sciences to refer to actual observational experience, a freedom of choice being left in that respect. Thus it is admitted that in principle the empirical sciences may become completely independent of the results of observations and experiments, the methodological role of which becomes incomprehensible in conventionalism. The connection between the empirical sciences and experience is confined to the fact that the basic statements of those sciences speak of observable phenomena, and as such are couched in empirical terms. That condition excludes metaphysical fantasy, but does not suffice to exclude fantasy expressed in an empirical language. This is probably no less grave a sin than that with which Prof. Popper reproached inductivism (that criticism being a misdirected one) and which is said to consist in obliterating the demarcation line between science and metaphysics. Here the demarcation line is obliterated between genuine empirical knowledge and something else which is not knowledge. Responsibility for that, in the opinion of the present author, rests with the too radical, anti-psychological tendencies revealed by Popper in his treatment of methodology as a branch of logic in the narrower sense of the term, that is, as a discipline concerned with statements solely from the point of view of the logical relations between them.[8]

4

We now come to the issue which is the focal point in the controversy
between deductivism and inductivism, namely, the criterion of acceptability of the non-basic empirical statements, in particualr statements of
universal generality, i.e., nomological statements, which are of exceptional importance in science. Following Prof. Popper, the present author
will confine herself to a discussion of the controversy with reference to
the nomological statements interpreted in this way.

Both parties agree that the acceptability of such statements is decided
by the method of justifying them. But whereas inductivism recommends
accepting the nomological statements which are sufficiently justified by
inductive inference, deductivism rejects induction as a method devoid
of any justifying value and generally useless from the point of view of
the needs of science, and at the same time points to another method, the
deductive one (also called the method of criticism of hypotheses or the
trial-and-error method), which is claimed to have none of the shortcomings that are said to disqualify the inductive method.

Professor Popper's arguments require a more detailed examination.
Let us begin with the justifying value. According to him induction cannot
be used as a justifying method mainly for two reasons.

First, because inductive inference goes in the direction which is
inconsistent with the direction of justifying inference. Here, observations
precede hypotheses, whereas in a justifying inference one starts with a
hypothesis being justified and only then one passes, as a result of inference, to observational statements, which are its consequences.

Second, because induction is a fallible inference from the logical
point of view: the conclusion of an induction is not deducible from its
premises and may prove false even if all the premises are true.

The method recommended by the deductivist program is in Prof.
Popper's opinion not subject to this criticism. The starting point here
always is some hypothesis, and the procedure consists in a critical
examination of the hypothesis being justified, that is, in drawing from
it by deduction logical consequences in the form of basic statements and
in investigating whether one of these consequences is not a false statement. If a false consequence is found, the given hypothesis is rejected
as false, again as a result of deductive inference. But if all the conse-

quences prove true and, moreover, satisfy certain additional conditions, the given hypothesis is accepted provisionally until a false consequence is found in the course of further investigations. It is claimed that the deductive character of the types of reasoning used in this method and the merely hypothetical acceptance of justified statements make the method correct from the point of view of logical validity. This, according to Prof. Popper, puts an end to the so-called problem of induction with which the theory of induction has been grappling at least since Hume without any prospects for its solution.[9]

Let us stop here for a moment. As for the first objection—a doubt must arise as to its conformity with the main tendency of Popper's which is expressed in the view that the validity of empirical theorems depends exclusively on their logical relation to the already accepted empirical theorems, and in particular, to basic sentences and on the conditions that these basic theorems satisfy. Is not there a misunderstanding here? Is not the idea rather that a reasoning *with the aim of justifying* — i.e. carried out in order to justify some theorem — always has as its starting point the theorem being justified and that, in this type of reasoning, the conclusion always precedes the premises with the help of which it is justified. This presumption is suggested by, e.g. the text on p. 31 of *The Logic of Scientific Discovery*. If this presumption were valid, the divergence between inductivism and deductivism in the matter under discussion would only be seeming.

And what about the validity of the second objection? The answer depends, above all, on the type of induction it is raised against and on the criteria of evaluation of its validity.

Now when referring to induction, Popper, as he explains it at the outset of his analysis,[10] has in mind its traditional interpretation, which he believes is extremely common up to this day among those who are in favor of induction. In the case of such an interpretation, "inference is called inductive if it consists in a transition from individual statements (sometimes also called singular statements) such as reports on observations or experiments to universal statements such as hypotheses or theories." Inference is here identified with 'transition' from statements to some other statements, and the peculiarity of inductive inference is seen in the type of the statements from which and to which that 'transition' takes place.

The vagueness of this formulations is well known. 'Transition' from some statements to some other statements, for instance from the statements $A_1, ..., A_n$ to the statement B, can be interpreted in at least three ways: either (1) we derive the statement B from the statements $A_1, ..., A_n$ as their logical consequence; or (2) we derive the statement B from the statements $A_1, ..., A_n$, already previously accepted, as their logical consequence, and we accept, besides, in the final phase of the whole procedure the statement B; or else (3) we accept the statement B simply because the statements $A_1, ..., A_n$ had been accepted before, 'acceptance' being interpreted either in the stronger sense of the word, marked by complete certainty about the truth of the statement in question or in the weaker sense, which covers hypothetical acceptance, without the certainty that the accepted statement is true.

Which of these interpretations is taken into account by Popper when he argues against induction? No comments to that effect are to be found in the text. His objections to induction seem to indicate that he usually understands 'transition' in the second sense above; this is the narrowest interpretation in which the term 'inference' does not differ in meaning from the term 'deductive inference' in its usual sense. In such an interpretation induction is in fact a type of inference that is incorrect from the logical point of view, and consequently does not contribute to the justification of statements obtained as a result of its application.

But this is not the only interpretation of the term 'induction' and not even the most common one. Among logicians another interpretation, corresponding to that described under point (3) above, is at least equally frequent. In such an interpretation of the term 'inference', induction is understood as a mental process consisting of hypothetically accepting a nomological statement on the basis of individual statements that are its particular cases.[11] It seems that in some parts of his analysis Popper means precisely such an interpretation of induction; at least it is only in the case of such an interpretation that some of the objections he raises can be explained.

The question arises whether, in the light of Prof. Popper's opinion, induction understood in such a way would have to be denied all justifying value. The answer to this question depends on whether the entailment of the conclusion by the premises is treated by Popper as an obligatory condition in every justifiable inference, or as a criterion of

justifiability for only those inferences in which the conclusion is accepted decidedly, i.e. those which claim to be infallible inferences. It is obvious that in the latter case induction understood in the way mentioned last is not subject to this objection, since the conclusion is accepted there only hypothetically. It is, on the other hand, subject to this objection, if we adopt the first assumption. Only in that case the method suggested by Popper does not present itself better in this respect. True, it is a deductive method – but not in the sense, in which deductive character might be considered to guarantee logical validity.

Attention has been drawn many times to the ambiguity of the term 'deduction'. It seems that the most frequent interpretations of the term are: (A) the traditional interpretation of deduction as 'transition from the general to the particular', where 'transition' may be interpreted either as in (1), or in (2), or in (3) in the specification given above (cf. p. 268); (B) a more general and modern interpretation – as 'transition from *ratio* to consequence' with analogous interpretational variants (we shall note them with the help of digits '1', '2', '3'). It is easy to notice that a deductive character of an inference guarantees its infallibility only if it is understood in the sense A.2, B.2, or B.3, since only those interpretations guarantee that the conclusion will be the consequence of premises (a reminder: we call a conclusion a statement accepted as a result of reasoning). Now, the method of criticizing hypotheses is a deductive method only in the senses A.1 and B.1, which distinguish deduction on the grounds of the chronological sequence of constituent steps of reasoning, and not on the basis of the logical relation that holds between the conclusion and premises.

Moreover, in case of the interpretation of induction under discussion the method of criticizing hypotheses is, as a matter of fact, nothing more but a particular variety of inductive inference, which satisfies certain special conditions. The controversy between deductivism and inductivism becomes then, to some respect at least, a purely verbal controversy arising from linguistic misunderstandings.

5

According to Popper, one of the most serious shortcomings of induction is the fact that it does not provide criteria that would make it possible to distinguish between hypotheses which have some scientific value and

those which are pseudoscientific and devoid of all cognitive value. The rules of inductive justification admit of the acceptance of a hypothesis as sufficiently justified simply if a certain number of facts that are in conformity with that hypotheses have been found and if no facts to the contrary have been encountered. These requirements are so weak that they can be satisfied with respect to almost every hypothesis. On the other hand, the deductivist rules are much more exacting: before a hypothesis is accepted, it must be examined critically, facts must be sought which would testify against it rather than in its favour, and arguments must be looked for which would speak against the hypothesis in question rather than against rival hypotheses. A hypothesis which does not stand up to such an examination does not deserve even a provisional acceptance in science.

At first, such an interpretation must seem astonishing. It would appear that the operation consisting in seeking confirmation of a hypothesis does not differ from the operation consisting in submitting that hypothesis to falsifying tests, since both operations are in fact reducible to drawing consequences from the hypothesis being verified and to examining whether those consequences are true statements, i.e., conforming with reality, or false statements.

But Professor Popper probably means not the mental procedure as such, but the conditions which are to be satisfied by the consequences obtained if they are to be given a justifying value. Now in Prof. Popper's opinion inductivism rests satisfied with the condition of truth. On the contrary, deductivism assumes that true consequences justify a hypothesis only if there exist 'proper (severe) tests', i.e., if the probability of the truth of these consequences is comparatively small in the light of existing knowledge, possible with a provisional assumption of falsity of the hypothesis being verified. Moreover, it assumes that a hypothesis may be considered sufficiently justified only, if it has been subject to honest attempts at falsification, taking into consideration all possible in this phase of science development proper tests and all 'serious' rival hypotheses.

In connection with such a conception of the matter, there may arise doubts as to whether it conforms with the state of things. The principle which ascribes to the facts that agree with the hypothesis being verified the greater justifying value the lesser the probability of the occurrence

of such a fact in the light of existing knowledge is often met with, and has been so before, also in the theory of induction, so that there is no need to involve deductivism in these matters.

Moreover, one might ask if the above presented postulates of deductivism are not actually present in the programme of eliminative induction, since the concern for the proper choice of justifying facts is particularly characteristic of eliminative induction and, since the rules governing this choice recommend taking into consideration both the positive and the negative cases. Popper foresees this objection. In this answer he emphasizes the distinctness of the basic guide lines: whereas, according to the laws of deductivism, one should put maximum energy to find facts such that would falsify the verified hypothesis, possibly confirming one of the rival hypotheses – according to the laws of eliminative induction one should seek for facts that would confirm the verified hypothesis and falsify its rival hypotheses.

Without questioning the validity of the above characteristics of eliminative induction on the grounds of the ordinary interpretation of Mill's canons and schemes – i.e. on the grounds of the interpretation, where the canon of agreement is used to justify theorems indicating the causes of the investigated phenomena understood as necessary conditions, and the canon of difference – to justify theorems indicating the causes understood as sufficient conditions, and the reasoning carried out according to these canons is treated as a deductive reasoning (in the sense distinguished on p. 269 under B.1) – it may be worth turning our attention to the possibility of a different interpretation which, it seems, makes eliminative induction similar to the method advocated by Popper. The different character of this interpretation can be seen especially in the suggestions concerning the interpretation of the term 'cause' in Mill's formulations: the cause in the case of the canon of agreement should be understood as a sufficient condition and in the case of the canon of difference – as a necessary condition. We shall demonstrate the course of reasoning on the example of the canon of agreement.

Let us suppose that we want to justify a hypothesis H stating that $(x)\,[A(x)\rightarrow B(x)]$, i.e., that every phenomenon of the type $A(x)$ is a sufficient condition of some phenomenon of the type $B(x)$ and that the rival hypotheses which deserve consideration are hypotheses of the type '$(x)\,[C(x)\rightarrow B(x)]$', '$(x)\,[D(x)\rightarrow B(x)]$', etc. In order to justify the

hypothesis H we should try to falsify it, i.e., in other words, we should look for facts that do not agree with it. The method in this case consists in deriving from this hypothesis of consequences such that their probability would be small under the provisional assumption of the truth of one of the 'serious' rival hypotheses and the falsehood of the hypothesis under investigation. If, e.g., it is true that the sufficient conditions of the phenomena of the type $B(x)$ are phenomena of the type $C(x)$ and not those of the type $A(x)$, we must derive from the hypothesis H consequences of the type: '$A(a) \cdot \bar{C}(a) \rightarrow B(a)$', '$A(a) \cdot \bar{D}(a) \rightarrow B(a)$' [12] etc. analogously, whose probability is small in the case of the given assumptions. If the consequences matched in this way are found to be true, they will confirm the tested hypothesis. In the opposite case the hypothesis will have to be rejected. Let us add that the application of the so-called joint method of agreement and difference makes also possible — as a result of analogous reasonings, taking as the starting point one after another the various rival hypotheses — the elimination of all the hypotheses taken into consideration except the tested one. For instance, the hypothesis '$(x)[C(x) \rightarrow B(x)]$' is falsified by the observation $C(a) \cdot \bar{A}(a) \cdot \bar{B}(a)$, for it is contradictory to the consequence of this hypothesis '$C(a) \cdot \bar{A}(a) \rightarrow B(a)$'.

In this approach the canons of eliminative induction simply indicate the method of finding the 'critical consequences', i.e. those that are most likely to falsify the tested hypothesis. The reasoning is inductive in character (in the sense given under A.2 on p. 269) — and not deductive, as it is the case with their ordinary interpretation — and hence it is fallible reasoning and, due to its nature, leading only to hypothetical conclusions. At the same time its course, it seems, conforms to the requirements of the deductivist rules.[13]

6

The concept of falsification, which in the deductivist system has played so important a role in the analysis described above (the concept of empirical statement, the theory of criticism of hypotheses), comes to the forefront again in Prof. Popper's analysis of the criterion of differentiation between rival hypotheses from the point of view of their cognitive value; this is the criterion which should guide the choice of that hypothesis which under given cricumstances is the most suitable for provisional

acceptance. Now according to Professor Popper, the greater or lesser cognitive value of the rival hypotheses depends on the degree of their falsifiability: that hypothesis which is easier to falsify is more valuable, and that hypothesis which is more difficult to falsify is less valuable. That criterion is supposed to be connected with the deductivists' concept of justification. Since the degree of justification of a hypothesis depends on the severity of the tests used, those hypotheses which admit of more severe tests are better. In Popper's opinion — which, let it be noted, is far from obvious — these are precisely the hypotheses which are easier to falsify and which are thus more exposed to disproval.

The question arises, of course, what does it mean to say that one hypothesis is easier to falsify than another? Let us consider the matter on some simple, specially chosen example. Suppose we have two rival generalizations. Namely, we do not know, after having observed a number of cases of malaria following mosquito stings whether all mosquitoes spread malaria or whether it is only the anopheles mosquito. Let us assume further that neither of these generalizations has been falsified by further observations. Which one is better: the first more general one, or the second, less general. According to Professor Popper's criterion the first generalization is more valuable, for it may be falsified easier: it is a *ratio*, not a consequence of the narrower generalization and hence it is false if, but not only if the narrower generalization is false. There are, somehow, more possibilities of falsifying a *ratio* than a consequence.

The example presented above, which, naturally, is only a particular case of a situation in which one may — according to Popper — compare some hypotheses from the point of view of their falsifiability is especially interesting also for some other reasons. Namely, it seems that one of the most relevant differences between the standpoint of Popper and the theory opposed by him reveals itself here, that it is only here that truly essential differences can be seen: while a deductivist chooses from among two rival hypotheses the more general one, as more liable to falsification, the inductivist — as is stressed by Popper himself — makes just the opposite choice. He adopts the hypothesis which is less risky and the transcending observational data, which hence is the least liable to falsification, for he is striving for knowledge as well justified as possible and, at the same time, identifies the degree of justification

with the degree of certainty. According to Popper a similar difference of opinion would demonstrate itself if choice were to be made
between two generalizations which have the same subject and predicates
differing in extension (e.g., 'Each anopheles mosquito spreads malaria'
and 'Each anopheles mosquito spreads infectious disease'): a deductivist
would consider better the generalization with the predicate narrower in
extension, i.e. the more determinate generalization, for it is the *ratio*
of the other and thus is more liable to falsification; the inductivist,
on the other hand, would choose the more cautious generalization, less
precise but, to his mind, better justified. Thus, according to Popper,
out of four rival generalizations differing with respect to their degree
of generality or degree of determinacy, the most valuable is the one
that is most general and, at the same time, most determinate, while on
the grounds of the inductive theory the most proper one is the least
general and the least determinate generalization.

In the eyes of Popper, that difference of opinion is an extremely important argument in favor of his view. He emphasizes that it is a fact
that the natural sciences strive for the most general and most determinate
laws and that as science progresses more such laws are introduced. For
instance, in mechanics Newton's law of gravitation long ago replaced
Kepler's laws, which have turned out to be its special cases. And if more
general and more precise laws are of a lesser value, as it is usually
assumed in the theory of induction, then why do we always strive to
formulate them? In the opinion of Professor Popper, the theory of
inductive justification leads just to such consequences, which are in
glaring disagreement with the actual state of science, whereas his own
idea adequately explains the trend in the natural sciences: in science we
strive for increasingly general and increasingly precise laws because
such laws can more easily be falsified, and that is precisely why they are
more valuable.

Is Professor Popper right in making his claim? What he says is
rather suggestive, but it is difficult to resist the impression that he simplifies the issue too much and consequently triumphs easily over his
adversary. For instance, there are two generalizations: 'All planets
attract one another with force inversely proportional to the square of
the distance between them' and 'All bodies attract one another with
force inversely proportional to the square of the distance between them';

the scientist tests those two generalizations exclusively with respect to planets (both will, of course, be confirmed ,since Mars, Venus, etc., are both planets and bodies) and then faces the choice between the less general and the more general. Should the scientist in such circumstances always choose the more general one, Popper's objection would be justified, since in view of the inductivist assumptions a more general law would not be as well justified with respect to given observations as a less general law with respect to *the same* observations. But it seems that in fact the situation is somewhat different. In the natural sciences, a more general law is accepted only when observations have been made which exceed the scope of applicability of the less general law and can be explained by the more general law, e.g., the universal law of gravitation replaced Kepler's laws only when Newton noticed that it is not only planets, but various other bodies as well, which gravitate. Thus the law of gravitation is justified on the strength of a different scope of observation than were Kepler's laws. If, moreover, such additional observations are very numerous and much differentiated, then in the light of the traditional theory of induction — just as in the light of Popper's theory — it may be said correctly that Newton's law — in spite of the fact that it is more general and perhaps even just because it is more general — is better justified than Kepler's laws. It might even be said, and not without justice, that as a rule less general (and also less precise) laws come to be replaced by more general (or more precise) laws only on the condition that the latter are better justified than were the former. And should it be really so, then we would have to question, at least in part, both the objections raised by Popper against induction and his own theory as well.

But, as we have seen, Popper not only blames the inductivists for not being able to explain the pursuit of possibly most general and precise laws, but also for the fact that, to be consistent, they should not aim at reaching empirical laws at all, contending themselves with collecting observations, if their aim is to obtain the possibly most certain theses. Could an inductivist find an answer to defend himself against this objection? It seems that he could — for he could rightly object to the interpretation of his standpoint given by Popper and assuming that the consideration of the degree of certainty is the only consideration that should be taken into account in building the empirical sciences. The

inductivist does not adopt this assumption at all. Thus he is not guilty
of inconsistency if, in introducing hypotheses, he takes into account
other considerations and recommends the choice of the most certain
hypothesis from among those only which satisfy certain additional conditions, e.g. they are suitable for the realization of the basic aims of the
empirical sciences — the explanation and prediction of phenomena, they
are sufficiently simple, etc. These additional postulates sufficiently justify
the postulate of aiming at general laws.

Moreover, it seems that a consistent deductivist would object to
drawing from his assumption those conclusions which Popper has drawn.
The theory of criticism of hypotheses makes the adoption of hypotheses
depend not only on the degree of their falsifiability, i.e., their relation
to possible basic statements, but also (and perhaps even principally) on
the degree of their justification, which is determined by the number,
variety, and severity of the tests used in their verification, that is, by
the number and kind of the accepted basic statements used as premises.
A greater degree of falsifiability does not necessarily accompany a
greater degree of justification. If it does, then the deductivist, too, will
not always recognize as better a statement that is more general and
more determinate compared with a less general and a less determinate
one. He will do so only when these additional conditions are satisfied,
conditions which — let it be remarked parenthetically — do not differ from
those by which the inductivist is guided in the choice of hypotheses.
Thus both of them, the deductivist and the inductivist alike, will probably make the same choice.

7

Such are Professor Popper's more important arguments intended to
discredit the inductive method and to pave the way for the method of
criticism of hypotheses, built on deductivist principles. It has been my
intention to demonstrate that Popper's endeavors have failed, either because his objections against induction were misdirected, or because they
referred only to some variations of the theory of induction, mainly
those that are most simplified and quite obsolete today, or because they
aimed not only at the theory of induction but also at the rival method
of criticism of hypotheses. Moreover, analysis has shown that if induc-

tion is interpreted in a certain manner that is rather common today, the method of criticism of hypotheses satisfies all the conditions imposed on induction and thus turns out to be just a special case of the latter.

Thus Professor Popper's investigations have failed in their principal purpose, yet they have nevertheless succeeded in bringing other valuable results. Such a wealth of detailed methodological recommendations as is contained in Popper's work is probably not to be found elsewhere; the same applies to the great number of pertinent and penetrating observations concerning the errors, shortcomings, and abuses that often mark actual inductive procedures, and to the strong emphasis laid on the need for caution and criticism even in the case of a provisional acceptance of hypotheses. It must be borne in mind that these are matters of primary importance both for research practice and for a rational attitude in collective life.

His critical analysis concerned with the various concepts of probability and with the endeavors to apply them in the theory of justification of hypotheses also deserves special mention. I have not raised these subjects here, since their discussion would take too much time and yet, so it seems, would not affect my conclusions. Besides, Professor Popper's opinions in these matters have undergone certain essential modifications that are to find expression in the second volume of *The Logic of Scientific Discovery*, already announced but not yet published.

REFERENCES

* First published in *Studia Filozoficzne* **1 (22)** (1961). Reprinted from a shortened version published in *Logic, Methodology and Philosophy of Science, Proceedings of the 1960 International Congress*, edited by E. Nagel, P. Suppes and A. Tarski, Stanford University Press, 1962. The completion to original version translated by E. Ronowicz.

[1] Karl R. Popper, *The Logic of Scientific Discovery*, London 1959.

[2] K. Popper, *op. cit.*

[3] It would be well to mention that in the vocabulary of the deductionist the term 'observable' in contrariety to 'observed' is a non-psychological term.

[4] A precise description, formulated only for general universal statements, (in short: for nomological statements, and in Popper's terminology for 'theories') is as follows: "A theory is to be called empirical or falsifiable if it divides the class of all possible basic statements unambiguously into the following two non-empty subclasses. First, the class of all those basic statements with which it is inconsist-

ent (...), we call this the class of the potential falsifiers of the theory; secondly the class of basic statements which it does not contradict" (*op. cit.*, p. 86).

[5] Although, as was pointed out by Popper (*op. cit.* p. 91), also all tautological statements result from basic sentences — e.g. on the basis of the formula such as: $q\to(p\to q)$, where this tautological statement should be substituted for q, and for p the conjunction of the basic sentences coming into play — they are not justified with the help of the basic statements, for they cannot be deduced from them without the vicious circle.

[6] Cf. e.g. H. Mehlberg, 'Positivisme et science', *Studia Philosophica* **III** (1948), § 11.

[7] K. Popper, *op. cit.*, pp. 34–36, 56 ff.

[8] K. Popper, *op. cit.*, pp. 30–32, 99 *et passim*. This is perhaps a good opportunity to mention that Prof. Popper is not always consistent in putting this into effect. For example, a certain breach is made by the assumption, important for deductivism, that the verification of a hypothesis has a justifying value only if it has been made in an honest and thorough way, with the desire to exhaust all the available tests that might be dangerous for the given hypothesis.

[9] K. Popper, 'Philosophy of Science', a personal report, in: *British Philosophy in the Mid-Century*, London 1957, pp. 33, 42, 182–3, 265–6.

[10] K. Popper, *op. cit.*, p. 27.

[11] Cf. e.g.: K. Twardowski, *Zasadnicze pojęcia dydaktyki i logiki* (*Basic Concepts of Didactics and Logic*), Lwów 1901; J. Hosiasson, 'Definicje rozumowania indukcyjnego' ('Definitions of Inductive Reasoning'), *Przegląd Filozoficzny* **31** (1928); K. Ajdukiewicz, *Zarys logiki* (*Outline of Logic*), Warszawa 1955; by the same author, *Główne zasady metodologii nauki i logiki formalnej* (*Basic Principles of the Methodology of Science and Formal Logic*) (script), Warszawa 1928; A. Lalande, *Vocabulaire technique et critique de la philosophie*, 5th edition, 1947; D. Runes, *The Dictionary of Philosophy*, N. York 1942; St. Jevons, *The Principles of Science*, new edition, N. York 1958; *Philosophical Writings of Peirce*, selected a. edited by J. Buchler, N. York 1955; S. Stebbing, *A Modern Introduction to Logic*, 4th ed. 1945, etc., etc.

[12] The direct consequence is the sentence: '$(x) [A(x)\cdot \bar{C}(x)\to B(x)]$', or '$(x) [A(x)\cdot \bar{D}(x)\to B(x)]$', etc.

[13] Cf. e.g., the characteristics of the method of criticizing hypotheses presented by J. Giedymin in the article 'Indukcjonizm i antyindukcjonizm' ('Inductivism and Anti-inductivism'), *Studia Filozoficzne* **2** (1959), p. 18, and accepted by Popper in his letter appended *in extenso* to the text of this article.

TADEUSZ KOTARBIŃSKI

CONCEPTS AND PROBLEMS IN GENERAL
METHODOLOGY AND METHODOLOGY OF
THE PRACTICAL SCIENCES *

The present discussion is an attempt to formulate and systematize problems of the methodology of the practical sciences. It will, therefore, be convenient for further discussion to establish in the first place the meaning of the word method[1] from which the term methodology is derived. Now, a method is always the method of some kind of human activity. Every activity is a kind of process. Every process is a kind of whole, i.e. a kind of complex object. Every complex object has a structure, which may be understood in either a broader, or a narrower sense. In the broader sense we call a structure the system of relations between the components of a whole; in the narrower sense—it is a set of components of a whole and the system of relations between them. We shall adopt for further discussions this notion of structure, narrower in range but richer in content and, since a given whole may be divided into components according to different criteria (for instance, a living organism may be conceived as composed of organs, or of its right and left parts, or of cells, etc.), whenever we use this term, we should indicate what principle of distinguishing the components of a given whole we have in mind. Thus, we may, for instance, consider the structure, i.e. construction—it is only the matter of what we call it—of a mature insect as consisting of a head, thorax and abdomen, the structure of a pen as consisting of a holder and pen, the structure of a subject-predicate sentence as consisting of subject, predicate and copula and, in another instance, we may, in our mind, divide the insect into tegument, entrails and limbs, the pen into molecules, the sentence into syllables, etc. The same applies to the components of a whole. Indeed, considering a bunch of flowers we may be interested in either the different species of plants in the bunch, or in the different colours of the flowers, and analysing

279

the structure of a clock we may be interested in, e.g., the differences in the size of the clock-hands: the minute hand, the hour hand, and the second hand, or the differences in the speed of their revolutions.

Activity is, I repeat, a certain process. Processes may be discussed as certain wholes and we may speak not only of the structure of a thing, but also, in a certain derivative sense, of the structure of processes, i.e. events, since events and processes are the same. In every process we can distinguish its component processes, e.g. by dividing it in our minds into stages (or phases) taking place in a sequence in time. A set of process components distinguished in this way and a system of relations of their succession in time, one after another, will be a certain structure of the process under discussion. There is a name for a structure like this: we call it the course of a process. Here is an example: the course of a flow of water in a given river: at first the water flows slowly and evenly over the smooth river-bed, then follows the turbulent flow over the cataracts, further on — a cascade, still further — the branching of its currents, etc. And if the process under discussion is an activity then the method of this activity is nothing else but its course plus its relation to the earlier preparation of some of its phases by others. Indeed, the method of a given activity is univocally defined by the construction of its component acts, the sequence of occurrence of these acts, and how the earlier ones prepare the way for the later ones. John and Peter reached the office they wanted in different ways. John came on foot, asked about the floor and room number at the information desk, went up by lift, and reported himself orally; Peter came by car, and, knowing the exact address went straight to the right place, up the stairs and showed his identity card: the component acts were different here and there were certain differences in the relations between the phases of the complex activity, apart from the sequence of relations in time, e.g. in John's case some of the component acts resulted from the information he had received and in Peter's case we have, at the end of the process, the relation of reference to a document, etc. Now, the method of a given activity is the course of its realization, if it has been an intended one, and if the person so acting has been conscious of its usefulness in other instances, and not only in this one, particular instance of its application. If, for instance, a team carrying out a surgical operation sterilizes the tools, applies anesthesia, cuts the body, puts stitches in the wound,

etc. — we shall be correct in stating that, not only has the operational process had a defined course and that the operation has been carried out in a certain defined way, but also that a certain established method of going about it has been applied.

Now what we know what a method is, the meaning of the term methodology [2] becomes clear, since methodology is — in our understanding — simply the science of methods. And, according to the title of the present discussion, we are to examine problems of the methodology of practical sciences. Thus, we shall have to consider the principle of distinguishing practical sciences from the others. We shall define practical sciences by reference to theoretical sciences. Every science whose basic aim is the acquisition of truths, i.e. finding the best substantiated answers to questions, and only such sciences, are called theoretical. Mathematics, physics, botany, history are examples of theoretical sciences. All other sciences are practical. Thus, every, and only such, science whose basic aim is something other than the acquisition of truths, [3] e.g. making a tool, or bringing about the recovery of a patient, or increasing a fortune, is practical. Every science, whether theoretical or practical is, of necessity, a compound activity, consisting of solving problems as well as other operations, such as manipulation of instruments, writing, etc.; it matters here, however, what eventually serves what: in the theoretical sciences these operations are basically subservient in relation to the act of solving the problem, whereas in the practical sciences solving problems is basically subservient in relation to the effort to construct something that is not an answer to a question. Naturally, before a constructional manipulation is undertaken in these cases it is preceded by an answer to the question what manipulation would lead to the creation of the intended object, but an attempt to acquire the answer to such a question is the means to attain an end, and not the end in itself. There are, at the same time, differences between the systems of activities from the point of view of the number of cognitive acts they involve, i.e. those which have as their main purpose the acquisition of a new truth. The more acts of this kind present in a given activity, the more scientific it is and, it is generally agreed, activities that are highly scientific, and whose aim is different from the acquisition of truths, are simply called practical sciences. Others must satisfy themselves with the name of craftsmanship or similar names, and the

boundary between craftsmanship, art and practical sciences is, because of their very nature, fluid. Shoemaking is correctly assigned to craftsmanship, surgery to the practical sciences, but if somebody was not sure whether the watchmaker's skill should be classified as a craft or a practical science, I should not be surprised.

Having established the meanings of words we can now attempt a look at the sphere of inquiries that is collectively called the methodology of practical sciences. A basic division of methods suggests itself here: into general methods which find application in all sciences, and thus, in the practical sciences but not only there – in the theoretical sciences as well, and into methods peculiar to the practical sciences, which are only applicable there. As examples of methods of the first type we may quote, for instance: the trial and error method (consisting of an attempt to carry out the intended operation and another, different attempt in the case of failure, etc., until the desired effect is achieved), the method of maximum simplification (by selecting the simplest possible means), the method of collective work division and many others.

These methods should be contrasted with those that are applied only when the basic aim of the activity is something other than an answer to some question, something different from the acquisition of some truth, e.g. if we have in mind mainly making a tool, building, or bringing about the recovery of a patient, or performing a sonata. Here again the need arises for differentiating the general methodology of practical sciences and the detailed methodologies of particular practical sciences. We shall, therefore, say a few words about the specific features of the method common to all the practical sciences as such. So, above all, the last, final operation in the practical sciences is always some kind of effort to shape the material, different from the cognitive effort, e.g. applying pressure to an outside body, dislocation, transmutation. Moreover, whereas in case of the search for a truth its content, i.e. its essential feature, can never be known in advance and reveals itself fully only after the investigation has been completed, so that the definite relevant result of this investigation cannot, because of its nature, be reasonably planned, in case of, e.g. efforts to find a medicine for a given disease, it is the essential feature of the intended product, namely, its effectiveness in this disease, that must, by its very nature, be planned in advance. Hence the preculiarity of the method of practical science: it relies

on remodelling the subject matter according to the planned pattern thus equipping it with the features relevant for the intended product. And the success of this remodelling is conditioned by the method peculiar to this type of operation, and which we shall call designing.[4] This is a search auxiliary to the main purpose of finding an answer to the question: what conditions attainable by way of action are essential in the composition of the subject matter in order to fulfil a hitherto non-occurrent set of conditions which are desired for some reason or other. Another form of designing is a search for conditions whose addition to the set of conditions already existing would suffice to generate the desired product.

However, the practical sciences are diverse and different methods are being used in their various branches. It is easy to make a natural distinction in the case of engineering, therapeutics, public administration, for instance. The most striking here are the restrictions imposed upon the use of certain methods resulting from moral or legal principles. Thus, engineers use experimental methods in search of new tools, whereas in therapeutic and administrational activities experimenting is severely limited since we hesitate to attempt operations that might expose people to dangerous effects. A high degree of freedom in this respect in case of living organisms other than human beings clearly offers wider possibilities of experimenting in medicine than in social activities. There are also considerable differences between the differentiated branches of the practical sciences from the point of view of the methods of measurement used. In the sphere of operations on inanimate matter measurement is flourishing and being greatly improved, while in dealing with animate matter in the form of living organisms (plant cultivation and animal breeding, therapeutics) only some parameters may be subject to measurement, especially if they go beyond the scalar system; and it is much more difficult to achieve a high degree of accuracy here; and, finally, measurement is gaining ground in the humanities and social fields, but with great difficulty.

There exists, therefore, a possibility of making distinctions between the methods applied in the sciences according to the range of their applicability: we distinguish the most general methods, relevant in all the sciences, methods of an intermediary degree of generality, characteristic of the whole range of practical sciences, in contrast with the methods

characteristic of the theoretical sciences in particular, and, finally, special methods characteristic for particular branches of the practical sciences.[5]

A question arises as to the degree of relevance of such distinctions. The answer to this question depends on access or lack of access to the postulate of adequacy[6] of general scientific propositions. This postulate, already outlined clearly in Aristotle's *Analytics*, and put forth again some seventy years ago by Petrażycki (independently, I believe, of Aristotle and the followers of his idea such as Petrus Ramus, Francis Bacon), calls for aiming at such subject-predicate general propositions of the type: 'every *A* is *B*', whose subject includes all individuals of which *B* is true. It can then also be said that 'every and only *A* is *B*'. And classes *A* and *B* are then mutually reversible. Petrażycki[7] calls propositions like these adequate. Here are some classical examples of adequate propositions: every triangle is a plane figure whose sum of internal angles equals two right angles, every physical body is a gravitating object. For instance, the proposition that every isosceles triangle has this sum of internal angles would not be adequate (since there are non-isosceles triangles and they, too, have a sum of internal angles equal two right angles). Likewise, the proposition that all stones gravitate would not be adequate since not only stones gravitate. But what is it that speaks in favour of the postulate of adequacy? At least two reasons. First, it guarantees economy of time, space and words in the exposition of a theory when, for instance, instead of attributing the feature separately to isosceles triangles, and then, separately to others, we attribute it to all triangles at once. Secondly, only after we have justified the general adequate proposition we do grasp the objective dependency of the relevant features of the predicate on the relevant features of the subject: indeed, the feature of being subject to gravitation is objectively bound not to the 'stony character' of the subject, but to its character as a physical body. By respecting the postulate of adequacy we become more resistant to suggestions to accept too narrow dependencies, for instance, regarding the cause of specialization. One who hears, for instance, a well established and valid, but very inadequate proposition that progress of production in economic units requires division of labour, is inclined to associate this tendency of development with some specifically economic motives that play an important role peculiar to the production

of so-called material goods which can be subjected to pecuniary comparative evaluation. But this conjecture would naturally be too narrow since specialization becomes a necessary component of any activity, as requirements become more exacting concerning the quality of work and products and the complexity and difficulty of the tasks involved increases. Only an adequate formulation of a proposition stating that specialization is indispensable gives prospects of a rational elucidation of its genesis.

Following the above arguments the reader might get the impression that we are defending here some deep and fundamental alienation of the problems that belong to a certain part of the methodology of the practical sciences and the problems peculiar to the methodology of the theoretical sciences. This is not so, for at least two reasons. First, every practical science is a long sequence of activities which, true, serve a different basic purpose than the justification of a new truth, but they also involve a great many component activities of a theoretical character. In short, to create something rationally one must learn a lot. And, if we isolate in our minds cognitive procedures we employ with this aim in view, we shall find, restricting ourselves to the period during which they were used, that their basic aim, among the aims that were pursued within that period of time, was the acquisition of an answer to a definite question. It appears, therefore, that before an expert in a practical science begins to design his intended product, and even more so, before he begins the manipulations necessary to realize the designed product, he has to solve theoretical problems and, naturally, in all those instances he will have to resort to the methods peculiar to the theoretical sciences. We shall not deal here with their dissection since that would go beyond the topic assigned by the title. Secondly, there is something that knits the practical and the theoretical sciences closely together. Indeed, the activity of designing itself is, after all, a kind of theoretical operation. True enough, this operation is, as regards the content of the guiding question, clearly a different one from other theoretical operations, but its common feature with them is that the problem here is the answer to a definite question about adequate conditions (necessary? sufficient?) that would have to be fulfilled, for the desired thing to be created. It follows from the above that complete preparation of the methodologist in the practical sciences must include problems that belong to the methodology

of the theoretical sciences; for instance, the recommendation that before the possibility of joint realization of two given conditions is checked experimentally one has to make sure that their synthesis in not excluded for *a priori* reasons, e.g. means of moving between street and flat based solely on a staircase must be excluded *a priori* when designing a multi-story block of flats for old people.

On the other hand, numerous phases of activity of the theoretical sciences possess a practical character. These are, e.g. such manipulative operations as assembling and transformation of the various components of the external material and, besides, also some mental activities of a practical character. For instance, whenever we deliberate on selecting from accepted axioms those which we can use as the simplest possible base to achieve the result that is derivable from them in the hypothet-ical-deductive system with the help of accepted rules, we are solving a practical problem, since we are concerned, in connection with this phase, mainly with the most economical course of work, and not with finding answers to definite questions. Thus, also in the pursuit of the theoretical aim we have to use methods peculiar to the practical sciences in certain phases.

Concluding this attempt to make distinctions between the types of methods used in the practical sciences—a problem auto-reflective in character, namely the question as to whether methodology as such is, or is not a practical science ... I believe that it is a practical science, since its basic aim is the realization of optimal methods, and this is something different from discovering new truths, although, it is true that in realizing optimal methods we must answer the question as to which methods are optimal, and activities involved in trying to answer this question are, in relation to the period of time they take, a preparatory theoretical science.

REFERENCES

* First published in *Studia Filozoficzne* **1 (74)** (1972). Translated by E. Ronowicz.
[1] T. Kotarbiński, 'O pojęciu metody' ('On the Concept of Method'), Warszawa 1957, *Zeszyty Wydziału Filozoficznego Uniwersytetu Warszawskiego* **1**; by the same author: 'De la notion de méthode', *Revue de Métaphysique et de Morale* **2** (1957).
[2] T. Kotarbiński, 'Treść i zakres pojęcia metodologii' ('The Content and Exten-sion of the Concept of Methodology'), *Myśl Filozoficzna* **4** (1956).

[3] A similar view is held by J. Zieleniewski. Cf. e.g. *Studia Filozoficzne* **3/4** (1963) reviews, p. 272; also his work *Organizacja zespołów ludzkich* (*Organization of Human Groups*), 3rd edition, 1967, p. 18.

[4] K. Tuchel, 'Zum Verhältnis von Kybernetik, Wissenschaft und Technik', *Akten des XIV. Internationalen Kongresses für Philosophie*, *Wien*: 2.–9. *September 1968*, Wien 1968, "Zwar bilden theoretisch-wissenschaftliche Kenntnisse heute in der Regel eine Voraussetzung des technischen Konstruierens, aber um zu einem neuen technischen Gebilde zu gelangen, reicht die Anwendung oder Verwendung dieser Kenntnisse nicht aus. Vielmehr muss das an einem Zweck orientierte schöpferische Vorausdenken hinzukommen, um die neue Gestalt eines technischen Gebildes entstehen za lassen" (p. 583).

[5] Two specializations of a kind are being developed at present in different countries: general methodology and the methodology of engineering or, in other words, of technique or technology.

General methodology is developed under the name of General Systems Theory, mainly in the U.S.A by A. Simon and also included in the sphere of cybernetics as initiated by N. Wiener, the problems of which are part of the problems of general methodology, concentrating around relatively isolated systems, feed-back, feeding and information. Cybernetics is being developed at present in other countries as well, including the Soviet Union and Poland, where its exposition has achieved a high degree of formalization in the works of H. Greniewski and M. Mazur.

France traditionally is the source of general methodology. We have in mind here certain Cartesian ideas. T. Kotarbiński, 'Idée de la méthodologie génerale', *Travaux de IX-e Congrès International de Philosophie* (*Congrès Descartes*), volume 4, Paris 1937. – By the same author: 'Les origines de la praxéologie', Academie Polonaise des Sciences, Centre Scientifique à Paris, Conférences, fascicule 58, Warszawa, PWN, 1965. They contain information on such methodological thinkers as C. Dunoyer, M. Martin, and, above all, E. Espinas. The idea of methodology, understood most generally under this name and in this meaning for the first time in France, was expressed in the collective work edited by R. Caude and A. Moles, entitled *Méthodologie, vers une science de l'action*, Gauthier-Villars, Paris 1964. This is a very inventive work.

In Poland, general methodology is being developed under the name of praxiology. T. Kotarbiński, *Traktat o dobrej robocie*, Ossolineum, Łódź 1955 (English translation: *Praxiology: An Introduction to the Sciences of Efficient Action*, Oxford and Warsaw 1965). – T. Pszczołowski, *Zasady sprawnego działania* (*Principles of Efficient Action*), 4th edition, Wiedza Powszechna, Warszawa 1967 – J. Zieleniewski, *Organizacja zespołów ludzkich, wstęp do teorii organizacji i kierownictwa* (*Organization of Human Groups, An Introduction to the Theory of Organization and Management*), 3rd edition, 1967. – T. Wójcik, 'Przedmiot, cel, zadania prakkseologii' ('The Subject, Aim, Tasks of Prakseology'), *Problemy Organizacji* **1**, PWE Warszawa 1963. – Z. Kleyff, 'Science, Technology and Economics as Integral Parts of Productive Processes', *Akten des XIV. Internationalen Kongresses für Philosophie*, *Wien*: 2–9 *September* 1968, Wien 1968. – H. Stonert, 'Analiza

logiczna pojęć nauki o działaniu' ('Logical Analysis of the Concepts of Science on Action'), *Prakseologia* **27** (1967). — By the same author, 'Charakterystyka twierdzeń nauk praktycznych w aspekcie metodologicznym' ('Characteristics of the Propositions of the Practical Sciences—Methodological Aspect') *Prakseologia* **28** (1967).

Earlier the idea of general methodology was expressed in a very original way in the work of A. Bogdanow, entitled *Tektologia* (*Tectology*), the first part of which was published in the year 1912. This work, which for a long time remained unnoticed, has recently gained the interest of methodologists, being studied now from the historical point of view. Also E. Słucki's pioneer article entitled 'Beitrag zur formal-praxeologischen Grundlegung der Oekonomik', which was published in *Zapiski Wydziału Społeczno-Ekonomicznego Uniwersytetu w Kijowie 1962*, passed unnoticed at first. Some information on methodological works concerning the practical sciences can be found in A. P. Ogurtsov's article: 'Praktika kak filosov-skaya problema' ('Practice as a Philosophical Problem'), *Voprosy filosofii* **7** (1967) 91–105.

The content of the above-mentioned *Akten des XIV. Internationalen Kongresses für Philosophie* demonstrates that there has been a considerable development of the problems of specific methodology of technology (understood as the art of engineers) in different countries. This is shown by works such as: J. Agassi, *The Logic of Technological Development*; Skolimowski, *On the Concept of Truth in Science and in Technology*; Walentynowicz, *On Methodology of Engineering Design*; Zieleniewski, *Why "Cybernetics and the Philosophy of Technical Science" only*; Zworykin, *Technology and the Laws of its Development*. See: *Akten des XIV. Internationalen Kongresses für Philosophie*, Wien: 2.–9. September 1968. The fact that the authors, while trying to give examples or definitions of such methods, actually solve more general problems often strikes the reader of works studying methods specific to engineering: they either give methods applicable in all activities, or methods which can also be applied in fields other than engineering, viz. in all practical sciences as such. A. Adam's *Philosophie der Technik* in the above mentioned *Akten* ... can be an example of this. We have the following attempt at a definition of technology there: "Wir konzipieren den Begriff Technik als die zeitliche (*t*), örtliche (*r*), und sachliche (*x, y*) Umwandlung von Etwas (*x*), in Etwas (*x, y*), allgemein $x(t_1, r_1) \rightarrow y(t_2, r_2)$, wobei auch $x = y$, $t_1 \geqslant t_2$, $r_1 = r_2$ zulässig ist" (p. 479).

Mention is due to the following articles in the field of the methodology of engineering: B. Walentynowicz, 'O istocie działalności inżynierów' ('On the Essence of the Activities of Engineers') *Przegląd Elektrotechniczny* **5** (1969). — W. Gasparski, 'Przyczynek do pojęcia techniki' ('A Contribution to the Notion of Technology'), *Zagadnienia Naukoznawstwa* **V** (1969).

[6] T. Kotarbiński, 'Z dziejów pojęcia teorii adekwatnej' ('On the History of the Concept of Adequate Theory') *Przegląd Filozoficzny* **3** (1937). Reprint: *Wybór pism* (*Selected papers*), vol. 2, Warszawa 1958. — By the same author, 'Petrażyckiego koncepcja twierdzenia adekwatnego na tle dawniejszych doktryn pokrewnych

(Petrażycki's Concept of Adequate Proposition on the Basis of Earlier Related Doctrines), *Z zagadnień teorii prawa i teorii nauki Leona Petrażyckiego*, Warszawa 1969.

[7] L. Petrażycki, *Vvedenie v izuchenie prava i nravstvennosti (An Introduction to the Study of Law and Morality)*, S.-Peterburg 1903, 3rd ed. 1908.

JERZY ŁOŚ *

THE FOUNDATIONS OF A METHODOLOGICAL
ANALYSIS OF MILL'S METHODS **

1. I shall present in this paper two systems which are fragments of physical language and its metalanguage, and which are to be used for the purposes of a methodological analysis of Mill's methods. However, this is not the only possible application of the formal results presented here; they can also be used to obtain precise definitions of numerous concepts of the methodology of empirical sciences. Some of those definitions can be found in Sections 13, 18, 23 and 24. The basic aim of this paper, however, is to lay the foundation of the analysis of Mill's methods, and all the problems discussed here are connected with it.

But is a methodological analysis of Mill's methods necessary? Mill formulated his methods in the middle of the 19th century (1843), i.e., over one hundred years ago. Much has been said about them since, every handbook has a separate chapter on them, but nobody, with the exception of a few (Nicod, Ajdukiewicz), has been able to say more or

* This paper originated in 1947 as my graduation thesis in philosophy at the Maria Curie-Skłodowska University in Lublin. It was written under the supervision of Professors Narcyz Łubnicki and Jerzy Słupecki, to whom I owe my permanent gratitude.

It is a little embarrassing to read one's own paper after almost thirty years, and still more embarrassing to prepare it again for publication. I would certainly not share now either its young author's sharp criticism of Mill or his enthusiasm for formalisms. In spite of that I have not changed the paper much for the new edition, presenting it as a translation and not as a new version. Improving a little the way of writing logical symbols, I have retained both methods of writing formulas, that of Łukasiewicz in the language and that of Peano–Russell in the metalanguage. When I was working on this paper, I considered the application of both symbolisms very convenient. I have also left intact some other naiveties, in so far as they do not lead to misunderstanding. As an example let me mention the 'Theorem of applicability' (Sec. 14), whose proof shows more than the theorem says. I hope the reader will accept this as an apology.

291

say it better than Mill. On the other hand, it is known that his methods are neither ideal nor complete. On the contrary, it is known that they are not applicable in their classical form; they are used in practice, but probably in a somewhat different form. Moreover, Mill formulated a hypothesis (or rather a theorem) about the dependence of his methods on the principle of causality. The problem raised by this hypothesis has not been adequately resolved yet. We do not even have at our disposal a formally correct language on the grounds of which that dependence could be worked out. Yet, since the times of Mill, the methodology of sciences, inductive as well as deductive, has progressed enough to make it possible to construct such a language and to formulate in it the problems connected with Mill's methods. This is my aim in the present paper. At the end I consider some problems concerning Mill's methods in their seemingly classical formulation. As the results of those considerations are negative for the methods, I also present some ways for their retrieval.

1. THE METHODS IN MILL'S FORMULATION

2. Mill considers his inductive logic as being closely related to experience.

Induction — he says — can be summarily defined as the generalization from experience.[1]

Mill took over this empirist attitude from the English empiricists, and directly from his father James St. Mill, from whom he also inherited his interest in economics. However, it would be erroneous to think that Mill, following the former empiricists, did not see any connection between induction and deduction. Such an opinion is likely to persist since, with respect to Mill's views on deduction, we know best his criticism of the syllogism, i.e., of the almost unique form of deductive reasoning studied by the logic of his times.

Every induction — Mill clearly says — may be thrown into the form of a syllogism.[2]

This linking of induction with deduction was Mill's greatest achievement. It resulted in the view, long accepted and apparently shared to a certain degree by Mill himself, that assertions justified by using his methods are as infallible as assertions justified by using the methods of logic. Where did Mill see the guarantee of this infallibility?

Mill clearly distinguishes between the procedure called proper induction and the rules of proceeding formulated by himself; in his main book he repeatedly denounces the former as a procedure which does not properly justify its conclusions. Some of his formulations may even suggest that he did not consider that procedure to be an induction. For, as has been said above, he says that every induction can be given the form of a syllogism, while a reasoning following proper induction cannot be given that form; at the same time, reasonings following his rules can be formulated not only as syllogisms; Mill even indicates a general premise serving all such reasonings. It is the principle of causality.

"If this be actually done, (i.e., when inductive reasoning assumes the form of a syllogism — J. Ł.) the principle which we are now considering, that of the uniformity of the course of nature, will appear as the ultimate major premise of all inductions" — says Mill.[2]

Thus Mill assigns the main role in induction to the principle of the uniformity of the course of nature, i.e., to the principle of causality. In formulating it Mill encounters various difficulties, and he does not explain what such a syllogism should be like; but his aim is clear and simple: to reduce induction to deduction.

3. The justification of laws is not the only field in which Mill intends to make use of deduction. His dream is to create a deductive system of the laws of nature. Knowing that many natural laws are interrelated by logical consequence, Mill thinks that the final form of the natural sciences should be a system in which, from a few basic laws justified, with as much certainty as possible, by induction, all the other laws would be inferred by deductive reasoning.[3] But the laws which would serve as a sort of axioms of the natural system (Mill calls them the laws of nature, as distinct from the empirical laws which are less strongly justified) should be justified particularly powerfully; with respect to that Mill once says that justification by using any of his methods is sufficient, and, at another place, that those laws should be justified by using the method of difference.

The following fragment concerns the latter possibility; "I answer (the question what we call the law of nature and what — an empirical law — J. Ł.) that no generalization amounts to more than an empirical

law when the only proof on which it rests is that of the Method of Agreement. For it has been seen that by that method alone we never can arrive at cause".[4]

Laws justified by other methods, particularly by proper induction, are at most empirical laws. This distinction is considered by Mill as very relative, depending indeed on our knowledge. Mill considers as the main task of science the raising of empirical laws to the dignity of laws of nature through proper justification.

4. In his "System of Logic" Mill formulated five methods of inductive procedure. They are: 1) the method of agreement, 2) the method of difference, 3) the joint method of agreement and difference, 4) the method of concomitant variation, 5) the method of residues. These methods are well known and are presented under the name of Mill's methods in almost every handbook of logic. Mill does not treat all the methods equally. As follows from the preceding section, in Mill's opinion the method of difference has the greatest power of justification. The classification is thus epistemological with respect to the likelihood of conclusions. But from the point of view of methodology the first two methods are the most important; the other three can be reduced to them, and are only their modifications.

We shall present here only the first two methods, trying to follow Mill's idea as faithfully as possible. This will serve us later as a basis for further discussion:

"First Canon. If two or more instances of the phenomenon under investigation have only one circumstance in common, the circumstance in which alone all the instances agree, is the cause (or effect) of the given phenomenon".[5]

"Second Canon. If an instance in which the phenomenon under investigation occurs, and an instance in which it does not occur, have every circumstance in common save one, that one occurring only in the former; the circumstance in which alone the two instances differ, is the effect, or the cause, or an indispensable part of the cause, of the phenomenon".[6]

This is the way in which Mill formulates his rules. Let us note the difference in his conclusions. The first method, that of agreement, discovers 'the cause of the phenomenon', while the second method

discovers 'the cause, or an indispensable part of the cause, of the phenomenon'. When we confront the first method with the excerpt quoted in the first section of this paper, we shall find a consistent interpretation difficult. This results from the ambiguity of the term 'cause' as used by Mill.[7] Mill would like to call the cause of a phenomenon its necessary and sufficient condition; but tending to such a formulation he does not succeed in the analysis of the necessary, sufficient and counteracting conditions, thus creating chaos and causing ambiguity. It seems that in the first canon Mill uses the term 'cause' to denote the sufficient but not necessary condition of a phenomenon. He knows that the method of agreement gives us nothing less than the cause, and the method of difference—nothing more than the cause. Such an opinion seems to be supported by the formulation of the second canon, as well as by the following fragment, saying what can and what cannot be eliminated.

"The Method of Agreement stands on the ground that whatever can be eliminated is not connected with the phenomenon by any law. The Method of Difference has for its foundation that whatever cannot be eliminated is connected with the phenomenon by a law".[8]

Mill illustrates his methods with several examples. He quotes them as a proof of the claim that he is not their author but their discoverer. But the schemata used by Mill in order to present the idea contained in the methods are more instructive than the examples. Those schemata are also well known, as they have entered the handbooks together with the methods as their best explanation. They show a kind of coincidences from which it is possible to infer causality by using the methods proposed. For the method of agreement, the schema is:

1)

	t	$t+\tau$
I	xyz	a
II	xyw	a
III	xwz	a

The rows marked with Roman numerals correspond to consecutive observations; the letters x, y, z denote all the phenomena occurring at a moment t, corresponding to the first observation; the letters x, y, w and x, w, z in the second and third observation have analogous mean-

ings. The letter a denotes the phenomenon examined, occurring every time at a moment later by τ than t (τ is a constant value). As is easy to see, the conclusion of three such observations according to the first method is that x is the cause of a.

A similar schema for the method of difference is as follows:

2)

	t	$t+\tau$
I	xyz	a
II	yz	

The conclusion following the second method is: x causes a. These schemata are very instructive. First of all, they clearly show what Mill calls "factors accompanying the phenomenon". This does not mean any simultaneous factors, but earlier ones, what agrees with Mill's view of the relation between cause and effect, according to which the cause should precede its effect at least in the sense that the beginning of the cause precedes the begining of the effect. Second, the schemata show that Mill's primary intention is to look for the causes of phenomena. Mill does not restrict himself to these two schemes: in other parts of this work he gives schemata which do not show so clearly the tendency to look for the causes.[9] These schemata are the following:

3)

t	$t+\tau$
xyz	abc
xyw	abd
xwz	adc

for the rule of agreement,

4)

t	$t+\tau$
xyz	abc
yz	bc

for the rule of difference.

Mill seems to see no difference between the first and the second type of those schemata. Why did such a penetrating investigator neglect this difference? This seems due to the fact that Mill constantly neglected the counterdomain of the relation of causality — the effect. Both in analysing causality and in formulating the principle of causality Mill always

deals with the cause, hardly ever with the effect; as a result we see that he overlooked the difference between the schemata illustrating his methods.

In the course of further discussion, when formulating the methods precisely, we shall use the schemata 3) and 4), in spite of the fact that Mill's original text suggests that schemata 1) and 2) reflect the author's intention better. As an excuse let us point out that schemata 3) and 4) seem to be much more general than 1) and 2).

2. THE STRUCTURE OF PHYSICAL LANGUAGE

5. The starting point of our discussion about physical language will be Professor K. Ajdukiewicz's[10] distinction of 3 types of rules of inference which occur in such a language. They are axiomatic, deductive and empirical rules. Leaving out the former two, which are known from the methodology of deductive sciences, we shall examine the latter in detail. Those rules, characteristic for physical language, are distinct from the other two types in that they are based on relations between empirical data and sentences (deductive rules are based on relations between sentences only). They allow us to accept a sentence if appropriate empirical data are present. The fact that we use such rules seems to be indubitable. The majority of the sentences we use in everyday life are justified directly or indirectly by such rules.

The introduction of empirical rules, however, makes the study of the properties of language much more difficult. It is not possible to elaborate such rules to a degree sufficient to prove the consistency of the language or to carry out any other demonstration analogous to those known from the methodology of formalized languages (that is one in which empirical rules do not prevail). This is because deductive rules are defined by means of structural relations between sentences or classes of sentences and sentences. A similar definition for an empirical rule is not possible because the domain of such a relation consists not of expressions but of empirical data. However, in order to enable ourselves to study at least in part a language containing empirical rules, we must characterize the class of sentences which can serve as the counter-domain of the relation on which such rules are based. The sentences belonging to that class, i.e., sentences which might be directly justified by using the empirical rule, are named differently. Carnap and Neurath

call them protocol sentences (*Protokollsätze*), and Popper calls them
basic sentences (*Basissätze*); in Poland it is generally accepted to call
them observational sentences. Let us retain this terminology.

The concept of observational sentences, which originated in the
so-called *Vienna Circle*, immediately gave rise to misunderstandings
and controversies among the main representatives of that *Circle*.
Many works have been published on this subject.[11] Do observational
sentences speak about relations between things, or do they only describe
the contents of observations? Do observational sentences belong to physical language or not? Are there time and space coordinates in observational sentences, and what role do they play there?

Without attempting an analysis of all these questions, which would
lead us too far, we shall present a concept of observational sentences
which seems to be closest to Popper's position; the latter has also been
attracting Carnap recently.

6. According to this concept observational sentences speak about things,
they belong to physical language and contain time and space coordinates
concerning the fact described by a given sentence. Popper gives the
following examples of observational sentences:

1. "It is thundering in the 13th district of Vienna on the 10th of June,
   1933, at 5.15 p.m.".
2. "There is a raven in the space-time region k", or generally:
3. "Such-and-such an event is occurring in the region k".[12]

Let us have a look at these examples. Each of them is a sentence
in which there is a time and space determination and an utterance:

1. 'it is thundering',
2. 'there is a raven',
3. 'Such-and-such an event is occurring'.

These utterances are not sentences in the normal sense. They can become
sentences only after a time and space determination is added to them.
They are occasional functions. Their truth or falsity, i.e. their logical
value, depends on the time and place at which they are uttered, or on
the time and space determination by adding which they are made into
sentences. They are thus incomplete expressions; they may be completed
either by adding a situation [13] or by adding to them arguments determining the time and place.

Such incomplete expressions have already been considered by philosophers. Meinong dealt with them,[14] investigating incomplete names; this led to incomplete sentences, for which the principle of contradiction was supposed not to be valid. Twardowski[15] devoted to them a short paper in which he refused to assign them the character of sentences. Finally, Smolka, in a lecture delivered in Lwów in 1920,[16] proposed to use them in interpreting the intermediary values between truth and falseness in many-valued logics. It is interesting to note that our concept does indeed lead to a type of many-valued logic, a direction of research which unfortunately we shall have to leave out of this paper.[17]

Let us call an 'event' what happened somewhere and at some time, and a 'phenomenon' — a defined type of events. We shall say, for example, that the fact that a fuse burned out in my flat on the 1st of January, 1947, at 10.00 p.m., is an event; and the class of such events, consisting in the burning out of fuses, the class of all the burn-outs of fuses, is a phenomenon.

To every event we can assign a sentence describing it; this will be an observational sentence concerning that event. On the other hand, to every phenomenon we can assign an utterance which is not precisely a sentence. Thus we assign the following descriptive sentence to the event of the previous example:

    4. 'On the 1st of January, 1947, at 10.00 p.m., a fuse burned out in my flat',

and the following expression to the phenomenon:

    4'. 'a fuse burned out'.

The sentences 1, 2, 3 and 4 state that, at appropriate spatio-temporal points, the phenomena described by 1', 2', 3', and 4' were realized. Let us call the expressions 1', 2', 3', and 4' occasional functions, remembering that we do not mean here utterances containing occasional expressions, as 'I am eating', but sentences where time and space determination is somehow lacking.

7. Let us distinguish from among occasional functions those whose logical values do not depend on time and place, but only on time. We shall call them temporal functions. We shall then say that the expression 4' is an occasional function in the wider sense; while the expression 'a fuse burned out in my flat' is a temporal function only.

Temporal functions describe phenomena a little more precisely than occasional functions in the wider sense do. We introduce the concept of temporal function only in order to restrict to them our further discussion concerning Mill's methods and thus to simplify it.

Now let 'p' be any temporal function, and let 't' denote any moment. Further on we shall use the expression 'p occurs at the moment t' or 'p is realized at the moment t'. We shall assume that 'p' is of the semantic category of sentence and that all observational sentences have the form 'p occurs at the moment t'. It should be kept in mind that this condition is necessary for a statement to be an observational sentence, but not sufficient.[18] Further on we shall assume that if, in a sentence of the form discussed, there is a true sentence in place of 'p' (for example the tautology 'if p then p'), then the whole sentence is true no matter what moment 't' denotes.

For a temporal function 'p', let us denote by T_1 the set of moments at which 'p' occurs (we shall then call the set T_1 the range of the function 'p'). 'Non-p' is then a new function, occurring at all those and only those moments at which 'p' does not occur. Let us denote by T_1' the range of its realization. Now, if 'q' is a temporal function with the range of realization T_2, 'p or q' is a new temporal function with the range of realization $T_1 \cup T_2$; 'if p, then q' is a new temporal function with the range of realization $T_1' \cup T_2$. The temporal function 'p or non-p' has the range of realization $T_1 \cup T_1'$, i.e., it is realized at every moment. A temporal function which is realized at every moment is a true sentence; a temporal function which is never realized is a false sentence. It is easy to show that a substitution of temporal functions into a theorem (tautology) of the two-valued propositional calculus results in a true sentence.

3. THE PRECISE FORMULATION OF MILL'S METHODS

8. Of course, when presenting his methods, Mill did not know the requirements of modern methodology, and for this reason he did not indicate what their position in the language should be, i.e., whether those methods are theorems or rules of inference, or whether they are only informal suggestions indicating procedures convenient in experimental research, originating from a generalization of concrete cases. Historically, the

latter possibility seems the most plausible. It is supported by the numerous examples given by Mill to illustrate his methods, and also by many fragments of his work in which he tried to encourage the reader to use those methods. However, we shall consider the methods as rules of inference. We have to do so because of Mill's ambition that the infallibility of those rules should be accepted on the grounds of the principle of causality. We shall thus regard Mill's methods as inference rules of the physical language.

9. The method known as the method of agreement can be formulated as follows:

α) For any classes of moments Z_1, Z_2 and any classes of temporal functions A_1, A_2, if

1) Z_2 is the class of all moments which are later than the moments in Z_1 by a fixed period of time τ;

2) A_1 is the class of all those temporal functions which occur at any moment of the class Z_1;

3) A_2 is the class of all those temporal functions which occur at any moment of the class Z_2,

then for any two other classes of moments Z_3, Z_4 if

4) Z_4 is the class of all moments which are later than the moments in Z_3 by the same fixed period of time τ,

5) every temporal function of A_1 occurs at every moment of the class Z_3,

then

6) every temporal function of the class A_2 occurs at every moment of the class Z_4.

Let us now recall schema 3) from Sec. 4. The three rows of the first column correspond to observations at three moments belonging to the class Z_1, and the lines of the second column correspond to observations at moments belonging to the class Z_2, which are later by τ than the moments belonging to the class Z_1. In our schema the phenomenon 'x' (or the class of which 'x' is the only member) is the class A_1; indeed: x is the only element occurring in every line of the first column. For analogical reasons 'a' is the only element of the class A_2.

We thus see that our rule includes schema 3) of Sec. 4. It is more general than that schema to the extent that:

I. it allows for any number of observations, since it contains no restrictions with respect to the power of the classes Z_1 and Z_2. Schema 3) contains such a restriction for technical reasons only. Let us remark that the original wording of the method of agreement does not contain such restrictions;

II. the power of the classes A_1 and A_2 is not limited, either. From the previous considerations it follows that we cannot afford such limitation; namely, if 'p' belongs to A_i then, for any 'q', 'p or q' also belongs to A_i ($i = 1, 2$).

10. While in the method of agreement the classes Z_1 and Z_2 could be unrestricted in power, the corresponding classes in the method of difference must necessarily consist of only two members:

β) For any moments z_1, $\tilde{z}_1$, z_2, $\tilde{z}_2$ and any classes of temporal functions A_1, A_2, if

1) z_2 and $\tilde{z}_2$ are moments later than the moments z_1 and $\tilde{z}_1$, respectively, by a fixed period of time;

2) A_1 is the class of all temporal functions occurring at the moment z_1 but not at the moment $\tilde{z}_1$;

3) A_2 is the class of all temporal functions which occur at the moment $\tilde{z}_1$ but not at the moment $\tilde{z}_2$;

then for any classes of moments Z_3, Z_4, if

4) Z_4 is the class of moments which are later than the moments in Z_3 by the same fixed period of time τ;

5) every temporal function of A_1 occurs at every moment of the class Z_3;

then

6) every temporal function of the class A_2 occurs at every moment of the class Z_4.

4. THE AXIOMS OF A FRAGMENT
OF PHYSICAL LANGUAGE

11. We shall now formulate a set of axioms of a fragment of physical language, in order to be able to use the expression 'p occurs at the moment t' in this language. In Professor K. Ajdukiewicz's theory of meanings the axioms given here could play the role of axiomatic rules,

determining the sense of the expression 'occurs at the moment', which we shall further on abbreviate by U.

Let us denote:

1. variables of the semantic category of sentences by $p_1, p_2, p_3, \ldots$,
2. variables of the semantic category of moments by $t_1, t_2, t_3, \ldots$,
3. variables of the semantic category of periods of time by $\tau_1, \tau_2, \tau_3, \ldots$

We shall read the formula

(I) $\qquad Ut_1 p_1$

as: 'p_1 occurs at the moment t_1', and the expression

(II) $\qquad U\delta t_1 \tau_1 p_1$

as: 'p_1 occurs at a moment later than t_1 by τ_1'.

In constructing our axioms we shall use the notation of Łukasiewicz,[19] placing the functors before the arguments, as in formulas (I) and (II).

Our system will be based on the two-valued propositional calculus with quantifiers and on the system of functional calculus without quantifiers for functional variables.[20]

12. We shall now give the list of axioms.

(A$_1$) $\qquad EUt_1 Np_1 NUt_1 p_1$.

This axiom reads: to say that the negation of p_1 holds at the moment t_1 is equivalent to saying that it is not true that p_1 holds at the moment t_1. This is a free interpretation of our axiom in everyday language. Further on, while formalizing our system as exactly as possible, we shall give such free interpretations of its formulas in order to make their understanding easier.

(A$_2$) $\qquad CUt_1 Cp_1 p_2 CUt_1 p_1 Ut_1 p_2$.

If at the moment t_1 the implication 'if p_1, then p_2' holds, then the fact that p_1 holds at the moment t_1 implies that p_2 holds at the moment t_1.

(A$_3$) $\qquad Ut_1 CCp_1 p_2 CCp_2 p_3 Cp_1 p_3$.

(A$_4$) $\qquad Ut_1 Cp_1 CNp_1 p_2$.

(A$_5$) $\qquad Ut_1 CCNp_1 p_1 p_1$.

As has been demontrated by Łukasiewicz,[21] the three propositional formulas in the axioms (A_3)–(A_5) which follow the letters 'Ut_1' form a complete and independent set of axioms of the two-valued propositional calculus. With the axioms (A_3)–(A_5) and accepted rules of inference we can prove within the system any sentence of the form '$Ut_1\alpha$', where α is any theorem of the propositional calculus.

(A_6) $\qquad C\varPi t_1\, Ut_1 p_1 p_1\,.$

If p_1 holds at every moment, then we can accept p_1 as a theorem of the system.[22]

We shall explain the following three axioms after having given the definitions:

(A_7) $\qquad \varPi t_1\, \varPi\tau_1\, \varSigma t_2\, \varPi p_1\, EU\delta t_1\, \tau_1\, p_1\, Ut_2\, p_1\,.$

(A_8) $\qquad \varPi t_1\, \varPi\tau_1\, \varSigma t_2\, \varPi p_1\, EU\delta t_2\, \tau_1\, p_1\, Ut_1\, p_1\,.$

(A_9) $\qquad \varPi t_1\, \varSigma p_1\, \varPi t_2\, EUt_2\, p_1\, \varPi p_2\, EUt_1\, p_2\, Ut_2\, p_2\,.$

13. We shall now give several definitions.

(D_1) $\qquad Ip_1\, p_2 \overset{\text{def}}{=} \varPi t_1\, EUt_1\, p_1\, Ut_1\, p_2\,.$

This is the definition of the extensional identity of temporal functions. The consequence of this definition is that in order to prove that two temporal functions (i.e., two phenomena) are not identical we must indicate two moments at which one of them holds and the other does not.

(D_2) $\qquad \rho t_1\, t_2 \overset{\text{def}}{=} \varPi p_1\, EUt_1\, p_1\, Ut_2\, p_1\,.$

This is the definition of the identity of two moments.

(D_3) $\qquad \pi t_1\, t_2 \overset{\text{def}}{=} \varSigma\tau_1\, \rho\delta t_1\, \tau_1\, t_2\,.$

This is the definition of the succession of two moments. The moment t_2 occurs after the moment t_1 if and only if there is an interval after which we get to t_2 from t_1.

(D_4) $\qquad v\tau_1\, t_1\, t_2 \overset{\text{def}}{=} \rho\delta t_1\, \tau_1\, t_2\,.$

This is the definition of a ternary relation: t_2 succeeds t_1 after an interval of time τ_1.

By introducing these definitions the axioms (A_7)–(A_9) can be considerably simplified:

(A$'_7$) $\Pi t_1 \Pi \tau_1 \Sigma t_2 \rho \delta t_1 \tau_1 t_2$.

(A$'_8$) $\Pi t_1 \Pi \tau_1 \Sigma t_2 \rho t_1 \delta t_2 \tau_1$.

The axiom (A_7) requires that the operation 'δ' be feasible in the set of moments, (A_8) requires that the inverse operation also be feasible. Both these axioms jointly require that the number of values which can be substituted for the variables of the semantic category of moments be infinite; and this, together with the definitions, implies that the number of temporal functions must be infinite. A more detailed discussion of these questions would lead us beyond the scope of this paper. However, in order to explain them shortly and in a simple way, let us remark that the formula

(T_1) $CK v \tau_1 t_1 t_2 v \tau_1 t_1 t_3 \rho t_2 t_3$ $((D_1), (D_4))$

stating that the operation 'δ' is single-valued in the set of moments, is a theorem of the system.

Finally, the axiom (A_9) will assume the form:

(A$'_9$) $\Pi t_1 \Sigma p_1 \Pi t_2 E U t_2 p_1 \rho t_1 t_2$.

This axiom, called the axiom of the clock, says that to any moment we can assign a temporal function occurring only at that moment. Let us note that, if it were not so, we would not be able to use a clock. Let us suppose, restricting ourselves to only twelve hours, that we measure time with a clock the short hand of which can take 12 positions and the long hand — 60 positions. 3 o'clock 45 minutes is then characterized by two temporal functions, one is: 'the short hand has taken position 3' and the other: 'the long hand has taken position 45'. The temporal function satisfying the axiom (A_9) with 't_1 equal to 3.45' is the conjunction of both these temporal functions.

The following is an immediate consequence of axioms (A_1) and (A_9):

(T_2) $\Pi t_1 \Sigma p_1 \Pi t_2 E U t_2 p_1 N \rho t_1 t_2$.

Let us have a look at some other interesting consequences of our axioms and definitions:

(T_3) $CUt_1 p_1 Ut_1 Ap_1 p_2$ $((A_1)-(A_5))$.

(T_4) $EUt_1 Kp_1 p_2 KUt_1 p_1 Ut_1 p_2$ $((A_1)-(A_5))$.

(T_5) $CK\rho t_1 t_2 v\tau_1 t_1 t_3 v\tau_1 t_2 t_3$ $(D_2), (D_5))$.

(T_6) $\rho t_1 t_1$ (D_2).

(T_7) $E\rho t_1 t_2 \rho t_2 t_1$ (D_2).

(T_8) $C\rho t_1 t_2 C\rho t_2 t_3 \rho t_1 t_3$ (D_2).

14. In order to show the consistency of our system of axioms it is enough to point out that the formula

(N) $EUt_1 p_1 p_1$

implies, with the aid of the propositional calculus with quantifiers and of the functional calculus, all the axioms (A_1)–(A_9) of the system discussed. Thus we cannot obtain any contradiction within this system, because the expression '$Ut_1 p_1$' can be interpreted as the assertion of the expression 'p_1'.

Let us further notice that (N) implies

(FA_6) $Cp_1 \Pi t_1 Ut_1 p_1$,

which is the reverse of axiom (A_6). It follows immediately that our system, reinforced by adding the expression (FA_6) to the axioms, remains consistent.

Let us now suppose that we have, in the reinforced system, any proper temporal function, for example 'it is thundering' (we now leave out the question of space determination). Since it is a proper temporal function, it is true that

(G_1) $K\Sigma t_1 Ut_1$ (it is thundering) $\Sigma t_2 Ut_2 N$ (it is thundering).

On the other hand, after substituting 'it is thundering' for 'p_1', we get as a consequence of (FA_6):

(G_2) $C\Sigma t_1 NUt_1$ (it is thundering) N (it is thundering).

From (G_1) and (G_2) we obtain

(G_3) N (it is thundering).

Obviously an analogical procedure will prove that

(G_4) (it is thundering)

and we shall have a contradiction within the system.

The above contradiction has arisen by adding the expression (G_1) to the system which states that 'it is thundering' is a proper temporal function (i.e., the range of realization of the function 'it is thundering' does not include all the moments and is not empty). We therefore reject the expression (FA_6), and we must examine whether it is not possible to get a contradiction from axioms (A_1)–(A_9) through a similar procedure. We shall call this question the question of applicability. The expression (FA_6) is not applicable because it leads to a contradiction if any proper temporal function is contained in the system.

Let L be a line (e.g. the set of real numbers) and let P be the family of all subsets of L. Further, let I be the set of closed intervals in L.

We consider the propositional variables (p) as ranging over P, the variables of the moments of time (t) as ranging over L, and finally the variables of the periods of time (τ) as ranging over I.

By $\emptyset$ we denote the empty set, $p_1 \cup p_2$ is the set-theoretical union of the sets p_1, p_2 in P, p_1' is the complement of p_1 in L, $t_1 \in p_1$ means as usual that the point t_1 belongs to the set p_1.

To every formula of our system an operation in L, P, I will be assigned in the following manner:

(i) $Np_1 = p_1'$;

(ii) $Cp_1 p_2 = p_1' \cup p_2'$;

(iii) $Ut_1 p_1 = \begin{cases} L \text{ if } t_1 \in p_1, \\ \emptyset \text{ otherwise}; \end{cases}$

(iv) $\delta t_1 \tau_1$ is the point on L which is the right end-point of the interval τ_1 shifted so that its left end-point is t_1;

(v) $\Pi p_1 \varphi(p_1) = \begin{cases} L \text{ if } \varphi(x) = L \text{ or every set } x \in P, \\ \emptyset \text{ otherwise}; \end{cases}$

(vi) $\Pi t_1 \varphi(t_1) = \begin{cases} L \text{ if } \varphi(x) = L \text{ for every point } x \in L, \\ \emptyset \text{ otherwise}; \end{cases}$

(vii) $\Pi \tau \varphi(\tau) = \begin{cases} L \text{ if } \varphi(x) = L \text{ for every interval } x \in I, \\ \emptyset \text{ otherwise}. \end{cases}$

It is easy to show that this interpretation converts every axiom of the system under consideration into a function with values in P which is identically equal to L. This is a hereditary property, i.e., if the premises have it, then the consequence has it too. Since by (i) if a formula has this property, then its negation cannot have it, the system is consistent.

It is now easy to prove the following theorem, which says that the system is applicable.

THEOREM OF APPLICABILITY. *Let A be a class of sentences of the form 'Utp' with constants at the places of 't' and 'p'. The system based on axioms* (A_1)–(A_9) *and all sentences of the class A is inconsistent if and only if there exist two sentences* '$Ut_1 p_1$' *and* '$Ut_2 p_2$' *in A such that the sentences*

$$\text{`}pt_1 \, t_2\text{' and `}Ip_1 Np_2\text{'}$$

are both consequences of the axioms (A_1)–(A_9).

This theorem says that in our physical language we can obtain a contradiction by applying the empirical rule in one way only.

5. THE METASYSTEM OF PHYSICAL LANGUAGE

15. As we have seen in Secs. 9 and 10, in order to formulate Mill's methods we must make use of classes of sentences and, moreover, of some concepts of a semantic nature. It is known that in order to do so, if we are to avoid contradictions, we must turn to a metalanguage. There are two methods of metalinguistic study described in literature. The first is the arithmetization of a system, the second is the axiomatic method.[23] We shall use the axiomatic method for our purposes, but our approach will be fundamentally different from those set up previously for the purposes of syntax and semantics, because it is to serve methodological purposes.

In Tarski's metalanguage, serving to define the concept of truth,[24] there are names of formulas of the language and their counterparts as expressions of the same semantic category as in the language; as a result every linguistic expression has its 'translation' in the metalanguage. Speaking freely, the consequence of this is that in the metalanguage we can speak simultaneously of formulas of the language and of the objects referred to by those formulas. By using this framework Tarski arrives at the definitions of semantic concepts, e.g., that of the class of true sentences.

For our purposes we shall reverse Tarski's method, and introduce the class of true formulas and some other concepts of syntactic nature as primitive concepts; they will occur in axioms. By using them, through the names of formulas we shall speak in the metasystem of the objects those formulas refer to, i.e., of moments, of events which correspond to elementary sentences of the system, and of phenomena which correspond to temporal functions.

It would be possible to translate, in Tarski's sense, into the metalanguage only the quantifier-free formulas of the propositional calculus. Other formulas of the language will only have their transpositions in the metalanguage, i.e., sentences asserting that a given formula is true. A difficulty arises here, as we cannot exclude the possibility of the existence of synonyms in the language, for example two different names for the same moment, or two temporal functions differently denoted but identical in the sense of (D_1). For example let 't_1' and 't_2' be names of two different moments. It may happen that '$\delta t_1\ \tau_1$' is the name of the same moment as 't_2'; then '$\rho \delta t_1\ \tau_1\ t_2$' will be true; but the statement '$\rho \delta t_1\ \tau_1$' = 't_2' will be false in the metasystem.

As a matter of fact, in our further discussion this difficulty concerns only the semantic category of moments. In order to avoid it, we shall give several definitions of concepts analogous to ordinary relations between classes but concerning only the classes of names of moments.

We shall take as the basis of the metasystem the language of the set theory, and thus we shall use expressions such as '$\in$' – the symbol of membership, '$\subset$' – the symbol of inclusion of one class into another, '$=$' – the symbol of equality and so on. Moreover, we shall use the constant terms of the propositional calculus, this time writing them between arguments, according to the Peano–Russell notation.

16. As primitive notions of our metasystem we will have:

V — the set of true formulas;

Fp — the set of syntactically correct formulas of the semantic category of sentences;

Ft — the set of syntactically correct formulas of the semantic category of moments;

$F\tau$ — the set of syntactically correct formulas of the semantic category of periods of time.

Moreover, we adopt the following list of names for single signs of the language:

Sign of the language:	Its name in the metalanguage:
C, N, E, K, I	C, N, E, K, I
U	U
$t_1, t_2, t_3, \ldots$	$t_1, t_2, t_3, \ldots$
$p_1, p_2, p_3, \ldots$	$p_1, p_2, p_3, \ldots$
$\tau_1, \tau_2, \tau_3, \ldots$	$\tau_1, \tau_2, \tau_3, \ldots$
Π, Σ	P, S
δ, ρ, ν	d, r, υ

In order to facilitate the reading of formulas we shall use special characters to denote variables restricted to particular sets. Thus $a, a_1, a_2, \ldots$ will be used for individual variables ranging over Fp and $A, A_1, A_2, \ldots$ for set-variables ranging over subsets of Fp. Analogously $z, z_1, z_2, \ldots$ and $Z, Z_1, Z_2, \ldots$ are used for variables restricted to Ft and subsets of Ft, respectively, and o will be a variable ranging over $F\tau$. Let us notice that, for instance, a, z and o are of the same semantic category (in the metalanguage), i.e., that of individuals. The writing convention is adopted only for the readers' convenience. In order to make the reading still easier, we shall use restricted quantifiers,[25] writing the formulas

$$\forall x\, x \in Y \to \varphi(x); \quad \forall X\, X \subset Y \to \varphi(X);$$

$$\exists x\, x \in Y \cdot \varphi(x); \quad \exists X\, X \subset Y \cdot \varphi(X);$$

in the form

$$\forall x \; \varphi(x); \quad \forall X \; \varphi(X);$$
$$\underset{Y}{} \qquad \underset{Y}{}$$

$$\exists x \; \varphi(x); \quad \exists X \; \varphi(X).$$
$$\underset{Y}{} \qquad \underset{Y}{}$$

If the shape of the bound variable indicates the set $Y \; (= Fp, Ft$ or $F\tau)$ to which it has been restricted, then the symbol of the set under the quantifier can be omitted.

Taking advantage of the universal order of signs in the syntactically correct formulas of the language, we form their structural-descriptive names by ordinary succession of the names of signs. We do not need here, as Tarski did in [16], any special functor corresponding to the succession of two symbols. We simply preserve in the structural-descriptive names the same succession as in the formula they refer to.

Thus, for instance,

$$Pt_1 \; Up_1 \; d \text{ is the name of } \varPi t_1 \; Up_1 \; \delta,$$

which is not a syntactically correct formula, but

$$Pt_1 \; Sp_1 \; Pt_2 \; EUt_2 \; p_1 \; rt_1 \; t_2$$

is the name of

$$\varPi t_1 \; \varSigma p_1 \; \varPi t_2 \; EUt_2 \; p_1 \; \rho t_1 \; t_2,$$

in which we recognize the axiom (A'_9).

In order to form the system of the metalanguage we shall adopt some axioms which will be described below by groups only.

First of all

(I) $\qquad V \subset Fp$

is an axiom stating that true formulas are among syntactically correct formulas.

The second group of axioms consists of syntactic rules of forming formulas of the language. Thus for instance we will have here the axiom

(II) $\qquad \underset{Ft}{\forall} z \underset{F\tau}{\forall} o \; dzo \in Ft,$

which describes the way of forming formulas of the semantic category of moments with the aid of the function 'δ'.

Further, we adopt as axioms all sentences saying that all the axioms and definitions of the language are true. For instance the sentences

(III)
$$CPt_1\, Ut_1\, p_1\, p_1 \in V,$$
$$Ert_1\, t_2\, Pp_1\, EUt_1\, p_1\, Ut_2\, p_1 \in V$$

are both among the axioms because the first says that (A_6) is true and the second that the equivalence of the definiens and the definiendum of (D_2) is true.

Finally, we adopt as axioms all the rules of inference holding in language. We will thus have in this group, for instance,

(IV) $\quad Ca_1\, a_2 \in V \cdot a_1 \in V \to a_2 \in V$

since this expresses the rule of modus ponens.

Having adopted the above axioms we can of course obtain in our metasystem every sentence asserting that a theorem of the system of the physical language under consideration is true (belongs to V). If α is, for instance, a name of a particular formula of the semantic category of moments, then we can prove in the metalanguage

(P) $\quad Sp_1\, Pt_2\, EUt_2\, p_1\, r\alpha t_2 \in V.$

Although we can prove in the metalanguage every sentence of the form (P), we cannot prove the general sentence

(TM$_1$) $\quad \forall z\, Sp_1\, Pt_2\, EUt_2\, p_1\, rzt_2 \in V,$

as well as we cannot prove

(TM$_2$) $\quad \forall z_1\, \exists a \forall z_2\, EUz_2\, arz_1\, z_2 \in V.$

The truth of both sentences results from the above considerations.[26]

In order to obtain in the metasystem this type of theorem we should adopt additional rules of inference allowing to replace names of variables by variables and names of quantifiers by quantifiers relative to the appropriate class. Were such rules adopted, the sentences (TM$_1$) and (TM$_2$) would become provable in the metasystem.

17. As is well known, in order to put a domain of science in an axiomatic form we ought to begin with the definition of the syntactically correct formulas, and then to establish rules of inference. Those rules

must include axiomatic rules, i.e., rules requiring that some sentences be accepted without proof. We usually define the syntactic correctness of formulas and express the rules of inference in natural, everyday language. But, to be exact, they should be expressed in the language in which we can speak about the constructed language, i.e., in the metalanguage. Thus, if we want to construct a formal language, we must already be in possession of another language, which will serve as the metalanguage for the newly constructed one.

It is characteristic (and it seems to be the only thing making the procedure reasonable) that the language which we start from — let us call it the constitutional metalanguage, or shortly the metalanguage K — does not have to be richer than the constructed one either in the number of semantic categories or in the number of logical types. The metalanguage K can be a language with a low number of logical types even for a language with an infinite number of logical types.

The metalanguage constructed in the preceding section is precisely such a metalanguage K for the fragment of physical language presented in the preceding chapter. As we see, it lacks some semantic categories that can be found in the language, for example the semantic category corresponding to the functor 'U'.

Let us now consider how we could extend our metalanguage in order to pass from the metalanguage K to a richer metalanguage. It is known that some languages are incomplete. This means that there exist in such languages sentences not containing free variables such that neither they nor their negations are theorems. The fragment of physical language presented in the preceding chapter also has this property. For instance

(n) $\Sigma p_1 \Sigma t_1 \Sigma t_2 K U t_1 p_1 U t_2 N p_1$

does not satisfy the interpretation of the first proof of consistency given in Sec. 14, and its negation does not satisfy the interpretation of the second proof of consistency given in that section. This sentence is therefore independent.

Now if, besides the axioms of the metalanguage described in the preceding section, we adopt as an axiom

(nn) $S p_1 S t_1 S t_2 K U t_1 p_1 U t_2 N p_1 \in V$,

then such a metasystem would no longer be the metasystem K for the fragment of physical language given in the preceding chapter. In particular, it would lose the property that 'V' in its axioms can be interpreted as a class of provable formulas.

We shall call a rule of inference of the language a sentence of the metalanguage of the form

(∗) $\forall a \; \varphi(a) \rightarrow a \in V$,

i.e., having the form of a conditional statement saying that every formula satisfying $\varphi(\cdot)$ is true. Also any other formula equivalent to a formula of form (∗) will be called a rule of inference. We see for instance that axiom (IV) in Sec. 16 is a rule of inference, because one of its equivalent forms is

$$\forall a (\exists a_1 \; Ca_1 \; a \in V \cdot a_1 \in V) \rightarrow a \in V.$$

In every language some rules of inference must be assumed at the outset, in forming the language; others are their consequences. We shall call them induced rules of inference. Most precisely, a rule of inference is an induced rule for a given language if it is a theorem of the metalanguage K for that language.[27]

We shall further show that Mill's methods are induced rules with respect to the axioms and rules of the fragment of physical language given in the preceding chapter.

18. We shall now give some definitions which will help in the final formulation of the problem. Not all of them are relevant to our question, but they are meant, additionally, to facilitate the reader's understanding of the character of the system and of the metasystem.

(DM$_1$) $a \in D \overset{\text{def}}{=} Pt_1 \; Ut_1 \; a \in V$.

We read this as follows: a is a descript if and only if it is true that a holds at every moment. If we wanted to understand our physical language as describing an isolated system, we would say that the class D is the class of sentences determinating that system. The fact that a must be the name of an expression of the semantic category of sentences results from axiom (I) of the metasystem and from the axioms describing the manner of forming formulas (group (II))

(DM$_2$) $a \in W \overset{\text{def}}{=} St_1 \; Ut_1 \; a \in V \cdot St_1 \; Ut_1 \; Na \in V$.

This reads: a is a variant if and only if there is a moment at which a holds and there is a moment at which a does not hold. W is thus the class of proper temporal functions, i.e., functions which are neither identically true nor false (cf. the example in Sec. 14).

$$(\text{DM}_3) \quad a \in \Delta z \overset{\text{def}}{=} a \in W \cdot U z a \in V.$$

Δz is the class of variants (i.e., something like a class of phenomena) which occur at the moment z. Relativizing our language again to an isolated system, we can say that Δz fully describes the state of the system at the moment z.

$$(\text{DM}_4) \quad z_1 \approx z_2 \overset{\text{def}}{=} r z_1 z_2 \in V.$$

This is the definition of the specific identity of two names of moments. The relation defined here is obviously different from that defined in (D_2) in Sec. 13. However, from the very form of definition (DM_4) we immediately see that far-reaching analogies will hold between these two relations and that we will have, in the metasystem, analogies of (T_6)–(T_8).

$$(\text{TM}_3) \quad z \approx z.$$

$$(\text{TM}_4) \quad z_1 \approx z_2 \cdot \leftrightarrow \cdot z_2 \approx z_1.$$

$$(\text{TM}_5) \quad z_1 \approx z_2 \cdot z_2 \approx z_3 : \rightarrow \cdot z_1 \approx z_3.$$

$$(\text{DM}_5) \quad Z_1 \mathrel{\text{Cl}} Z_2 \overset{\text{def}}{=} \bigvee_{z_1} \underset{z_2}{\exists} z_2 \quad z_1 \approx z_2.$$

This is the definition of specific inclusion of two classes of names of moments.

$$(\text{DM}_6) \quad Z_1 \sim Z_2 \overset{\text{def}}{=} Z_1 \mathrel{\text{Cl}} Z_2 \cdot Z_2 \mathrel{\text{Cl}} Z_1.$$

This is the definition of the specific identity of two classes of names of moments.

It follows from (TM_3)–(TM_5) that

$$(\text{TM}_6) \quad Z \sim Z,$$

$$(\text{TM}_7) \quad Z_1 \sim Z_2 \cdot \leftrightarrow \cdot Z_2 \sim Z_1,$$

$$(\text{TM}_8) \quad Z_1 \sim Z_2 \cdot Z_2 \sim Z_3 : \rightarrow \cdot Z_1 \sim Z_3.$$

$$(DM_7) \quad z_1 \overset{o}{\to} z_2 \overset{\text{def}}{=} voz_1 z_2 \in V,$$

$$(DM_8) \quad Z_1 \overset{o}{\Rightarrow} Z_2 \overset{\text{def}}{=} \underset{z_1}{\forall} z_1 \underset{z_2}{\exists} z_2 \; z_1 \overset{o}{\to} z_2 \cdot \underset{z_2}{\forall} z_2 \underset{z_1}{\exists} z_1 \; z_1 \overset{o}{\to} z_2.$$

The relation defined corresponds to that of the succession of two moments; the first definition is analogous to (D_4); the second definition has as a consequence the theorem:

$$(TM_9) \quad Z_1 \overset{o}{\Rightarrow} Z_2 \cdot Z_3 \overset{o}{\Rightarrow} Z_4 : \to \cdot Z_1 \cup Z_3 \overset{o}{\Rightarrow} Z_2 \cup Z_4.$$

Let us note that, by means of the rules, the axioms and the definitions (DM_4) and (DM_7), we can prove

$$(TM_{10}) \quad z_1 \overset{o}{\to} z_2 \cdot z_1 \overset{o}{\to} z_3 : \to \cdot z_2 \approx z_3,$$

which is an analogon of (T_1).

It is easy to derive from (TM_{10}) and the definitions (DM_6) and (DM_8) the following theorems, which are rather important for the considerations to follow:

$$(TM_{11}) \quad Z_1 \overset{o}{\Rightarrow} Z_2 \cdot Z_1 \overset{o}{\Rightarrow} Z_3 : \to \cdot Z_2 \sim Z_3,$$

$$(TM_{12}) \quad Z_1 \overset{o}{\Rightarrow} Z_2 \cdot Z_1 \sim Z_3 : \to \cdot Z_3 \overset{o}{\Rightarrow} Z_2.$$

We shall give one more definition:

$$(DM_9) \quad a \in \theta Z \overset{\text{def}}{=} \underset{z}{\forall} z \; a \in \varDelta z.$$

This definition has a very clear-cut meaning, inherited from definition (DM_3). For any set of moments Z, θZ is the class of variants which occur at every moment belonging to Z.

6. FORMAL ANALYSIS OF MILL'S METHODS

19. By using the concepts defined so far it is easy to formulate in the metalanguage Mill's method of agreement:

$$(KM_1) \quad Z_1 \overset{o}{\Rightarrow} Z_2 \cdot \theta Z_1 = A_1 \cdot \theta Z_2 = A_2 : \cdot \to \cdot \forall Z_3 \; \forall Z_3 \; Z_4 \overset{o}{\Rightarrow} Z_4 \cdot$$

$$\cdot A_1 \subset \theta Z_3 : \to \cdot A_2 \subset \theta Z_4.$$

(KM$_1$) is of course equivalent to

(KM$_2$) $\quad Z_1 \overset{o}{\Rightarrow} Z_2 \cdot \rightarrow \cdot \forall Z_3 \forall Z_4 \quad Z_3 \overset{o}{\Rightarrow} Z_4 \cdot \theta Z_1 \subset \theta Z_3 : \rightarrow \cdot$

$\qquad \theta Z_2 \subset \theta Z_4$.

Comparing (KM$_1$) with the rule α) of Sec. 9, we see that they both express the same. Using the notation of α), we may write (KM$_1$) (neglecting quantifiers) as

$$1) \cdot 2) \cdot 3) \cdot : \rightarrow \cdot 4) \cdot 5) : \rightarrow \cdot 6).$$

The method of difference can be written as follows ('\' means the set difference):

(KM$_3$) $\quad z_1 \overset{o}{\rightarrow} z_2 \cdot \tilde{z}_1 \overset{o}{\rightarrow} \tilde{z}_2 \cdot \Delta z_1 \backslash \Delta \tilde{z}_1 = A_1 \cdot \Delta z_2 \backslash \Delta \tilde{z}_2 = A_2$

$\qquad :: \rightarrow \cdot \forall Z_3 \forall Z_4 \cdot Z_3 \overset{o}{\Rightarrow} Z_4 \cdot A_1 \subset \theta Z_3 : \rightarrow \cdot A_2 \subset \theta Z_4$.

The comparison of (KM$_3$) with the rule β) (Sec. 10) should not present any difficulties to the reader.

20. In order to prove (KM$_1$), or its equivalent form (KM$_2$), we first have to prove some lemmas. They will be a little more complicated than those we learned previously. The proofs will be presented in a non-formal way, since otherwise they would become completely non-intuitive.

(TM$_{13}$) $\quad Z_2 \mathrel{\subsetneq} Z_1 \cdot \rightarrow \cdot \theta Z_1 \subset \theta Z_2$.

The proof will be carried out by *reductio ad absurdum*. Let us suppose that (TM$_{13}$) is false; therefore $Z_2 \mathrel{\subsetneq} Z_1$ and there exists a temporal function $a \in Fp$, such that $Uz_1 \, a \in V$ for all $z_1 \in Z_1$, but for some $z_2 \in Z_2$, $Uz_2 \, a \notin V$. This follows from the definitions (DM$_3$) and (DM$_9$). Since $z_2 \in Z_2$ and $Z_2 \mathrel{\subsetneq} Z_1$, by (DM$_5$) there exists a $z_3 \in Z_1$ such that $z_2 \approx z_3$. But this contradicts the existence of the temporal function a, whose properties show that $z_2 \approx z_3$ is false for every $z_3 \in Z_1$ (we use here (DM$_4$) and suitable axioms of the metasystem). This ends the proof.

(TM$_{14}$) $\quad \theta Z_1 \subset \theta Z_2 \cdot \rightarrow \cdot Z_2 \mathrel{\subsetneq} Z_1$.

In order to prove this let us note that the following two propositions are

immediate consequences of the axioms and definitions of the meta-system:

$$(TM_{15}) \quad \underset{Ft}{\forall} z_1 \underset{Fp}{\exists} a \underset{Ft}{\forall} z_2 \, EUz_2 \, aNrz_1 \, z_2 \in V \, .$$

$$(TM_{16}) \quad \underset{Ft}{\forall} z_1 \underset{Fp}{\exists} a \underset{Ft}{\forall} z_2 \, a \in \varDelta z_2 \cdot \leftrightarrow \cdot (z_1 \approx z_2)' \, .$$

The first one is a metasystem analogon of (T_2), the second one is an immediate consequence of the former and of definitions (DM_3) and (DM_4).

Let us proceed now to the proof of (TM_{14}). If $Z_2 \subset Z_1$ does not hold, then by (DM_5) there exists a $z_2 \in Z_2$ such that $z_2 \approx z_1$ does not hold for any $z_1 \in Z_1$. Let us consider an $a \in Fp$ which occurs at exactly those moments $z_3 \in Ft$ for which $z_3 \approx z_2$ does not hold. By (TM_{16}), such an a does exist. Since $z_3 \approx z_2$ is false for all $z_3 \in Z_1$, for all such z_3 it will be true that $Uz_3 \, a \in V$. We conclude that (by (DM_3) and (DM_9)) $a \in \theta Z_1$. However, we see that $Uz_2 \, a \in V$ does not hold for that particular z_2 which we have chosen previously. Finally $a \in \theta Z_2$ is not true and neither is $\theta Z_1 \subset \theta Z_2$, which ends the proof.

$$(TM_{17}) \quad Z_1 \subset Z_2 \cdot \leftrightarrow \cdot Z_1 \cup Z_2 \sim Z_2 \, .$$

Indeed, if $Z_1 \subset Z_2$ is not true, then by (DM_5) there exists a $z_1 \in Z_1$ such that, for all $z_2 \in Z_2$, $z_1 \approx z_2$ does not hold. Since $z_1 \in Z_1$, we have $z_1 \in Z_1 \cup Z_2$, which implies that $Z_1 \cup Z_2 \sim Z_2$ does not hold.

Let us now suppose that $Z_1 \cup Z_2 \sim Z_2$ does not hold. By (DM_3) we have anyway $Z_2 \subset Z_1 \cup Z_2$; thus in this case $Z_1 \cup Z_2 \subset Z_2$ does not hold. By (DM_5) we may find a $z_1 \in Z_1 \cup Z_2$ such that $z_1 \approx z_2$ is false for any $z_2 \in Z_2$. From (TM_3) it follows that $z_1 \in Z_1$, because otherwise $z_1 \approx z_1$ would be false. Finally we conclude that $z_1 \in Z_1$ has the property that $z_1 \approx z_2$ for every $z_2 \in Z_2$, which implies that $Z_1 \subset Z_2$ is not true, as required for the proof.

$$(TM_{18}) \quad Z_1 \overset{o}{\Rightarrow} Z_2 \cdot \rightarrow \cdot \forall Z_3 \forall Z_4 \; Z_3 \overset{o}{\Rightarrow} Z_4 \cdot Z_3 \subset Z_1 : \rightarrow \cdot Z_4 \subset Z_2 \, .$$

In order to prove this, let us assume

1. $Z_1 \overset{o}{\Rightarrow} Z_2$;

2. $Z_3 \overset{o}{\Rightarrow} Z_4$;

3. $Z_3 \subset Z_1$.

We deduce step-by-step:

4. $Z_3 \cup Z_1 \sim Z_1$, by 3 and (TM_{17});

5. $Z_3 \cup Z_1 \overset{o}{\Rightarrow} Z_4 \cup Z_2$, by 2, 1 and (TM_9);

6. $Z_1 \overset{o}{\Rightarrow} Z_4 \cup Z_2$, by 4, 5 and (TM_{12});

7. $Z_4 \cup Z_2 \sim Z_2$, by 6, 1 and (TM_{11});

8. $Z_4 \subsetneq Z_2$, by 7 and (TM_{17}).

Conclusion 8 is the final conclusion in (TM_{18}), which ends the proof.

From (TM_{18}), applying (TM_{13}) and (TM_{14}), we immediately obtain (KM_2), which is equivalent to (KM_1). We may thus say that we have proved the method of agreement (see Sec. 19).

21. Let us observe that unfortunately the result obtained in the preceding section is trivial. Namely, it follows from (TM_{13}), (TM_{14}) and (TM_{18}) that whenever we deduce by the method of agreement a causal link between classes $A_1 = \theta Z_1$ and $A_2 = \theta Z_2$, these classes will have the property that all the events in A_1 occur only at moments of Z_1 and all the events in A_2 occur only at moments of Z_2. In other words, the situation A_1 can never happen again, and so A_2 cannot happen again.

In order to make this argument more specific, let us take a closer look at the method of difference, written in a slightly modified form:

(KM_4) $\quad z_1 \overset{o}{\to} z_2 \cdot \tilde{z}_1 \overset{o}{\to} \tilde{z}_2 \cdot \Delta z_1 \backslash \Delta \tilde{z}_1 = A_1 \cdot \Delta z_2 \backslash \Delta \tilde{z}_2 = A_2 \cdot$

$\qquad \cdot Z_3 \overset{o}{\Rightarrow} Z_4 \cdot A_1 \subset \theta Z_3 ::: \to \cdot A_2 \subset \theta Z_4 .$

We shall show that if the assumptions of (KM_4) are satisfied by some $z_1, z_2, o, \tilde{z}_1, \tilde{z}_2, A_1, A_2$ and Z_3, Z_4, then Z_3 consists of the only element z_1 [28] and Z_4 consists of the only element z_2. Indeed, according to the axiom of the clock (resp. to its translation into the metalanguage, see group (III) of axioms in Sec. 16) there exists a temporal function $a_1 \in Fp$ which occurs only at the moment z_1 and another temporal function a_2 which occurs only at the moment z_2. We see that $a_1 \in A_1$

and $a_2 \in A_2$; thus $a_1 \in \theta Z_1$ and $a_2 \in \theta Z_2$. But, by the definition of a_1, if Z_1 contains any z such that $z \sim z_1$ does not hold, then $a_1 \in \theta Z_1$ can not be true. The same method of proof applies to z_2 and Z_2.

22. The reasoning above yields a proof of (KM_4), and thus also of (KM_3), in the metasystem. Both this proof and the preceding proof of (KM_1) show that Mill's methods are induced rules with respect to the axioms and rules of the fragment of our physical language, and in order to obtain them it is not necessary to assume any principle of causality. Unfortunately they also show that the causal laws that can be obtained by those methods are only a description of what has already happened during our observations, and the situation cannot occur again so that the law will never be applied again. It is a consequence of the axiom of the clock.

Nevertheless, it seems to be a fact that Mill's methods are being applied, not only in scientific activities but also in everyday life, and the fact that we use a clock does not seem to interfere. We simply observe some phenomena and we assume that the occurrence of the observed phenomena is not influenced by the occurrence of other phenomena which are not observed. In other words, we restrict ourselves to a certain isolated system.

Let us now try to interpret our fragment of physical language as concerning such an isolated system, and let it be a periodical system with respect to the time measured with a clock outside the system. As we know from physics, such systems do not exist, but we also know that such fictitious systems are often used in physics in order to carry out various reasonings. We may suppose that our system (we shall call it system U) is a pendulum in motion, suspended in an empty room and not subject to any external influence.

We shall now read the formula

(I) $U t_1 p_1$

as: 'p_1 occurs at the moment t_1 in the system U'. Here 'p_1' is a temporal function corresponding to the phenomenon occurring in the system U, while the value of 't_1' is read on the clock which is outside the system.

It is easy to check that our system of axioms is valid under such an interpretation. The meaning of some terms, however, particularly those

denoting relations between moments, both in the language and in the metalanguage, will be changed. Indeed, let us denote the period of the oscillations of our pendulum by τ_1. If we carry out the observations at any moment t_1 and at a moment later by τ_1, we arrive at the conclusion that

(i) $\rho t_1 \, \delta t_1 \, \tau_1$

but, on the grounds of the definitions (D_3), we arrive also at the conclusion that

(ii) $\pi t_1 \, \delta t_1 \, \tau_1$.

We thus see that our definitions of relations between moments have lost their former sense. According to the intuitions now adopted, we thus have to interpret the relation

(iii) $\rho t_1 \, t_2$

as 't_1 is a moment equivalent to the moment t_2 modulo the period of the system U_{11}', and the relation

(iv) $\nu \tau_1 \, t_1 \, t_2$

as: 'the moment t_2 is equivalent modulo the period of the system U to the moment by τ_1 later than t_1'.

With this in mind we shall be able to read the method of agreement analogously to the rule (Sec. 10), modifying only the corresponding points 1) and 4), for instance the point 1), as follows:

1') Z_2 is a class of moments equivalent modulo the period of the system to moments which are later than the moments in Z_1 by a fixed period of time.

In this interpretation Mill's methods enable us to obtain some applicable laws concerning the system U. But we could obviously obtain the same laws without those rules, because of their tautological character, which we have demonstrated.

23. The attempt to save the validity of Mill's methods, as presented in the preceding section, does not seem satisfactory. Indeed, the applicability range of the rules understood in this way would be so narrow that they would completely lose their present character. We might

try to improve them in still another way, by rejecting the axiom of the clock and reinforcing the fragment of physical language with an axiom corresponding to the principle of causality. However, this method, certainly the closest to Mill's concept, is difficult to accept because the axiom of the clock is our only protection against the metaphysical and extrasensory understanding of time. But if we decided to reject this axiom, it would not be difficult to find a new one, suitable for our purposes.

It can easily be shown that, in order to prove the method of agreement in an appropriate metasystem K, it is enough to adopt the axiom

$$(A_{10}) \quad \Pi p_1 \, \Pi \tau_1 \, \Pi t_1 \, C U \delta t_1 \, \tau_1 \, p_1 \, \Sigma p_2 \, K U t_1 \, p_2$$

$$\Pi t_2 \, C U t_2 \, p_2 \, U \delta t_2 \, \tau_1 \, p_1 \, ,$$

which may be called the principle of causality. In order to demonstrate both the method of agreement and the method of difference, we must adopt the axiom

$$(A_{11}) \quad \Pi p_1 \, \Pi \tau_1 \, \Pi t_1 \, \Sigma p_2 \, E U \delta t_1 \, \tau_1 \, p_1 \, U t_1 \, p_2 \, ,$$

which may be called the strong principle of causality.[29]

24. But it seems that the most suitable way to save the validity of Mill's methods is to extend our fragment of physical language. Let us remark that we can do that in two ways:

1. By introducing variables of space coordinates, i.e., basing our language on the primitive expression

$$(I') \quad U t_1 \, x_1 \, y_1 \, z_1 \, p_1$$

which can be read: 'p_1 occurs at the moment t_1 in a place with coordinates x_1, y_1, z_1'.

2. By introducing variables for the semantic category of the functor 'U' (we could denote those variables by $U_1, U_2, \ldots$) and reading the expression

$$(II') \quad U_1 \, t_1 \, p_1$$

as 'p_1 occurs in the system U_1 at the moment t_1'. The system U_0 satisfying the condition

$$(i) \quad \Pi U_1 \, C U_1 \, t_1 \, p_1 \, U_0 \, t_1 \, p_1$$

could be called the 'world', and we could then adopt the axiom of the clock for this system. Further, we could consider the functors defined by (D_2)–(D_4) but for U_0 as the equality of two moments and as time succession.

It is difficult to predict which of the two possibilities presented will enable us to formalize the intuitions of Mill's methods more easily (possibly it will be necessary to extend the fragment of physical language in both ways), but as the second method seems to be closer to our previous considerations, I shall try to outline the appropriate further procedure for it.

We can say that the system U_1 contains the system U_2 if

(ii) $\qquad \Pi t_1 \, \Pi p_1 \, C U_2 \, t_1 \, p_1 \, U_1 \, t_1 \, p_1 \, .$

We can see that the system U_0 contains all the systems. We shall call U_1 the non-clock system if

(iii) $\qquad \Sigma t_1 \, \Sigma t_2 \, K N \rho t_1 \, t_2 \, \Pi p_1 \, E U_1 \, t_1 \, p_1 \, U_1 \, t_2 \, p_1$

(the relation ρ being defined in the system U_0); in the opposite case we shall call U_1 the clock system.

It is clear that for the clock system Mill's methods are tautologically satisfied, but the situation is different for the non-clock system. It should be examined what assumptions are necessary in order that Mill's methods be valid also in such systems, and what relations the two systems must satisfy in order that the law demonstrated by using Mill's methods for one of them be also valid in the other. It seems that a sufficient condition is for the first system to be contained in the second. But is it a necessary condition?

25. The problems outlined in the last section seem to be relevant for the methodological analysis of Mill's methods. The formal material presented in this paper forms a basis for their solution. It seems, however, that these are not the only problems which can be considered by using this formalism. Problems concerning the question of time, so important for the methodology of empirical sciences, problems concerning the testability of observational sentences and many others – can also be considered by using the framework presented in this paper.

REFERENCES

** First published in *Annales UMCS* (1947). Translated by S. Wojnicki.

[1] Mill [10], Vol. 1, Book 3, Chapter 3, p. 343.

[2] Mill [10], Vol. 1, Book 3, Chapter 3, p. 345.

[3] As to Mill's system of laws of nature, see Stejnbarg [15].

[4] Mill [10], Vol. 2, Book 5, Chapter 16, p. 43.

[5] Mill [10], Vol. 1, Book 3, Chapter 8, p. 430.

[6] Mill [10], Vol. 1, Book 3, Chapter 8, p. 431.

[7] This equivocality is also pointed out by Stejnbarg [15].

[8] Mill [10], Vol. 1, Book 3, Chapter 8, p. 432.

[9] Mill [10], Vol. 1, Book 3, Chapter 8, pp. 434–5.

[10] Ajdukiewicz [1], Sec. 5, pp. 113 ff, also Ajdukiewicz [2].

[11] Carnap [4], Secs. 3 and 6, Carnap [5], Neurath [12], Popper [13].

[12] Popper [13], pp. 48 and 58.

[13] This is Ajdukiewicz's opinion, presented by Kokoszyńska [7], p. 32, particularly footnote 1.

[14] Meinong's opinion on this question is presented by Łukasiewicz [8], pp. 121 ff.

[15] Twardowski [19].

[16] Smolka [14].

[17] See J. Łoś, 'Many-valued logics and formalization of intentional functions' (in Polish), *Kwartalnik Filozoficzny* **17** (1948) 59–78.

[18] If 'p occurs at the moment t' is an observational sentence then for instance, following Popper, we would not consider the sentence 'non-p occurs at the moment t' as observational; see Popper [13], p. 59.

[19] The principles of this notation are presented in Łukasiewicz [9].

[20] The propositional calculus with quantifiers is presented in Łukasiewicz [9], the functional calculus without quantifiers for functional variables (the so-called first-order functional calculus) — in Hilbert–Ackermann [6].

[21] Łukasiewicz [9].

[22] The axioms (A_1)–(A_6) together with rules of inference allow us to prove any theorem of the propositional calculus with quantifiers without using the propositional calculus as a primitive system.

[23] The axiomatic method is discussed in detail by Tarski [16]; see also Tarski [17].

[24] Comp. Tarski [16], pp. 21 ff.

[25] We introduce relativized quantifiers following Mostowski [11], p. 19.

[26] The axioms imply both (TM_1) and (TM_2) in the sense of Tarski, see Tarski [18], but neither (TM_1) nor (TM_2) is a formal consequence of axioms.

[27] Concerning induced rules, comp. Ajdukiewicz [1], pp. 155 ff.

[28] More precisely we should say: if $z \in Z_3$ then $\{z\} \sim Z_1$. The same applies to Z_4 and z_2.

[29] (A_{10}) and (A_{11}) are counterparts of the so-called principles of determinism, cf. Ajdukiewicz [1], pp. 285 ff.

BIBLIOGRAPHY

[1] Kazimierz Ajdukiewicz, *Główne zasady metodologii nauk i logiki formalnej* (*The Basic Principles of the Methodology of Sciences and of Formal Logic*), Lectures edited by M. Presburger, Wydawnictwa Koła Matematyczno-Filozoficznego Słuchaczów Uniw. Warszawskiego, Vol. 15, Warszawa 1928 (litographed).

[2] Kazimierz Ajdukiewicz, 'Sprache und Sinn', *Erkenntnis* **4**, 100–138.

[3] Kazimierz Ajdukiewicz, 'Naukowa perspektywa świata' ('The Scientific View of the World'), *Przegląd Filozoficzny* **37**.

[4] Rudolf Carnap, 'Die physikalische Sprache als Universalsprache der Wissenschaft', *Erkenntnis* **2**, 432–465.

[5] Rudolf Carnap, 'Über Protokollsätze', *Erkenntnis* **3**, 215–228.

[6] D. Hilbert, und W. Ackermann, 'Grundzüge der theoretischen Logik', *Die Grundlehren der Mathematischen Wissenschaften* **27**, Berlin 1928.

[7] Maria Kokoszyńska, 'O różnych rodzajach zdań' ('On Various Kinds of Sentences'), *Przegląd Filozoficzny* **43**.

[8] Jan Łukasiewicz, *O zasadzie sprzeczności u Arystotelesa* (*On Aristotle's Principle of Contradiction*), Kraków 1910.

[9] Jan Łukasiewicz, *Elementy logiki matematycznej* (*Elements of Mathematic Logic*), Lectures edited by M. Presburger, Wydawnictwa Koła Matematyczno-Fizycznego Słuchaczów Uniwersytetu Warszawskiego, Vol. 18, Warszawa 1929 (litographed).

[10] John Stuart Mill, *A System of Logic, Ratiocinative and Inductive*, Longmans, London 1865.

[11] Andrzej Mostowski, *Logika matematyczna* (*Mathematical Logic*), Monografie Matematyczne, Vol. 18, Warszawa–Wrocław 1948.

[12] Otto Neurath, 'Protokollsätze', *Erkenntnis* **3**, 204–214.

[13] Karl Popper, 'Logik der Forschung', *Schriften zur wissenschaftlichen Weltauffassung* **9**, Vienna 1935.

[14] Franciszek Smolka, 'Paradoksy logiczne a logika trójwartościowa' ('Logical Paradoxes and Three-Valued Logic'). The author's review of the lecture delivered in Lvov on July 2nd, 1920, *Ruch Filozoficzny* **5**, 171.

[15] Dina Stejnbarg, 'Pojęcie prawa przyrodniczego u J. St. Milla' ('J. St. Mill's Concept of Natural Law'), *Przegląd Filozoficzny* **34**.

[16] Alfred Tarski, 'Der Wahrheitsbegriff in den formalisierten Sprachen', *Studia Philosophica* **I** (1936). Originally published in Polish in 1933.

[17] Alfred Tarski, 'O ugruntowaniu naukowej semantyki' ('On the Foundation of Scientific Semantics'), *Przegląd Filozoficzny* **39**.

[18] Alfred Tarski, 'O pojęciu wynikania logicznego' ('On the Concept of Logical Consequence'), *Przegląd Filozoficzny* **39**.

[19] Kazimierz Twardowski, *O tak zwanych prawdach względnych. Rozprawy i artykuły filozoficzne* (*On So-Called Relative Truths. Philosophical Papers and Articles*). Collected and edited by disciples, Lwów 1927.

JERZY ŁOŚ

SEMANTIC REPRESENTATION OF THE PROBABILITY OF FORMULAS IN FORMALIZED THEORIES*

In probability theory, or rather in its foundations, there has long been a trend in favour of identifying events, i.e., objects to which probability is ascribed, with formulas of certain theories. Without adducing arguments in favour of that idea I shall confine myself to mentioning its principal representatives, namely J. M. Keynes, J. Nicod, H. Jeffreys, H. Reichenbach, R. Carnap, and in Poland J. Łukasiewicz and K. Ajdukiewicz.

It is, of course, formally possible to ascribe probability to formulas, since formulas form a Boolean algebra (the term 'sentences' is sometimes used instead of 'formulas', but I shall not use it since I want to make a distinct difference between sentences and sentential functions). But it does not seem that the interpretation of formulas of a language as events is always the same. Moreover, it seems that at least two interpretations can be distinguished, the confusion of which occasionally leads to errors. One such error has been discovered when Nicod's works were studied in my seminar in the Polish Academy of Sciences' Institute of Philosophy and Sociology in the academic year 1958/9. Roughly speaking, it consists in that Nicod confuses the probability of appearance of a causal relationship between two phenomena with the probability of existence of such a relationship. It seems that we have to do here with two quite different probabilities. In the first case we are concerned with a kind of a sentential function in the form of the implication: $A(x) \rightarrow B(x)$, the question of what is the probability of that relationship being the question of what is the probability (or frequency) of drawing by lot (or obtaining) such an individual (object, moment, point in space – according to interpretation) which would satisfy that sentential function. In the second case we are concerned with the probability of the world

327

we live in possessing a certain characteristic (namely that A is the cause of B). It would be difficult to say in what sense the term 'probability' might be used here, since we in no case draw at random the world we live in, but that is not the point. The point is that we face the necessity of making a distinction between the probability of a sentential function, understood as the probability of its being satisfied by elements chosen in a certain way, and the probability of a sentence whose truth depends not on such and such elements, but on the whole of the relationships among those objects which form the universe of discourse, in a word, on the model in question.

In less general considerations than those which refer to the principle of causality the issue loses its metaphysical aspect of 'drawing a world at random'. Let us consider the following example. Suppose we investigate the theories of the ordering of a set by the relation $<$. Let us reflect, how to interpret the statement that the sentential function $x_1 < x_2$ has the probability $\frac{1}{2}$: $P(x_1 < x_2) = \frac{1}{2}$. Apparently this means (in the frequency interpretation) that by drawing, in a given way, the elements x_1 and x_2 from a given ordered set we obtain elements which in one half of the cases will satisfy the sentential function in question.

But it is obvious that although $P(x_1 < x_2) = \frac{1}{2}$, nevertheless $P(\prod_{x_1} \prod_{x_2} x_1 < x_2) = 0$, and that because the formula with the probability of which we are now concerned is a sentence, and moreover a sentence that is false in any ordered set.

But what about the probability of the sentence $\sum_{x_1} \prod_{x_2} x_1 \leqslant x_2$?

If the ordered set in question has been determined then that sentence, which expresses the existence of the least element in that set, will be true or false in that set, and hence will have the probability 1 or 0.

It seems possible for that sentence to have a probability other than 1 or 0, e.g., $P(\sum_{x_1} \prod_{x_2} x_1 \leqslant x_2) = \frac{1}{2}$. But in such a case we must imagine that there is a given class of ordered sets and a given way of drawing its elements. The probability of the sentence $\sum_{x_1} \prod_{x_2} x_1 \leqslant x_2$ is the probability of drawing from that class such an ordered set in which the least element exists.

In this way we have as it were two probabilities, one for sentential function, when – given a certain model – we draw elements and inquire

whether they satisfy that function, and the other for sentences, when we draw a model (in the example above: an ordered set) and inquire whether the sentence is true in that model.

Of course, nothing prevents us from applying the first case to sentences or the second case to sentential functions. But then in the first case all the sentences will have probabilities equal to 1 or 0, and in the second case the sentential functions will have the same probabilities as their generalizations (covering of all variables by universal quantifiers). Thus, we shall have for instance $P(x_1 < x_2) = P(\prod_{x_1} \prod_{x_2} x_1 < x_2) = 0$.

What an interpretation is then to be given to formulas if these two extreme cases are to be avoided? It seems that there is a middle course. We may interpret the probability of a formula as the probability of its being satisfied by a certain sequence of elements obtained by double drawing: first we draw a model in accordance with a probability given for the class of models, and next from that model, also in accordance with a probability given in that model, we draw a sequence of elements.

For that procedure we must have a class of models $\{M_t\}_{t \in T}$ and a probability, to be symbolized μ, in that class, or rather in the set T, for the models belonging to the class concerned are indexed by elements of that set. Finally, we must have a probability in the class of sequences of every model M_t; let that probability be symbolized v_t. Then every formula α, whether a sentence or a sentential function, has its probability $v_t(\alpha)$ in the model M_t. By fixing α and changing t in the set T we obtain changing values of $v_t(\alpha)$ (if α is a sentence, these values will be only 0 and 1), and thus we have to do with a function in the set T in which the probability μ is defined.

This is a random variable (certain conditions of measurability must be satisfied here) for which we can compute the expected value $E_\mu(v_t(\alpha))$. That expected value is a number which depends only on α.

Let us put

(0) $\qquad P(\alpha) = E_\mu(v_t(\alpha))$.

In this way we define a certain probability in the set of all formulas α.

In order to explain certain details of this way of defining the probability of formulas, and at the same time to impart precision to the concepts involved, we shall refer to a simplified example.

Let the class of models consist exactly of four models: $\{\mathbf{M}_i\}$, $i=1, 2,$ 3, 4. Every model, as usual, consists of the set A_i and the relations $R_j^{(i)}$ which are interpretations for the primitive signs of the theory: $\mathbf{M}_i = \langle A_i, R_1^{(i)}, R_2^{(i)}, \ldots \rangle$. Let the symbol S stand for the set of the formulas belonging to our theory, and $A_i^{\omega_0}$ for the set of sequences of the elements of the set A_i, i.e., an element x of $A_i^{\omega_0}$ is an infinite sequence $x = \langle x_1, x_2, \ldots \rangle$ where all x_j are elements of A_i.

For every α belonging to S, let $\sigma_i(\alpha)$ stand for the set of those sequences from $A_i^{\omega_0}$ which satisfy α; let it further be supposed that in every set $A_i^{\omega_0}$ there is given the probability $\bar{v}_i$, which is anyhow defined for all the sets $\sigma_i(\alpha)$.

Let us put $v_i(\alpha) = \bar{v}_i(\sigma_i(\alpha))$. In this way the probabilities in S are defined. It can easily be verified by the formula

(1) $$\sigma_i(\alpha \vee \beta) = \sigma_i(\alpha) + \sigma_i(\beta),$$

where $\vee$ symbolizes disjunction, and $+$ the addition of sets. If the conjunction $\alpha \wedge \beta$ is contradictory then the sets $\sigma_i(\alpha)$ and $\sigma_i(\beta)$ are disjoint, and the assumption that $\bar{v}_i$ is a probability function leads to additivity:

(2) $$v_i(\alpha \vee \beta) = v_i(\alpha) + v_i(\beta).$$

Other conditions required of probability (that $v_i(\alpha)$ should range between 0 and 1, and that $\bar{v}_i(\alpha)$ should be 1 for tautologies) result directly from the assumption that v_i is a probability.

Let it now be supposed that the probability μ is defined in the set of the indices of models, i.e., in our case, in the set of the numbers 1, 2, 3, 4. In this case, since the set is finite (it consists of four elements), this means simply that the numbers from 1 to 4 are correlated with four non-negative numbers $\mu(1)$, $\mu(2)$, $\mu(3)$ and $\mu(4)$ whose sum is 1. Hence every $v_i(\alpha)$ becomes a random variable of the parameter i. The expected value of that random variable is computed by taking the average of its values, weighted by the values of probability:

(3) $$E_\mu(v_i(\alpha)) = \mu(1)v_1(\alpha) + \mu(2)v_2(\alpha) + \mu(3)v_3(\alpha) + \mu(4)v_4(\alpha).$$

The expected value — not only in this case, but in general — is additive, which means that the expected value of a sum is the sum of the expected values. Hence if we take two expressions, α and β, belonging to S, and

on the strength of (3) compute the expected value of the function $v_i(\alpha) + v_i(\beta)$, we obtain

(4) $\qquad E_\mu\big(v_i(\alpha) + v_i(\beta)\big) = E_\mu\big(v_i(\alpha)\big) + E_\mu\big(v_i(\beta)\big).$

From this and from (2) it follows that when the conjunction $\alpha \wedge \beta$ is contradictory we have

(5) $\qquad E_\mu\big(v_i(\alpha \vee \beta)\big) = E_\mu\big(v_i(\alpha) + v_i(\beta)\big) = E_\mu\big(v_i(\alpha)\big) + E_\mu\big(v_i(\beta)\big).$

The expected value of (3) no longer depends on i, as can be seen from the right side of that formula. For fixed probabilities μ, v_1, v_2, v_3, v_4 it depends only on α, so that we may put

$\qquad P(\alpha) = E_\mu\big(v_i(\alpha)\big).$

The function of a formula, when so defined, is, as can easily be verified, a probability. In particular, if $\alpha \wedge \beta$ is contradictory, then

$\qquad P(\alpha \vee \beta) = P(\alpha) + P(\beta)$

results from (5).

In order to give more precision to the example in question let us assume that we consider theories of a densely ordered set, and let the models $\mathbf{M}_i$ be: 1. the open segment $A_1 = (0, 1)$; 2. the closed segment $A_2 = [0, 1]$; 3. the segment $A_3 = (0, 1]$, open on the left and closed on the right; 4. the segment $A_4 = [0, 1)$, closed on the left and open on the right. All those segments are ordered by the ordinary relation 'lesser than'. (Since we consider theories with one primitive concept, the models include only one relation which interprets them.) Let it further be assumed that $\mu(1) = 1/8$, $\mu(2) = 3/8$, $\mu(3) = 2/8$, $\mu(4) = 2/8$, and finally that the probabilities $\bar{v}_i$, defined in the sets of sequences of numbers from the corresponding segment A_i are probabilities connected with such a choice: there is such a way of drawing numbers from a segment that the probability of drawing a point from every subsegment is equal to the length of that subsegment, and in order to draw a sequence we perform infinitely many independent draws with replacement. This way of defining probability may appear inexact, but in fact it is quite precise. In mathematics probability thus defined is called product probability of Lebesgue's measure in the product $A_i^{\omega o}$. It can easily be demonstrated that

for every probability $\bar{v}_i$ there is $\bar{v}_i\left(\sigma_i(x_1\leqslant x_2)\right)=v_i(x_1\leqslant x_2)=1/2$, whence

$$P(x_1\leqslant x_2)=E_\mu\left(v_i(x_1\leqslant x_2)\right)$$

$$=1/8\cdot 1/2+3/8\cdot 1/2+2/8\cdot 1/2+2/8\cdot 1/2=1/2\,.$$

But of course $\sigma_i\left(\prod_{x_1}\prod_{x_2} x_1\leqslant x_2\right)$ is an empty set for every $i=1,2,3,4$, for $\prod_{x_1}\prod_{x_2} x_1\leqslant x_2$ is a false sentence in every model. Hence it follows that

$$\bar{v}_i\left(\sigma_i(\prod_{x_1}\prod_{x_2} x_1\leqslant x_2)\right)=v_i\left(\prod_{x_1}\prod_{x_2} x_1\leqslant x_2\right)=0\,.$$

On the other hand, the sentence $\sum_{x_1}\prod_{x_2} x_1\leqslant x_2$, which states the existence of a least element, is true in the second and the fourth model, and false in the first and the third. Thus $\sigma_i\left(\sum_{x_1}\prod_{x_2} x_1\leqslant x_2\right)=A_i^{\omega 0}$ for $i=2,4$, and $=0$ for $i=1,3$. Consequently, $v_i\left(\sum_{x_1}\prod_{x_2} x_1\leqslant x_2\right)=1$ for $i=2,4$, and $=0$ for $i=1,3$. On computing the probability P we obtain

$$P\left(\sum_{x_1}\prod_{x_2} x_1\leqslant x_2\right)=0\cdot 1/8+1\cdot 3/8+0\cdot 2/8+1\cdot 2/8=5/8\,.$$

Thus we see that both sentential functions and sentences can have here probabilities other than 0 and 1.

In the case of an infinite set T, and hence of an infinite set of models $\{M_t\}_{t\in T}$, the computation of the expected value $E_\mu(v_t(\alpha))$ is not so simple as in the formula (3). It is expressed by the abstract integral

$$(6)\qquad E_\mu\left(v_t(\alpha)\right)=\int_T v_t(\alpha)\mu(dt),$$

and for the existence of that integral it suffices that the integrated function $v_t(\alpha)$ (which is anyhow bounded) should be measurable, that is, that for every number l the set of those t in T for which $v_t(\alpha)<l$ should have a definite probability μ. The condition is not trivial, since probability in infinite sets T is usually defined not for all the subsets of T, but only for a certain field of such subsets.

An additional condition for the existence of the expected value (6) is that the probability μ should be defined over a denumerably additive field of subsets of T and should itself be denumerably additive. Hence it must be so that if μ is defined for each of the sets $X_1, X_2, \ldots$ (*ad inf.*),

then it is also defined for their sum $\overset{\infty}{\underset{i=1}{\bigcup}} X_i$, and moreover if the sets $X_1, X_2, \ldots$ are pairwise disjoint, then

$$\mu\left(\bigcup_{i=1}^{\infty} X_i\right) = \sum_{i=1}^{\infty} \mu(X_i).$$

The probabilities $\bar{v}_i$ need not satisfy these conditions; it suffices if they satisfy the condition of additivity for two pairwise disjoint sets.

What has been stated above and explained by examples is a certain semantic method of defining probability of formulas. It is semantic for we start from probability in models and in sets of models and then define the probability of formulas by making use of the semantical concepts. In order to avoid misunderstandings we shall describe that method once more, this time in a purely formal fashion.

Let S be the set of all formulas built of given constant predicates and such that individual variables are the only free variables. Let further $\{\mathbf{M}_t\}_{t \in T}$ be the class of models interpreting formulas belonging to S, and for every t belonging to T let v_t be a probability defined in the set of infinite sequences of elements of the model $\mathbf{M}_t$ and such that for every α belonging to S the set $\sigma_t(\alpha)$ of those sequences which satisfy α has a definite probability. Let finally μ be a probability in T, denumerably additive and defined over a denumerably additive field, such that for a given α from S, $v_t(\alpha) = \bar{v}_t(\sigma_t(\alpha))$ as a function of t is measurable with respect to μ. With these assumptions, the function P of the formula α, defined as the expected value of the function $v_t(\alpha)$ with respect to the probability μ : $P(\alpha) = E_\mu(v_t(\alpha))$, is a probability in S.

Let us now pass to the probabilistic intuitions connected with the procedure described above. 'Probabilistic' means not connected with the fact that the events under consideration are formulas. For that purpose let us imagine that we have the bag W with cubes marked with the numbers 1, 2, 3, 4, each cube with only one number, and further four boxes U_1, U_2, U_3, U_4, containing balls marked with the letters a, b, c, d, also each ball with only one letter. In this scheme, which is known as the 'bag and boxes scheme', when we know the probabilities of drawing the cubes marked with numbers and the probabilities of drawing from the various boxes balls marked with letters, we can compute the probability of drawing a ball marked with a given letter, assuming that we

first draw a cube from the bag W, and next a ball from that box which bears the number that marks the cube that has been drawn first.

If we draw 'honestly' both from the bag and from the boxes, i.e., if the probability of drawing a given letter or a given number depends on what a given box or the bag contains, then from the data pertaining to the experiment — just from what is contained in the bag and the boxes — we can compute the probability of 1° that a given number will be drawn from the bag; let these probabilities be $\mu(1)$, $\mu(2)$, $\mu(3)$, $\mu(4)$; 2° that a given letter will be drawn from the box U_t, where t stands for one of the numbers 1, 2, 3, 4; let these probabilities be $v_t(a)$, $v_t(b)$, $v_t(c)$, $v_t(d)$. These latter are conditional probabilities. To compute the probabilities of drawing the letter $x(=a, b, c, d)$ in the whole experiment we must resort to the formula

(7) $$P(x)=\mu(1)v_1(x)+\mu(2)v_2(x)+\mu(3)v_3(x)+\mu(4)v_4(x)$$

which is analogous with the formula (3).

The probability of drawing a or b is given by the formula:

$$P(a \text{ or } b)=\mu(1)[v_1(a)+v_1(b)]+\mu(2)[v_2(a)+v_2(b)]+$$
$$+\mu(3)[v_3(a)+v_3(b)]+\mu(4)[v_4(a)+v_4(b)].$$

Let it be noted that nothing changes in these considerations if it is assumed that some boxes contain only balls marked with the letters a and b, and the others, only balls marked with the letters c and d. In such a case for some boxes $v_i(a)+v_i(b)=1$ and $v_i(c)+v_i(d)=0$, and for the others $v_i(a)+v_i(b)=0$ and $v_i(c)+v_i(d)=1$. The probability $P(a \text{ or } b)$ is then the sum of those $\mu(i)$ for which $v_i(a)+v_i(b)=1$. It need be neither 0 nor 1, although the relative probability: that of drawing a or b from a given box always is 0 or 1.

We have considered the determination of the probabilities of drawing balls with appropriate numbers on the basis of the knowledge of the principles of drawing and the knowledge of what the bag and the boxes contain. Let us now consider a problem which is reverse in a sense: can the bag and the boxes be made to contain such cubes and balls, respectively, that the probabilities of drawing balls marked with the various letters should have the values determined in advance: $P(a)=w_1$, $P(b)=w_2$, $P(c)=w_3$, $P(d)=w_4$? Of course, $w_1+w_2+w_3+w_4$ must be 1,

and $w_i \geqslant 0$, for $i = 1, 2, 3, 4$, but if it is so, the problem can be solved without difficulty. It suffices so to adjust the balls contained in the boxes that the balls marked with the letter a should be w_1, the balls marked with the letter b should be w_2, etc. (w_i must be rational numbers, for otherwise only an arbitrarily exact approximation can be obtained — this difficulty will be disregarded here as inessential for further considerations).

As result we obtain what we want to have, regardless of what the bag contains during the first draw.

Since the problem is easy to solve, let us make it a little more complicated. Let be required that the boxes 1 and 3 contain only balls marked with the letters a and b, and the boxes 2 and 4, only balls marked with the letters c and d; in other words, let $v_i(a) + v_i(b)$ be 1 for $i = 1, 3$, and let it be 0 for $i = 2, 4$.

This can be done, too, even so that the numbers 3 and 4 may not be represented at all in the bag, so that $\mu(3) = \mu(4) = 0$. Let it be noted that under such conditions it follows that $P(1) = P(a) + P(b) = w_1 + w_2$ and $P(2) = P(c) + P(d) = w_3 + w_4$. From the formula (7) we obtain

$$w_1 = P(a) = \sum_{i=1}^{4} \mu(i) v_i(a) = \mu(1) v_1(a) = (w_1 + w_2) v_1(a). \quad \text{Hence} \quad v_1(a)$$

$$= \frac{w_1}{w_1 + w_2}$$ is the sufficient and necessary condition. Other probabilities are computed in an analogous manner.

When computing probabilities of formulas we have to do with a procedure which is quite similar to the box scheme described above, although it is much more complicated. But like in the box scheme we have double drawing, first of the model — which corresponds to drawing a cube from the bag — and then of a sequence of elements from the model drawn by lot — which in turn corresponds to drawing a ball from a box. As in the case of the box scheme we may here pose the question: given the probability P for the expression S of a certain theory, can we so select the models $\{\mathbf{M}_t\}_{t \in T}$, the probability μ in the set T, and the probabilities v_t for sequences of elements from the models $\mathbf{M}_t$, that the formula (0) should hold, i.e., that

$$P(\alpha) = E_\mu(v_t(\alpha))$$

should hold for every formula α from S?

Note that we are in a similar situation as in the case of the box scheme, when it was required that certain boxes should contain exclusively balls marked with the letters a and b, and the others, only balls marked with the letters c and d. However the models $\mathbf{M}_t$ be selected, sentences (formulas without free variables) will be true or false in those models, so that for the sentences α we shall have either $v_t(\alpha)=1$ or $v_t(\alpha)=0$.

The answer to the question posed above is in the affirmative: it is possible so to select the models $\mathbf{M}_t$ and the corresponding probabilities μ and v_t.

But before that answer in the affirmative is formulated as a theorem, let the formal properties of formulas and probability be examined. Let S be the set of formulas. There is no need to assume that the formulas belonging to S are elementary, that is, that they include only individual variables. It must be assumed, however, that all the formulas in S are of a certain fixed type, but only such in which the variables of higher types are quantifier-bound.

There is no need to explain the first assumption: the point is that it should be permitted to connect formulas belonging to S by sentential connectives without going outside the set S. The second assumption makes us possible to confine ourselves to drawing a sequence of elements of models (the probability v_t) without the need of drawing, for instance, subsets. The latter would hold if the formulas belonging to S would include variables ranging over sets. It seems that the question: what is the probability of the formula $\alpha(A)$?, where A is a set, means the same as the question: what is the probability of drawing the set A which satisfies α?

Now let Z stand for the set of sentences belonging to S (i.e., formulas without free variables), and let $Cn(X)$ stand for the set of consequences that can be deduced from the subset X of S. Remember that 1° Both S and Z are closed under sentential connectives, in particular alternation, conjunction and negation; 2° We call a system such a set X for which $Cn(X)=X$; 3° Every system which includes a formula also includes all the substitutions of that formula; 4° For every system X holds the formula $Cn(X \cap Z)=X$.

Let now X_0 be a consistent system $(X_0 \neq S)$. By probability in S, related to X_0, we mean the function $P(\alpha)$, defined for α belonging to S,

with values ranging between 0 and 1, equal to 1 for the formulas belonging to X_0, and such that if the conjunction of two formulas, α and β, is contradictory in the sense that its negation belongs to X_0 $((\alpha \wedge \beta)' \in X_0)$, then for their disjunction we have $P(\alpha \vee \beta) = P(\alpha) + P(\beta)$.

It can be seen from the above that if P is a probability related to X_0, and X_1 is a system contained in X_0, then P also is a probability related to X_1.

Now let P be a probability related to X_0, and let F stand for the set of those formulas α belonging to S for which $P(\alpha) = 1$. Obviously, X_0 is included in F, but although F need not be a system (for instance, F need not be closed under substitutions, it may be so that $P(\alpha(x_1)) = 1$ and $P(\alpha(x_2)) < 1$), yet there is a greatest system included in F, namely the system $Cn(F \cap Z)$. Of course, P is a probability related to that system.

As can be seen, from the formal point of view nearly every probability may be treated as related to the various systems (unless $Cn(F \cap Z)$ is the system of tautologies); if it is so, what is the meaning of relating P to X_0 and not to some other X_1, included in X_0?

The meaning of relating it precisely in such a way will be explained by an example. Let X_1 be included in X_0, which is in turn included in F, and let α be a sentence that belongs to X_0 but not to X_1. If we treat P as related to X_0, then we think that whatever model be drawn, α will be true in it. But if we treat P as related to X_1, we do not exclude the falsehood of α in a model of the class from which we draw; the only provision is that the subclass of those models in which α is false must have probability equal to 0.

Thus, the relating of probability to a certain definite system X_0 restricts its interpretation by drawing of models and from models more than does the computation of that probability.

If $S(\mathbf{M}_t)$ stands for the set of the true formulas in the model $\mathbf{M}_t$ then if we want to interpret, in the class of models $\{\mathbf{M}_t\}_{t \in T}$, the probability P related to X_0, we should require that X_0 be included in every $S(\mathbf{M}_t)$, but not more, so that the intersection of all $S(\mathbf{M}_t)$ should give exactly X_0. In the case of elementary theories such a class of models can always be found for a given system. This is confirmed by Gödel's theorem on the completeness of the first-order functional calculus, which—in a formulation that suits our purpose well—says that: for every consistent system X there is a model $\mathbf{M}$ such that $X \subset S(\mathbf{M})$.

Godel's theorem does not hold for non-elementary theories and there-fore if we want to obtain, for a given probability P related to X_0, a class of models in which that probability can be interpreted properly, we must accept the condition that there exists for X_0 such a class of models $\{M_t\}_{t \in T}$ that X_0 is exactly the common part of all the $S(M_t)$. A system which has that property will be called ω-regular. Note that a system X_0 for which there exists no such model M that $X_0 \subset S(M)$ certainly is not ω-regular (such a system is called ω-inconsistent), but the existence of one such model does not ensure ω-regularity; hence ω-regularity is a stronger property than ω-consistency.

Let now X stands for an arbitrary system including X_0, to which the probability P is related (X may include F or not). As we know, we have $Cn(X \cap Z) = X$. Let the set $X \cap Z$ of all the sentences from X be arranged as the sequence $\zeta_1, \zeta_2, \ldots$, and let z_n stand for the conjunction of the first n sentences of that sequence. The sequence of probabilities, $P(z_1), P(z_2), P(z_3), \ldots$, is certainly non-increasing, and hence it may converge to zero or to some number greater than zero. In the latter case we have $P(z_n) \geqslant \varepsilon > 0$ for all n. But at the same time we have

$$\bigcup_{n=1}^{\infty} Cn(z_n) = Cn(X \cap Z) = X.$$

It seems therefore that should we like to extend the probability P so that it should cover systems as well, we would have to impart to the system X the probability ε. Here the difficulty emerges for if that prob-ability is understood as a drawing of models then this would have to mean that the drawing of such a model M_t in which every formula belonging to X would be true, i.e., $X \subset S(M_t)$, has the probability $\varepsilon > 0$; yet, if the given theory is not elementary, it may happen that such models do not exist at all, in other words, that the system X is ω-inconsistent. We must protect ourselves against such a possibility by imposing upon P a condition which, in the case of $P(z_n) \geqslant \varepsilon > 0$, would guarantee the existence of an appropriate model. In order to avoid the formation of the sequence ζ_i and the conjunctions z_n, let us formulate that condition as follows:

(C)   If there exists such an $\varepsilon > 0$ that for every sentence ζ belong-ing to the system X we have $P(\zeta) \geqslant \varepsilon$, then X is not an ω-in-consistent system.

A probability that satisfies the condition (C) will be called continuous. Note again that the condition (C) is essential only in the case of non-elementary theories. In elementary theories there are no ω-inconsistent systems, and hence every probability is continuous.

These introductory considerations may now be followed by the final formulation of the theorem on the semantic representation of probability.[1]

THEOREM. For every continuous probability in S, related to an ω-regular system X_0, and for every family of models $\{M_t\}_{t \in T}$ such that for every system Y, complete and ω-consistent and including X_0, there is such a t in T that $Y = S(M)_t$, there exist

1° denumerably additive probability μ defined for the subsets of T;

2° probabilities v_t, each defined for the subsets of the set $A_t^{\omega o}$ which consists of sequences of elements of the model M_t such that if we symbolize by $\sigma_t(\alpha)$ the set of those sequences from $A_t^{\omega o}$ which satisfy α we have for every α from S:

$$P(\alpha) = E_\mu\big(v_t(\sigma_t(\alpha))\big) = \int_T v_t\big(\sigma_t(\alpha)\big)\mu(dt).$$

The proof of that theorem does not involve difficulties from the mathematical point of view. Given the family $\{M_t\}_{t \in T}$, we first define the measure for subsets so that to every sentence ζ from Z we ascribe the set $T(\zeta)$ of those t for which ζ is true in M_t. Next we put $\mu(T(\zeta)) = P(\zeta)$ and in this way the probability μ for the sets $T(\zeta)$ is given. Availing ourselves of the continuity of P and of the assumption concerning the family $\{M_t\}_{t \in T}$ we extend μ as to become denumerably additive probability (Kolmogorov's theorem on the extension of measure). Finally, μ being already given, we determine v_t by means of the Radon–Nikodym theorem.

This outline does not tantamount to a proof, but the mathematical apparatus to be used in this connection is unfortunately too complicated for a complete proof to be given here.

Two important remarks must be made here.

First, the probability μ is unique, the probabilities v_t are not unique, but the family $\{v_t\}_{t \in T}$ is unique 'almost everywhere'. This means that if both μ, $\{v_t\}_{t \in T}$ and μ', $\{v_t'\}_{t \in T}$ satisfy the thesis of the new theorem, then $\mu = \mu'$, and by drawing from the set T according to the probability

μ we have the probability 0 of drawing such a t that $v_t \neq v_t'$ (in other words: $\mu \{t \in T: v_t \neq v_t'\} = 0$).

Second, throughout all these considerations we have been making the tacit assumption that we have to do with an ordinary theory, in which the set S is denumerable. This is a very natural assumption, which was not questioned for many years, but in recent times investigations have, for various reasons, covered theories in which the set S is non-denumerable.

The reservation must be made in advance that for such theories our theorem cannot be demonstrated with the methods outlined above. The Radon–Nikodym theorem simply does not suffice to determine the probabilities v_t. Still worse, I am inclined to believe that for non-denumerable theories the theorem given above is just not true.

The theorem given above amounts to what is stated in the title of the present paper: a semantic representation of the probability of formulas in formalized theories. It also enables us to make a natural transition from probability defined over formulas to a probability connected with a random drawing of a point from a set. In that case it is sometimes said that we have to do with 'Kolmogorov's scheme'; this refers to such an interpretation of probability which fall under Kolmogorov's well-known axiom system.

Thus, this is a path from logical probability to Kolmogorov's scheme. It has long been known that that path can be covered in various ways. The way shown in the present paper is confined to semantics, and hence to the conceptual apparatus most closely connected with logic. This fact seems to indicate that this is a proper path to follow.

REFERENCE

* First published in *Studia Logica* **XIV** (1963). Reprinted from *The Logico-Algebraic Approach to Quantum Mechanics*, Vol I, C. A. Hooker (ed.), D. Reidel Publishing Company, Dordrecht-Holland, 1975.

[1] The full proof of this theorem may be found in J. Łoś, 'Remarks on Foundations of Probability, Semantical Interpretation of the Probability of Formulas', *Proceedings of the International Congress of Mathematicians 1962*, Stockholm 1963, pp. 225–229.

SEWERYNA ŁUSZCZEWSKA-ROMAHNOWA

CLASSIFICATION AS A KIND OF DISTANCE FUNCTION. NATURAL CLASSIFICATIONS *

1. INTRODUCTION

There exists a close analogy between the theoretical procedure of classifying a domain of objects and the physical operation of segregational ordering of a collection of bodies, which in the simplest case consists in first forming a number of heaps of the elements of the collection to be ordered and then in arranging these heaps so as to form clusters of heaps and eventually also clusters of clusters of heaps, etc.

Indeed, it is possible to consider a classification (and so it is often considered) as a plan of such a segregational ordering either of the whole classified domain, provided this domain is a collection of bodies succeptible to such ordering, or of some subclasses of this domain.

In view of this relation between both procedures it seems natural to conceive of classification as establishing a *distance* between components of each pair of elements of the classified domain.

Let us assume, for example, that a certain classification may be presented in the form of the following diagram:

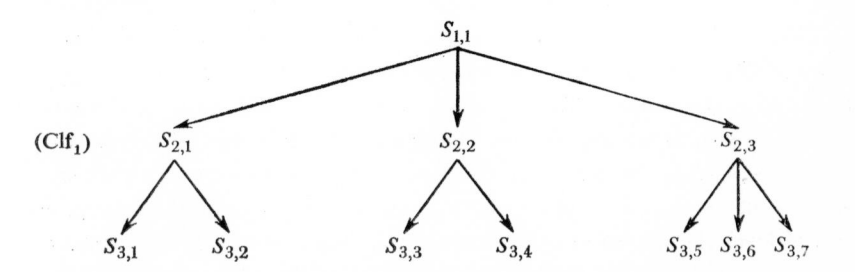

This diagram is to be interpreted as follows. The classified domain, or the *space* of the classification Clf_1, as it will be referred to, is the class $S_{1,1}$. This class has been first divided into disjoint and adding

up to $S_{1,1}$ classes $S_{2,1}$, $S_{2,2}$, $S_{2,3}$; a similar division of each of the latter results in classes $S_{3,1}$, ..., $S_{3,7}$, which form the last level of Clf_1.

The classification Clf_1 distinguishes three kinds of pairs of elements of the space $S_{1,1}$: pairs x, y such that both x and y belong to the same class of the last level of the classification (e.g. $x \in S_{3,5}$ and $y \in S_{3,5}$), pairs x, y such that x and y belong to different classes of the last level but to the same class of the second level (e.g. $x \in S_{3,5}$, $y \in S_{3,6}$, $x \in S_{2,3}$, $y \in S_{2,3}$), and pairs x, y such that the only class within Clf_1 to which both x and y belong is $S_{1,1}$.

The distance established by the classification Clf_1 between the elements of the pairs of the first kind is the smallest — let us denote it conventionally by 0; the distance corresponding to pairs of the second kind is greater — this will be denoted by 1; the distance corresponding to pairs of the third kind and denoted by 2 is the greatest.

Let $x \mathbin{|\overline{Clf_1}|} y$ stand for *the distance between x and y as determined by* Clf_1. Then the following holds: $x \mathbin{|\overline{Clf_1}|} y = 0$ *if and only if there exists a class $S_{3,i}$ such that $x \in S_{3,i}$, and $y \in S_{3,i}$.*

$x \mathbin{|\overline{Clf_1}|} y = 1$ *if and only if $x \mathbin{|\overline{Clf_1}|} y \neq 0$ and there exists a class $S_{2,i}$ such that $x \in S_{2,i}$ and $y \in S_{2,i}$.*

$x \mathbin{|\overline{Clf_1}|} y = 2$ *if and only if $x \mathbin{|\overline{Clf_1}|} y \neq 0$ and $x \mathbin{|\overline{Clf_1}|} y \neq 1$ and $x \in S_{1,1}$ and $y \in S_{1,1}$.*

$x \mathbin{|\overline{Clf_1}|} y$ is therefore a function which assumes the values 0 or 1 or 2 for different pairs of elements of the space of the classification Clf_1. A function $x \mathbin{|\!-\!\!-\!\!-\!|} y$ analogously determined for an arbitrary n-level classification (provided we have a proper definition of such a classification) will assume values $0, ..., n-1$.

Two different n-level classifications Clf_i and Clf_j having the same space differ from each other in that the corresponding functions $x \mathbin{|\overline{Clf_i}|} y$ and $x \mathbin{|\overline{Clf_j}|} y$ assume different values at least for some pairs of elements of this space. Similarly, if one segregational ordering of a collection of bodies, e.g. the distribution of all books of a library on various shelves, in various bookcases and in various rooms (which would correspond to a 4-level classification of the collection of the books of this library), is replaced by another, there will be pairs of books whose elements were located on the same shelf or in the same bookcase or in the same room in the original ordering, whereas in the new one they will be placed on different shelves or in different cases or in different rooms.

If classification is approached as a procedure establishing certain distances between the elements of the classified domain, then a certain explication of the concept of *natural classification* suggests itself. According to this explication *a classification Clf_i of a given class of things S is a natural classification of this class, if the elements of S separated with respect to Clf_i by a smaller distance than some others are also in reality closer to each other than those latter.* In other words *if for arbitrary elements x, y, z, r of the class S — if $x \overline{|_{Clf_i}|} y < z \overline{|_{Clf_i}|} r$, then x and y are in reality closer to each other than are z and r.*

What is meant, however, by *closer in reality.* It may mean, e.g. differing less from each other in such and such a respect (for example, separated from each other by a smaller distance in a given property space), or related by closer affinity, or — when S is a set of numbers — having a smaller absolute difference.

From these examples it is apparent that the concept of natural classification must be relativized. It would be impossible at this moment of our analysis to give precise definitions. Let us accept for the time being the following characterization.

A classification Clf_i is a natural classification of a class S with respect to the magnitude of the qualitative difference of such and such a type between the elements of S, or with respect to the magnitude of the degree of the affinity between the elements of S, etc. if for all elements x, y, z, r of the class S, if $x \overline{|_{Clf_i}|} y < z \overline{|_{Clf_i}|} r$, then the magnitude of the qualitative difference of this type between x and y is smaller than the magnitude of such a difference between z and r, or x and y are related by closer affinity than z and r, etc.

The following are two artificial but simple examples of natural classification in the explicated sense.

EXAMPLE I. Let the following diagram represent the genealogy of the third generation of the male offsprings of an individual $\mathscr{J}$ on paternal line

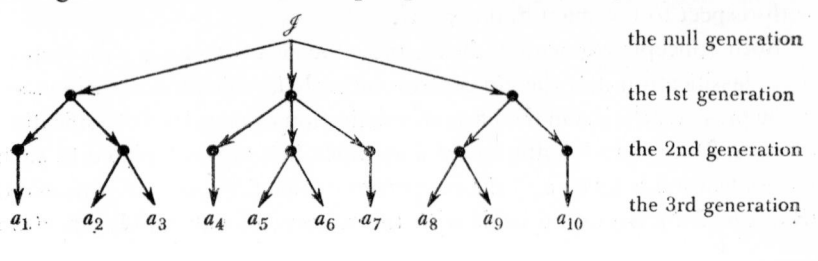

the null generation

the 1st generation

the 2nd generation

the 3rd generation

a_1 a_2 a_3 a_4 a_5 a_6 a_7 a_8 a_9 a_{10}

Let now $S_{1,1}$ be the class of individuals $\{a_1, ..., a_{10}\}$. Let the function $d_{gen}(x, y)$, (to be read as: the degree of affinity between x and y), assume values in accordance with the formula $d_{gen}(x, y) = 3 - r$, where r is the numeral of the latest generation from the null to the 3rd in which x and y have a common ancestor (we assume that each male individual is his own ancestor on paternal line).

It is easy to verify that the following classification of the class $\{a_1, ...$ ..., $a_{10}\}$ is natural, in the sense explicated above with respect to the magnitude $d_{gen}(x, y)$, in other words with respect to the degree of affinity between the elements of the set of the third generation of the offsprings of $\mathscr{I}$.

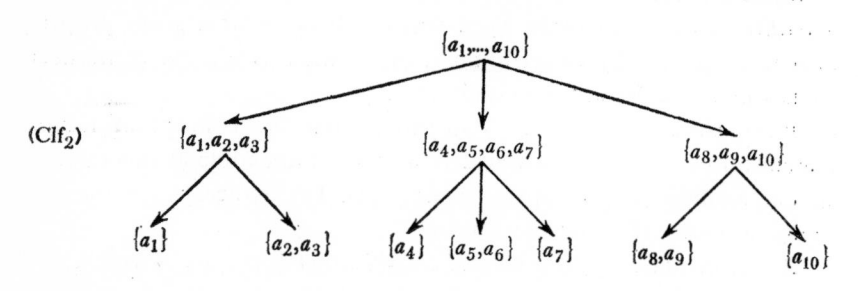

EXAMPLE II. Let us consider the following classification:

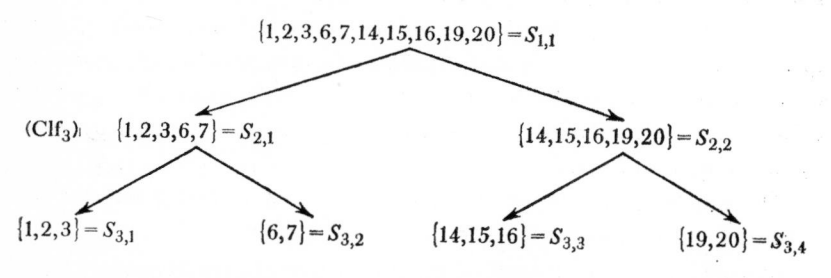

It is easy to see that Clf_3 is a natural classification of the class $S_{1,1}$ with respect to the magnitude $|x - y|$.

Both concepts explicated above, the concept of distance with respect to a classification and the concept of natural classification are discussed in the present article. On the basis of a definition of an n-level classification (Section 2) a general definition of a distance function correlated to such a classification is given and the properties of this function are considered (Sections 3–5). Section 6 deals with the concept of natural classification

with respect to a magnitude defined in the classification space. Besides the definition of this concept the author formulates here a condition necessary and sufficient for the existence of an n-level classification, natural with respect to a given magnitude. The last section discusses two other concepts of natural classification definable in terms of the basic concept introduced in Section 6.[1]

The idea behind this formal treatment was to contribute to the elucidation of formal-logical principles of empirical classification. It makes namely precise a view on classification which has suggested itself in a natural way to systematists, i.e. to those who undertake to solve concerete, empirical classificatory problems.

So, for instance, when discussing various classifications they often use the concept of distance with respect to a given classification between the elements of the classified domain and consider whether to the diversification with respect to this distance corresponds a real diversification of this domain with respect to properties in which they are interested in the given context. Accordingly, the criticism of classifications given by systematists may often be interpreted as showing that the classification in question is unnatural in some respect, and the progress in classifying – as a transition from a classification which in the light of acquired knowledge turned out to be unnatural, to a natural one.

To support what has been said above, here is a rather typical example from the history of biological systematics. The example comes from Cuvier's treatise *Sur un nouveau rapprochement à établir entre les classes qui composent le règne animal*, which was important for the development of methods of biological classification. Now, in this treatise the author criticises a certain classification of the animal kingdom, the classification which – so he says – he made use of before. In this classification we find first a dichotomy of the animal kingdom into the *vertebrata* and the *animals without backbone*, then these are subdivided into a number of groups, known as classes: the vertebrates are subdivided into *mammalia, aves, pisces* and *reptilia*. The other group into: *mollusca, insecta* etc. The defect of this classification, according to Cuvier, consists in that it treats as classes groups of differents levels, "so that, e.g. the class of the *mollusca* so far as the importance of its main characteristics and the diversification of animals belonging to it are concerned, is almost equivalent to the whole group of the *vertebrata*".[2]

Let us note that the above comment of Cuvier may be formulated equivalently as follows. Within the class of *mollusca* belonging to the last level of the discussed classification there are animals not less differentiated than are animals belonging to different classes of the last level within the *vertebrata*. Using the introduced symbolism and designating by Clf_4 the classification criticized by Cuvier we may say: there exist elements of the space of Clf_4: x, y, z, u such that $x \mid \overline{Clf_1} \mid y = 0$, $z \mid \overline{Clf_1} \mid u = 1$, but the difference between x and y is not less than that between z and u, which means that the classification Clf_4 is not natural.

As a result of this criticism Cuvier proposes a new classification of the kingdom of animals, in which he first divides this kingdom into four *phyla* (the *vertebrata*, the *mollusca*, the *arthropoda* and the *coelenterata*), and then divides each of these *phyla* into a number of classes.

The term *natural classification* itself and the distinction between natural and artificial classification were introduced, so it seems, by systematists of the XVIII c., when scientific methods of classifying the domain of living beings were formed. All concepts of natural classification discussed by methodologists and philosophers originated with systematists.[3] J. S. Mill, for instance, took over from biologists the pre-Darwinian, morphological concept of natural classification; Venn introduced into his methodology the Darwinian concept of natural classification as genealogical classification.[4]

As may be seen from the above historical outline, there exist several concepts of naturalness of classification. The concept introduced in this article, scil. the basic concept explicated in Section 6, is so general that several other concepts of natural classification may be treated as its particular cases.

The formulation of such a general concept of natural classification would be impossible without having a generalized concept of distance, i.e. the concept of a *distance function*, and without assuming that the likeness or rather its converse, i.e. the qualitative difference may be conceived as a *distance* (in the generalized sense) between the differing objects.

A classification mirroring the diversification of a given class with respect to such and such a magnitude will be used below, instead of the phrase *a classification of a given class natural with respect to the given magnitude*. In accordance with this terminology the classification Clf_2,

for example, mirrors the diversification of the class $\{a_1, \ldots, a_{10}\}$ with respect to the function $d_{gen}(x, y)$. The expression *mirrors* suggesting the idea of a quasi-semantical relation between the given classification and a certain state of things within the classified domain is used here intentionally, scil. with reference to the terminology of systematists which has a similar connotation. Instead of distinguishing between natural and artificial classifications they sometimes distinguish in a similar sense between *classifications revealing real differences* and *classifications which artificially differentiate things.*

   Is, however, this quasi-semantical terminology justified from the logical point of view? The following remarks may be suggested. The naturalness of the classification Clf_2, for instance, is a conformity relation holding between the quantity $x\,|_{\overline{Clf_2}}|\,y$ and the quantity $d_{gen}(x, y)$. This relation is such that the latter quantity increases with the increase of the former. The former however is, in a sense, a certain element of the logical structure of the classification Clf_2. This is to be understood as follows. If we map by means of an arbitrary one-one relation the space of the classification Clf_2 on itself, so that to every element x of this space there corresponds some element x' of the same space, then each of the classes $S_{i,j}$ of the classification Clf_2 is transformed into a class $S'_{i,j}$, and the system of all classes $S_{i,j}$ forming the classification Clf_2 is transformed into a corresponding system of the classes $S'_{i,j}$ forming the classification Clf'_2. Now, all the *distances* are invariant under any such transformation of Clf_2 into a classification Clf'_2 having the same logical structure as Clf_2. In other words, *for all x, y belonging to the space common to both Clf_2 and Clf'_2,* $x\,|_{\overline{Clf_2}}|\,y = x'\,|_{\overline{Clf_2}}|\,y'$. The quantity $d_{gen}(x, y)$ is, of course, not so related to the structure of the classification Clf_2. It is an empirical quantity. The naturalness of Clf_2 is hence a conformity relation holding between a quantity forming, so to say, a part of the logical structure of Clf_2 and a quantity which is independent of this structure.

2. THE CONCEPT OF CLASSIFICATION

By a classification we shall mean a finite matrix (double sequence) of classes fulfilling certain (to be specified below) conditions. This is one of the meanings of the term classification suggested by the commonly used tree-diagrams of classification. According to this meaning a clas-

sification corresponding to the tree-diagram Clf_1 discussed in the preceding section is identical with the following double sequence of classes

$$S_{1,1}$$
$$S_{2,1}\, S_{2,2}\, S_{2,3}$$
$$S_{3,1}\, S_{3,2}\, S_{3,3}\, S_{3,4}\, S_{3,5}\, S_{3,6}\, S_{3,7}$$

Before specifying the conditions distinguishing classifications among all finite matrices of classes we shall explain the notation applied.

Let $\mathfrak{M}$ be a finite matrix of classes with n rows. The letters r, $\bar{r}$, $\bar{\bar{r}}$ will be used as variables the range of values of which is the class of natural numbers: $[1, \ldots, n]$ (i.e. the class of the numerals of the rows of $\mathfrak{M}$); $i(r)$ will be used to denote the number of elements of the rth row of $\mathfrak{M}$; the letters h^r, $\bar{h}^r$, $\bar{\bar{h}}^r$ will be used as variables the range of values of which is the set of natural numbers: $[1, \ldots, i(r)]$ (the set of the numerals of the elements of the rth row of $\mathfrak{M}$); by $[r, h^r]$ we shall mean the ordered pair the first element of which is the number r, the second the number h^r (thus $[r, h^r]$ denotes the respective index pair of $\mathfrak{M}$); by $\{r, h^r\}$, on the other hand, we shall mean the h^rth element of the rth row of $\mathfrak{M}$, i.e. the class appearing in this place.[5]

Df. 2.1. A matrix of classes $\mathfrak{M}$ with n rows is a *classification* if and only if for all r, h^r, $\bar{h}^r$,

    I. $\{r, h^r\} \neq \Lambda$;

    II. $h^r \neq \bar{h}^r \to \{r, \bar{h}^r\} \cap \{r, \bar{h}^r\} = \Lambda$;

    III. $i(1) = 1$;

    IV. *if* $r < n$, then there are $i(r)$ natural numbers: $k_1^r, \ldots, k_{i(r)}^r$ such that

    a. $i(r+1) = k_1^r + \ldots + k_{i(r)}^r > i(r)$ (hence for at least one of those numbers $k_{h^r}^r$, $k_{h^r}^r > 1$) ;

    b. $\{r, 1\} = \{r+1, 1\} \cup \ldots \cup \{r+1, k_1^r\}$,
    $\{r, 2\} = \{r+1, k_1^r+1\} \cup \ldots \cup \{r+1, k_1^r+k_2^r\}$,

    . . . .

$$\{r, i(r)\} = \{r+1, (\sum_{j=1}^{i(r)-1} k_j^r)+1\} \cup \ldots \cup \{r+1, \sum_{j=1}^{i(r)} k_j^r\}.$$

T. 2.1. $i(1) = 1 < i(2) < \ldots < i(n)$.

(Conditions III and IV of Df. 2.1. According to T.2.1 at least one of the branches of any but the last level of the tree diagram of a classification forks into at least two branches of the next level. Hence of the diagrams

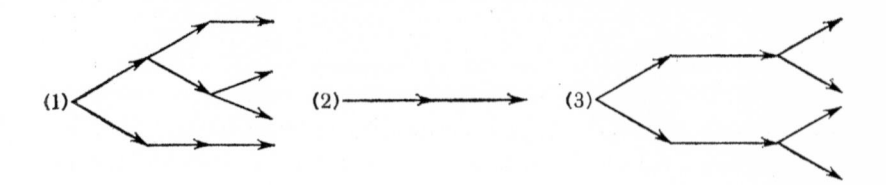

only (1) may be a diagram of a classification).

T. 2.2.   *For all* r, $\{r, 1\} \cup \ldots \cup \{r, i(r)\} = \{1, 1\}$.

(Conditions III and IV of Df. 2.1. According to T.2.2 every branch of any but the last level of a diagram of any classification has a prolongation in the next level. Hence of the diagrams

only (1) may represent a classification according to the meaning given here to this term.)

Df. 2.2.   The class $\{1, 1\}$ will be called the space of the classification $\mathfrak{M}$, shortly sp'$\mathfrak{M}$.

T. 2.3.   *For every index pair* $[r, h^r]$ *such that* $r > 1$, *there is an index pair* $[r-1, h^{r-1}]$ *such that* $\{r, h^r\} \subset \{r-1, h^{r-1}\}$.

PROOF. Let $r \in [2, \ldots, n]$. Then $r-1 < n$ and $r-1$ fulfils the antecedent of Df. 2.1, condition IV. Let $k_1^{r-1}, \ldots, k_{i(r-1)}^{r-1}$ be the numbers corresponding according to Df. 2.1, cond. IV to the $(r-1)$st row of $\mathfrak{M}$. Let further h^r be any of the numbers $1, \ldots, i(r)$. One of the following $i(r-1)$ cases occurs.

(1) h^r is one of the numbers $1, ..., k_1^{r-1}$ and then $\{r, h^r\}$ $\subset \{r-1, 1\}$.

(2) h^r is one of the numbers $k_1^{r-1}+1, ..., k_1^{r-1}+k_2^{r-1}$, and then $\{r, h^r\} \subset \{r-1, 2\}$,

. . . .

$(i(r-1))h^r$ is one of the numbers $\left(\sum\limits_{j=1}^{i(r-1)-1} k_j^{r-1} \right)+1, ...$

$..., \sum\limits_{j=1}^{i(r-1)} k_j^{r-1}$, and then $\{r, h^r\} \subset \{r-1, i(r-1)\}$.

Thus in any case there is a pair $[r-1, h^{r-1}]$ such that $\{r, h^r\} \subset \{r-1, h^{r-1}\}$.

T. 2.4.   *For every index pair* $[r, h^r]$, *for every* $\bar{r} < r$, *there is a pair* $[\bar{r}, h^{\bar{r}}]$ *such that* $\{\bar{r}, h^{\bar{r}}\} \subset \{r, h^r\}$. (This is a consequence of T. 2.3.)

T. 2.5.   *For every index pair* $[r, h^r]$ *of any row of* $\mathfrak{M}$, *for every index pair* $[\bar{r}, h^{\bar{r}}]$ *such that* $\bar{r} < r$

$$\{r, h^r\} \subset \{\bar{r}, h^{\bar{r}}\} \vee \{r, h^r\} \cap \{\bar{r}, h^{\bar{r}}\} = \Lambda.$$

PROOF. Let $1 \leqslant \bar{r} < r \leqslant n$ and let $[r, h^r]$ be any index pair of the rth row of $\mathfrak{M}$. According to T. 2.4 $\{r, h^r\}$ is included into one of the classes of the $\bar{r}$th row of $\mathfrak{M}$. According to Df. 2.1 $\{r, h^r\}$ is not empty whereas the product of any two classes of the rth row is empty. From this it follows that $\{r, h^r\}$ is included into exactly one of the classes of the rth row and has no common elements with any other.

T. 2.6.   *For every* $r < n$ *there are index pairs* $[r, h^r]$, $[r+1, h^{r+1}]$, $[r+1, \bar{h}^{r+1}]$ *such that*

     I. $h^{r+1} \neq \bar{h}^{r+1}$,

     II. $\{r+1, h^{r+1}\} \subset \{r, h^r\}$,

     III. $\{r+1, \bar{h}^{r+1}\} \subset \{r, h^r\}$.

(T.2.1, T.2.3.)

T. 2.7.   *For every index pair* $[r, h^r]$ *such that* $r < n$, *there is a pair* $[r+1, h^{r+1}]$ *such that* $\{r+1, h^{r+1}\} \subset \{r, h^r\}$. (Df. 2.1, condition IV.)

T. 2.8. *For every* $x \in \text{sp}'\mathfrak{M}$, *for every* r, *there is exactly one index pair* $[r, h^r]$ *such that* $x \in \{r, h\}^r$. (Df. 2.1, condition II, 2.2.)

3. THE LINES OF THE ELEMENTS OF THE SPACE OF A CLASSIFICATION

Let $\mathfrak{M}$ be any classification with n rows, x—any element of $\text{sp}'\mathfrak{M}$.

Df. 3.1. The n-element sequence $\mathfrak{C}$ of index pairs of $\mathfrak{M}$ such that for every r the rth element of $\mathfrak{C}$ is the only index pair $[r, h^r]$ fulfilling the condition $x \in \{r, h^r\}$, will be called the line of x in $\mathfrak{M}$ or shortly $\mathfrak{L}_{\mathfrak{M}}'x$. (In case it is known which classification is considered the index $\mathfrak{M}$ may be omitted.)

Let x, y be any elements of $\text{sp}'\mathfrak{M}$, l_r^x, l_r^y—the second element of the rth pair of $\mathfrak{L}'x$ or $\mathfrak{L}'y$ rsepectively. The lines $\mathfrak{L}'x$ and $\mathfrak{L}'y$ may then be noted as follows

$$\mathfrak{L}'x : [1, l_1^x], \dots, [r, l_r^x], \dots, [n, l_n^x],$$

$$\mathfrak{L}'y : [1, l_1^y], \dots, [r, l_r^y], \dots, [n, l_n^y].$$

T. 3.1. *There is an index pair* $[r, h^r]$ *such that*

I. $h^r = l_r^x = l_r^y$, i.e. $[r, h^r] = [r, l_r^x] = [r, l_r^y]$,

II. *for every* $\bar{r} < r$, $l_{\bar{r}}^x = l_{\bar{r}}^y$,

III. *for every* $\bar{\bar{r}} > r$, $l_{\bar{\bar{r}}}^x \neq l_{\bar{\bar{r}}}^y$.

PROOF. Let us note first that

(1) $[1, l_1^x] = [1, l_1^y]$ and

(2)   for every r, if $l_r^x = l_r^y$ then for every $\bar{r} < r$, $l_{\bar{r}}^x = l_{\bar{r}}^y$.

(1) is an immediate consequence of the Definitions 2.1 and 3.1. As regards (2) let us assume that r satisfies the condition $l_r^x = l_r^y$ and that $\bar{r} < r$ Then (a) $\{r, l_r^x\} = \{r, l_r^y\}$; hence (b) $x \in \{r, l_r^y\}$; but (c) $x \in \{\bar{r}, l_{\bar{r}}^x\}$; hence (d) $\{r, l_r^y\} \cap \{\bar{r}, l_{\bar{r}}^x\} \neq \Lambda$. (d) together with T. 2.5 implies (e) $\{r, l_r^y\} \subset \{\bar{r}, l_{\bar{r}}^x\}$; but (f) $y \in \{r, l_r^y\}$; (e) together with (f) yields: $y \in \{\bar{r}, l_{\bar{r}}^x\}$. The index pair $[\bar{r}, l_{\bar{r}}^x]$ is hence the only index pair $[\bar{r}, h^{\bar{r}}]$ such that $y \in \{\bar{r}, h^{\bar{r}}\}$. Hence $l_{\bar{r}}^x = l_{\bar{r}}^y$.

(1) and (2) being assumed the proof of T. 3.1 is immediate.

From (1) it follows that the class of numbers $r \in [1, ..., n]$ such that for a certain h^r, $h^r = l_r^x = l_r^y$, is not empty. Let r' be the greatest of such numbers. Hence there is a $h^{r'}$ such that $h^{r'} = l_{r'}^x = l_{r'}^y$, which means that the index pair $[r', h^{r'}]$ fulfils the condition I of T. 3.1; according to (2) $[r', h^{r'}]$ fulfils also the condition II; III must also be fulfilled in view of the assumption defining r'.

It is obvious that any two index pairs of $\mathfrak{M}$ fulfilling I, II, III of T. 3.1 are identical.

Df. 3.2.   The only index pair $[r, h^r]$ fulfilling I, II, III of T. 3.1 will be called the *meet of the lines* $\mathfrak{L}^rx$ *and* $\mathfrak{L}^ry$, and the number r the *level of the meet of* these lines.

Let us note that in case the lines $\mathfrak{L}^rx$ and $\mathfrak{L}^ry$ are identical, their meet is the pair $[n, h_n] = [n, l_n^x] = [n, l_n^y]$, and that if n is the level of the meet of two lines, these lines are identical.

4. THE FUNCTION $\ulcorner x \mid_{\overline{\mathfrak{M}}} \mid y \urcorner$[6]

Let $\mathfrak{M}$ be any classification with n rows. We shall introduce now the concept of the *distance in* $\mathfrak{M}$ between any two elements x, y belonging to the space of $\mathfrak{M}$. This distance will be denoted by $x \mid_{\overline{\mathfrak{M}}} \mid y$.

Df. 4.1.   For every $x, y \in \mathrm{sp}^r\mathfrak{M}$, $x \mid_{\overline{\mathfrak{M}}} \mid y = n - w$, where w is the level of the meet of $\mathfrak{L}^rx$ and $\mathfrak{L}^ry$.

The above definition assigns to every pair x, y of elements of $\mathrm{sp}^r\mathfrak{M}$ exactly one value $n - w$. $\ulcorner x \mid_{\overline{\mathfrak{M}}} \mid y \urcorner$ is hence a function defined in $\mathrm{sp}^r\mathfrak{M}$. Let us consider the properties of this function.

T. 4.1.   *The class of all values of* $\ulcorner x \mid_{\overline{\mathfrak{M}}} \mid y \urcorner$ *for* x, $y \in \mathrm{sp}^r\mathfrak{M}$, *is the class of natural numbers*: $[0, ..., n-1]$; *in other words*
I. *for any* $x, y \in \mathrm{sp}^r\mathfrak{M}$, $x \mid_{\overline{\mathfrak{M}}} \mid y \in [0, ..., n-1]$, *and*
II. *for every* $v \in [0, ..., n-1]$, *there exists an* $x \in \mathrm{sp}^r\mathfrak{M}$ *and an* $y \in \mathrm{sp}^r\mathfrak{M}$ *such that* $x \mid_{\overline{\mathfrak{M}}} \mid y = v$.

PROOF. Part I of T. 4.1 is an immediate consequence of Df. 4.1 and the obvious fact that for all $x, y \in \mathrm{sp}^r\mathfrak{M}$, the level of the meet of $\mathfrak{L}^rx$ and $\mathfrak{L}^ry$ is one of the numbers $1, ..., n$. Part II, however is to be proved.

Let us distinguish two cases: $v=0$ and $v>0$, i.e. $v \in [1, ..., n-1]$. Suppose $v=0$. According to Df. 2.1 sp'$\mathfrak{M}$ is not empty. Let x be any element of sp'$\mathfrak{M}$. The level of the meet of the lines $\mathfrak{L}'x$ and $\mathfrak{L}'x$ is n; hence $x \mid_{\overline{\mathfrak{M}}} \mid x = n - n = 0$; hence there exists an $x \in$ sp'$\mathfrak{M}$ and an $y \in$ sp'$\mathfrak{M}$ such that $x \mid_{\overline{\mathfrak{M}}} \mid y = 0$.

Let us assume now that $v \in [1, ..., n-1]$ and that $n-v=r$; hence $r \in [1, ..., n-1]$. Let further $[r, h^r]$, $[r+1, h^{r+1}]$, $[r+1, \overline{h}^{r+1}]$ be such index pairs of $\mathfrak{M}$ that (1) $h^{r+1} \neq \overline{h}^{r+1}$, (2) $\{r+1, h^{r+1}\} \subset \{r, h^r\}$ and (3) $\{r+1, \overline{h}^{r+1}\} \subset \{r, h^r\}$. (The existence of such index pairs may be assummed in view of T. 2.6.) Let now x and y be any elements of $\{r+1, h^{r+1}\}$ and $\{r+1, \overline{h}^{r+1}\}$ respectively. We may assume this since both classes are according to Df. 2.1 not empty.

Let us consider the lines $\mathfrak{L}'x$ and $\mathfrak{L}'y$. In view of (2) and (3) $x \in \{r, h^r\}$ and $y \in \{r, h^r\}$; hence $h^r = l_r^x = l_r^y$. In view of (1) however $l_{r+1}^x \neq l_{r+1}^y$. The pair $[r, h^r]$ is hence the meet of $\mathfrak{L}'x$ and $\mathfrak{L}'y$; hence $x \mid_{\overline{\mathfrak{M}}} \mid y = n - r = v$. The proof of T. 4.1 is thus complete.

(Let us note that part II of T. 4.1 would not be true if the condition $i(r+1) > i(r)$ were cancelled from Df. 2.1. It would be possible then to define a classification corresponding for instance to the diagram:

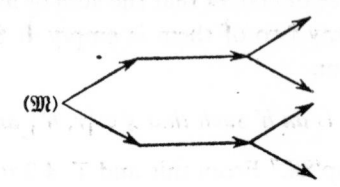

Such a classification would statisfy condition I of T. 4.1 since obviously for all $x, y \in$ sp'$\mathfrak{M}$, $x \mid_{\overline{\mathfrak{M}}} \mid y \in [0, 1, 2, 3]$, but there would not exist a pair x, y of elements of sp'$\mathfrak{M}$ such that $x \mid_{\overline{\mathfrak{M}}} \mid y = 2$.)

T. 4.2. *For all $x, y \in$ sp'$\mathfrak{M}$, for every r,*

(1) $\begin{cases} x \mid_{\overline{\mathfrak{M}}} \mid y \leqslant n-r \text{ if and only if there is an } h^r \text{ such that} \\ x \in \{r, h^r\} \text{ and } y \in \{r, h^r\}. \end{cases}$

PROOF. Let r be one of the numbers $1, ..., n$, h^r any of the numbers $1, ..., i(r)$, x, y—any elements of sp'$\mathfrak{M}$ such that

(2) $x \in \{r, h^r\}$ and $y \in \{r, h^r\}$.

Let us consider the lines $\mathfrak{L}^\iota x$ and $\mathfrak{L}^\iota y$. It follows from (2) that

(3) $l_r^x = l_r^y = h^r$.

Let m be the level of the meet of $\mathfrak{L}^\iota x$ and $\mathfrak{L}^\iota y$. It follows from (3) that

(4) $m \geqslant r$ (see T. 3.1 and D. 3.2)

hence

(5) $x \mid_{\overline{\mathfrak{M}}} y = n - m \leqslant n - r$.

Thus we have shown that the right side of the equivalence (1) implies the left side. A similar reasoning shows that the left side implies the right side. T. 4.2 is equivalent to the following statement. *For every value $n-r$ of the function $\ulcorner x \mid_{\overline{\mathfrak{M}}} y \urcorner$ the two relations*

(1) $\ulcorner x \mid_{\overline{\mathfrak{M}}} y \leqslant n - r \urcorner$ *and*

(2) $\ulcorner$ *there is an h^r such that $x \in \{r, h^r\}$ and $y \in \{r, h^r\}$ $\urcorner$*
are identical in sp'$\mathfrak{M}$.

Let us note now that according to Df. 2.1 every row of any classification is such a sequence of classes that the sum of all these classes is sp'$\mathfrak{M}$ and the product of any two of them is empty. It follows from this that for every r the relation

$\ulcorner$ *there is an h^r such that $x \in \{r, h^r\}$ and $y \in \{r, h^r\}$ $\urcorner$*

is an equivalence in sp'$\mathfrak{M}$.[7] From this and T. 4.2 it follows that

T. 4.3. *any of the relations*

$\ulcorner x \mid_{\overline{\mathfrak{M}}} y \leqslant n - 1 \urcorner$, ..., $\ulcorner x \mid_{\overline{\mathfrak{M}}} y \leqslant n - n \urcorner$ *is an equivalence in* sp'$\mathfrak{M}$.

Moreover, since the numbers of elements of any row of $\mathfrak{M}$ is finite,

T. 4.4. *any of the families of abstraction classes of the relations listed in T. 4.3, i.e. any of the families*

$\mathfrak{A} (\ulcorner x \mid_{\overline{\mathfrak{M}}} y \leqslant n - 1 \urcorner$, sp'$\mathfrak{M}$), ..., $\mathfrak{A} (\ulcorner x \mid_{\overline{\mathfrak{M}}} y \leqslant n - n \urcorner$, sp'$\mathfrak{M}$), *is finite.*

It is also apparent that

T. 4.5. *for every r, the rth row of $\mathfrak{M}$ is a permutation of the family* $\mathfrak{A}\,(\ulcorner x \mid_{\overline{\mathfrak{M}}} y \leqslant n-r \urcorner,\ \mathrm{sp}`\mathfrak{M})$.

From T. 4.1 and T. 4.2 follows:

T. 4.6. *For every value $n-r$ of the function* $\ulcorner x \mid_{\overline{\mathfrak{M}}} y \urcorner$, $x \mid_{\overline{\mathfrak{M}}} y$
$= n-r$, *if and only if there is an h^r such that $x \in \{r, h^r\}$ and $y \in \{r, h^r\}$ and, moreover, for every $\bar{r} > r$, for every $h^{\bar{r}}$, either x or y is not an element of $\{\bar{r}, h^{\bar{r}}\}$. In other words, $x \mid_{\overline{\mathfrak{M}}} y = n-r$ if and only if r is the greatest of the numerals of such rows of $\mathfrak{M}$ in which x and y belong to the same class.*

(Let us note that of the relations $\ulcorner x \mid_{\overline{\mathfrak{M}}} y = n-r \urcorner$ only $\ulcorner x \mid_{\overline{\mathfrak{M}}} y = n-n \urcorner$ is an equivalence. Consider the following example. Let $\mathfrak{M}$ be a classification corresponding to the following diagram:

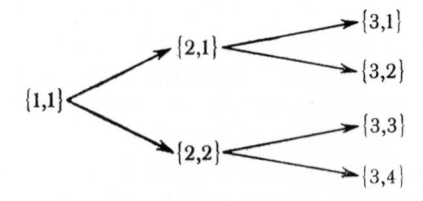

Let further $x \in \{3, 3\}$, $y \in \{3, 3\}$, $z \in \{3, 2\}$. Then $x \mid_{\overline{\mathfrak{M}}} z = 2$, $z \mid_{\overline{\mathfrak{M}}} y = 2$, whereas $x \mid_{\overline{\mathfrak{M}}} y = 0$. The relation $\ulcorner x \mid_{\overline{\mathfrak{M}}} y = 2 \urcorner$ is therefore not transitive and hence not an equivalence.)

T. 4.7. *For all $x, y, z \in \mathrm{sp}`\mathfrak{M}$,*

    I. $x = y \rightarrow x \mid_{\overline{\mathfrak{M}}} y = 0$,

    II. $x \mid_{\overline{\mathfrak{M}}} y = y \mid_{\overline{\mathfrak{M}}} x$,

    III. $x \mid_{\overline{\mathfrak{M}}} y \geqslant 0$,

    IV. $x \mid_{\overline{\mathfrak{M}}} y + y \mid_{\overline{\mathfrak{M}}} z \geqslant x \mid_{\overline{\mathfrak{M}}} z$ (condition of triangle).

PROOF. Conditions I, II and III are immediate consequences of Df. 4.1. Very easy is also the proof of IV.

It follows namely from T. 4.3 that for any $v \in [0, ..., n-1]$, that is for any value of $\ulcorner x \mid_{\overline{\mathfrak{M}}} y \urcorner$, the relation $\ulcorner x \mid_{\overline{\mathfrak{M}}} y \leqslant v \urcorner$ is an equivalence in $\mathrm{sp}`\mathfrak{M}$. Let $x, y, z$ be any elements of $\mathrm{sp}`\mathfrak{M}$, and let us put

$x \mid_{\mathfrak{M}} \mid y = d$, $y \mid_{\mathfrak{M}} \mid z = d'$. Either (1) $d \geqslant d'$ or (2) $d' \geqslant d$. In case (1) $x \mid_{\mathfrak{M}} \mid y \leqslant d$ and $y \mid_{\mathfrak{M}} \mid z \leqslant d$, and consequently $x \mid_{\mathfrak{M}} \mid z \leqslant d$. In case (2) $x \mid_{\mathfrak{M}} \mid y \leqslant d'$ and $y \mid_{\mathfrak{M}} \mid z \leqslant d'$, and hence $x \mid_{\mathfrak{M}} \mid z \leqslant d'$. But d and d' being values of the function $\ulcorner x \mid_{\mathfrak{M}} \mid y \urcorner$ are both positive numbers (cond. III). Hence both in case (1) and in case (2) $x \mid_{\mathfrak{M}} \mid z \leqslant d + d' = x \mid_{\mathfrak{M}} \mid y + y \mid_{\mathfrak{M}} \mid z$.

(In the above proof only two properties of the function $\ulcorner x \mid_{\mathfrak{M}} \mid y \urcorner$ have been made use of: (1) all values of $\ulcorner x \mid_{\mathfrak{M}} \mid y \urcorner$ for arguments belonging to sp‘$\mathfrak{M}$ are non negative numbers; (2) each of the relations $\ulcorner x \mid_{\mathfrak{M}} \mid y \leqslant v \urcorner$, where v is any value of $\ulcorner x \mid_{\mathfrak{M}} \mid y \urcorner$ for $x, y \in$ sp‘$\mathfrak{M}$, is transitive. Hence any function $f(x, y)$ defined in a given class S and satisfying the conditions (1) and (2) fulfils also the condition of triangle. One may ask whether any function satisfying (1) and the condition of triangle fulfils also the condition (2). The following example shows that this is not the case. Let S be the set of natural numbers $[1, \ldots, 10]$. The values of the function $\ulcorner |x - y| \urcorner$ for arguments taken from S are all non negative. This function fulfils also the condition of triangle. Condition (2) however is not fulfilled, since for instance $|1 - 2| \leqslant 1$, $|2 - 3| \leqslant 1$ but $|1 - 3| > 1$.)

> Df. 4.2. A function $\ulcorner d(x, y) \urcorner$ defined in a given class S and assuming for arguments belonging to S only numeral values is a *quasi distance* in S if and only if for all $x, y, z \in S$,
>
> I. $x = y \to d(x, y) = 0$,
>
> II. $d(x, y) \geqslant 0$,
>
> III. $d(x, y) = d(y, x)$,
>
> IV. $d(x, y) + d(y, z) \geqslant d(x, z)$.

(We shall also use in the sequel, in Sections 6 and 7, the concept of *distance in a given class S*. The definition of this last concept may be obtained from Df. 4.2 by replacing condition I by the following: $d(x, y) = 0$ if and only if $x = y$.)

Once Df. 4.2 is assumed T. 4.7 may be expressed as follows:

The function $\ulcorner x \mid_{\mathfrak{M}} \mid y \urcorner$ *is a quasi distance in* sp‘$\mathfrak{M}$.

Let $\mathfrak{M}$ be a classification whose rows are: $\mathfrak{R}_1, \ldots, \mathfrak{R}_{n-1}, \mathfrak{R}_n$. The sequence $\mathfrak{R}_1, \ldots, \mathfrak{R}_{n-1}$ is obviously also a classification. Let us denote

it by $\mathfrak{M}'$. It is easy to show that both classifications have the same space and moreover that

T. 4.8. *for all* $x, y \in \mathrm{sp}\,'\mathfrak{M} = \mathrm{sp}\,'\mathfrak{M}'$,

$$x \mid_{\overline{\mathfrak{M}}} \mid y > 0 \rightarrow x \mid_{\overline{\mathfrak{M}}} \mid y = x \mid_{\overline{\mathfrak{M}'}} \mid y + 1.$$

(In case $x \mid_{\overline{\mathfrak{M}}} \mid y = 0$, the equality $x \mid_{\overline{\mathfrak{M}}} \mid y = x \mid_{\overline{\mathfrak{M}}} \mid y$ is of course not excluded.)

T. 4.9. *If two classifications* $\mathfrak{M}$ *and* $\mathfrak{M}'$ *are satisfying the conditions*

  I. $\mathrm{sp}\,'\mathfrak{M} = \mathrm{sp}\,'\mathfrak{M}'$ *and*

  II. *for all* $x, y \in \mathrm{sp}\,'\mathfrak{M}$, $x \mid_{\overline{\mathfrak{M}}} \mid y = x \mid_{\overline{\mathfrak{M}'}} \mid y$,
  then they have the same number of rows and the corresponding rows of both classification differ at most in the order of their elements (T. 4.1, T. 4.5).

Df. 4.3. Of two classifications satisfying the above conditions I and II we shall say that they are *equivalent* or that they *classify in the same way their common space.*

5. CLASSIFYING FUNCTIONS

Df. 5.1. A function $\ulcorner d(x, y) \urcorner$ defined in the class S will be said to *determine an **n**-level classification of S* if and only if there is a classification $\mathfrak{M}$ with **n** rows such that

  I. $S = \mathrm{sp}\,'\mathfrak{M}$ and

  II. for all $x, y \in S$, $d(x, y) = x \mid_{\overline{\mathfrak{M}}} \mid y$.

T. 5.1. *A function* $\ulcorner d(x, y) \urcorner$ *defined in the class* S *determines an **n**-level classification of* S *if and only if for all* $x, y \in S$, *for all* $v \in [0, ..., n-1]$,

  I. $d(x, y) \in [0, ..., n-1]$,
  II. *there is a* $z \in S$ *and a* $u \in S$ *such that* $d(z, u) = v$,
  III. *the relation* $\ulcorner d(x, y) \leqslant v \urcorner$ *is an equivalence in* S,
  IV. *the family of abstraction classes* $\mathfrak{A}(\ulcorner d(x, y) \leqslant v \urcorner, S)$ *is finite.*

PROOF. If the function $\ulcorner d(x,y)\urcorner$ determines an n-level classification of S then the function $\ulcorner d(x,y)\urcorner$, the number n and the class S are satisfying the right side of T. 5.1. (Df. 5.1, T. 4.1, T. 4.3, T. 4.4.).

Less obvious is the converse of this: if $\ulcorner d(x,y)\urcorner$, n and S are satisfying the right side of T. 5.1 then $\ulcorner d(x,y)\urcorner$ determines an n-level classification of S. We shall prove this by induction relatively to n.

Let us assume that the function $\ulcorner d(x,y)\urcorner$ the number 1 and the class S are satisfying the right side of T. 5.1. It follows from this assumption that for all $x,y \in S$, $d(x,y)=0$, and that there is an $x \in S$ and an $y \in S$ such that $d(x,y)=0$. Hence S is not empty. Let $\mathfrak{M}$ be a classification having exactly one row and consequently exactly one element: $\{1,1\}=S$. Obviously (1) $S=\mathrm{sp'}\mathfrak{M}$ and (2) for all $x,y \in S$, $d(x,y)=x\,|_{\overline{\mathfrak{M}}}|\,y=0$, which means that $\ulcorner d(x,y)\urcorner$ determines a 1-level classification of S. Thus we have shown that for every function $\ulcorner d(x,y)\urcorner$, for every class S, if $\ulcorner d(x,y)\urcorner$, S and the number 1 are satisfying the right side of T. 5.1, they are also satisfying the left side.

Let us assume now that for every function $\ulcorner d(x,y)\urcorner$ and every class S if $\ulcorner d(x,y)\urcorner$, S and the natural number l are satisfying the right side of T. 5.1: then $\ulcorner d(x,y)\urcorner$ determines an l-level classification of S. We shall prove that if this is the case then for every function $\ulcorner d(x,y)\urcorner$ and every class S if $\ulcorner d(x,y)\urcorner$, S and the number $l+1$ are satisfying the right side of T. 5.1, then $\ulcorner d(x,y)\urcorner$ determines an $(l+1)$-level classification of S.

Let us assume that $\ulcorner d(x,y)\urcorner$, S and $l+1$ are satisfying the right side of T. 5.1. We shall define a new function $\ulcorner d'(x,y)\urcorner$ by assuming that for any $x,y \in S$.

$$d(x,y)=0 \to d'(x,y)=0,$$

$$d(x,y)>0 \to d'(x,y)=d(x,y)-1.$$

It is easy to show that the function $\ulcorner d'(x,y)\urcorner$, the class S and the number l are satisfying the right side of T. 5.1. Hence according to the assumption of the inductive proof there is a classification $\mathfrak{M}$ with l rows such that $S=\mathrm{sp'}\mathfrak{M}$ and for all $x,y \in S$, $x\,|_{\overline{\mathfrak{M}}}|\,y=d'(x,y)$.

Let $\mathfrak{M}'$ be such a classification, $\mathfrak{R}_1, \ldots, \mathfrak{R}_l$—the lst, the 2nd, ..., the lth row of $\mathfrak{M}'$, respectively; let moroever d_0, d'_0, d_1 stand for $\ulcorner d(x,y)=0\urcorner$, $\ulcorner d'(x,y)=0\urcorner$, $\ulcorner d(x,y)\leqslant 1\urcorner$, respectively. The last, i.e. the

*l*th row of $\mathfrak{M}'$:

$(\mathfrak{R}_l)$ $\{l, 1\}, \ldots, \{l, h^l\}, \ldots, \{l, i(l)\}$

is a permutation of the family $\mathfrak{A}(\ulcorner x \mid_{\mathfrak{M}'} y = 0 \urcorner, S)$, i.e. the family $\mathfrak{A}(d_0', S)$.

The relation d_0' is however identical in S with the relation d_1, since for all $x, y \in S$, $d'(x, y) = 0$ if and only if $d(x, y) \leqslant 1$. Hence

(1) $\mathfrak{R}_l$ is a permutation of the family $\mathfrak{A}(d_1, S)$.

Further, since d_0 is an equivalence in S and moreover for all h^l, $\{l, h^l\} \subset S$,

(2)   for all h^l, d_0 is an equivalence in $\{l, h^l\}$.

Let us note now that (a) $d_0 \subset d_1$; (b) $\bigcup\limits_{j=1}^{i(l)} \{l, j\} = S$; and (c) for all h^l and all $\bar{h}^l$, $h^l \neq \bar{h}^l \rightarrow \{l, h\} \cap \{l, \bar{h}^l\} = \Lambda$. (a), (b), (c) together with (1) and (2) imply

(3) $\bigcup\limits_{j=1}^{i(l)} \mathfrak{A}(d_0, \{l, j\}) = \mathfrak{A}(d_0, S)$.

At last, since (according to the assumption stating that $\ulcorner d(x, y) \urcorner$, S and the number $l+1$ are satisfying the right side of T. 5.1) there is a pair x, y of elements of S such that $d(x, y) = 1$.

(4)   at least one of the families $\mathfrak{A}(d_0, \{l, h^l\})$ has more than one element and hence differs from the unit family ($\{l, h^l\}$).

Let now $\mathfrak{R}_{l+1}$ be a sequence of classes which may be obtained from the sequence of families

$$\mathfrak{A}(d_0, \{l, 1\}), \ldots, \mathfrak{A}(d_0, \{l, h^l\}), \ldots, \mathfrak{A}(d_0, \{l, i(l)\})$$

by replacing each of the families $\mathfrak{A}(d_0, \{l, h^l\})$ by an arbitrary permutation of its elements.

From the assumptions defining $\mathfrak{R}_{l+1}$ together with (1), (2), (3), (4) and the fact that $\mathfrak{M}'$ is a classification whose space is S it follows that the double sequence of classes

$(\mathfrak{M}'')$ $\mathfrak{R}_1, \ldots, \mathfrak{R}_l, \mathfrak{R}_{l+1}$

is a classification having S as its space.

I will prove now that for all $x, y \in S$,

$$d(x, y) = x \, |_{\overline{\mathfrak{M}''}}| \, y .$$

First let us note that

(5)　　for all $x, y \in S$, $d(x, y) = 0$ or $d(x, y) = 1$ or $d(x, y) > 1$.

Let x, y be such elements of S that $d(x, y) = 0$. Then x and y are both elements of the same class of the sequence $\mathfrak{R}_{l+1}$, since $\mathfrak{R}_{l+1}$ is a permutation of the family $\mathfrak{A}(d_0, S)$. But x and y are also satisfying the equality $x \, |_{\overline{\mathfrak{M}'}}| \, y = 0$, since $\mathfrak{R}_{l+1}$ is the last row of $\mathfrak{M}''$. Hence

(6)　　for all $x, y \in S$, if $d(x, y) = 0$ then $d(x, y) = x \, |_{\overline{\mathfrak{M}''}}| \, y$.

Suppose now that x and y are elements of S for which $d(x, y) = 1$ holds. Then $d'(x, y) = d(x, y) - 1 = 0$; hence also $x \, |_{\overline{\mathfrak{M}'}}| \, y = 0$; hence x and y are both elements of the same class of the row $\mathfrak{R}_l$; but x and y do not appear both as elements in the same class of the row $\mathfrak{R}_{l+1}$, for were this the case, $d(x, y) = 0$ would hold, contrary to assumption. Hence l is the numeral of the last of such rows of $\mathfrak{M}''$ in which x and y belong to the same class, and consequently $x \, |_{\overline{\mathfrak{M}''}}| \, y = 1$. Hence

(7)　　for all $x, y \in S$, if $d(x, y) = 1$ then $d(x, y) = x \, |_{\overline{\mathfrak{M}''}}| \, y$.

Suppose at last that x and y are elements of S such that $d(x, y) > 1$. Then $d'(x, y) = d(x, y) - 1 > 0$. Hence also $x \, |_{\overline{\mathfrak{M}'}}| \, y > 0$. From this together with T. 3.8 (it is easy to see that $\mathfrak{M}$ and $\mathfrak{M}'$ are satisfying the assumption of this theorem) it follows that

$$x \, |_{\overline{\mathfrak{M}''}}| \, y = x \, |_{\overline{\mathfrak{M}'}}| \, y + 1 .$$

But also

$$x \, |_{\overline{\mathfrak{M}'}}| \, y = d'(x, y) .$$

Hence

$$x \, |_{\overline{\mathfrak{M}''}}| \, y = d'(x, y) + 1 .$$

The definition of $\ulcorner d'(x, y) \urcorner$ and the assumption $d(x, y) > 1$ imply that

$$d(x, y) = d'(x, y) = 1 , \quad \text{hence}$$

(8)　　for all $x, y \in S$, if $d(x, y) > 1$ then $x \, |_{\overline{\mathfrak{M}''}}| \, y = d(x, y)$.

From (5), (6), (7) and (8) it follows that for all $x, y \in S$, $d(x, y) = x \,|\overline{\underset{\mathfrak{M}''}{}}|\, y$. The proof of T. 5.1 is thus completed.

Let S be any set, $\ulcorner d(x, y) \urcorner$ — a function determining an **n**-level classification of S, $\mathfrak{M}$ — any classification.

Df. 4.2. $\mathfrak{M}$ is *connected with* $\ulcorner d(x, y) \urcorner$ if and only if

I. $S = \mathrm{sp}\,{}^{\text{‘}}\mathfrak{M}$ and

II.   for all $x, y \in S$, $d(x, y) = x \,|\overline{\underset{\mathfrak{M}}{}}|\, y$.

It is easy to see that

T. 5.2.   *any two classifications which are connected with* $\ulcorner d(x, y) \urcorner$ *are equivalent, i.e. are classifying in the same way the class S.*

6. NATURAL CLASSIFICATIONS

Let S be any non empty class, $\mathfrak{M}$ — a classification such that $S = \mathrm{sp}\,{}^{\text{‘}}\mathfrak{M}$, $\ulcorner d(x, y) \urcorner$ — any distance in S.

Df. 6.1. $\mathfrak{M}$ *mirrors the diversification of S relatively to* $\ulcorner d(x, y) \urcorner$ if and only if for all $x, y, z, u \in S$,

$$x \,|\overline{\underset{\mathfrak{M}}{}}|\, y < z \,|\overline{\underset{\mathfrak{M}}{}}|\, u \to d(x, y) < d(z, u).$$

(Examples illustrating the above definition have been given in Section 1.)

The concept of natural classification corresponding to Df. 6.1 may also be explained by way of another equivalent definition. Before formulating this definition, which will have here the status of theorem, two auxiliary definitions must be given.

Let S, $\mathfrak{M}$, **n**, $\ulcorner d(x, y) \urcorner$, V be a non empty set, a classification having S as its space, the number of rows of $\mathfrak{M}$, any distance in S, the set of all values of $\ulcorner d(x, y) \urcorner$ for $x, y \in S$, respectively.

Let us distinguish now an **n**-element sequence of subsets of V:

$V_0, \ldots, V_{n-1}$ by assuming that

for all $v \in V$, for all $i \in [0, \ldots, n-1]$, $x \in V_i$, if and only if there is a pair x, y of elements of S such that $d(x, y) = v$ and $x \,|\overline{\underset{\mathfrak{M}}{}}|\, y = i$.

Df. 5.2. The sequence of subsets of V fulfilling the above condition will be called the *correlate of the compound*[8] $\langle S, \mathfrak{M}, \ulcorner d \cdot (x, y) \urcorner \rangle$.

Examples. Let S be the class of numbers: $\{1, ..., 10\}$, $d(x, y)$—the function $\ulcorner |x-y| \urcorner$, $\mathfrak{M}$—the following classification.

$$\{1, ..., 10\},$$
$$\{1, ..., 5\}, \{6, ..., 10\},$$
$$\{1, 2, 3\}, \{4, 5\}, \{6, 7\}, \{8, 9, 10\}.$$

In this example V is the class numbers: $\{0, ..., 9\}$, V_0 is the class of all values of the function $\ulcorner |x-y| \urcorner$ for x, y belonging to the same class of the third row of $\mathfrak{M}$, hence $V_0 = \{0, 1, 2\}$; V_1 is the class of all values of this function for x, y belonging to the same class of the second row of $\mathfrak{M}$ but to different classes of the third row, hence $V_1 = \{1, 2, 3, 4\}$; the last element of the correlate under consideration $V_2 = \{1, ..., 9\}$.

In the case, however, of the second example of a natural classification given in Section 1, the correlate of the compound $\langle S_{1,1}, Clf_3, \ulcorner |x-y| \urcorner \rangle$ is:

$$V_0 = \{0, 1, 2\}, \quad V_1 = \{3, 4, 5, 6\}, \quad V_2 = \{7, ..., 19\}.$$

The theorem which follows lists certain common properties of any correlates of any compounds. The symbols, $\mathfrak{M}$, n, $\ulcorner d(x, y) \urcorner$, V, V_0, ..., V_{n-1} will have the meaning explained above.

T. 6.1.   I.   *For all $i \in [0, ..., n-1]$, $V_i \neq \Lambda$,*

II. $\displaystyle\bigcup_{j=1}^{n-1} V_j = V$,

III. *for all $x, y \in S$, for all $i \in [0, ..., n-1]$,*

$$x |_{\overline{\mathfrak{M}}}| y = i \rightarrow d(x, y) \in V_i.$$

(The proof of all three theorems is immediate. Let us note that the proof of I involves T. 3.1, II.)

And here is the second auxiliary definition.

Df. 6.2. The finite sequence of classes of numbers $V_0, ..., V_r$ (r is any natural number) is an *increasing partition* of the class of numbers V, if and only if

I. $\displaystyle\bigcup_{i=0}^{r} V_i = V$,

II. for all $i, j \in [0, \ldots, r]$, if $i < j$ then for all $v \in V_i$, for all $v' \in V_j$, $v < v'$.

The examples given above show that for certain compounds $\langle S, \mathfrak{M},$ $\ulcorner d(x, y)\urcorner \rangle$, the correlate of the compound is not an increasing partition of the class V of all values of $\ulcorner d(x, y)\urcorner$ for $x, y \in S$. The following theorem is however true.

T. 6.2. *The classification $\mathfrak{M}$ mirrors the diversification of the class S relatively to $\ulcorner d(x, y)\urcorner$ if and only if the correlate of the compound $\langle S, \mathfrak{M}, \ulcorner d(x, y)\urcorner \rangle$ is an increasing partition of V.*

PROOF. Let us assume that $\mathfrak{M}$ mirrors the diversification of S relatively to $\ulcorner d(x, y)\urcorner$, i.e. that for all $x, y, z, u \in S$,

(1) $x\,|_{\overline{\mathfrak{M}}}|\,y < z\,|_{\overline{\mathfrak{M}}}|\,u \rightarrow d(x, y) < d(z, u)$.

Let i, j be such elements of $[0, \ldots, n-1]$, v, v' — such numbers, that $i < j$, $v \in V_i$, $v' \in V_j$. From $v \in V_i$ it follows that there is a pair x, y of elements of S such that $d(x, y) = v$ and $x\,|_{\overline{\mathfrak{M}}}|\,y = i$. Let x, y be such a pair. Analogously let z, u be such a pair of elements of S that $d(z, u) = v'$ and $z\,|_{\overline{\mathfrak{M}}}|\,u = j$. Obviously $x\,|_{\overline{\mathfrak{M}}}|\,y < z\,|_{\overline{\mathfrak{M}}}|\,u$. This and (1) implies: $d(x, y) < d(z, u)$, i.e. $v < v'$. Hence for all $i, j \in [0, \ldots, n-1]$, if $i < j$ then for every $v \in V_i$, for every $v' \in V_j$, $v < v'$. From this and T. 6.1, II it follows that the correlate of the compund $\langle S, \mathfrak{M}, \ulcorner d(x, y)\urcorner \rangle$ is an increasing partition of V. Thus we have shown that from the left side of T. 6.2 follows the right side.

Let us assume now that the correlate of the compound $\langle S, \mathfrak{M},$ $\ulcorner d(x, y)\urcorner \rangle$ in an increasing partition of V, i.e. that

(2)   For all $i, j \in [0, \ldots, n-1]$, if $i < j$, then for all $v \in V_i$, $v' \in V_j$, $v < v'$.

Let further x, y, z, u be such elements of S that $x\,|_{\overline{\mathfrak{M}}}|\,y < z\,|_{\overline{\mathfrak{M}}}|\,u$. We put $x\,|_{\overline{\mathfrak{M}}}|\,y = i$, $z\,|_{\overline{\mathfrak{M}}}|\,u = j$. Obviously $i, j \in [0, \ldots, n-1]$, and $i < j$; moreover $d(x, y) \in V_i$, $d(z, u) \in V_j$. Hence according to (2)

$d(x, y) < d(z, u)$. Hence for all x, y, z, $u \in S$, $x \mid_{\overline{\mathfrak{M}}} \mid y < z \mid_{\overline{\mathfrak{M}}} \mid u \rightarrow$ $\rightarrow d(x, y) < d(z, u)$.

Thus we have shown that from the right side of T. 6.2 follows the left side. The proof of T. 6.2 is complete.

T. 6.3. *If $\mathfrak{M}$ mirrors the diversification of S relatively to $\ulcorner d(x, y) \urcorner$ then for every $i \in [0, ..., n-1]$, for all $x, y \in S$, $d(x, y) \in V_i$ $\rightarrow x \mid_{\overline{\mathfrak{M}}} \mid y = i$.*

PROOF. Let us assume that $\mathfrak{M}$ mirrors the diversification of S relatively to $\ulcorner d(x, y) \urcorner$, that $i \in [0, ..., n-1]$, and that x, y are elements of S such that $d(x, y) \in V_i$. It follows from this that there is a pair z, u of elements of S such that $d(x, y) = d(z, u)$ and $z \mid_{\overline{\mathfrak{M}}} \mid u = i$. We assume that z, u is such a pair and put $x \mid_{\overline{\mathfrak{M}}} \mid y = j$. Suppose $i \neq j$. From this and the assumption that $\mathfrak{M}$ mirrors the diversification of S relatively to $\ulcorner d(x, y) \urcorner$ it follows that $d(x, y) \neq d(z, u)$ contrary to what is assumed regarding z, u. Hence $i = j$. Hence T. 6.3.

From T. 6.1 and T. 6.3 it follows that if $\mathfrak{M}$ mirrors the diversification of S relatively to $\ulcorner d(x, y) \urcorner$ then for any $i \in [0, ..., n-1]$, for all $x, y \in S$,

$$d(x, y) \in V_i \equiv x \mid_{\overline{\mathfrak{M}}} \mid y = i.$$

This implies that for any $i \in [0, ..., n-1]$, the relation $\ulcorner d(x, y) \in V_i \urcorner$ is identical in S with the relation $\ulcorner x \mid_{\overline{\mathfrak{M}}} \mid y = i \urcorner$; hence also any of the relations $\ulcorner d(x, y) \in \bigcup_{j=0}^{i} V_j \urcorner$ $(i \in [0, ..., n-1])$ is identical in S with the corresponding relation $\ulcorner x \mid_{\overline{\mathfrak{M}}} \mid y \leqslant i \urcorner$. This together with T. 4.3 and T. 4.4 yields

T. 6.4. *If $\mathfrak{M}$ mirrors the diversification of S relatively to $\ulcorner d(x, y) \urcorner$ then for any $i \in [0, ..., n-1]$,*

I. *the relation $\ulcorner d(x, y) \in \bigcup_{j=0}^{i} V_j \urcorner$ is an equivalence in S, and*

II. *the family $\mathfrak{A}(\ulcorner d(x, y) \in \bigcup_{j=0}^{j} V_j \urcorner, S)$ is finite.*

Let S be any class, $\ulcorner d(x, y) \urcorner$ —any distance in S, n—any natural number (not 0), V—the class of all values of $\ulcorner d(x, y) \urcorner$ for $x, y \in S$.

T. 6.5. *If there exists a classification $\mathfrak{M}$ with **n** rows such that S $= \mathrm{sp}\ \mathfrak{M}$, and $\mathfrak{M}$ mirrors the diversification of S relatively to $\ulcorner d(x, y) \urcorner$, then there exists an **n**-element increasing partition of V: $V_0, ..., V_{n-1}$, such that for any $i \in [0, ..., n-1]$,*

    I. $V_i \neq \Lambda$,

    II. *the relation* $\ulcorner d(x, y) \in \bigcup_{j=0}^{i} V_j \urcorner$ *is an equivalence in S*,

    III. *the family* $\mathfrak{A}(\ulcorner d(x, y) \in \bigcup_{j=0}^{i} V_j \urcorner, S)$ *is finite.*

T. 6.5 is an immediate consequence of T. 6.1 and T. 6.4. Making use of T. 5.1 we show easily the truth of the following converse of T. 6.5.

T. 6.6. *If there exists an **n**-level increasing partition of V: $V_0, ..., V_{n-1}$, such that for any $i \in [0, ..., n-1]$, the conditions I, II, III appearing in T. 6.5 are fulfilled, then there exists a classification $\mathfrak{M}$ with **n** rows such that $\mathfrak{M}$ mirrors the diversification of S relatively to $\ulcorner d(x, y) \urcorner$.*

PROOF. Let $V_0, ..., V_{n-1}$ be an increasing partition of V such as postulated in the antecedent of T. 6.6. Let further $\ulcorner Clf(x, y) \urcorner$ be a function for which the following condition holds:

(1) $\begin{cases} \text{for all } x, y \in S, \text{ for all } i \in [0, ..., n-1], \\ Clf(x, y) = i \text{ if and only if } d(x, y) \in V_i. \end{cases}$

Condition (1) assignes to every pair x, y of elements of S exactly one value $Clf(x, y)$, since (according to our assumption stating that $V_0, ..., V_{n-1}$ is an increasing partition of V)

$$\bigcup_{i=0}^{i} V_i = V \text{ and } V_i \cap V_j = \Lambda \text{ (for } i \neq j \text{ and } i, j \in [0, ..., n-1]).$$

$\ulcorner Clf(x, y) \urcorner$ fulfils moreover the following condition. For all $x, y \in S$, for any $i \in [0, ..., n-1]$,

    I. $Clf(x, y) \in [0, ..., n-1]$,

    II. there is a pair z, u of elements of S such that $Clf(z, u) = i$,

    III. the relation $\ulcorner Clf(x, y) \leqslant i \urcorner$ is an equivalence in S,

    IV. the family $\mathfrak{A}(\ulcorner Clf(x, y) \leqslant i \urcorner, S)$ is finite.

(This is an obvious consequence of (1) and the assumptions made regarding $V_0, ..., V_{n-1}$.)

It follows from this that the function $\ulcorner Clf(x, y) \urcorner$, the class S and the number n are satisfying the right side of T. 5.1. Hence

(2) $\ulcorner Clf(x, y) \urcorner$ determines an n-level classification of S.

Further from (1) and the assumption stating that $V_0, ..., V_{n-1}$ is an increasing partition of V it follows that

(3) $\begin{cases} \text{for all } x, y, z, u \in S, \\ Clf(x, y) < Clf(z, u) \rightarrow d(x, y) < d(z, u). \end{cases}$

From (2), (3) and the Definitions 5.1 and 6.1 it follows that there exists a classification $\mathfrak{M}$ with n rows such that $\mathfrak{M}$ mirrors the diversification of S relatively to $\ulcorner d(x, y) \urcorner$. This ends the proof of T. 6.5.

7. TWO FURTHER CONCEPTS OF NATURAL CLASSIFICATION

According to the definition of classification given in Section 2, the last row of a classification may be a sequence of unit classes. If for example to the classification Clf_2 considered in Section 1 the row: $\{a_1\}, ..., \{a_{10}\}$ is added, the double sequence of classes thus obtained is a classification according to Df. 2.1 and even a classification which just as Clf_2 mirrors the diversification of the class $\{a_1, ..., a_{10}\}$ relatively to $d_{gen}(x, y)$.

Now however we shall change the concept of classification so as to exclude such matrices of classes from its extension.

Appendix to Df. 2.1. A double sequence of classes $\mathfrak{M}$ with n rows is a classification *only if* at least one of the classes of the nth row of $\mathfrak{M}$ contains more than one element.

An obvious consequence of this is the following theorem.

> *For every classification* $\mathfrak{M}$, *there is a pair* x, y *of elements of* sp$\mathfrak{M}$ *such that* $x \neq y$ *and* $x \mid_{\overline{\mathfrak{M}}} \mid y = 0$.

Let us also note that this change of the concept of classification implies a corresponding change of the concept of a *function determining an n-level classification* of a given class. The proper definition of this last concept is T. 5,1. In order to adjust T. 5.1 to the now changed concept

of classification, to the conditions postulated for $\ulcorner d(x, y) \urcorner$ by the right side of T. 5.1, the following condition must be added. *There is a pair x, y of elements of S such that $x \neq y$ and $d(x, y) = 0$.*

Turning again to the example Clf_2 given in § 1 let us note that if we cancel in Clf_2 the last, i.e. the third row, or leave the third and cancel the second, the classification thus obtained mirrors the diversification of the class $\{a_1, \ldots, a_{10}\}$ relatively to $d_{gen}(x, y)$. It seems however natural to say that in either case the new classification mirrors this diversification less exactly than the original classification Clf_2. Examples of this kind justify the introduction of the concept of a classification *exactly mirroring* the diversification of a given class relatively to a given function.

This will be the first of the new concepts of natural classification defined in this section. First however the auxiliary concept of an enlargement of a given classification must be introduced.

Let $\mathfrak{M}$ be any classification with n rows, $\mathfrak{M}'$ any classification with m rows, r — any of the numbers $1, \ldots, n$

Df. 7.1. $\mathfrak{M}'$ *is an enlargement of* $\mathfrak{M}$ *in the place after the rth row if and only if*

I. $m = n + 1$,

II. for every $\bar{r} \in [1, \ldots, n]$
if $\bar{r} \leqslant r$ then the $\bar{r}$th row of $\mathfrak{M}$ and the $\bar{r}$th row of $\mathfrak{M}'$ are identical;
if $\bar{r} > r$ then the $\bar{r}$th row of $\mathfrak{M}$ and the $(\bar{r}+1)$st row of $\mathfrak{M}'$ are identical.

(In other words, $\mathfrak{M}$ may be obtained from $\mathfrak{M}'$ by cancelling the $(r+1)$st row of $\mathfrak{M}'$ and changing appropriately the indexes of rows appearing after the $(r+1)$st.)

Df. 7.2. $\mathfrak{M}'$ *is an enlargement of* $\mathfrak{M}$ *by* 1 *row if and only if there is an* $r \in [1, \ldots, n]$ *such that* $\mathfrak{M}'$ *is an enlargement of* $\mathfrak{M}$ *in the place after the rth row.*

As a consequence of these definitions we get

T. 7.1. *For any classifications* $\mathfrak{M}$ *and* $\mathfrak{M}'$ *if* $\mathfrak{M}'$ *is an enlargement of* $\mathfrak{M}$ *by* 1 *row then* sp'$\mathfrak{M}$ = sp'$\mathfrak{M}'$.

Let $\mathfrak{M}$ be any classification with n rows, r — any of the numbers $1, \ldots, n$, $\mathfrak{M}'$ — any enlargement of $\mathfrak{M}$ in the place after the rth row, S — the common space of $\mathfrak{M}$ and $\mathfrak{M}'$. Let further $m(x, y)$ and $m'(x, y)$ denote (for any $x, y \in S$) the level of the meet of the lines $\mathfrak{L}_{\mathfrak{M}}{}^{\prime}x$ and $\mathfrak{L}_{\mathfrak{M}}{}^{\prime}y$ or $\mathfrak{L}_{\mathfrak{M}'}{}^{\prime}x$ and $\mathfrak{L}_{\mathfrak{M}}{}^{\prime}y$ respectively.

T. 7.2.   *For every* $x, y \in S$

$$\text{I. } m(x, y) < r \to m'(x, y) = m(x, y),$$

$$\text{II. } m(x, y) = r \to m'(x, y) = r \lor m'(x, y) = r + 1,$$

$$\text{III. } m(x, y) > r \to m'(x, y) = m(x, y) + 1 \geqslant r + 2.$$

(This is an obvious consequence of the definitions involved.)

T. 7.3.   *For all* $x, y, z, u \in S$,

$$m(x, y) > m(z, u) \to m'(x, y) > m'(z, u)$$

(A consequence of T. 7.2).

T. 7.4.   *If* $\mathfrak{M}'$ *mirrors the diversification of* S *relatively to* $\ulcorner d(x, y) \urcorner$
then $\mathfrak{M}$ *mirrors the diversification of* S *relatively to* $\ulcorner d(x, y) \urcorner$
(where $\ulcorner d(x, y) \urcorner$ *is an arbitrary distance in* S).

PROOF. Assume that $\mathfrak{M}'$ mirrors the diversification of S relatively to $\ulcorner d(x, y) \urcorner$, i.e. that for all $x, y, z, u \in S$

$$(1) \qquad (n+1) - m'(x, y) < (n+1) - m'(z, u) \to d(x, y) < d(z, u).$$

Let now x, y, z, u be any elements of S such that $x |_{\overline{\mathfrak{M}}}| y < z |_{\overline{\mathfrak{M}}}| u$, i.e. that $n - m(x, y) < n - m(z, u)$. But $m(x, y) \in [1, \ldots, n]$ and $m(z, u) \in [1, \ldots, n]$. Hence $m(x, y) > m(z, u)$. Hence according to T.7.3, $m'(x, y) > m'(z, u)$. But $m'(x, y) \in [1, \ldots, n+1]$ and $m'(z, u) \in [1, \ldots, n+1]$.

Hence

$$(n+1) - m'(x, y) < (n+1) - m'(z, u).$$

Hence according to (1) $d(x, y) < d(z, u)$.

Thus we have shown that if (1) then for all $x, y, z, u \in S$, $x |_{\overline{\mathfrak{M}}}| y < z |_{\overline{\mathfrak{M}}}| u \to d(x, y) < d(z, u)$. Hence T. 7.4.

Let $\mathfrak{M}$ be any classification whose space is S, $\ulcorner d(x, y) \urcorner$ — any distance in S.

Df. 7.3. $\mathfrak{M}$ *mirrors exactly the diversification of* S *relatively to* $\ulcorner d(x, y) \urcorner$ *if and only if*

I. $\mathfrak{M}$ mirrors the diversification of S relatively to $\ulcorner d(x, y) \urcorner$ and

II. There does not exist a classification $\mathfrak{M}'$ such that $\mathfrak{M}'$ is an enlargement of $\mathfrak{M}$ by 1 row, and $\mathfrak{M}'$ mirrors the diversification of S relatively to $\ulcorner d(x, y) \urcorner$.

The concept of natural classification corresponding to Df. 6.1 has in certain cases a geometrical meaning. Let P_1, P_2 be any measurable properties such as for instance *weight* and *height*. Let S be a class of objects such that to every element of S a definite degree of the property P_1 and of the property P_2 may be assigned. Then to every $x \in S$ corresponds a point x' of the two-dimension $\{P_1, P_2\}$-space and for any $x, y \in S$, the measure of the qualitative difference between x and y relatively to P_1 and P_2 is the distance of the points x' and y' in this space. Let us denote this distance by $d(x, y)$.

Let us assume now that there exists a two-level classification of S mirroring the diversification of this class relatively to $\ulcorner d(x, y) \urcorner$ and let

$(\mathfrak{M})$ $\{1, 1\} = S,$
$\{2, 1\}, \ldots, \{2, i(2)\}$

be such classification. Then the following holds.

(1) $\begin{cases} \text{For every } x, y, z, u \in S, \\ x \mid_{\overline{\mathfrak{M}}} \mid y = 0 \wedge z \mid_{\overline{\mathfrak{M}}} \mid u = 1 \to d(x, y) < d(z, u). \end{cases}$

This condition (1) which states that $\mathfrak{M}$ mirrors the diversification of S relatively to $\ulcorner d(x, y) \urcorner$ may in the case under consideration be expressed in another way.

Let S' be the class of all points of the $\{P_1, P_2\}$-space corresponding to elements of S. An equivalent statement of (1) is as follows.

> *For any* $x', y', z', u' \in S'$, *if* x' *and* y' *are points corresponding to elements of the same class the second row of* $\mathfrak{M}$ *while* z' *and* u' *are points corresponding to elements of two different classes of this row, then* x' *lies closer to* y' *than* z' *to* u'.

From this it follows that S' divides into clusters of points such that the distance of any two points belonging to the same cluster is less than the

distance of any two points belonging to different clusters. Only in case S' may be so partitioned, there exists a two-level classification of S mirroring the diversification of this class relatively to $\ulcorner d(x, y) \urcorner$.

A class of points of a property space may however divide into distinctly separated clusters although the above condition is not fulfilled. Let us also note that the systematists are sometimes considering families of classes which are delineated in the following way. Any two elements belonging to the same class of the family are either more alike than any two elements belonging to different classes of this family or else one of these elements is connected with the other by a sequence of intermediate forms (also belonging to the same class) so that any two neighbours in the sequence are more alike than any two elements of different classes of the family. Classifications corresponding to such families seem also 'natural'. The concept of natural classification covering such cases must however be weeker than the concept corresponding to Df. 6.1.

This weeker concept of natural classification will be defined now, however only for classifications with finite spaces.

Let $S, \ulcorner d(x, y) \urcorner, V$ be any finite class, any distance in S, the class of all values of $\ulcorner d(x, y) \urcorner$ for arguments belonging to S respectively.

By $\bar{d}(x, y)$ for any $x, y \in S$, we shall mean the smallest of the numbers $v \in V$ such that

> I.   there exists a finite sequence of elements of S: $a_1, \ldots, a_m$ such that $a_1 = x$, $a_m = y$ and
>
> II. for any $i \in [1, \ldots, m-1]$, $d(a_i, a_{i+1}) \leqslant v$.

From the assumption that S is finite it follows that the class of elements of V fulfilling the above condition is also finite. This class is not empty since $v = d(x, y)$ is its element. $\ulcorner \bar{d}(x, y) \urcorner$ is hence a function defined in S. Let $\bar{V}$ be the class of all values of $\ulcorner \bar{d}(x, y) \urcorner$ for $x, y \in S$. The following theorem may be easily proved.

T. 7.4. *For all $x, y \in S$, for all $\bar{v} \in \bar{V}$,*

> I. $\bar{d}(x, y) = 0 \equiv x = y$,
>
> II. $\bar{d}(x, y) = \bar{d}(y, x)$,
>
> III. $\bar{d}(x, y) \leqslant 0$,
>
> IV. $\bar{d}(x, y) \leqslant v \wedge d(y, z) \leqslant v \rightarrow \bar{d}(x, z) \leqslant \bar{v}$,
>
> V. $\bar{d}(x, y) + \bar{d}(y, z) \leqslant \bar{d}(x, z)$.

As a consequence of T. 7.4 we have

T. 7.5.   I. $\ulcorner \bar{d}(x, y) \urcorner$ is a *distance in S*;
II. *for all* $\bar{v} \in V$ *the relation* $\ulcorner \bar{d}(x, y) \leqslant \bar{v} \urcorner$ *is an equivalence in S.*

Df. 7.4. $\mathfrak{M}$ *mirrors weakly the diversification of S relatively to* $\ulcorner d(x, y) \urcorner$ *if and only if* $\mathfrak{M}$ *mirrors (as understood by Df. 6.1) the diversification of S relatively to* $\ulcorner \bar{d}(x, y) \urcorner$.

T. 7.6.   *If there exists a classification* $\mathfrak{M}$ *with* ***n*** *rows such that* $\mathfrak{M}$ *mirrors weakly the diversification of S relatively to* $\ulcorner d(x, y) \urcorner$ *then the class* $\bar{V}$ *of all the values of* $\ulcorner \bar{d}(x, y) \urcorner$ *for* $x, y \in S$ *contains at least* ***n*** *numbers different from 0.*

PROOF. Let $\mathfrak{M}$ be a classification with ***n*** rows such that $\mathfrak{M}$ mirrors (according to Df. 6.1) the diversification of S relatively to $\ulcorner d(x, y) \urcorner$. Then

(1)   for all $x, y, z, u \in S$, $x \mid_{\overline{\mathfrak{M}}} \mid y < z \mid_{\overline{\mathfrak{M}}} \mid u \rightarrow \bar{d}(x, y) < \bar{d}(z, u)$.

Moreover (see T. 3.1 and the appendix to Df. 2.1) for every $i \in [0, ...,$ $n-1]$ there is an $x \in S$ and an $y \in S$ such that $x \neq y$ and $x \mid_{\overline{\mathfrak{M}}} \mid y = i$. Let x, y be any elements of S such that $x \neq y$ and $x \mid_{\overline{\mathfrak{M}}} \mid y = 0$. We put $\bar{d}(x, y) = a_1$. Since $x \neq y$, $a_1 \neq 0$. By way of a similar reasoning we find the numbers $a_2, ..., a_n$ – all different from 0. According to (1) $a_1, ..., a_n$ is an increasing sequence. Hence $\bar{V}$ contains at least ***n*** numbers different from 0.

T. 7.7.   *If the class* $\bar{V}$ *of all values of* $\ulcorner \bar{d}(x, y) \urcorner$ *for* $x, y \in S$ *contains at least* ***n*** *numbers different from 0, then there exists a classification* $\mathfrak{M}$*, with* ***n*** *rows such that* $\mathfrak{M}$ *mirrors weakly the diversification of S relatively to* $\ulcorner d(x, y) \urcorner$.

PROOF. Suppose that V contains at least ***n*** numbers different from 0. But for all $\bar{v} \in \bar{V}$, $\bar{v} \geqslant 0$, hence

(1) $\bar{V}$ contains at least ***n*** positive numbers.

Since S is finite, $\bar{V}$ is also finite. Let us denote by a the greatest of the numbers contained in $\bar{V}$. Obviously $a > 0$ and moreover for all $\bar{v} \in \bar{V}$, $0 \leqslant v \leqslant a$. According to (1) $\bar{V}$ contains at least $n-1$ numbers v such that

$0 < v < a$. Let $a_1, \ldots, a_n = a$ be an increasing sequence of positive elements of $\overline{V}$. According to T. 7.5.

(2)   for any $i \in [1, \ldots, n]$ the relation $\ulcorner \overline{d}(x, y) \leqslant a_i \urcorner$ is an equivalence in S.

Also, since S is finite

(3)   for every $i \in [1, \ldots, n]$ the family $\mathfrak{A}(\ulcorner \overline{d}(x, y) \leqslant a_i \urcorner, S)$ is finite.

Let us define the function $\ulcorner Clf(x, y) \urcorner$ by assuming that for any $x, y \in S$,

(4)
$$
\begin{cases}
Clf(x, y) = 0 \equiv \overline{d}(x, y) \leqslant a_1, \\
Clf(x, y) = 1 \equiv a_1 < \overline{d}(x, y) \leqslant a_2, \\
\cdots\cdots\cdots\cdots\cdots\cdots\cdots\cdots \\
Clf(x, y) = n - 1 \equiv a_{n-1} < \overline{d}(x, y) \leqslant a_n.
\end{cases}
$$

From this definition it follows that

(5)
$$
\begin{cases}
\text{for all } x, y \in S, \text{ for every } i \in [0, \ldots, n-1], \\
Clf(x, y) \leqslant i \equiv \overline{d}(x, y) < a_{i+1}.
\end{cases}
$$

The following is also true:

(6)   There is a pair x, y of elements of S such that $x \neq y$ and $Clf(x, y) = 0$.

From (2)–(6) and the definition of the sequence $a_1, \ldots, a_n$ it follows that

(7)   The function $Clf(x, y)$ determines an n-level classification of S.

(See T. 5.1 and the remarks commenting the appendix to Df. 2.1.) According to (4) however the following also holds.

(8)   For all $x, y, z, u \in S$, $Clf(x, y) < Clf(z, u) \rightarrow \overline{d}(x, y)$

$< \overline{d}(z, u)$.

From (7) and (8) it follows that there exists a classification $\mathfrak{M}$ with n rows such that $\mathfrak{M}$ mirrors the diversification of S relatively to $\ulcorner \overline{d}(x, y) \urcorner$, i.e. such that $\mathfrak{M}$ mirrors weakly the diversification of S relatively to $\ulcorner d(x, y) \urcorner$. The proof of T. 7.7 is complete.

REFERENCES

* First published in *Studia Logica* **XII** (1961).

[1] The concept of natural classification considered in the present article is a relative concept. Philosophers and methodologists are however often using an absolute concept of natural classification. By a natural classification they mean a classification based on essential affinities and differences of the objects classified. See for example V. Lenzen, 'Procedures of Empirical Sciences', *International Encyclopedia of Unified Science* **I**, p. 31–35, The University of Chicago Press, 1938. As regards this absolute concept the fundamental problem is that of the nature of essential affinities and differences. This problem of essence has been studied in Poland by prof. Roman Ingarden. In this article it is not considered.

[2] Cited from F. Dannemann, *Erläuterte Abschnitte aus den Werken hervorragender Naturforscher*, Leipzig 1902, p. 247.

[3] Here is how J. Venn in his *Principles of Empirical Logic*, London 1907, describes the ideas of those systematists "who have spoken in favour of a 'natural' system of classification". "They considered themselves, says Venn, to be in some way following Nature in their scheme of arrangement, and to be making their dispositions in such a way that the things which should stand nearest each other in their scheme, should be those which were actually most closely related" (p. 333//334)... They thought that "on such a scheme those things were placed in close proximity which are actually 'related to' or in 'affinity with' each other" (p. 337). Let us note that Venn is using here our concept of the distance within a given classificatory system between the things belonging to the classification space.

[4] J. Venn: *l. c.*, p. 338–341.

[5] I will be also using: a) the functors of the propositional calculus $\rightarrow$, $\vee$, $\wedge$, $\equiv$ as shorthands for 'if ... then', 'or', 'and', 'if and only if', respectively; b) the following symbols of the calculus of classes: $\cup$, $\cap$, $\supset$, ε, Λ (null class); c) the sign of identity '$=$'.

[6] The corner quotation marks in the formula '$\ulcorner x \overline{\overline{m}} \vert y \urcorner$' are meant to inform that what is spoken of is the corresponding function and not its value for the arguments x, y. This value will be denoted by '$x \overline{\overline{m}} \vert y$'. A similar notation will be used below in the case of relations. Thus for example the formula '$\ulcorner x < y \urcorner$' names the relation of *being smaller than* whereas '$x < y$' states that x is smaller than y.

[7] i.e. is reflexive, symmetric and transitive in sp $\mathfrak{M}$.

[8] By 'the compound $\langle S, \mathfrak{M} \ulcorner d(x, y) \urcorner \rangle$' I mean the ordered triple the first element of which is the class S, the second – a classification $\mathfrak{M}$ (whose space is S), the third – the distance $\ulcorner d(x, y) \urcorner$ (which is defined in S).

W. MEJBAUM

THE PHYSICAL MAGNITUDE AND EXPERIENCE *

INTRODUCTION

The title of this paper determines too large a subject. I am interested here in the problems concerning the operationist concept of physical magnitude: according to this concept, the sense of the terms denoting physical magnitudes is determined by description of respective measurement methods. The neopositivistic methodology — by which I mean the views of Carnap and his followers — has accepted the operationist ideas. According to the neopositivistic position, physical magnitudes are partly definable on the basis of observational terms. The definition of physical magnitude consists of a system of postulates, which have the form of general implications; and their antecedents are formulated in observational language. Each postulate of such a system determines a method of measurement of the considered magnitude. For short, we shall call these postulates 'operational definitions'.

The neopositivistic model of knowledge is based on the dichotomy of analytic and synthetic sentences. In our case operational definitions are analytic sentences, while physical laws connecting two or more physical magnitudes are synthetic ones. Also observational statements which describe the measurements carried out and their results are called synthetic sentences, as well as atomic sentences assigning to a given object a certain value of the considered physical magnitude. The latter sentences are accepted as consequences of observational statements and of operational definitions (this enumeration of synthetic sentences is of course far from being complete).

The dichotomy of analytic and synthetic sentences and that of observational and theoretical terms expresses a certain view on the dynamics of knowledge. The foundations of our knowledge are based on directly

decidable synthetic statements formulated in observational language. Arbitrary operational definitions enrich the language by completing it with theoretical terms—among other things, with names of physical magnitudes. On the grounds of observational statements and of operational definitions the truth or falsity of sentences containing theoretical terms is decided, and, among other things, physical laws may be falsified.

I am convinced that the neopositivistic theory of knowledge has some essential merits. It provides a very simple scheme of cognitive processes, and it is generally accepted that the development of any branch of knowledge should start with possibly simple idealizations (I think it becomes particularly evident in the case of the Marxist concept of relative truth). Of course, if the idealization should be effective, it must *idealize*, i.e. stress some objective properties of the object of study. I think that the neopositivistic theory fulfils this requirement.

It is a fact that every branch of science enriches the natural language (or that of another branch of science) with new specific terms; it does so either by modifying the meaning of words of the natural language or by introducing new words. It is also a fact that we can understand the meaning of a new term only by grasping its relations with meaning of words already known. Thus it seems rational to proceed (in the theory of knowledge) to a stratification of language, distinguishing two or more levels. From this point of view the distinction between observational and theoretical terms and the search for analytic sentences to connect them is a useful step.

Nevertheless, the model knowledge thus constructed appears to be too primitive. It is not true that the sense of observational terms remains unchanged during the development of the theory. On the other hand, in my opinion, the latter statement may be maintained as a simplifying assumption, i.e. as a conscious idealization. But the situation is more complicated in the case of the semantic relations between the theoretical terms and those considered as observational. Gradually it becomes clear that it is impossible, in a set of general statements including a given theoretical term, to distinguish a subset of falsifiable hypotheses (synthetic statements) and subset of conventions defining the meaning of the term (analytic statements), where the second subset would be sufficiently rich to explicate the sense of the considered term. Moreover, the alleged operational definitions sometimes appear to be statements

more falsifiable than some others from higher levels of the theory. If it is true — with which I agree — that for the purpose of falsification of a physical theory it is necessary to have a measurement technique which proved infallible in the course of experiments, then it is also true that a theory is necessary to set up a measurement technique. The recognition of these facts led some neopositivists — I think here of Mehlberg — to assign a 'semi-definitional' function to all, or almost all, general statements of physics: in some experimental situations every statement of physics — traditionally considered as synthetic — can play the role of a convention determining the meaning of one of its terms. But it should be added here — which Mehlberg does not do — that also operational 'definitions' are only semi-definitions, and that in some experimental situations they lose the immunity of empirically unquestionable sentences and are as falsifiable as all the hypotheses. Thus instead of the black and white schema of convention and experience, analysis and synthesis, we get a grey one: every statement is half-analytic and half-synthetic.

I believe that such a situation is the crisis of neopositivism. At the same time it is, in my opinion, the crisis of the theory of knowledge as a science. The latter statement needs some more explanation (explanation — not justification as the subject in question could hardly be settled on several pages of print. Everything I say in this introduction should be taken as a declaration; I do so because I think that the knowledge of the author's general theoretical position will make it easier to evaluate the text and to select the main problems from the marginal ones).

In my opinion there should be a twofold evaluation of neopositivism from the Marxist point of view. First, it is a philosophy following Hume's patterns, and as such it evokes the well-known objections concerning all versions of subjectivism. I shall omit this part of the problem. On the other hand, neopositivism is a philosophy which has outlined a detailed theory of knowledge, where, for the first time in the history of that science, the formal apparatus was indispensable. Such is the lasting contribution of neopositivism to science. In a critique of the neopositivistic model of knowledge it is not enough to negate the theoretical basis, which made the authors choose some particular solutions. It is then necessary to construct another model of knowledge — a more complex one — and to show on this basis that the neopositivistic model is not satisfactory.

I do not believe it to have been accidental that it was precisely sub-
jectivistic philosophy which gave rise to the first utilizable idealization
of the cognitive process. A subjectivist has the courage of simplifying
because he does not realize that he is doing so. His world is poor: it
does not contain causality (there are only regular successions of phenomena), it includes no unobservable objects and hidden parameters, and
finally, it does not have any theory which would explain the nature,
but only general hypotheses allowing the deduction of predictions. In such
a situation it is easier to admit that the theory of knowledge is impossible
and that the subject of research cannot undergo any schematization.

Saying that the crisis of neopositivism is that of the scientific theory
of knowledge I mean among other things the revival of conventionalism
and pragmatism, which takes place in the Western philosophical thought.
The contemporary subjectivist has already understood that knowledge
is too complex to be presented in dichotomies: analytic-synthetic,
observational-theoretic or — let us add — descriptive-emotive. He then
gives up all conceptual distinctions. He restricts himself to stating that,
in the course of development of knowledge, our beliefs are falsified or
modified, where the beliefs include religious superstitions and moral
norms as well as the axioms of logic and mathematics, physical theories
and singular sentences about facts. This constitutes a regression both
from the ideological point of view (because we agree with neopositivism
that religion is something different from science and that the will to
know is not the same as the will to believe) and from that of the perspectives of a detailed theory of knowledge. The conviction that every
statement is both analytic and synthetic, evaluating and descriptive,
is a useless assumption.

To criticise neopositivism on the grounds that it introduces the concepts of analytic sentences and observational terms, although such
sentences and terms do not exist in reality, is to me the same as the
possible criticism of the kinetic gas theory for the introduction of the
concept of perfect gas, although such gases do not exist.

In this paper, I shall outline a simplifying model of procedures, of
accepting or of falsifying statements which contain terms denoting
physical magnitude. In Section 7 I deal with some detailed questions
concerning the neopositivistic solution of the problem and with some
aspects of the logical theory of partial definitions. The last part of the

paper (Section 8) deals with the more general question of the variability of meanings of theoretical terms; the previous analysis of problems of physical magnitudes is only an exemplification. Throughout the paper I use a very simplified logical notation, I omit quantifiers wherever it does not lead to misunderstandings. The same symbols indicate extralogical constants and variables, for example the symbol $F(x)$ stands for a propositional formula in some places and for a singular sentence in others. The correct interpretation is indicated by the context.

1. MEASUREMENT AND METHOD OF MEASUREMENT

Let the terms F_i and G denote respectively the physical magnitudes f_i and g. We shall introduce formulas $F_i(y, k, \alpha)$ and $G(x, k, \beta)$, which read: 'α is a numerical value of a physical magnitude f_i for an object y at the time k', and 'β is a numerical value of a physical magnitude g for an object x at the time k'. We shall be interested in the physical statements of the form

$$(1.1) \qquad W_i(x, y, k) \to \big(F_i(y, k, \alpha) \cdot (\beta = \varphi_i(\alpha)) \to G(x, k, \beta)\big),$$

where $W_i(x, y, k)$ is a propositional formula, and $\varphi_i(\alpha)$ — a mathematical function. We shall call the antecedent of the implication (1.1) *physical conditions*, and the consequent of (1.1) — *physical dependency*. We consider the set of objects satisfying the formula $W_i(x, y, k)$ as the range (or the limits of applicability) of a physical dependency. This range has been determined by the terms of the postulate of the form (1.1). (For the sake of simplicity we shall not take into explicit consideration that such a range is variable in time.)

A physical dependency may bind more than two magnitudes. Therefore, besides (1.1) we must consider the scheme

$$(1.2) \qquad W_{ij}(x, y, y', k) \to \big(F_i(y, k, \alpha) \cdot F_j(y', k, \alpha') \cdot (\beta = \varphi_{ij}(\alpha, \alpha'))$$
$$\to G(x, k, \beta)\big)$$

and analogical schemes for four or more magnitudes. Moreover, in further considerations we shall restrict ourselves to dependencies binding two magnitudes. The restriction is not essential as all the conclusions of this paper — formulated for postulates of the form (1.1) — may be

easily generalized for postulates of the form (1.2) as well as for postulates binding more that three magnitudes.

Every statement of the form (1.1) defines a *method of determining* the physical magnitude g. To determine the value β of the magnitude g for the body x at the time k it is necessary and sufficient to see that for a body y, the time k and the number α of the formulas $W_i(y, x, k)$, $F_i(y, k, \alpha)$ are satisfied. The value β then equals $\varphi_i(\alpha)$. Such is the procedure considered here as the method of determining a given magnitude.

In further considerations we shall often use a symbol P_i to denote the i-th term of a sequence of postulates of the form (1.1), and the symbol M_i to denote the appropriate method of determining the magnitude concerned by the postulates.

Parallel to the term 'method of determining the physical magnitude g', we shall use the terms 'method of measurement of the magnitude g' and 'identifying test for the magnitude g'. We consider the methods of measurement as a subclass of the methods of determining physical magnitudes.

It seems that in natural language the term 'method of measurement' is used in the following cases:

a) when the experimentator can make the bodies x and y meet the conditions $W_i(x, y, k)$, without modifying (or without any significant modification) of the value of the physical magnitude g for the body x;

b) when the experimentator can directly decide whether the number α satisfies the formula $F_i(y, k, \alpha)$.

The conditions a) and b) are fulfilled for all commonly used measuring instruments. $F_i(y, k, \alpha)$ is then the function 'the instrument y indicates the number α at the time k'. Doubt may arise whether the relation F_i determined by this function should be considered a physical magnitude. The question has no real significance for us. Let us note only that to consider the indications of instruments as physical magnitudes is conform to the terminology used by some specialists; it may be useful to cite Landau's position, formulated in his *Quantum Mechanics*.[1]

Measurements should be distinguished from the operations which — for lack of a better term — might be called *computings*. Let us consider the definition of density as quotient of mass by volume. This definition refers to the method of determining density: to determine the density of a body, the value of its mass should be divided by the value of its

volume. Such a procedure could hardly be called measurement, if the actual meanings of words are to remain unchanged. For this reason we have introduced the general term 'method of determining a physical magnitude', which includes methods of measurement as well as computings such as that of density.

Of course, before calculating density it is necessary to determine the values of volume and mass. The calculation needs therefore other previous measurements or calculations, supplying the respective data. In natural language measurement often denotes the whole procedure in the course of which two or more measurements or calculations might be distinguished. In scientific literature a distinction is often made between *direct* measurements and *indirect* measurements. In the course of the former the value of the measured magnitude is identified with the indication of the instrument; the latter consists in utilizing directly measured values of some physical magnitudes $g_1, g_2, ..., g_n$ to calculate the value of the magnitude g_{n+1}. Such classification of measurements can be found in Mehlberg's book.[2]

We shall call identifying operations both the direct and indirect measurements and calculations. An identifying operation is then for us a concrete application of any method of determining a physical magnitude. We shall adopt the following partial explanation of the newly introduced concept:

(1.3)   'An identifying operation for a magnitude g for a body x was carried out by the method M_i' if:

       a) it had been established that, for a body y at the time k and for the numbers α and β, the formula

$$F_i(y, k, \alpha) \cdot (\beta = \varphi_i(\alpha))$$

       is satisfied;

       b) on those grounds the number β was considered as the possible value of the physical magnitude g for the body x.

The explanation needs modifying in the case of an identification test concerning the dependency of more than two physical magnitudes (instead of the number α we then have a set of numbers).

Let me emphasize that from the point of view adopted here, the result of the identifying operation the number β does not always have

to be identified with the value of the measured magnitude. For example, it may happen that the result of the measurement differs so much from the previous evaluations that we are inclined to reject it as erroneous. For this reason we have written in the point b) of the explanation: 'the number β was considered as the possible value of the physical magnitude g for the body x', instead of 'it was accepted that the number β represents the value of the physical magnitude g for the body x'.

We shall introduce some further necessary concepts. We shall call the body x *object of operation*. In the case when the considered identifying operation can be called measurement according to linguistic tradition, the body y is a *measuring instrument*. The number α is then the *indication of the instrument* in the case when body y is an instrument. We shall call the number β *result of the identifying operation* (or result of measurement or calculation).

We can say that an identifying operation (a measurement or calculation) carried out by the method M_i is *correct* (in relation to that method[3]) if and only if the conditions $W_i(x, y, k)$ are fulfilled for the body x and the appropriate body y at the time k. We can say that the result of an identifying operation is *adequate*, if and only if it equals the value of the measured magnitude, that is, if x is the object and β the result of the operation — when the sentence $G(x, k, \beta)$ is true.

We can say that the method M_i is *correct* if and only if the corresponding postulate P_1 is true. We can see that, if the method M_i is correct and if the operation carried out with it is also correct, then the result of the operation must be adequate. Indeed, sentences of the form $G(x, k, \beta)$ are consequences of the postulate P_i and of corresponding sentences of the forms $W_i(x, y, k)$, $F_i(y, k, \alpha)$ and $\beta = \varphi_i(\alpha)$.

From now on, we shall interpret in two ways the variable k of the postulates of the form (1.1) and (1.2): either as a time variable, as we have said at the beginning, or as a discrete variable ranging over the set of measurements or operations carried out in the course of an experiment. Of course, all the statements true under the first interpretation will also remain true under the second.

It should be pointed out that the adopted understanding of correctness of both the method of determining a physical magnitude and of the operation carried out by this method allow to be ' correct' an operation carried out by using an incorrect method. Therefore correct (in

relation to a given method) measurements may give inadequate results: in cases when the method is incorrect. The explication (1.3) is not meant as a definition of the concept of identifying operation. We shall not attempt to find the definition at all; we only want to give it a sense definite enough to be able to employ the concepts of the *result of operation* and of *correctness of operation*.

It should be emphasized that from the point of view adopted here the experimentator's intentions are essential to the measurement (the same is true for any identifying operation). The measurement is then a social or psychological phenomenon rather than a purely physical one. The fact of interaction between the instrument and the object is not in itself a measurement: the thermometer behind the window does not measure the temperature if there is nobody to read its indications. Neither does a man carry out a measurement if he looks at the thermometer without knowing how to use it. On the other hand, we shall consider as measurement the determination of a physical magnitude by using an improperly scaled instrument or one which is out of order; but such a measurement will not be a correct one. A man not knowing that a clock has stopped and using it carries out a measurement; so does a child who plays with an imitation of a clock, and believes that the toy is a real clock. Of course, such measurements are evidently incorrect.

We are then rejecting the definition of Landau and Lifshic who say: "It should be strongly emphasized that we do not mean a process of 'measurement' occurring in the presence of a physicist-observer. In quantum mechanics we call measurement any process of interaction between classical and quantum objects, occurring independently of any observer" [4]. An interaction of two objects, whatever their nature, does not correspond to a measurement for us, except for the case when there is someone who believes the existence of such an interaction, and who wants to get some information about one of the objects by studying the state of the other. Landau's definition makes it impossible to distinguish between correct and incorrect measurements, as well as adequate and inadequate results. This is a shortcoming of the definition, as these distinctions exist in linguistic use (possibly under the form of other words). It is generally accepted that a measurement can involve an error, and definite controlling procedures are used to eliminate errors. Even for the purpose of analysis of these procedures, Landau's under-

standing of the concept of measurement would not be convenient for the theory of knowledge.

All the remarks presented above are simplifying. Parallel to postulates of the form (1.1) and analogical postulates for a greater number of physical magnitudes, more complex postulates may be considered; as, for example, postulates determining probabilistic dependencies between two or more physical magnitudes. We shall, however, restrict ourselves in principle to the analysis of postulates of the form (1.1) and of corresponding indentifying operations.

2. POSTULATES OF UNIQUENESS AND DIVERGENCE OF RESULTS

We adopt as true for every physical magnitude F the postulate

$$(2.1) \qquad F(x, k, \alpha) \cdot F(x, k, \alpha') \to (\alpha = \alpha').$$

We shall call the postulates of the form (2.1) *postulates of uniqueness*.

We say that two identifying operations have given *divergent results* when one of these operations leads to the assertion of a sentence $G(x, k, \beta)$ and the other to the assertion of a sentence $G(x, k, \beta')$, while $\beta \neq \beta'$. It is clear that the acceptance of divergent results of measurements makes a system of knowledge contradictory on the grounds of postulates of uniqueness.

In practice two results of measurements are seldom directly divergent, for seldom do two different measurements concern the same object at the same moment. More often results of measurements appear to be *indirectly divergent* on the grounds of some additional continuity assumptions. By *continuity assumptions* I mean here statements of the forms

$$(2.2) \qquad \prod_{k, k' \in T} \left(F_i(x, k, \alpha) \equiv F_i(x, k', \alpha) \right),$$

$$(2.3) \qquad \prod_{x, x' \in K} \left(F_i(x, k, \alpha) \equiv F_i(x', k, \alpha) \right),$$

where T is a finite period of time, and K—a set of objects.[5] Continuity assumptions of the form (2.2) may be in some cases justified by the preservation laws concerning some physical magnitudes.

For instance, two different results of measurements of density of two different samples of mercury are directly divergent on the grounds of the assumption that all samples of mercury have the same density. In this case, the set K contains all the bodies being mercury.

3. THE PRINCIPLE OF GENETIC ARGUMENTATION

Deductive reasoning is assigned to every identifying operation; the premises of that reasoning are: singular sentences of the forms $W_i(x, y, k)$ and $F_i(y, k, \alpha)$, and a postulate P_i characterizing the test used. The conclusion of the reasoning is the sentence $G(x, k, \beta)$. For short, we shall call the premises of the reasoning premises of the operation and its conclusion – the conclusion of operation.

In the case of divergence of results of two operations it must be assumed that at least one of their premises is false (in the case of indirect divergence a corresponding continuity assumption should be taken into consideration). Thus the disclosure of a divergence of results makes it necessary to question the premises adopted at the beginning.

We can say that the disclosure of the divergence of the results of two or more operations falsifies the conjunction of all the premises of these operations. But none of the premises, considered separately, is falsified yet. Although we know that one of them is false, we do not know which one.

We shall adopt the following *principle of genetic argumentation*:

(3.1)   If the results of identifying operations are divergent, it is necessary to find the error made in the course of these operations.

Genetic argumentation may lead us to reject any of the premises of the identifying operation, i.e. both the singular sentences $W_i(x, y, k)$ or $F_i(y, k, \alpha)$ and the postulate P_i. It may also challenge – in the case of indirect divergence of results – the adopted continuity assumptions.

The procedure of finding an error made in the course of identifying operations is slightly different in the case of challenging the truth of the singular premises $W_i(x, y, k)$ or $F_i(y, k, \alpha)$, or that of the general postulate P_i. We shall present successively the procedures used in each of these cases.

4. THE CONTROL OF PREMISES OF THE FORM $F_i(y, k, \alpha)$

Let us first consider direct measurement. In this case a premise of the form $F_i(y, k, \alpha)$ is an observational sentence stating the indication of the instrument. In the course of controlling procedure this premise can be challenged in two ways:

1. by considering the possibility of an imprecise reading of the indication of the instrument,

2. by considering the possibility of a totally erroneous reading of the indication of the instrument (for instance in the case of sensory illusion or hallucination).

Let us consider the first possibility.

The scale of an instrument determines its maximal precision of measurement. Therefore the theorem stating the indication of the instrument should be formulated as

$$(4.1) \qquad \sum_{\zeta \in P_{\alpha \delta_i^0}} F_i(y, k, \zeta),$$

where α is an approximate indication of an instrument, δ_0^i — the smallest division of the scale and $P_{\alpha \delta_i^0}$ — the set of numbers belonging to the interval $\langle \alpha - \delta_0^i, \alpha + \delta_0^i \rangle$. Thus, if we consider the inexactitude of an instrument, instead of a premise of the form $F_i(y, k, \alpha)$ we must use one of the form (4.1). Moreover, we must also reformulate the conclusion of direct measurement, and instead of the form $G(x, k, \beta)$ we get a theorem of the form

$$(4.2) \qquad \sum_{\zeta' \in P_{\beta \delta_i}} G(x, k, \zeta'),$$

while the value of the error δ_i may be determined mathematically as we know the function

$$(4.3) \qquad \zeta' = \varphi_i(\zeta).$$

In the simplest case — when the appropriate postulate P_i makes us assume that the value of the measured magnitude is equal to the indication of the instrument — the error δ_i will be equal to the value of the division δ_i^0.

By taking instrument inexactitude into consideration we may eliminate the divergence of results of direct measurements provided this

divergence does not exceed the value of $2\delta_i$. In further considerations we shall take it for granted that this operation has been carried out: when speaking of divergence of results of identifying operations we shall therefore mean a divergence which cannot be eliminated by considering the inexactitude of instruments scales. For simplicity of notation we shall still refer to sentences of the forms $F_i(y, k, \alpha)$ and $G(x, k, \beta)$, with the reservation, however, that in more precise formulation they should take form of (4.1) or (4.2).

In some cases it should be assumed that the inexactitude of measurement exceeds the value indicated on the scale. Such is the situation, for instance, when the instrument indicator oscillates chaotically; oscillations of that sort appear during measurements carried out by some sensitive galvanometers. (As we know from theory, the oscillations can be sometimes explained by thermodynamical fluctuations. The galvanometer's maximal sensitiveness determined in this way is of the order of 10^{-12} mA.)[6] Another case is that of measurements of magnitudes which change with time: the experimentator may then assign to some moment of time an indication read a little earlier or later.

In all such cases it should be assumed that the inexactitude of the reading of the instrument's indication exceeds δ_i^0. Therefore, the result of measurement may be characterized by an error exceeding δ_i; after an evaluation of this error the conditions of the operations carried out should undergo an additional correction.

It can be seen that in explaining divergences of results by considering the inexactitude of the reading of an instrument's indication one uses information about the construction and the functioning of the instrument, as well as about the experimentator's behaviour. This information consitutes the premises of genetic argumentation. An important fact is that the value of the error of measurement, resulting from an inexactitude in reading the instrument's indication, is estimated independently of the value of the divergence of results of measurements carried out with this instrument. Therefore it is not always possible to explain a divergence of results by considering errors of that type.

Let us now consider the possibility of a totally erroneous reading of an instrument's indication. Analysing such errors we must consider the process of sense perception as an identifying operation. Its premises are singular statements ascertaining that the observer had the impres-

sion the instrument was indicating a given value, and that the observation was carried out in appropriate conditions; another premise is the general postulate stating that in those conditions the result of observation conforms with reality. All these premises may be challenged, and the most doubtful is the assumption that the observation was carried out in appropriate conditions. In many situations the result of observation may be divergent from the instrument's real indication: bad light, too short time of observation, lack of concentration of the observer, or various pathological changes in the observer's organism. The disclosure of such situations makes it possible to explain the origin of the error; in effect, both the sentence stating the result of observation and the results of identifying operations for which that sentence is a premise become unreliable and should be rejected.

In practice an observation is usually considered as incorrect when its result is clearly divergent from those of several other observations, and when it could hardly be assumed that the object of observation had undergone changes in the meantime. I have here in mind the common case when among several readings of an instrument's indications one diverges from the others, e.g., after having taken somebody's temperature one first reads 37°C, then 39°C, and then again several times 37°C. As we could hardly assume that the volume of mercury had gone up and then fallen down in the short period between the first and the third observation, we agree to treat the second one as incorrect. The premise of genetic argumentation is in this case the rather unprecise conviction that the conditions in which the series of observations was carried out guarantee that *generally* the results should be adequate, although sporadically something else may occur.

All the remarks presented above concerned direct measurement, i.e. one for which the premise of the form $F_i(y, k, \alpha)$ is an observational statement referring to the instrument's indication. To reject this premise it is necessary to challenge the correctness of the observation as a result of which it had been accepted. Let us now consider the rejection of the premise of the form $F_i(y, k, \alpha)$ in the case when the given observation is a calculation.

In the case of calculations the premise of the form $F_i(y, k, \alpha)$ is a statement ascertaining the result of an earlier identifying operation. To reject this premise it is, then, necessary to challenge the correctness

of that earlier operation. As a result, we must challenge and reject either one of the premises P_j or $W_j(x, y, k)$ for one of the earlier operations, or the observational premise of the form $F_j(y, k, \alpha)$ for direct measurement, which begins the considered series of operations. The latter possibility has already been dealt with; the former will be the subject of the two following sections.

5. THE CONTROL OF CORRECTNESS ASSUMPTIONS OF IDENTIFYING OPERATIONS

We shall call the premise of the form $W_i(x, y, k)$ *correctness assumption* of a given identifying operation (in relation to the method M_i). The correctness assumptions of an identifying operation will be dealt with by using a concrete example: let it be the temperature measurement with a normal mercury thermometer. The correctness assumption may then be formulated in the sentence: 'a *correct* measurement of the temperature of the body x was carried out by using the mercury thermometer y'. The sentence contains an enigmatic term, i.e. the term 'correct'. This peculiarity does not disappear after stylistic changes. Of course, another word may be substituted for 'correct', e.g. 'normal' or 'proper'. We can thus say that the thermometer functioned *properly*, and the external conditions were *normal*. Someone else can say that the thermometer was *not out of order*, or simply that it was *good*. All these terms actually have a normative character. The question arises whether and how they could be eliminated.

To eliminate the normative terms found in correctness assumptions or measurements, the assumptions should be formulated in a purely physical language, describing as exhaustively as possible the state of the objects involved in the process of measurement. The thermometer construction should then be exhaustively described, and the conditions determined in which the thermometer indication would be identical to the value of the temperature of the environment. By applying this procedure we shall obtain the list of conditions essential for the correctness assumption of the measurement to be true. As a result, a sequence of terms will be obtained, which we shall call *postulates of correctness of measurement*, for the method M_i. In the simplest case it will be the

sequence of implications

(5.1) $W_i(x, y, k) \to V_{ij}(x, y, k),$ $= 1, 2, ..., m,$

where each of the formulas $V_{ij}(x, y, k)$ determines one of necessary conditions of correctness of measurement. We can also assume that the conjunction of all the necessary conditions $V_{ij}(x, y, k)$ yields a *sufficient condition* of correctness of measurement for the given method; instead of a conjunction of implications (5.1) we could then accept the equivalence

(5.2) $W_i(x, y, k) \equiv V_{i1}(x, y, k) \cdot V_{i2}(x, y, k) ... V_{im}(x, y, k).$

Let us give an example of some conditions of correctness of temperature measurement by using a mercury thermometer:

a) From the moment of beginning the measurement (i.e. the contact of the thermometer with the object) a certain period of times would elapse before the reading of the thermometer's indication. This period is necessary for the equalization of the temperatures of the instrument and of the object;

b) the object's temperature should be constant during the measurement. A mercury thermometer (as well as other instruments based on the principle of thermal dilatability) is not suitable for measuring rapidly variable temperatures;

c) there should be no circumstances which would make impossible the equalization of the temperatures of the instrument and of the object. If the measurement is carried out by using an outdoors thermometer, the instrument should be situated in the shadow. For similar reasons the thermometer should not be held in hand, unless we want to know the hand's temperature;

d) the calorific capacity of the object should greatly exceed that of the thermometer;

e) a medical thermometer should be shaken down before the beginning of the measurement;

f) the temperature of the object should not go beyond the limits fixed by the thermometer's scale;

g) the tube and the bulb should not be broken, the column of mercury should be continuous, the tube should be strongly fixed to the scale. The tube should have a constant diameter.

It can be easily seen that even for a very primitive instrument the list of conditions of correctness of measurement is hard to complete. The task would get still harder if we wanted to eliminate the term 'temperature' from the conditions of correctness of temperature measurement (which is an operationist requirement).

Practically, the physicist does not have any established list of conditions of measurement correctness. In the course of preparations to the measurement an analysis is carried out of the possible disturbances which might make the measurement inadequate; appropriate preventive measures are then taken. In the case when, for some reasons, the result of measurement seems hardly acceptable, another possibly detailed analysis of the circumstances is carried out. The list of correctness conditions is practically completed anew in every case of non-trivial identifying operation.

The assumption of measurement correctness is in no case an observational statement. Only some conditions of measurement correctness can be established by direct observation, and even then we do not always proceed in this way. It is certainly possible to check up directly whether the thermometer's tube is not broken, if the amperometer's circuit has not been interrupted or if all the pieces of a watch are properly set up. But usually all these details are not checked, or at least not all of them — unless the result or measurement appears to be unacceptable; in that case a more thorough control is carried out.

Other conditions of measurement correctness are not subject to direct testing at all. The calorific capacity of a thermometer and of the measurement's object may be determined only by measuring, while the corresponding identifying operations are more complex than temperature measurement. To ascertain that a primitive test was properly applied it would be in many cases necessary to use tests much more complicated. Let us for example consider the evaluating of the pH of a solution by using a litmus-paper. If one wanted to check up whether the object one is dealing with is indeed a litmus-paper, one would have to proceed to operations more complicated than the primitive test to which such a paper is needed. The same situation occurs in the case of the question whether the liquid in a thermometer bulb is indeed mercury.

Certainly, the latter question may be turned down on the grounds that it does not matter whether there is mercury or another liquid in the ther-

mometer — if only the liquid's volume and temperature are approximately connected by linear dependence, and if the thermometer is appropriately scaled. Then, however, in order to guarantee the correctness of the measurement one should first check up whether the liquid in the thermometer does indeed fulfil that requirement. By analogy, one may turn down the question whether the paper dealt with it litmus-paper by formulating instead the question if the object considered — whatever might it be from the chemical point of view — has the property of changing its colour when pH = 7. It is clear that also in this case the measurement correctness assumption can by no means be considered an observational statement.

As a result we have to assume that measurement correctness assumptions are almost always (or always) accepted without sufficient justification. We are not able to test (it is at least practically impossible) whether a measurement has been correct, we can only *assume* it provisionally. For that reason, should any inconsistencies of results be discovered, the assumptions of correctness of the considered operations are the first to be questioned. When it is done, the falsification of premises of an identifying operation consists generally in the falsification of its correctness assumption, or of the correctness assumptions of earlier operations (in cases when the considered operation is a calculation).

It can be seen that measurements (or any identifying operations) may be carried out only when a certain special cognitive attitude is taken up; we shall call that attitude here *correctness presumption* of an observation or measurement. The result of an identifying operation may be considered adequate if only it is not contradictory with other operations' results and if correctness conditions have been respected. On the other hand, to reject a result as inadequate it is not sufficient to state its inconsistency with other results; it is necessary to attempt additionally to find the *origin* of the fault, i. e. to say *explicite* which of the correctness conditions has not been fulfilled.

Correctness presumption may be reckless to a various degree. In the course of every identifying operation one always preliminarily checks up whether some of its correctness conditions are fulfilled. In the case of measuring temperature with a normal thermometer one always makes sure that the instrument one is using has the characteristic features

of a thermometer—otherwise one could not declare (even an incorrect one) that a measurement has been carried out. Further on one checks up whether the thermometer is not broken. One waits until the instrument's temperature becomes equal to that of the object. Nevertheless, no matter how precise and detailed should these preparations be, an incorrectness of measurement can never be totally excluded. We always assume more than we know.

Measurement correctness presumption is turned down if a divergence of results is stated. In the course of controlling the identifying operation's premises, we are inclined to accept any—however little justified—hypothesis which might explain the stated divergence. It should be pointed out, though, that according to the principle of genetic argumentation an identifiying operation's premise should not be rejected without any reason—say, by random choice. Moreover, the genetic explanation should be *confirmable* in the course of later operations: the application of safety measures indicated by the explanation should prevent a possible repeated divergence of results.

6. THE CONTROL OF MEASURING METHODS

The control of a measuring method (or any identifying test) is a procedure tending to falsify and to modify a postulate P_i characteristic for the considered method. An important thing is that in physics a measuring method and the attached postulate are never totally rejected. The result of a controlling procedure is only a certain modification of the previously accepted postulate P_i—either by accepting more rigorous correctness conditions, or by correcting the dependence which is the consequent of the postulate.

We shall call *adequacy postulate* for a given identifying test a postulate P_i attached to that test. Let us rewrite once more the implication (1.1), which we consider as a typical scheme of the adequacy postulate:

$$(6.1) \qquad W_i(x, y, k) \rightarrow \big(F_i(y, k, \alpha) \cdot (\beta = \varphi_i(\alpha)) \rightarrow G(x, k, \beta)\big).$$

Let us also rewrite the scheme of correctness postulates (5.2)

$$(6.2) \qquad W_i(x, y, k) \equiv V_{i1}^{\mathbb{F}}(x, y, k) V_{i2}(x, y, k) \dots V_{im}(x, y, k).$$

Let us first consider a modification of an identifying test by limiting the range of its applicability. We proceed to such a modification when a factor

appears which has not been previously taken into consideration, and which disturbs the measurement; in the case of the influence of such a factor, the dependence in the consequent (6.1) may not be realized. The list of measurement correctness conditions should therefore be completed. Let us denote by $W_i'(x, y, k)$ the range of applicability of the modified test and by $V_{im+1}(x, y, k)$ the additional measurement correctness conditions. We then have:

(6.3) $W_i'(x, y, k) \rightarrow (F_i(y, k, \alpha) \cdot (\beta = \varphi_i(\alpha)) \rightarrow G(x, k, \beta)),$

(6.4) $W_i'(x, y, k) \equiv V_{i1}(x, y, k) \cdot V_{i2}(x, y, k) \dots V_{im}(x, y, k) \times$

          $\times V_{im+1}(x, y, k).$

In this way the inadequacy of some results of the primary test M_i may be explained in all those cases when the test was applied to bodies which did not fulfil the conditions $V_{im+1}(x, y, k)$. On the grounds of the new test M_i these results appear to be incorrect.

The problem may be formulated as follows: A physical dependence Z_i is given

(6.5) $F_i(y, k, \alpha) \cdot (\beta = \varphi_i(\alpha)) \rightarrow G(x, k, \beta).$

We must find such a sentential formula $W_i^*(x, y, k)$ that the postulate

(6.6) $W_i^* (x, y, k) \rightarrow (F_i(y, k, \alpha) \cdot (\beta = \varphi_i(\alpha)) \rightarrow G(x, k, \beta))$

will be true. From this point of view the sentential formulas $W_i(x, y, k)$ and $W_i'(x, y, k)$ are successive approximations of the formula $W_i^*(x, y, k)$. The postulates

(6.7) $W_i^* (x, y, k) \equiv W_i(x, y, k),$

          $W_i^* (x, y, k) \equiv W_i'(x, y, k),$

are testable hypotheses concerning the searched range of acceptability of the dependence (6.5). The adequacy postulate (6.1) may be considered as the partial definition of the predicate W_i^*.

This point of view corresponds to the practice of natural language. The simplest answer to the question: what does 'good' thermometer *mean*? is: it is a thermometer which shows α degrees if and only if the temperature is α degrees. Or rather the utterance 'a *correct* measurement of temperature was carried out with a mercury thermometer' means

'a temperature measurement was carried out with a mercury thermometer *constructed* and *used in such a way* that the thermometer's indication must be equal to the value of temperature for the object of measurement'. *How* should the thermometer be constructed and *in what way* should the measurement be carried out so that the result's adequacy be guaranteed—will be determined by experience. The history of measuring techniques is the history of efforts to construct the best possible instruments, providing increasing precision of measurement. But when we speak of progress in measuring techniques, we thereby assume some objective criterion of progress. We assume that some of the physical magnitudes measured in the 20th century are identical to those measured in the 19th; the progress consists in that we have learned to measure them better.

The concepts of 'correct measurement', 'good instrument' or 'normal physical conditions' have a specific descriptive and evaluating character. In every period of the evolution of knowledge they have an empirical content, determined by the correctness conditions known in that period. On the other hand, these conditions are chosen with regard to a certain aim; we want that in every case they are fulfilled also the considered physical dependence be realized. For this reason the empirical content of the concepts considered is changing; it undergoes successive corrections in the course of the evolution of physics.

An elasticity of the relation between theory and experience is thus obtained. The range of applicability of a theory is not *a priori* fully determined, it is determined gradually. Therefore it seldom happens that a theory is totally rejected. The essence of a theory are physical dependences, and not postulates stating that these dependences are realized in certain conditions. The question 'is a dependency Z realized in conditions W?' has an additional character to the question 'in what conditions is a dependence Z satisfied?'. These conditions should be investigated through successive approximations W, W', W'', $W^{(n)}$.

We have suggested at the beginning of this section that an identifying test may be modified in two ways. We have shortly described above the modification through limiting the range of applicability of a given test, i.e. through replacing previous conditions $W(x, y, k)$ with new ones $W'(x, y, k)$. Let us now consider the question of modifying a physical dependence constituting the consequent of a corresponding postulate (6.1). Our present question is: 'what physical dependence is satisfied

under conditions $W_i(x, y, k)$?'. Or else, if it is known that under conditions $W_i(x, y, k) \cdot V_{im+1}(x, y, k)$ the dependence should not be modified, the question may be: 'what physical dependence is satisfied under conditions $W_i(x, y, k) \cdot \sim V_{im+1}(x, y, k)$?'. We shall consider several typical answers to these questions.

1. *The influence of the approximative character of physical dependences.* Many physical dependences are realized only under very rigorously determined conditions, which practically occur very seldom or not at all. There also are physical dependences which are never precisely realized, under any conditions. Let us consider, for example, the dependence of Boyle–Mariotte, which is precisely realized only for perfect gases, while such gases do not exist.

In such cases the application of a corresponding identifying test gives a result charged with an error. Test control allows to determine the maximum value of the error under the conditions $W_i(x, y, k)$. We shall denote the error by δ_i and call it *admissible measurement error* under the conditions $W_i(x, y, k)$. By taking the admissible measurement error into consideration we obtain the following modification of the adequacy postulate (6.1)

$$(6.8) \qquad W_i(x, y, k) \to \left(F_i(y, k, \alpha) \cdot (\beta = \varphi_i(\alpha)) \to \sum_{\zeta \in P_{\beta \delta_i}} G(x, k, \zeta)\right).$$

The admissible measurement error should not be mistaken for the error connected with an imprecise reading of the instrument's indication (Section 4). The imprecise reading of an instrument's indication may nevertheless be considered a special case of admissible measurement error, i.e. the admissible error of an observation carried out under normal conditions by a normal observer.

Admissible measurement errors for intermediary measurements are determined by using Gauss' error calculus. It is then necessary to know the addmissible errors for the successive operations constituting the given intermediary measurement.

Let us take for example an ordinary watch of a common brand. The owner can tell the time with the exactitude to one second – a greater exactitude is impossible because of the character of the instrument's scale. Practically, though, the owner can reasonably tell the time with an exactitude not exceeding 1 minute, as he is conscious of a certain irregularity of the watch's movement.

As another example we can consider the case of a man measuring the time of duration of a process with a chronometer. We must assume that the admissible measurement error exceeds the inexactitude of the instrument, as it is possible that the observer (considered here as part of the instrument) has stopped the watch a little too early or too late. That is why in athletics competitions the timing is done by several umpires. To minimalize the error some additional primitive statistic methods are used, extreme results are left out and the average is calculated from the others. Such a method is worth closer consideration.

Under what assumptions can we reasonably consider that the average value for a series of measurement results differs from the real value by a number smaller than the error δ_i, taken as admissible for a single measurement carried out under the given conditions? How much can we reduce the error by carrying out longer and larger series of measurements? It is clear that it would be enough to assume that the probability of error is a decreasing function of its absolute value. In such a case we might expect that the average value of a sufficiently long series of measurements will be unrestrictedly close to the real value. The latter statement, however, is not practically accepted. This shows that the assumption it is based on is *implicite* considered as too strong.

If five people meet and each of them has a not very well regulated watch, they will be wise in leaving out the extreme indications (especially if those differ distinctly from the others) and in calculating the average from the others. The same holds for measurements carried out in sport competitions. Nobody believes, though, that in order to determine the time exactly (say, with the exactitude to 1 second) it would be sufficient to calculate the average of the indications of a hundred or more watches.

2. *Random error of measurement.* In cases when the probability of error is a decreasing function of its absolute value we can call the error of measurement *random error*. The arithmetic mean for a long series of results approaches then the real value.

Instead of the adequacy postulate (6.1) a certain modified postulate may then be considered; its consequent is a probabilistic dependence. On the grounds of the dependence and of the known value f_i we can then determine the probability (or the probability density) of whether the measured magnitude g will assume one of the possible values.

3. *Systematic errors of measurement.* We shall say that the results obtained by the test M_i under the conditions $W_i(x, y, k)$ are charged with a *systematic error* if under these conditions instead of the dependence (6.5) another dependence Z_i' is satisfied:

(6.9) $F_i(y, k, \alpha) \cdot (\beta = \varphi_i'(\alpha)) \rightarrow G(x, k, \beta).$

The systematic error can be eliminated by taking instead of the adequacy postulate (6.1) the modified postulate

(6.10) $W_i(x, y, k) \rightarrow (F_i(y, k, \alpha) \cdot (\beta = \varphi_i'(\alpha)) \rightarrow G(x, k, \beta)).$

The considered test M_i is thereby modified.

In particular cases it may appear that the systematic error is constant for the whole range of values f_i and g. The modified adequacy postulate will then take the form:

(6.11) $W_i(x, y, k) \rightarrow (F_i(y, k, \alpha) \cdot (\beta = \varphi_i(\alpha) + \delta_i) \rightarrow G(x, k, \beta)),$

where δ_i is the systematic error committed in applying the primary test M_i based on the postulate (6.1). In other cases the value of the systematic error will depend on the value of the magnitude f_i.

Another practically important case is the situation when the value of the systematic error becomes dependent on a parameter which is the value of a physical magnitude distinct from both f_i and g. Instead of the postulate (6.1) we then take the postulate

(6.12) $W_i(x, y, k) \cdot V_{im+1}(x, y, k, \alpha') \rightarrow (F_i(y, k, \alpha) \cdot (\beta = \varphi_i'(\alpha, \alpha'))$

$\rightarrow G(x, k, \beta)),$

while we consider the additional measurement correctness conditions $V_{im+1}(x, y, k, \alpha')$ as realized when the parameter assumes the value α'. Let us now suppose that there exists such a value φ that

(6.13) $\prod_\alpha (\varphi_i'(\alpha, \zeta) = \varphi_i(\alpha)),$

therefore, that

(6.14) $W_i(x, y, k) \cdot V_{im+1}(x, y, k, \zeta) \rightarrow (F_i(y, k, \alpha) \cdot (\beta = \varphi_i(\alpha))$

$\rightarrow G(x, k, \beta)).$

As we can see, the modification of the physical dependence in the pos-

tulate (6.1) leads in this case to a repeated restriction of the range of the unmodified dependence (the dependence (6.5)).

Let us consider the example of length measurement by means of an optional ruler. Such a measurement is charged with a systematic error depending on temperature. Taking this into consideration we can:

a) determine the value of the temperature in which the measurement result equals the real length of the object,

b) determine the correction which should be made in the case when the temperature is higher or lower; the value of the correction will of course be different for different temperatures.

Let us return to the question of admissible measurement error, which we dealt with in point 1. A measurement may be charged with a systematic error and a random one, and both can occur simultaneously. If we do not know the correction necessary to eliminate systematic errors and if we ignore the probability distribution of random errors, we are not able to determine the precise value of the measured magnitude, even on the base of an unrestrictedly long series of measurement results. In such a situation we must accept the determination of the admissible error for the given conditions and we must consider that the real value is situated in a certain interval. Thus by replacing the postulate (6.1) with the postulate (6.8) we obtain a tentative solution; for a further progress we can either consider probabilistic dependences which include random error, or postulates of the form (6.10) and (6.12) which eliminate systematic errors. It is risky to use primitive statistic procedures (calculations of the average, leaving out extreme results) as long as there is no evidence of the random character of the errors. In case of overlapping of random and systematic errors we can speak of a specified probability of a certain number value of the error only when there is a determined probability of the fulfilment of additional conditions $V_{im+1}(x, y, k, \alpha')$. Even in that case, though, the probability distribution will generally be unsymmetrical, and hence the mean value of a long series of results may distinctly diverge from the real value.

It is clear that in the course of a procedure aiming at modifying an identifying test we must always have at disposal another test, which we consider correct. We than always have a *questionable* test and a *control* one. We consider the results of the control test as strictly adequate,

that is equal to the real values of the measured magnitude. On these grounds we can test the adequacy of the results of the questionable test. If these are not adequate, we evaluate the measurement error. We then modify the questionable test by either limiting its range of applicability or changing the appropriate physical dependency. We eventually check if the modified test gives results corresponding to those of the control one; if it does, we can consider the procedure finished. It will be slightly different in the cases of determining the values of systematic errors and the probabilities of accidental errors. Nevertheless, the basic scheme remains unchanged; it is necessary to compare the results of the questionable test with those of the control one and to explain the causes of the divergences observed.

On what grounds, however, can we decide *which* test is to be the control one and *which* to question? We shall represent here the view which may be summarized in two points:

a) the choice of the control test is at first optional. Every identifying test may be submitted to control on the basis of any other test for the same magnitude;

b) the choice later proves to have been proper or not depending on the effectiveness of further procedure. If it is possible to find an acceptable modification of the questionable test, the correctness assumption of the control test may be considered as confirmed. If it is not possible to find such a modification, the assumption should be considered as *falsified*. The control test should then be changed; the new control test may be either one previously questioned or any test not considered before.

What has been said in point (a) means that *prima facie* all identifying tests are equivalent, in the sense that each of them may be questioned on the basis of any other test. It is in this sense that we shall further speak of the *basic equivalence* of identifying tests. We do not intend to defend this view at any price: we agree that there are some specific identifying tests which always assume a controlling function and are never questioned. The corresponding adequacy postulates play then the role of definitions of some physical magnitudes on the basis of others. We mean here in the first place such undoubtedly definitional dependences as the identity of acceleration with the first derivative of velocity. But if we leave out the indubitable definitions, the other identifying

tests are as a rule equivalent in the sense presented above. It particularly applies to the majority of methods of measurement (direct and indirect) as well as the corresponding adequacy postulates. There are possibly some exceptions; for instance, the operational definition stating that if the extremities of two adjacent bars are coincident, then the bars are equal in length. Another exception might be Einstein's definition of simultaneity.

The view presented in point (b) is based on the *principle of genetic argumentation* (Section 3). We have said that in the case when the adequacy of the result of an identifying operation is questioned, it is necessary to show what kind of error has been committed, that is to reject *explicitely* the premise of the operation; then additional arguments must be found which would prove that the rejected premise is indeed false, and eventually it is necessary, in the course of further research, to confirm the explanation proposed. In our present case, as the general premise of an identifying operation has been questioned—that is, the corresponding adequacy and correctness postulates—a genetic explanation above all requires a modified postulate. Indeed: on the basis of new adequacy and correctness postulates it is possible to determine in what cases the previous postulates led to inadequate results, and to evaluate the degree of thus arising errors. The empirically stated divergence of results of the control and questioned tests should then be explained on the basis of the modification adopted. This holds for both the circumstances in which the divergence appeared and for its numerical value.

Further on—as we have already said—it should be checked whether the control test and the questioned one do not give divergent results.

This procedure clearly limits the free choice of modifications. As long as we consider a certain finite series of already carried out operations, we have indefinitely many propositional functions describing circumstances in which the questioned test has been used—i.e. functions which were realized when the result of the questioned test was divergent from that of the control test, and which were not realized when both tests were in agreement. Each of these functions may serve to formulate additional correctness conditions, limiting the range of applicability of the test. But if we proceed to control examination, it will appear that although the additional conditions are fulfilled, it is not possible

to achieve the agreement of the two tests. This means that the proposed modification does not meet our requirements and that we have to go on to further research.

Of course, these difficulties may be avoided if only the restriction of the range of applicability of the questioned test is radical enough. In each case it is possible to define such correctness conditions that their fulfilment would prove a logical or physical impossibility. This — practically useless — solution will, however, be ruled out if it is required that the range of applicability of the modified test includes circumstances in which a systematic agreement of both (the questioned and the control) primary tests was observed.

Let us note on the other hand that the restriction of the range of applicability of a test may be the object of further explanation procedure. A postulate modified by *limiting the range* (additional correctness conditions) may be justified by a postulate modified by *changing a physical dependence*: a postulate of the form (6.14) justified by a postulate of the form (6.12). In this sense Van der Waals' dependence is confirmed with a satisfactory exactitude only for certain intervals of pressures and temperatures. Such a procedure is closely related to the principle of correspondence widely recognized by physicists.

In some cases the explanation procedure goes even further. We refer to physical theory and the related causal model of phenomena. We might ask *why* does Van der Waals' law rather than that of Boyle–Mariotte apply to real gases; answering, we may quote the supposition that real gas is constituted by molecules of a non-null volume and the supposition that these molecules are bound with attraction forces. Similarly, one might ask *why* Rayleigh–Jeans' dependence, which determines the distribution of radiation of an perfectly black body, is not satisfied for small wave lengths, while Planck's dependence is satisfied for the total range of variables. The answer is given by the quantum theory (it is worth noting that, without assuming the quantum theory structure of light, Planck's formula could not have been discovered; it is too complicated to be guessed through interpolation of empirical data).

We cannot undertake here an analysis of explanation procedures used in physics. We only express the opinion that these procedures are submitted to rigorous rules which may render ineffective the ex-

planation procedure. If two tests are given, M_i and M_j, it may happen that

1. the control of test M_i on the basis of test M_j is effective, in the sense that it is possible to find a modified test M_i' such that the corresponding adequacy postulates fulfil the requirements for a physical law,

2. the control of tests M_j on the base of test M_i is not effective, in the sense that it is not possible to find a modified test M_j' fulfilling the said requirements.

Let us consider again the example of measuring length with a ruler. As we have said, there occur systematic errors depending on temperature. These errors may be evaluated theoretically – by reference to previously established regularities of thermal dilatation of solids – and empirically determined by using (for example) any optical method of length measurement.

This method then assumes the role of a control test, and the explanation procedure is effective. Let us now try to assume that the measurement by using a measure gives adequate results for every interval of temperatures (in other words, let us assume that the coefficient of linear expansion of our measure equals 0). Let us choose the object determined for the carrying out of a series of measures, which can be for example the edge of a mass characteristic for its considerable thermic capacity. We then heat the measure up to 100°C and carry out a measurement, then we cool it down to 0°C and carry out another measurement, and we repeat this several times. It then appears that the length of the edge measured changes according to the state of the measure: we would have to assume that the action on *one* body (the measure which we heat up or cool down) results in changing the length of *another* body (the mass). We shall moreover compare this result with the eventual results of measurements of the length of the edge of the mass by using other measures or optical methods. It will then appear that heating or cooling the standard measure entails not only a change of the length of the object of measurement, but also that of the length of other measures. It will also appear that the changing of the temperatures of the mass and of the measure may influence the course of light rays – if we want to explain the would-be inadequacy of the results of measurement obtained by using optical methods.

This example illustrates very well what we understand by non-effectiveness of some genetic argumentation. If a wrong control test is chosen it proves impossible to find modifications of other tests, at least if one does not want to reject such principles of physical thinking as that of close influence or to give up applying mathematical formalism to physics.

It should also be pointed out that the possibility must be considered that both the control of test M_i on the basis of test M_j and the control of test M_j on the basis of test M_i are not effective, and in which the same negative result is obtained by questioning both tests M_i and M_j on the basis of any other test for the considered magnitude. Let us assume that in such a case we should consider that each of the tests M_i and M_j concerns another physical magnitude. We would then deal with a sort of 'dissociation' of the magnitude g into 2 different magnitudes g_i and g_j. To be exact we would have to say that the magnitude g — measured by using primary tests M_i and M_j — *does not exist*, while there are two or more different magnitudes which had been for some time erroneously identified.

Following some neopositivist concepts of physical magnitude (which we shall deal with in Section 7), a necessary and sufficient condition of accepting the existence of a magnitude denoted by the term G is the consistency of the system of adequacy postulates for that term on the grounds of a set of individual sentences describing the carried out identifying operations and their results. In other words, the necessary and sufficient condition for the recognition of the existence of a physical magnitude would be the agreement of all the results of correct identifying operations for that magnitude. On the other hand, according to the concept constructed in the present paper, the necessary condition for the existence of a physical magnitude is the possibility of eliminating any divergence of results of appropriate identifying operations if the requirements of genetic argumentation are fulfilled.

We shall not try to decide whether in the history of physics there were cases of 'dissociating' a physical magnitude as a result of the non-effectiveness of efforts tending to establish appropriate tests. A good example might be the differentiating, in the theory of gases, between specific heat by constant volume and specific heat by constant pressure.

Let us now consider the case when both the control of test M_i on the basis of test M_j and the control of test M_j on the basis of test M_i prove to be effective. We have at first the pair of postulates

$$(6.15) \qquad W_i(x, y, k) \to \big(F_i(y, k, \alpha) \cdot (\beta = \varphi_i(\alpha)) \to G(x, k, \beta)\big),$$

$$(6.16) \qquad W_j(x, y, k) \to \big(F_j(y, k, \alpha) \cdot (\beta = \varphi_j(\alpha)) \to G(x, k, \beta)\big).$$

Let us suppose that, as a result of the control of test M_j on the basis of test M_i, we obtain

$$(6.17) \qquad W_i(x, y, k) \to \big(F_i(y, k, \alpha) \cdot (\beta = \varphi_i(\alpha)) \to G(x, k, \beta)\big),$$

$$(6.18) \qquad W_j(x, y, k) \to \big(F_j(y, k, \alpha) \cdot (\beta = \varphi_j'(\alpha)) \to G(x, k, \beta)\big)$$

(test M_i and the corresponding adequacy postulate remain unchanged by these assumptions). Let us additionally suppose that there is such a function $\psi(\ldots)$ that

$$(6.19) \qquad \varphi_j'(\alpha) = \psi\big(\varphi_j(\alpha)\big).$$

It is then clear that the control of test M_i on the basis of test M_j is also effective and gives as result

$$(6.20) \qquad W_i(x, y, k) \to \big(F_i(y, k, \alpha) \cdot (\beta = \psi^{-1} \cdot (\varphi_i(\alpha))) \to G(x, k, \beta)\big),$$

$$(6.21) \qquad W_j(x, y, k) \to \big(F_j(y, k, \alpha) \cdot (\beta = \varphi_j(\alpha)) \to G(x, k, \beta)\big).$$

(The symbol $\psi^{-1}(\ldots)$ denotes here a function reciprocal to $\psi(\ldots)$).

It can be seen that there is no possibility of deciding empirically which of the pairs of postulates ((6.17–18) or (6.20–21)) should be accepted. The choice of one of the pairs is then a question of convention. By making a choice we determine the scale of the considered magnitude.

In some cases we obtain the criterion for choosing the scale of physical magnitude by requiring that the considered magnitude be additive. It concerns such magnitudes as mass, length, energy or speed. The additivity postulate determines the scale with the exactitude to linear transformation. We consider additivity postulates as part of the adequacy postulates for a given physical magnitude. Appropriate identifying

tests are assigned to them, as to other adequacy postulates. If the values of a given physical magnitude are known for all the parts of an object, it is possible to calculate on the basis of these postulates the value assumed by the examined magnitude for the whole object.

The scale of other magnitudes is sometimes determined by giving the definitional dependences linking them with already scaled up magnitudes. Temperature scale ought to be dealt with separately. As the definition of the temperature as mean kinetic energy of molecules did not practically determine any effective measuring method the question arose concerning the choice of the so called thermometric body. On a clearly conventional basis it was decided to choose hydrogen. This meant accepting as a definition that the dependence of pressure upon temperature (with constant volume) was linear for hydrogen. It determines the so called empirical temperatures scale with the exactitude to linear transformation. In order to eliminate the remaining unprecision additional values have been accepted: of 100° for the temperature of boiling water and the value of 0° for that of water freezing-point (Celsius scale).

From the cognitive point of view we have here an interesting fact of giving a quasi-definitional status to a law which we independently know to be only relatively true. On the empirical scale temperature will assume values slightly different from the value of temperature on the scale determined by statistic definition, but for certain pressures and temperatures the differences are so small that they might be left out. Nevertheless the statistic temperature definition is the basic fact, decisive for determining the range of applicability of the gas thermometer. Thus the situation is as follows: we *accept* that, for hydrogen, the dependence of pressures on temperature is linear in that interval of pressures and temperatures in which—as it is already *known*—this dependence does not differ much from linear. If we were to say that fully lawful definitions play the role of king with respect to other statements, we might compare the role of definitional determining of the empirical temperatures scale to regency.

The choice and definition of the unit of measure of a physical magnitude are parts of the procedure aiming at determining the scale of this magnitude. We shall not deal with this problem in the present paper, leaving it for a separate one.

7. THE QUESTION OF DEFINABILITY OF PHYSICAL MAGNITUDES ON THE GROUNDS OF OBSERVATIONAL TERMS

The neopositivistic theory of knowledge considers the names of physical magnitudes theoretical terms definable on the grounds of observational terms. The proposed definitions do not differ substantially from the adequacy postulates considered in this paper. They have the form of reduction sentences; the antecedent gives a description of the application of a method or measuring test, the consequent states that under thus determined conditions the result of measurement is *ex definitione* equal to the value of the measured magnitude.

Two questions should be distinguished in the criticism of the neo-positivistic view. First: can the postulates of adequacy and of measurement correctness (no matter how formulated) be considered arbitrary definitions of a physical magnitude. Second: if the answer to the first question is positive, then do these definitions have an operational character, that is, are they indeed definitions of theoretical terms only on the grounds of observational terms. Let us first consider the second question, without questioning (provisionally) the hypothesis which it includes.

A possible (partial) definition of a physical magnitude may either be the adequacy postulate

$$(7.1) \qquad W_i(x, y, k) \rightarrow \left(F_i(y, k, \alpha) \cdot (\beta = \varphi_i(\alpha)) \rightarrow G(x, k, \beta) \right),$$

or the statement (on the basis of the correctness postulate (5.2))

$$(7.2) \qquad V_{i1}(x, y, k) \cdot V_{i2}(x, y, k) \dots V_{im}(x, y, k)$$

$$\rightarrow \left(F_i(y, k, \alpha) \cdot (\beta = \varphi_i(\alpha)) \rightarrow G(x, k, \beta) \right).$$

The non-operational character of the postulate (7.1) is clear. It is not possible to find through a single direct observation (cf. Section 5) whether the measurement has been *correct*, whether the physical conditions during the measurement were *normal*, whether the instrument functioned *correctly* and whether it was well graduated. The terms emphasised in the preceding sentence are in no case observational ones, no matter how liberally should the word 'observational' be treated.

Let us then consider the statement (7.2). The assumption of measurement correctness — undoubtedly non-observational — does no longer occur here. Instead we have in the antecedent the enumeration of cor-

rectness conditions known in a given period. As we know, these conditions are very numerous, even for primitive measuring methods. Some of them may be formulated in observational language. Others, however, include undoubtedly theoretical terms. For instance, it is beyond doubt that the term 'pressure' must be considered as theoretical, as the term 'temperature' is considered to be theoretical. And even such a simple definitional test as that permitting to attribute the temperature of 100°C to boiling water needs a qualification that during the measurement the air pressure equals 1 atm.

The difficulty may be changed by a requirement to replace the condition 'the pressure is 1 atm.' by a description of methods and results of measurement which will allow to decide that the condition is fulfilled (that is, by an alternative of antecedents of possible operational definitions of pressure). Realizing such a requirement, we would indeed consistently formulate monstruously complicated definitions for the simplest magnitudes.

What is more relevant the construction of a really operational definition of physical magnitude seems impossible even if we were to leave out the difficulties connected with the multitude of measurement correctness conditions and the theoretical terms necessary for their formulation (by the way, these terms should also obtain operational definitions). The definitions of theoretical terms necessary to define the considered magnitude must refer to still other theoretical terms, which in turn will want defining. If it is indeed so, the antecedent of the investigated definition will always include theoretical terms and we shall never approach the operational ideal. If we consider the fact that the number of logically independent theoretical terms is finite in a given period of time, we may reasonably presume that a persistent search of an operational definition of physical magnitude must eventually lead to a vicious circle in defining. This inconvenience appears in the fact that in order to graduate a thermometer we must dip it into boiling water by normal pressure, while in order to verify empirically whether the pressure is normal we must use a barometer which functions correctly in normal temperature.

This confused situation must be very unpleasant to those empiricists who are convinced that everything we know of the world should be demonstrated on the grounds of observational statements and of arbi-

trarily adopted terminological conventions. This probably explains why the philosophers who have given a determined shape to Bridgman's general idea by identifying operational definitions with statements of the form (7.1) or (7.2) have not noticed that such a concretized view is undefendable. It is astonishing that for instance Mehlberg, who pays so much attention to operational definitions, is satisfied with formulations like

(7.3)   "If a thermometer has been applied to the object x for some time then, by definition, the body x has the temperature T if and only if the thermometer reads T."[7]

In reality it is very frequent that a thermometer indicates a number different than the real value of the temperature, so that the expression (7.3) — if we are to speak seriously — requires a substantial correction. Indeed, some pages later (*op. cit.*, p. 237) Mehlberg does propose a modification of the definition (7.3) because of the need to take into consideration thermal fluctuations:

(7.4)   "If a thermometer is applied for some time to objects of sufficiently large class then, by definition, almost every object of this class has the temperature T if the thermometer applied to it reads T and, conversely, almost every thermometer applied to such an object endowed with the temperature T will read T."

The sensitivity limit for a gas thermometer — determined by thermal fluctuations — is of the order of $10^{-12}°K$ (cf. Leontowicz, *op. cit.*, p. 93). Errors caused by fluctuations can be left out, as they are much below the level of other errors, unavoidable in the case of the considered measuring method (the same holds for eventual quantum effects). But it should be remembered that the fact alone of applying a thermometer to an object is not sufficient to determine the probability of adequacy of the result.

Let us now consider the first of the above-mentioned questions. It is about whether systems of postulates of the form (7.1) or (7.2) may be treated as arbitrary definitions of physical magnitudes — we are now leaving out the question of their operational or non-operational character. A positive answer implies that all the terms needed to describe

the correctness conditions $V_{ij}(x, y, k)$ $(j=1, 2, ..., n)$ have meanings established independently of the meaning of the considered physical magnitude g. By that assumption the postulates of measurement correctness

$$(7.5) \qquad W_i(x, y, k) \equiv V_{i1}(x, y, k) \cdot V_{i2}(x, y, k) ... V_{im}(x, y, k)$$

will be arbitrary definitions of the term 'correct' in contexts 'a correct measurement of physical magnitude g has been carried out by using method M_i'. Statements of the form (7.2) $(i=1, 2, ..., n)$ will be arbitrary definitions of the term denoting magnitude g. And finally, adequacy postulates of the form (7.1) will have the character of short stylizations of definitions (7.2), equivalent to the latter on the grounds of definition (7.5).

In order to discuss the logical problems connected with that concept, we shall further on employ a modified and simplified formulation of adequacy postulates. Instead of postulates of the form (1.1) we shall consider postulates

$$(7.6) \qquad W_i(k, x) \rightarrow \big(F_i(k, x, \alpha) \equiv G(k, x, \alpha)\big), \qquad i=1, 2, ..., n$$

adopting the following interpretation of the symbols used:
— the symbol $W_i(k, x)$ stands for the formula: the k-th correct identifying operation was carried out for magnitude g for body x by using test M_i;
— the symbol $F_i(k, x, \alpha)$ stands for the formula: value α has been obtained as a result of the k-th identifying operation for magnitude g for body x by using test M_i;
— the symbol $G(k, x, \alpha)$ stands for the formula: in the course of the k-th identifying operation for magnitude g for body x by using test M_i magnitude g has assumed the value α for body x.[8] As it can be seen now only the symbol $F_i(k, x, \alpha)$ obtains an interpretation substantially different from the previous one.

We achieve further simplification of the formulation by accepting that we are interested only in divergence of results which arise as a result of correct application of two different identifying tests to the same object for the same moment of time. In this case we can leave

out the variable k from the formulation and consider the system of postulates

$$(7.7) \qquad W_i(x) \rightarrow \big(F_i(x, \alpha) \equiv G(x, \alpha)\big), \qquad i = 1, 2, \ldots, n.$$

Such a simplification does no harm as long as it concerns the question whether adequacy postulates can be considered as partial arbitrary definitions of the magnitude considered.

The system (7.7) proves to be contradictory (on the grounds of the postulates of uniqueness) in the case when the formula

$$(7.8) \qquad W_i(x) \cdot W_j(x) \cdot F_i(x, \alpha) \cdot F_j(x, \beta), \qquad \alpha \neq \beta$$

is satisfied. The difficulty connected with the view that the system (7.7) is an arbitrary definition of magnitude g arises, as the negation of every sentence of the form (7.8) is logically implied by that system; and the sentences of the form (7.8) do not contain the defined term and are true or false independently of the definition we accept for the term G. Reversing this way of reasoning we must agree that the system (7.7) will be falsified if only we empirically state the truth of a sentence of the form (7.8), and thus it should be accepted that arbitrary definitions may be subject to falsification as well as other statements of physics.

Let us now consider a certain attempt to eliminate the mentioned difficulty by reformulating the system of postulates (7.7). We mean here the opinion that every measuring method determines another physical magnitude. In symbolic notation we would then have to introduce the sequence of physical magnitudes $g_1, g_2, \ldots, g_n$ and the corresponding formulas $G_1(x, \alpha)$, $G_2(x, \alpha)$, $\ldots G_n(x, \alpha)$. We would then have a sequence of partial definitions

$$(7.9) \qquad W_i(x) \rightarrow \big(F_i(x, \alpha) \equiv G_i(x, \alpha)\big), \qquad i = 1, 2, \ldots, n.$$

This system is no longer falsifiable by sentences of the form (7.8). On the other hand, statements of the form

$$(7.10) \qquad G_i(x, \alpha) \equiv G_j(x, \alpha), \qquad i \neq j$$

establishing relations of magnitudes g_i and g_j are falsifiable.

Note that the postulates (7.7) were created by simplifying (leaving out certain variables) postulates of the form (7.1). If we go back to postulates (7.6) it will immediately become clear that it is not enough

to eliminate a possible divergence of results of two or more measurements carried out by using *different* methods; it is also necessary to take into consideration the case of direct or indirect divergence of results obtained by using the same method (for instance when a simultaneous measurement has been carried out by using two or more instruments). By explicitly taking into consideration the variable k ranging over the set of identifying operations, instead of the system (7.9) we obtain the system

$$(7.11) \quad W_i(x, k) \to \big(F_i(k, x, \alpha) \equiv G_i(x, k, \alpha)\big), \quad i = 1, 2, ..., n.$$

It is clear that each of the definitions forming the system (7.11) has testable consequences, as (on the grounds of postulates of uniqueness) from each of them it results that two measurements of magnitude g_1, carried out for the same object and the same moment of time must give identical results. To eliminate this we would have to accept that every measuring instrument determines another physical magnitude. This result seems too paradoxical, however, to be treated seriously.

The paradox consists in that the object of discussion has been changed. We were searching a definition establishing the meaning of the term denoting physical magnitude g. Instead we have obtained a sequence of definitions (7.9) establishing the meaning of some other terms, denoting magnitudes $g_1, g_2, ..., g_n$. Whatever could be said about these new definitions then does not matter to us (even the amusing detail that each of the definitions (7.9) — as it is seen after writing it in the form (7.11) — is charged with exactly the same inconvenience as definition (7.7): it has empirically testable consequences).

The problem appears again if we ask about the relations of meaning between the term denoting magnitude g and those denoting magnitudes $g_1, g_2, ..., g_n$. For instance the following equivalence might be accepted — as an arbitrary definition:

$$(7.12) \quad G(x, \alpha) \equiv \big(G_1(x, \alpha) \vee G_2(x, \alpha) \vee ... \vee G_n(x, \alpha)\big).$$

It is inconvenient to the extent that equivalence (7.12) is falsifiable (on the grounds of postulates of uniqueness) by every statement of the form

$$(7.13) \quad G_i(x, \alpha) \, G_j(x, \beta), \quad \alpha \neq \beta_0.$$

It is now clear that by introducing magnitudes $g_1, g_2, ..., g_n$ we achieve nothing but a new formulation of the problem we dealt with when considering the definition (7.7). Even if we were to agree that every measuring method defines *some* physical magnitude, it would remain an open question how to define the physical magnitudes which a physicist talks *explicitly* about and which are always measurable by using several different measuring methods.

In recent literature on the subject the predominant view is that the falsifiability of the system of postulates (7.7) or of the definition (7.12) does not itself disqualify that system as an arbitrary definition of the term G. This is Carnap's position,[9] and in Polish literature that of Przełęcki[10] and Wójcicki.[11] These authors attempt to present arbitrary definition as the conjunction of two sentences, one of which (containing term G and not falsifiable) would be the so-called *analytic component* of the considered definition, and the other (not containing term G) — its *synthetic component*. Carnap speaks of respectively A-postulate and P-postulate.

The analytic and synthetic components of arbitrary definition may be distinguished in several ways, yielding non-identical pairs of sentences. If $D(G)$ stands for the considered arbitrary definition, then in each case the synthetic component may be the existential sentence

$$(7.14) \quad \sum_{\Phi} D(\Phi)$$

(where Φ is a predicate variable in the case when the defined term G is a predicate), and the analytic component — the sentence

$$(7.15) \quad \sum_{\Phi} D(\Phi) \rightarrow D(G).$$

As it can be easily seen, the conjunction of sentences (7.14) and (7.15) is logically equivalent to the definition $D(G)$.

It is worth stressing that the 'synthetic' component of an arbitrary definition may in special cases prove to be an analytic sentence. Indeed: it may happen that the existential sentence (7.14) is a logically true sentence.

As it has been demonstrated by Wójcicki, the implication (7.15) is the weakest possible analytic component of definition $D(G)$, in the sense that it logically results from any other sentence suitable to be

the analytic component of that definition. For this reason (as there are no other criteria) Wójcicki is inclined to treat the sentence (7.15) as the 'proper' analytic component of the considered definition. Carnap's position is identical; he does not even consider other alternatives.

Let us return to our system of postulates (7.7). Its 'proper' synthetic component will be the existential sentence

$$(7.16) \quad \sum_{\Phi} \prod_{i,x,\alpha} \left(W_i(x) \to \left(F_i(x,\alpha) \equiv \Phi(x,\alpha) \right) \right)$$

(variable i ranges over the sequence of numbers $1, 2, \ldots, n$). As it is easy to prove (I am leaving out the trivial demonstration) the sentence (7.16) is logically equivalent to the structure:

$$(7.17) \quad \prod_{x,\alpha} \left(\sim \left(W_i(x) \cdot W_j(x) \cdot F_i(x,\alpha) \cdot \sim F_i(x,\alpha) \right) \right), \quad i,j = 1, 2, \ldots, n.$$

Thus the synthetic component of the definition (7.7) is identical to the conjunction of all the sentences of system (7.17).[12] Hence on the grounds of postulates of uniqueness negations of sentences of the form (7.8) result immediately.

To shorten the formulation we shall introduce a symbol $R(x, \alpha)$ for the following sentential formula:

$$(7.18) \quad \prod_{i,j} \left(\sim \left(W_i(x) \cdot W_j(x) \cdot F_i(x,\alpha) \cdot \sim F_j(x,\alpha) \right) \right), \quad i,j = 1, 2, \ldots, n$$

The synthetic component of the definition (7.7) — i.e. the system of sentences (7.17) — may then be written in the form: $\prod_{x,\alpha} R(x, \alpha)$.

We shall now consider some detailed questions connected with the non-unique character of the distinction of the synthetic and analytical component of arbitrary definition. Wójcicki states that (*op. cit.*, p. 143) "the solution of this problem may sometimes — it seems — have a practical significance". I would not agree with this assumption. Besides, the author himself does not seem to pay much attention to it, as he finally arbitrarily pronounces himself in favour of the distinction ((7.14)–(7.15)).

Two theorems demonstrated by Wójcicki are essential to the solution of this question; the first says that all the possible synthetic components of arbitrary definition are logically equivalent, and the second — that all the possible analytic components of such a definition are logically equivalent on the grounds of any synthetic component. According to

the first theorem we might, after all, accept that (7.14) is a synthetic component of the definition $D(G)$. According to the second theorem any analytical component of the definition $D(G)$ is logically equivalent to the implication (7.15) on the grounds of the statement (7.14). It follows that, as long as we accept (7.14) as true, the choice of a sentence as the analytical component does not change anything in the range of applicability of the term G (it is obvious, as this range is simply determined by the definition $D(G)$). But in case when (7.14) is falsified, we have to reject the whole definition $D(G)$, i.e. we must also reject all the possible analytical components of this definition. In the case of the falsification of the synthetic component of definition $D(G)$ we see that the denotation of term G does not exist. This result does not depend on the choice of the analytical component of definition $D(G)$. As we can see, then, the choice of the sentence as analytical component of definition $D(G)$ (from among possible analytical components of this definition) does not have any practical significance. For this reason, speaking of analytical component of the definition $D(G)$, we shall always mean implication (7.15) — which, anyway, agrees with the decision made by Wójcicki in this matter.

A certain lack of clarity in this problem — seen in both Wójcicki's and others' works — results, in my opinion, from the lack of distinction between the choice of a sentence $D(G)$ as *arbitrary definition* of term G and the choice of one of the consequences of sentence $D(G)$ as *analytical component* of this definition. The first choice may have considerable cognitive significance. We shall exemplify this point:

1. Let the *arbitrary definition* of term G be not the system (7.7) but its analytical component. We have introduced the symbol $\prod_{x,\alpha} R(x, \alpha)$ to denote the synthetic component of the considered structure (cf. (7.18)), so we may write the analytical component in the form of a system of implications:

$$(7.19) \quad \prod_{x,\alpha} R(x, \alpha) \to (W_l(y) \to (F_l(y, \beta) \equiv G(y, \beta))), \quad l=1, 2, \ldots, n.$$

We decide that we consider now the system (7.19) as a definition of the term G (instead of the previously considered system (7.7)).

The definition (7.19) is non-falsifiable (contrarily to the definition (7.7)). For, as it can be easily seen, its synthetic component is a logically

true sentence. The ranges of applicability of the term G determined by the definitions (7.7) and (7.19) are identical. Let us notice, however, that in the case of falsifying the theorem $\prod_{x,\alpha} R(x, \alpha)$ the system (7.19) will be treated differently, depending on whether we consider it the definition of the term G or only the analytical component of another definition (i.e. the definition (7.7)). In the first case the system (7.19) remains a theorem in spite of the falsification of the sentence $\prod_{x,\alpha} R(x, \alpha)$.

It only appears to be vacuously satisfied, as we already know that the antecedent of every implication (7.19) is false. In the second case — as we already said when discussing the definition (7.7) — the system (7.19) is rejected as we reject the whole definition (7.7) of which it is a component. In the first case the term G remains (after the falsification of $\prod_{x,\alpha} R(x, \alpha)$) undoubtedly meaningful, in the second case — the problem is open to discussion.[14]

2. A system of implications

$$(7.20) \qquad R(x, \alpha) \rightarrow \left(W_l(x) \rightarrow \left(F_l(x, \alpha) \equiv G(x, \alpha)\right)\right), \qquad l = 1, 2, \ldots, n,$$

is one of the possible analytical components of the system (7.7). The implication (7.10) was considered in different formulation[14] by Prze-łęcki, for the case of two postulates (*op. cit.*, p. 107). The author was interested in any arbitrary definition of the form of the system of reduction sentences (and not in the question of definitions of physical magnitudes).

In case when the system (7.7) consists of only two postulates

$$(7.21) \qquad \begin{aligned} & W_1(x) \rightarrow \left(F_1(x, \alpha) \equiv G(x, \alpha)\right), \\ & W_2(x) \rightarrow \left(F_2(x, \alpha) \equiv G(x\,\alpha)\right), \end{aligned}$$

the analytical component of the form (7.21) will be the pair of implications

$$(7.22) \qquad \sim \left(W_1(x) \cdot W_2(x) \cdot \left(F_1(x,\alpha) \cdot \sim F_2(x,\alpha) \vee \sim F_1(x,\alpha) \cdot F_2(x,\alpha)\right)\right)$$

$$\rightarrow \left(\left(W_1(x) \rightarrow \left(F_1(x, \alpha) \equiv G(x, \alpha)\right)\right),$$

$$\sim \left(W_1(x) \cdot W_2(x) \cdot \left(F_1(x,\alpha) \sim F_2(x,\alpha) \vee \sim F_1(x,\alpha) \cdot F_2(x, \alpha)\right)\right)$$

$$\rightarrow \left(W_2(x) \rightarrow \left(F_2(x, \alpha) \equiv G(x, \alpha)\right)\right).$$

In Przełęcki's view, the choice of the implications (7.22) as the analytical component of the definition (7.21) may be justified by the following reasons: the implications (7.22) determine the same range of applicability of the term G as the definition (7.21); they have no empirically testable consequences (they are not falsifiable); the equivalence of both postulates (7.21) is maintained. The latter condition is not fulfilled, for instance by the implications

$$(7.23) \quad W_1(x) \rightarrow \big(F_1(x, \alpha) \equiv G(x, \alpha)\big),$$

$$W_2(x) \cdot \sim W_1(x) \rightarrow \big(F_2(x, \alpha) \equiv G(x, \alpha)\big),$$

which are the analytical components of the system (7.21), too. In this case the first of the postulates (7.21) has a somewhat predominant position; the results obtained using the test characterized by the second postulate are taken into consideration only in case when the first postulate has not been applied.

Note that, if the system (7.21) is an arbitrary definition of the term G the 'non-symmetry' of the system (7.23) does not constitute any shortcoming. Giving priority to one of the tests has no practical significance as long as both tests give consistent results. And in case of incompatibility of results—when it is necessary to establish the priority of one of the tests—the term G loses its sense anyway, as a result of the falsification of the definition (7.21). Thus, as long as the system (7.21) is the definition of the term G it does not matter which of the systems (7.22) or (7.23) we choose as the analytical component of that definition, because—according to Wójcicki's theorem discussed above—all the analytical components of an arbitrary definition are equivalent on its grounds.

The situation changes if we ask which of the systems (7.22) and (7.23) is more suitable to be the *definition* of the term G. In this case the system (7.22) has indeed the advantage of treating both tests as equivalent. I believe that the fundamental equivalence of all measuring methods is a fact which should be taken into consideration in any reasonable analysis of physical concepts. But for some other reasons neither the system (7.22) nor the structure (7.23) can be considered good definitions of physical magnitude.

Let us suppose that we have accepted the system of sentences (7.20) or—in the case of only two postulates—(7.22) as arbitrary definition of

the physical magnitude g. This definition remains a theorem also in case of falsification of the system of theorems (7.17), i.e., if we find that the utilized measuring methods may give divergent results in spite of strict observance of correctness conditions. Therefore, there is no reason against further application of these methods, if in one of the subsequent measurements only one of them is used. Similarly, if the use of two or more methods leads *in this case* to coherent results, then on the basis of the definition (7.20) we shall have to consider the measurement adequate. The definition (7.20) characterizes then the behaviour of a physicist who does not learn on his errors. The same is true of the definition (7.23).

Maybe — although I doubt it — there exist scientific concepts for which definitions of the form (7.22) or (7.23) would be adequate. In any case, these definitions are not suitable to introduce names of physical magnitudes.

The peculiarities of the definitions (7.22) and (7.23) result from the fact that the antecedents differentiating them from the system of postulates (7.7) are not separately quantified. The quantification of the antecedents of the definition (7.22) will turn it into a special case of the already discussed definition (7.19).

We can say generally that no arbitrary definition of the term G denoting a physical magnitude will be satisfactory if it formulates any criteria of applicability of the term G which would hold also in the case of falsification of the system (7.7). This is true as well for the definitions (7.22) and (7.23) and for any of the numerous similar solutions. It is impossible to decide *a priori* what system of postulates will be accepted to replace the falsified system (7.7). The question will be resolved in the course of search for the causative explanation of the divergence of results of the methods characterized by the system (7.7), i.e. on the basis of the results of additional empirical research. It is possible to predict in general what kind of investigations will be appropriate, but not their results.

We shall now ask what is the epistemological sense of the view that the system (7.7) or the definition (7.12) is an arbitrary definition with components distinguished in the way presented above. According to Ajdukiewicz's explanation [15] of the concept of 'arbitrary definition', "we call a sentence arbitrary definition or postulate of language if in

this language there is a terminological convention establishing that the terms of that sentence should symbolize objects which satisfy the sentence when they are put in place of the terms". Wójcicki gives an analogous definition of this concept; in the corresponding place he speaks of the denoting directive instead of the terminological convention — but this difference has, no doubt, no substantial significance. It is now absolutely obvious that in the language of physics there is no such terminological convention which would demand to understand the term G so that the system (7.7) be realized. The same can be said of the directive of denoting. Then the only epistemological sense which might be assigned to the theorem saying that (7.7) is an arbitrary definition would be that physicists behave as if they accepted the considered convention or directive. What will then be the difference between the behaviour of a physicist who considers (7.7) as an arbitrary definition and a physicist who does not — while both accept the system (7.7)?

According to the colloquial everyday sense of the terms 'arbitrary definition' or 'terminological convention' we would, in any case, be inclined to consider that arbitrary definition cannot be subject to falsification. Let us take two typical examples.

1. The arbitrary definition of bachelor ('a bachelor is an unmarried man') is non-falsifiable, because the only way to find whether someone is a bachelor is to find whether he is a man and whether is unmarried; thus the situation cannot arise in which a criterion would indicate that Mr. X is a bachelor without simultaneously indicating that Mr. X is unmarried.

2. The arbitrary definition of a centimeter ('a centimeter is one hundredth part of a meter') is not non-falsifiable in the sense discussed above. For we have rulers scaled in meters and ones scaled in centimeters; the situation can then arise in which a measurement carried out with a meter-scaled ruler gives the result 2, and a measurement carried out with a centimeter-scaled one — let's say 199. But the definitional character of the considered sentence appears in the fact that in the described situation everybody will say that one of the rulers is badly scaled, and nobody — that one meter equals 99,5 centimeters, contrarily to the previously accepted hypothesis.

In both cases the arbitrary definition is one of the many general theorems containing the considered term ('bachelor' or 'centimeter'), and

its special character consists in that it remains the accepted sentence in the case of falsification of the system of theorems; on the other hand another sentence of the system is rejected (thus showing the non-definitional, hypothetical status of the rejected sentence). Accepting this point of view we shall say that distinguishing the arbitrary definition of the term G in the investigated fragment of knowledge has an epistemological sense only if we simultaneously assign a non-definitional status to some other theorems containing the term G. By making such a distinction we formulate a definite hypothesis concerning the dynamics of knowledge.

Let us consider from that point of view the question of definability of physical magnitude. We agree that among general theorems containing the term G (which we now consider as the name of the chosen magnitude) there exist non-falsifiable sentences, i.e. sentences which the physicist will defend at the cost of serious modification of other theorems of the theory. Is this 'non-falsifiability' absolute — that is, can there really be *no situation* in which the physicist would consider such a sentence as falsified — is a question which we need not decide. We shall not, for instance, consider whether it is really impossible to conceive such a set of newly accepted facts which would make the physicists revise the theorem that acceleration is the first derivative of velocity. It is sufficient for us that this theorem is undoubtedly non-falsifiable in very many predictable situations. For that reason (and even leaving out the psychological fact that physicists consider the theorem as definitional) we can, in the epistemological reconstruction of physics, assign to it the status of arbitrary definition.

It is important that the set of theorems which (with more or less doubts) are suitable to assume the role of the arbitrary definition of a chosen magnitude g is rather small. It probably includes the postulates of uniqueness, obvious definitions, such as the definition of acceleration, maybe postulates of additiveness, maybe also the operational definition of simultaneity and the analogous definition of length.

We now have the following alternative:

a) We accept that the arbitrary definition of the term G is the conjunction of only those theorems which are practically unquestionable in the considered sense (postulates of uniqueness, etc.). Such a definition would however determine too narrow a range of applicability for the considered term. It appears, I believe, that the major part of the postu-

lates of measurement adequacy (and particularly for methods of indirect measuring) are falsifiable theorems; these postulates could not then belong to the system of theorems constituting the considered definition. The operationist concept of measuring terms denoting physical magnitudes is then undefendable.

b) We accept that the system of theorems which forms the arbitrary definition of the term G can include empirically questionable theorems. We construct the definition of the term G in such a way that each of the adequacy postulates representing the methods of determining the considered magnitude belongs to the system. We can thus give the term G a range of applicability conforming to scientific practice.

However, we meet two difficulties. The first is the question of the meaning of the term 'correct' in the context 'a correct measurement has been carried out' (cf. Section 5, and also considerations on pp. 429, 430). I think that an attempt to eliminate this term shows that in order to define a chosen theoretical term G it is necessary to accept that some other theoretical terms $G_1, G_2, ..., G_n$ (names of other physical magnitudes) have an already established meaning. As this difficulty is true for each of the considered terms, it appears that an attempt to define all the physical magnitudes leads to a vicious circle. We then have to either accept that some theoretical terms have the character of primary terms (which is against the neopositivistic concept of the definability of *all* the theoretical terms on the grounds of observational terms), or to accept that the system of postulates for a chosen term G describes both the meaning of this term and the meaning of other theoretical terms necessary to formulate the conditions of measurement correctness.

A more detailed analysis of this problem requires taking into consideration the physical knowledge as a whole — which seems to be premature in the present state of research. In that situation I can only restrict myself to the presentation of my views — which I have done above — without trying to find a sufficient argumentation.

In the epistemological model presented in this paper the theorems stating the correctness of an identifying operation are accepted on the grounds of provisional assumptions; such a procedure is effective no matter whether and how the terms can be defined which are necessary to formulate correctness conditions. It is sufficient to explain how a physicist is able at all to measure a magnitude although he realizes that,

in order to find whether the measurement has been correct, he would have to undertake a series of introductory identifying operations indicating the state of the instrument, that these operations would require, the use of other instruments the correctness of which might in turn be questioned and so on ad infinitum. By adopting the principle of presumption of correctness the physicist breaks the vicious circle; in my opinion, this fact should be taken into consideration in every efficient model of physical knowledge.

Let us consider the second of the signalized difficulties. As the arbitrary definition of the term G should include all the identifying tests for the considered magnitude, we must accept that any modification of any of the tests changes the definition. We cannot predict which test will undergo a modification, as it depends on the results of later investigations conducted according to the principle of genetic argumentation. We then have to accept that all the tests are basically equivalent (with the possible exception of some few tests based on obvious definitions — as the definition of acceleration). It follows that the arbitrary definition of the term G may be either a complete system of postulates of adequacy (7.7) or the analytical component of that system (7.19). Other solutions cause undesirable consequences: we a priori decide which test will be modified in the case of a divergence of results (while there are no factual arguments for such an anticipation), or else — if the system (7.20) is to be accepted as definition — we decide that in spite of a divergence of results obtained by using different tests, none of these tests will be modified (which assumption opposes the known cognitive facts).

If we accept, however, that the system (7.7) (or — which in this case does not make any substantial difference — the system (7.19)) is the arbitrary definition of the term G, it appears that an arbitrary definition is a falsifiable theorem, and that the falsification of any hypothesis concerning the magnitude g implies the falsification of the definition of that magnitude. This result is undesirable from the point of view of the introductory assumption that arbitrary definitions are — as compared to hypotheses — a relatively constant element of knowledge. We shall consider this question in more detail.

Let $T(G)$ be the conjunction of sentences $T_1(G), T_2(G), ..., T_n(G)$ chosen so that every theorem accepted in a given moment of time t_1 and essentially containing the term G is a consequence of $T(G)$. Let us

suppose that we decide to consider $T(G)$ the arbitrary definition of the term G. Let us also assume that the synthetic component of that definition is a falsifiable sentence. It follows that the definition $T(G)$ is also falsifiable. In the case of falsification of this definition we must *prima facie* reject it as a whole, so that all its consequences are also rejected, i.e. all the theorems containing the term G.

In reality, though, we do not encounter such a total rejection of a whole fragment of knowledge. In the case of falsification of the synthetic component of the definition $T(G)$ we do not reject all the sentences of the sequence $T_1(G), T_2(G), ..., T_n(G)$, but only one or a few of them. We thus obtain a certain new sequence of theorems containing the term G, for example: $T_1^*(G), T_2(G), ..., T_n(G)$. This sequence differs from the previous one by its first term: the theorem $T_1(G)$ has been replaced by the modified theorem $T_1^*(G)$.

Do we have to reject, in this situation, the view that $T(G)$ was the arbitrary definition of the term G? Not at all. We can still maintain that $T(G)$ was the arbitrary definition of the term G in the moment t_1 and at the same time accept that as this definition was falsified in a later moment t_2, on the basis of the terminological convention a new definition $T^*(G')$ was adopted which was equivalent to the conjunction of all the sentences of the sequence $T_1^*(G'), T_2(G'), ..., T_n(G')$. The term G' is not equivalent to the term G because the sentential formula $T^*(\Phi)$ is not logically equivalent to the formula $T(\Phi)$. If, then, in natural language we always use the same word — for example 'temperature' — we must assume that the sense of this word changes with every modification of the system of theorems in which it is used.

As we can see, to call arbitrary definition of the term G the complete system of theorems in which this term appears may be reduced, as a matter of fact, to the choice of a certain way of speaking about epistemological problems. We agree that scientific theories are falsified and replaced by others with the reservation that these theories are called arbitrary definitions of the contained terms, and the replacing of an old theory by a new one is for us a change of terminological convention. Instead of saying that experience has abolished the view that the atom is indivisible and has forced us to adopt the view that it is divisible, we could say that experience has forced us to reject one definition of the atom (from which it results that atoms are indivisible) and that we

have replaced it by a new arbitrary definition, determining a new concept 'atom$_1$'; it results in particular from this definition that atom is divisible. The history of knowledge is then the history of the development of concepts.

Our problem may now be formulated as follows: is the choice of a new arbitrary definition arbitrary? That is, in the case of falsification of the definition $T(G)$, can we adopt any — but non-falsifiable — new definition $T^*(G')$? From our point of view there is pratically no arbitrariness because of the principle of genetic argumentation. But in this case arbitrary definitions are not, in reality, chosen in an arbitrary way at all, and 'terminological conventions' are almost uniquely dictated by experience.

The system of postulates (7.7) does not, of course, axiomatize all the theorems of physics in which the defined term G occurs. Note, however, that every falsifiable physical law containing the term G must be represented by a postulate belonging to that system. Indeed: if the considered law is falsifiable, then there must exist measuring methods for all the concerned physical magnitudes. If we now measure all the considered magnitudes except a magnitude g, then on the grounds of these results and of the considered law we can calculate what value should be assumed by the magnitude g. Thus every law of physics concerning the magnitude g determines an identifying test for that magnitude; the corresponding postulate must therefore belong to the system (7.7). The falsification of any law of physics containing the magnitude g is then at the same time the falsification of the system (7.7), if we accept that this system as a whole is the arbitrary definition of the considered magnitude. When replacing the falsified law by another we must then introduce an appropriate correction of the system (7.7), i.e. adopt a new arbitrary definition for the new term G'.

We thus come to the conclusion that the theorem stating that the system (7.7) is, as a whole, the arbitrary definition of the term G does not say anything but the truism that in the case when different measuring methods give divergent results it is necessary to modify the system of postulates of adequacy and of correctness of measurement for these methods. No indication results there from as to how is the desired modification to be carried out. The basic equivalence of measuring methods is not questioned, as all the postulates are rejected *prima facie*

(as consequences of the falsified definition). In this situation I do not think that the question whether the system (7.7) is an arbitrary definition or not might be subject to substantial discussion, as we agree that this system undergoes successive modifications according to the rules of causative explanation of phenomena.

8. THE OBJECTIVITY OF PHYSICAL MAGNITUDES

On the grounds of the Marxist theory of knowledge it is accepted that physical concepts (including concepts of physical magnitudes) reflect objective properties of matter. For this reason we must reject the view of arbitrary choice of concept structure adopted in the sciences, an arbitrariness limited only by the requirement of simplicity of the constructed theory. I am inclined to think—according to the opinion of part of the Marxist philosophers (e.g. of Ładosz [16])—that concepts are divided into true and false, and that the Marxist category of relative truth may be applied to concepts. Maybe the controversy upon that subject has a terminological character. Nevertheless, if for some reasons we limit the range of the terms 'truth' and 'falsity' to sentences and theories, there arises the need to construct concepts similar to categories of *adequacy* (absolute and relative) and *inadequacy*; of course, the terms 'adequacy' and 'inadequacy' might be replaced by others (the adequacy of concepts—which we shall deal with below—has nothing to do with the adequacy of measurement result, discussed in the detailed part of this paper). The controversy over the question whether it is indeed necessary to differentiate adequate and inadequate (true and false) concepts is certainly not merely terminological.

Philosophers who consider the distinction useless usually say that instead of calling a concept true it is enough to speak of true existential proposition stating the existence of the denotatum of a given term. If we accept this, we must agree that by introducing the categories of true and false concepts we unnecessarily complicate the terminology. I do not intend, however, to defend the view that concepts are true resp. false in that primitive sense.

Meanings of words change with time. It could hardly be doubted that the meaning of the word 'atom' was different in the epoch of Demo-

critus from what it is in the 20th century. Similarly, the meaning of the word 'gold' has changed as a result of the development of contemporary chemistry, and later of nuclear physics. The meaning of the word 'tuberculosis' has changed since the discovery of Koch bacillus. As we see, in different stages of the cognitive process different concepts correspond to the same words.

Not only meanings, but also denotata of terms change with time. In Democritus's language, the term 'atom' denoted a class of undivisible objects, fitted with a sort of hooks; as such objects do not exist, we must agree that the denotatum of Democritus's 'atom' is an empty set. A similar empty set is the denotatum of the term 'atomic nucleus' if — after a relatively recent physical theory — we agreed to understand by 'atomic nucleus' a material object of definite dimensions, built from protons and electrons. But can we be sure that the denotatum of the term 'atomic nucleus', understood according to a newer theory as a system of protons and electrons, is not an empty set?

If meanings and denotata of terms are changeable, how is it possible to compare contemporary and past knowledge? Our predecessors not only said something else than we do, but they also spoke of *something else* than we do. What Democritus had to say about undivisible particles fitted with hooks is in no way comparable to what a contemporary physicist can say about divisible particles built of electrons and nucleons. What alchemistry said about philosophers' stone cannot be negated by chemistry, which says nothing about this stone. The opinion of popular medicine that dog fat is a good remedy for turberculosis cannot be negated by scientific medicine, as a succesor of Koch understands by 'tuberculosis' something else than a country quack. A doctor and a quack treat patients suffering different diseases; no wonder then that they use different means.

It is not our point here to multiplicate such paradoxes, but to find a way of eliminating them. As every stage of the historical process of knowledge is simultaneously a negation and a continuation of the preceding one, we should establish a reference system common to them all; the state of affairs which we intend to describe is that reference system. Thus there exists tuberculosis, which is a point of interest both for Koch and the quack, there exists the atom, which is the subject of consideration of both Democritus and nuclear physics. The word 'atom'

(or 'tuberculosis') always refers to the same object, which is not necessarily identical with the one denoted by the word in a given stage of development. The successive meanings and denotata of words are approximations to the real meanings and denotata, which are discoverable in the course of development of science.

The meaning of a theoretical term is then not something which can be arbitrary established, but something which should be investigated. The totality of theorems accepted in a given period of time and containing the considered term is a more or less successful attempt to explicate the real meaning of the term.

By saying that a concept is relatively true or adequate, we mean that this concept belongs to a sequence of successive approximations to the meaning of a corresponding term.

How can we, however, evaluate the cognitive value of a scientific theory, if we do not know the real meanings of the included terms but only approximate ones? In practice we consider as binding the meaning which we presently have. We must therefore agree not only that a considered theory has merely an approximate cognitive value, but also that any attempt to determine that value leads to provisional results, which may be modified in further course of history. Not only is the corpuscular theory of light merely relatively true, but so is every theorem determining the limits of application of the theory. Indeed, the corpuscular theory of light was evaluated differently before the discoveries of Planck and Einstein than it is evaluated on the grounds of contemporary quantum field theory.

We must agree that at the moment when a theory is presented it is impossible to find out what does this theory indeed speak of. It is only the further progress in research which enables us to define the subject of the investigations of earlier theories. It was only after Koch's discovery that the question could be answered about what was the disease which had been treated by using dog fat; and it is only on the grounds of contemporary chemistry that we learn what was the subject of the alchemists' investigations.

I do not think, however, that the real meanings of theoretical terms are so unprecise that there is no possibility of answering the questions 'what is atom?', 'what is 'tuberculosis'?', 'what is temperature?'. We can always answer the question 'what is tuberculosis?' in this way:

　　　— it is the disease (or one of the diseases) which in the 20th century
　　　　was supposed to be caused by bacteries called Koch bacilli.
Similarly, to the question 'what is temperature?', we can always answer:
　　　— it is the physical magnitude which in the 20th century was meas-
　　　　ured — among others — by using an instrument called thermo-
　　　　meter.

Or

　　　— it is the physical magnitude which was written T in the equation
　　　　of state for perfect gases.

Or else

　　　— it is the magnitude identified (in the same epoch) with the mean
　　　　kinetic energy assigned in a certain way to a set of molecules.

　　It is important that such explanations — if they are to be correct — use
not only the vocabulary of the corresponding science, but also that of
the history of this science. They can thus be expressed on the grounds
of the theory of knowledge or of the history of knowledge. The first
of the above-formulated meanings of the term 'temperature' possibly
corresponds to the intuitions which led to operationism (as, indeed, to
introduce a new magnitude into physics it is necessary to give at the
same time some — even imperfect — measuring method). But it is not an
'operational definition' in the sense of the word which was given to it
by the neopositivistic methodology.

　　A more exhausting consideration of the evolution of meanings of
theoretical terms would demand the choice of a definite theory of meaning — as we know, there is no unanimity on that subject. It is anyway,
clear that the answer to the question, what is the meaning of the word
'temperature', or what is temperature, must be relativized to the language in which the answer is to be given. We have agreed (Section 6)
that some dependences relating physical magnitudes, postulates determining the scale, and possibly also postulates of uniqueness and operational definitions of length and time — have the character of non-questionable theorems, i.e., in an epistemological reconstruction of physical
language they can be considered as theorems accepted according to the
same principles as arbitrary definitions. According to the presently
developed concept we should then accept that some meaning dependences already adopted for theoretical terms will remain valid also in

the future, so that they have a value for both the presently known approximate meanings of terms and their ideal real meanings. For example, we accept that although the meanings of the words 'mass' and 'density' are changeable, their mutual relation is not; every revision of the meaning of the term 'mass' entails a corresponding revision of the meaning of the term 'density' and vice versa. The changeability of meanings of theoretical terms becomes visible if, when asking 'what is temperature?' we demand the criteria of applicability of this term on the grounds of experimental physics (and not theoretical) or – going further – on the grounds of natural language (prescientific). It is clear – for any empiricism: neopositivistic or Marxist – that theoretical terms are introduced into language by accepting the theorems which bind them with observational terms.

The analysis of relations binding names of physical magnitudes to observational terms demonstrates not only the changeability of meanings of the terms belonging to the first group but also a specific 'imprecision' of these meanings. I mean here the dialectics – if I may say so – of the synthetic and analytic function of an identifying test for a physical magnitude and the appropriate adequacy postulate. If I accept that M_i is a control test, and M_j a questioned one, I am thereby 'narrowing' the meaning of the considered term G, accepting that it is determined not by the complete system of adequacy postulates for this term but by a subsystem obtained by leaving out the postulate P_j. In that context the postulate P_j is treated as a hypothesis, and the other postulates (if logically independent on the postulate P_j) assume the definitional function. We then deal not with one meaning of the term G, determined for the given epoch by the system of postulates U, but with a spectrum of meanings determined by the subsystems $U_1, U_2, ..., U_n$ obtained – respectively – by leaving out the postulates $P_1, P_2, ..., P_n$. In the course of investigation of a genetic explanation of divergences of results of identifying operations one of these postulates – let us call it U_k – is accepted to be non-questionable, and (if the procedure is effective) the postulate P_k is modified, and as a result the whole system U. If we now reject the operationist concept of physical magnitudes, we do so considering that there is no such subsystem U_k which

1° would consist of postulates fulfilling the requirements for operational definitions;

2° would constitute a sufficient basis for all controlling procedures questioning the postulates of the system U.

I do not want to say that there do not exist in physics postulates at the same time operational and non-questionable — to mention only the already dealt with definition of simultaneity. I only say that operational definitions are not necessary for empirical verifiability of theoretical theorems, because the mechanism of falsification and modification of theoretical theorems presented in this paper is also suitable to work, if none of the adequacy postulates for the considered term fulfil the requirements of an operational definition.

I should like to add here that I have found a similar concept of 'flickering' meanings in a book rather distant from the problems interesting me now: I mean the work of the British ethic Nowell-Smith.[17]

Finally, I should like to draw the reader's attention to a certain analogy observed in the analysis of meanings of theoretical terms on the one hand and in the analysis of meanings of proper names. A proper name — let it be 'John Smith' — denotes a definite spatio-temporal object. If I have to explain to somebody now who John Smith is and how he can be recognized from among other people, including people having the same name and Christian name, I must give some more information about him: give his description, address, place of birth, the names of his parents, maybe his profession, character, hobby, family relations or habits in which he indulges. A 'theory' of John Smith is thus obtained; theory the task of which is to distinguish the real John from someone who pretends to be him, or else to recognize John in someone who for some reasons is hiding his identity and pretends to be Paul, Peter or Andrew. This theory determines the provisional meaning of the term 'John Smith' in my language.

Let us now suppose that in the course of further research it appears that either the object satisfying my theory does not exist or that there are more than one such objects. We then have two possibilities:

1° to accept that if there is no object satisfying the theory of John Smith, then John Smith does not exist, and that if there are two objects satisfying the theory, then we must arbitrarily decide which of them is to be called John Smith;

2° to accept that the theory has been incomplete or in some details false. To find a new theory and to apply it not only to John's further

life, but also to verify on its grounds all information concerning John's past.

In my opinion, by accepting or rejecting the first of these solutions we resolve an epistemological problem analogous to that of accepting or rejecting the operationist conception of physical magnitudes. If we intend to assign meanings to physical magnitudes on the grounds of natural language we can do it only in the sense in which we can say that the description of John Smith determines the meaning of the proper name 'John Smith'.

I should not like the view presented here to be understood as an attempt to give to physical magnitudes (or to corresponding concepts) a reality in Plato's world of ideas. The objectivity of physical magnitudes — as I understood it — consists in that the world is so constructed that it forces the human intellect to choose a definite conceptual apparatus. The 'real' meaning of the word temperature or the 'true' concept of temperature is not something that can be seen or grasped by an act of intuition; it is something that is being constructed in the course of efforts aiming not so much to learn about temperature, but to learn about matter.

REFERENCES

* First published in *Studia Filozoficzne* **2** (**41**) (1965). Translated by S. Wojnicki.
[1] L. Landau and E. Lifšic, *Quantum Mechanics*, 2nd ed., part I, Chap. 7, Pergamon Press, Oxford 1965.
[2] H. Mehlberg, *The Reach of Science*, Chap. 17, Toronto 1958.
[3] In short, we shall sometimes refer to 'correct operations' taking it *implicite* for granted that they are correct in relation to a given method.
[4] *Op.cit.*, p. 13.
[5] The word 'continuity' is maybe not the best because of the discrete, as a rule, character of the set K; but I do not know of any better term.
[6] M. A. Leontowicz, *Fizyka statystyczna* (*Static Physics*), Warszawa 1957, p. 91.
[7] *Op. cit.*, p. 235.
[8] Perhaps it is worth stressing that we intend to understand symbol k in the sense that two different operations k_1 and k_2 may be carried out in the same moment of time (for instance by using a different instrument for each of these operations).
[9] R. Carnap, 'Beobachtungssprache und theoretische Sprache', *Dialectica* **12** (1958).
[10] M. Przełęcki, 'Pojęcia teoretyczne a doświadczenie' ('Theoretical Concepts and Experience'), *Studia Logica* **XI** (1961).

[11] R. Wójcicki, 'Analityczne komponenty definicji arbitralnych', ('Analytical Components of Arbitrary Definitions'), *Studia Logica* XIV (1963).

[12] It is essential to the proof that one of the consequences of system (7.17) is the structure

$$\prod_{x,\alpha} \left(W_i(x) \rightarrow \left(F_i(x,\alpha) \equiv \left(W_1(x) \cdot F_1(x,\alpha) \vee W_2(x) \cdot F_2(x,\alpha) \vee \ldots \right. \right.$$

$$\left. \left. \ldots \vee W_n(x) \cdot F_n(x,\alpha) \right) \right), \quad i=1,2,\ldots,n.$$

[13] It is questionable whether, in the case when the synthetic component of an arbitrary definition is false, the whole definition should be considered as a non-sensical utterance, or as a false sentence, and its analytical component — as a non-sensical utterance or as a true sentence. The first solution is supported by the argument that it is hardly possible to assign a determined meaning to a term which, as it has just been demonstrated, cannot be included in the language, because there exists no object being its denotatum.

[14] Przełęcki considers postulates of the form

$$(x)\left(\Phi_1 x \supset (Qx \equiv \Phi_2 x)\right),$$

$$(x)\left(\Phi_3 x \supset (Qx \equiv \Phi_4 x)\right),$$

in which the numerical variable α does not appear. This difference does not matter for the problem discussed further.

[15] K. Ajdukiewicz, 'Trzy pojęcia definicji' ('Three Concepts of Definition'), *Studia Filozoficzne* 5/8 (1958).

[16] J. Ładosz, *Wielowartościowe rachunki zdań a rozwój logiki (Many-Valued Propositional Calculi and the Development of Logic)*, § 23, Warszawa 1961.

[17] P. H. Nowell-Smith, *Ethics*, London 1954.

H. MORTIMER

PROBABILISTIC DEFINITION ON THE EXAMPLE
OF THE DEFINITION OF GENOTYPE *

The aim of this work is to show that for some theoretical terms, i.e. terms which are not definable by using only empirical terms, the possibility exists of formulating definitions which can be considered empirical in a weaker sense. This will be shown on the example of the definition of the term 'genotype' in Mendelian genetics.

Attempts have been recently made by several authors dealing with the methodology of empirical sciences to formulate the definition of this term by using only empirical terms. The search for such definitions is connected with the tendency to discover the so-called empirical sense of the theoretical terms occurring in empirical sciences, which as a rule aims at learning what results of observations constitute the criterion for accepting or rejecting sentences containing such terms. There is a well-known J. H. Woodger's work *Biology and Language*,[1] in which the definitions have been formulated of homozygous and heterozygous genotype, in such a way that besides logical terms there are only empirical terms in the definienses. However, these definitions have proved to be inadequate: they have been criticized by M. Przełęcki in *On the Concept of Genotype*,[2] where the author has shown the paradoxical consequences of these definitions. Przełęcki has also considered some possibilities of weakening Woodger's definitions by formulating them as Carnap's reduction sentences which, however, as the author has himself pointed out, can also be seriously objected to from the point of view of adequacy.

The definitions presented in this paper will be normal (i.e. equivalence) definitions; besides logical and empirical terms, the definienses will contain the mathematical term 'probability', and for this reason the definitions are called probabilistic. They can aspire to be empirical to the extent that they give some criteria (although mostly only probabilistic ones) which, at least in some situations, allow to accept or to reject on

the grounds of some observations hypotheses concerning the genotype of an organism.

The problem of the definition of genotype will be restricted here to a very simple theory, which is only a fragment of Mendel's genetics. This fragment can be considered as an extremely simplified explanation of some regularities in the inheritance of properties (thus, for instance, the influence of environment will be left out here, as well as the pheno- menon of mutation). Therefore the definition of genotype presented here can be considered as only a simplified outline of the definition of this concept in Mendel's genetics. But from the point of view of the aims of the present paper it is justified to restrict the considerations to the fragment of genetics discussed. As a matter of fact, among the axioms of the theory presented here numerous are not accepted generally, but only in definite situations. But there are also assumptions which are gen- erally accepted in Mendelian genetics, and they are the reason forcing the formulation of the empirical definition of genotype in the form of a probabilistic definition. These assumptions concern the random mechanism of inheriting genes.

It should be remarked that contemporary genetics indicates the direc- tion of search for the empirical definition of genotype which is com- pletely different from the one chosen by the above-mentioned authors of methodological works, and from that of this paper. This new direc- tion is based on the opinion that some substances contained in cell chromosomes and accessible to observation are identical with the genes postulated by Mendelian genetics. The development of research can therefore lead to the situation in which the term 'gene' will be the name of observable bodies, and then the definition of genotype by using this term will become an empirical one. This, however, belongs to the future. For the time being—it seems —the way of inheriting properties by the offspring is still an empirical criterion of distinguishing genotypes, as it was for Mendel.

1. THE EXPLANATORY CHARACTER OF MENDEL'S HYPOTHESIS

Before entering the subject of the paper it might be worth giving some introductory explanation about Mendel's hypothesis of the existence of genes and about its significance for the explanation of some observed regularities in inheriting properties.

Here are two examples of observed facts concerning the inheritance of properties.

A. The so-called Andalusian hens can have three colours of feathers: white, black or grey.[3] It has been observed that the offspring of black couples is always black, that of white couples is white and that of grey couples can be white, black or grey; it has been moreover noticed that if the set of the offspring of grey couples is numerous in the first generation, the proportions of white, black and grey individuals are close to 1/4, 1/4, 1/2. It has also been observed that the offspring of crosses between white and black individuals always have grey feathers (in the first generation), and the offspring of white (resp. black) individuals with grey ones is white (resp. black) and grey in proportion close to 1/2, 1/2.

B. Axolotls, a sort of American amphibians, can have two colours of the skin: white and black.[4] It has been observed that the offspring of white couples is always white, and that of black couples can be twofold in the first generation:

a) all the individuals are black, or

b) some individuals are black and others are white in proportion close to 3/4, 1/4.

Differences have also been observed among the first generation of the progeny of crosses of white and black axolotls:

a) all the individuals are black, or

b) some are white and some black in a proportion close to 1/2, 1/2.

The results of the observations described are typical for the situation among numerous species of animals and plants as to the regularity in the inheriting of certain properties. Examples analogical to the schemes of the two situations described here are always quoted in handbooks of genetics as representing the simplest schemes of the inheriting of properties. Mendel's observations concerning *Pisum sativum*, that is the observations which started genetics, were conform to scheme B.

In explaining the observed regularities in inheriting properties Mendelian genetics is based on the hypothesis of the existence of genes in the cells of organisms. It has been assumed that genes are placed in pairs along chromosomes (the so-called pairs of alleles) in an order definite and constant for the given species. The pair from a definite place influences a definite sort of properties (for example the colour of feathers);

genes from other places influence other properties (for example the size). The case in which it is assumed that there is exactly one pair of genes influencing properties of a given sort is the simplest and the theory presented below will be restricted to such a case.

The different observable properties of a certain sort, the so-called phenotypes of a given species, are explained in the theory by the fact that the genes 'responsible' for the properties of that sort are different. Thus, for example, to explain the differences in the colour of feathers among Andalusian hens it is assumed that these hens can have genes of two kinds: genes causing black feathers, or as the geneticians say genes for black colour of feathers, let us denote them by g_A, and genes for white colour of feathers, g_B. But the colour of feathers of an organism is decided by the composition of the pair of alleles of a given place. As there are two kinds of genes, three different compositions of a pair of alleles are possible: $g_A g_A$, $g_B g_B$ and $g_A g_B$. Such a composition in a pair of alleles causing a given property is called the genotype for that property. Andalusian hens then have three different genotypes for the colour of feathers, which determine this colour as follows: individuals with the genotype $g_A g_A$ are black, individuals with the genotype $g_B g_B$ are white and individuals with the genotype $g_A g_B$ have an intermediate grey colour. Genotypes composed of similar genes (e.g. $g_A g_A$) are called homozygous and those composed of different genes — heterozygous.

The observed regularities in inheriting phenotypes are explained as follows: when the cell divides before fertilization, creating two gametes, the pairs of alleles also divide and each allele goes to a different gamete. During fertilization the male and the female gametes unite to form a zygote, from which a new organism develops. The organism then inherits the genotype: one gene of a pair comes from the father and the other from the mother. The combination of genes in a zygote resulting from fertilization is supposed to be random. It is also assumed that inheriting father's and mother's genes are independent events. Thus inheriting of genes has the probabilistic scheme of drawing one ball from each of two boxes. The regularities in inheriting the colour of feathers among Andalusian hens can then be explained in a simple way: the crossing of black individuals must give exclusively black offspring, as in both such organisms all the gametes contain only genes

for black colour g_A. For the same reason the offspring of two white organisms must be white. In turn, the offspring of a white individual with a black one must be grey: all the female gametes have the gene g_A, and all the male ones – the gene g_B (or conversely), thus the zygote must have the mixed composition $g_A g_B$. The offspring of a cross of two grey individuals can have various phenotypes. In such a case there will be two kinds of male gametes and two kinds of female ones: one half with the gene g_A and the other with g_B. In this situation – according to the assumption of randomness and independence – the probability of the creation of a zygote cmposed $g_A g_A$ equals 1/4, that of a zygote composed $g_B g_B$ is the same, and the composition $g_A g_B$ has the probability 1/2. A similarly simple explanation can be given for the observed proportion of white and grey (black and grey) organisms in the progeny of crosses of white (black) organisms with grey ones.

In order to explain the situation of the case B an additional assumption is made explaining the fact that the species involved there has only two colours of the skin. It is namely assumed that genes for one of these two colours are dominant and the genes for the other one are recessive (geneticians then say that one of these properties is dominant and the other one recessive). In the case of axolotls it is assumed that the dominant genes are those for black colour and the recessive ones – for white colour. The recessive character of the latter consists in that when such a gene is combined with a dominant one in a pair of alleles, it does not influence the given property leaving the effect of the dominant gene unaltered. This means that an organism of heterozygous genotype has the dominant property, say, B, exactly as has an organism of homozygous genotype $g_B g_B$. It explains why crosses of black axolotls give various results: if a pair of black homozygotes is chosen, or a pair composed of a black homozygote with a heterozygote, then of course all the offspring must be black. But if a pair of heterozygotes is chosen (which do not differentiate in colour from homozygotes), it is easy to calculate that the probability of the genotype $g_A g_A$, i.e. for the white (recessive) colour, in the set of the offspring is 1/4; the remaining 3/4 will consist of homozygotes $g_B g_B$ in a proportion close to 1/4 and of heterozygotes $g_A g_B$ in a proportion close to 1/2 – i.e. black axolotls. Different result in crosses of black axolotls with white ones obtain a similarly simple answer: the pairs $g_A g_A$ and $g_B g_B$ must have exclusively heterozy-

gous and thus black offspring; while the pairs $g_A g_B$ and $g_A g_A$ can have both, black offspring ($g_A g_B$), and white ($g_A g_A$).

The father of genetics, Mendel, did not investigate the 'nature' of genes. In his theory the gene had been a hypothetical entity, the existence of which was postulated in order to explain the regularities observed in inheriting properties, as described above.

2. A RECONSTRUCTION OF A FRAGMENT OF MENDELIAN GENETICS

A certain fragment of Mendelian genetics will now be presented in the form of an axiomatized theory. It will be based on the genetic theory presented in an axiomatized form by J. Neyman in his *First Course in Probability and Statistics*, modified here according to the needs of the present paper.[5]

The presented theory will refer to a situation analogical to that described in the example B, that is, the set of considered organisms S (the universe of the theory) will contain two phenotypes with respect to the properties of the sort considered, and one of these properties will be dominant and the other recessive. In the situation described in example A the formulation of the empirical definition of genotype does not constitute any problem at all, as equivalences occur there between possessing a definite phenotype and an appropriate genotype. The problem of formulating such a definition arises only, if such a unique assignement of observable properties to genotypes is not assumed.

In formulating the axioms the following abbreviations will be used:

S — the set of organisms examined by the theory (a species or a subset of a species of organisms),

C_1 and C_2 — phenotypes (observable properties),

G_1, G_2 and G_3 — genotypes; G_1 will symbolize the set of organisms with a genotype homozygous for the property C_1, G_2 — the set of organisms homozygous for C_2, and G_3 — the set of heterozygous organisms,

F — the set of female organisms,

M — the set of male organisms,

$g_i F$ — the set of organisms which have inherited the gene for C_i from the mother (i.e.: $x \in g_i F \leftrightarrow \bigvee_y (y$ is the mother of x and x has inherited the gene for C_i from y)),

$g_i M$ — the set of organisms which have inherited the gene for C_i from the father (analogously as above),

FG_i — the set of organisms having mothers with the genotype G_i (i.e.: $x \in FG_i \leftrightarrow \bigvee\limits_y (y \in G_i$ and y is the mother of x)),

MG_i — the set of organisms having fathers with the genotype G_i (analogically as above).

2.1. The Axioms of the Theory

A.1.   a) $S = G_1 + G_2 + G_3$,

          b) $G_i \cdot G_j = \emptyset$ for $i \neq j$,

          c) $G_i \cdot F \neq \emptyset$ and $G_i \cdot M \neq \emptyset$ for $i = 1, 2, 3$.

This axiom determines the division of the universe S into three subsets of organisms having three different genotypes; the axiom, moreover, says that each of these genotypes is exemplified by some female and some male organisms.

A.2.   a) $P(g_i F/FG_i) = 1 = P(g_i M/MG_i)$ for $i = 1, 2$,

          b) $P(g_i F/FG_3) = \frac{1}{2} = P(g_i M/MG_3)$ $i = 1, 2$.

The probability that an organism has inherited the gene for the property C_i from the mother (resp. from the father), when the mother (resp. the father) has a homozygous genotype for C_i, equals one and the probability of such an event with respect to the heterozygous genotype of the father, or of the mother, equals 1/2.

A.3. $P(g_i F \cdot g_j M/FG_k \cdot MG_l) = P(g_i F/FG_k) \cdot P(g_j M/MG_l)$

          for $i, j = 1, 2$ and $k, l = 1, 2, 3$.

This axiom states that the inheritance of genes from the father and from the mother are independent events.

A.4.   a) $g_i F \cdot g_i M = G_i$ for $i = 1, 2$,

          b) $g_1 F \cdot g_2 M + g_2 F \cdot g_1 M = G_3$.

That is, an organism has a homozygous genotype if it has inherited similar genes from the mother and from the father, and a heterozygous

genotype if it has inherited different genes from the mother and the father.

The next four axioms will concern the subset Z of the set S which satisfies two conditions:

$$Z \cdot F \neq \emptyset \quad \text{and}$$

$$Z \cdot M \neq \emptyset.$$

Further on we shall denote by Z any subset of S which satisfies these two conditions, i.e. composed of organisms of both sexes. We shall also use the term 'progeny of the set Z in the degree n', in short $P^n(Z)$, which we shall define in terms of the three-argument relation 'x is direct offspring (child) of y and z', in short $xDyz$, as follows:

D. $x \in P^0(Z) \leftrightarrow x \in Z$,

$$x \in P^n(Z) \leftrightarrow \bigvee_y \bigvee_z \left[y \in P^{n-1}(Z) \wedge z \in P^{n-1}(Z) \wedge xDyz \right].$$

This means: $P^0(Z)$ is identical with Z; $P^1(Z)$ is the set of direct offspring of the organisms belonging to Z, etc.

A.5.   a) $P\left(FG_i/P^n(Z)\right) = P\left(G_i/F \cdot P^{n-1}(Z)\right)$,

         b) $P\left(MG_i/P^n(Z)\right) = P\left(G_i/M \cdot P^{n-1}(Z)\right)$,

         for $i = 1, 2, 3$ and $n \geqslant 1$.

This axiom states the independence of the possession of offspring from the genotype (organisms with any genotypes have equal probabilities of having progeny).

A.6. $P\left(FG_i \cdot MG_j/P^n(Z)\right) = P\left(FG_i/P^n(Z)\right) \cdot P\left(MG_j/P^n(Z)\right)$,

         for $i, j = 1, 2, 3$ and $n \geqslant 1$.

The Axiom 6 states the independence of genotypes of parents pairs; organisms of the species S do not show any preferences as to the partner's genotype and have equal probabilities of having progeny with partners of any genotypes.

A.7. $P\left(G_i/F \cdot P^n(Z)\right) = P\left(G_i/M \cdot P^n(Z)\right)$

         for $i = 1, 2, 3$ and $n \geqslant 1$.

The inheritance of the genotype is independent of sex.

A.8. $\quad P(G_i/FG_j \cdot MG_k \cdot P''(Z)) = P(G_i/FG_j \cdot MG_k)$,

for $i, j, k = 1, 2, 3$ and $n \geqslant 1$.

The inheritance of the genotype is independent of generation (i.e. it is identical in each generation on assumption of given genotypes of the parents).

The next axiom assigns appropriate phenotypes C_1 and C_2 to organisms of definite genotypes:

A.9. $\quad$ a) $G_1 = C_1^{\pi}$,

$\quad\quad$ b) $G_2 + G_3 = C_2$.

It is assumed here that property C_1 is recessive and C_2-dominant, as it can be found among both homozygous organisms G_2 and heterozygous G_3. Axioms 9 and 1 of course entail the division of the set S into subsets C_1 and C_2.

2.2. The Language of the Theory

We distinguish among the terms of the theory presented above:
   a) logical constants (with terms of the set theory),
   b) mathematical terms: symbols of real numbers and the symbol of probability P (defined for subsets of S so that $P(S) = 1$),
   c) theoretical terms of genetics: G_i (the symbol of genotype), g_i (the symbol of relation: x has inherited the gene for C_i from y),
   d) empirical terms of biology: C_i, F, M, x is direct offspring of y and z, $P''(Z)$, x is the mother of y, x is the father of y, S,
   e) individual variables ranging over the set S.

As we can see the terms $g_i M, g_i F, MG_i, FG_i$ are composed of theoretical and empirical terms — cf. the list of abbreviations before the axioms.

It is assumed about all the constant terms except the theoretical terms of genetics that they are interpreted independently of the axioms.

2.3. The Theorems of the Theory

We shall now present some consequences of axioms which will be used in further discussion:

T.1. $\quad$ a. $g_i F \neq \emptyset$ $\quad$ and $\quad$ $g_i M \neq \emptyset$ $\quad$ for $i = 1, 2$.

It results from the Axiom 1 (non-emptiness of genotypes) and from the equalities determined by A.4.

T.1.   b. $g_i F \cdot g_j F = g_i M \cdot g_j M = \emptyset$ for $i, j = 1, 2$ and $i \neq j$.

This means that it is not possible to inherit two different genes either from the mother or from the father. It results from A.1.b. and from the equalities determined by A.4; if an organism inherited two different genes from the father (or from the mother), it would have two different genotypes on the grounds of A.4, which is contradictory to A.1.b.

T.1.   c. $g_1 F + g_2 F = g_1 M + g_2 M = S$.

It results from the Axiom 1.a and from A.4.

Thus the Theorems 1a, b and c determine two divisions of the universe S: one into subsets $g_1 F$ and $g_2 F$ and the second into subsets $g_1 M$ and $g_2 M$, i.e. with respect to the kind of genes inherited from the mother and to the kind of genes inherited from the father.

The Theorem T.1 and the Axiom A.2.a give us the consequence:

T.2. $P(g_i F / FG_j) = P(g_i M / MG_j) = 0$ for $i, j = 1, 2$ and $i \neq j$

as on the grounds of T.1.b and c: $g_i F = S - g_j F$ for $i, j = 1, 2$ and $i \neq j$, the same being true for $g_i M$.

We shall now formulate the theorems concerning the inheritance from parents when their genotypes are given.

T.3.   a. $P(G_i / FG_k \cdot MG_l) = P(g_i F / FG_k) \cdot P(g_i M / MG_l)$

        for $i = 1, 2$ and $k, l = 1, 2, 3$,

     b. $P(G_3 / FG_k \cdot MG_l)$

        $= P(g_1 F / FG_k) \cdot P(g_2 M / MG_l) + P(g_2 F / FG_k) \cdot P(g_1 M / MG_l)$

        for $k, l = 1, 2, 3$.

The Theorem 3.a is an immediate consequence of the Axioms 4 and 3. The Theorem 3.b results from the Axioms 4 and 3 and from the Theorem T.1.b (as it requires the transformation of the probability of the sum into the sum of probabilities).

The Theorems T.3.a and b allow to calculate, by means of the Axiom 2 and the Theorem T.2, the probabilities of any genotypes for given

genotypes of the parents; thus, for example:

$$P(G_1/FG_1 \cdot MG_1) = 1 \quad \text{(T.3.a and A.2.a)},$$

$$P(G_1/FG_3 \cdot MG_3) = \tfrac{1}{2} \cdot \tfrac{1}{2} = \tfrac{1}{4} \quad \text{(T.3.a and A.2.b)},$$

$$P(G_3/FG_1 \cdot MG_2) = 1 \cdot 1 + 0 = 1 \quad \text{(T.3.b, A.2.a and T.2)}.$$

We shall now prove two theorems concerning the inheritance of genotypes in the first generation of the offspring of a given set Z, i.e. in the set $P^1(Z)$. Let us remind that the symbol Z stands for a subset of S which satisfies the conditions: $Z \cdot F \neq \emptyset$ and $Z \cdot M \neq \emptyset$, i.e. a subset containing both male and female organisms. Our theory allows to calculate the probabilities of the form $P(G_i/P^n(Z))$, for any high n, if only the distribution of genotypes is given in the sets $Z \cdot F$ and $Z \cdot M$. We shall now adopt the following notation for such a distribution:

$$P(G_i/Z \cdot F) = p_i' \quad (i = 1, 2), \qquad P(G_i/Z \cdot M) = p_i'' \quad (i = 1, 2),$$

$$P(G_3/Z \cdot F) = q' \quad \text{and} \quad P(G_3/Z \cdot M) = q''.$$

T.4.a. $\quad P(G_i/P^1(Z)) = (p_i' + \tfrac{1}{2}q') \cdot (p_i'' + \tfrac{1}{2}q'')$

for $i = 1, 2$.

T.4.b. $\quad P(G_3/P^1(Z))$

$$= (p_1' + \tfrac{1}{2}q') \cdot (p_2'' + \tfrac{1}{2}q'') + (p_2' + \tfrac{1}{2}q') \cdot (p_1'' + \tfrac{1}{2}q'').$$

PROOF. Axiom 1 determining the division of S into 3 genotypes entails that for a pair of organisms — we are now interested in pairs of parents — there are 9 possible combinations of their genotypes: $G_1 G_1$, $G_2 G_2$, $G_3 G_3$, $G_1 G_2$, $G_2 G_1$, $G_1 G_3$, $G_3 G_1$, $G_2 G_3$ and $G_3 G_2$; we then have the division of $P^1(Z)$ into 9 subsets with respect to the combinations of the parents' genotypes. Therefore, an event consisting in that an organism of the set $P^1(Z)$ has inherited the genotype G_i equals the sum of 9 events exclusive by pairs $FG_j \cdot MG_k \cdot G_i$. We then obtain the following equality for the probability of the event G_i in the first generation of the set Z:

$$P(G_i/P^1(Z)) = \bigcup_{j=1}^{3} \bigcup_{k=1}^{3} P(G_i \cdot FG_k \cdot MG_j/P^1(Z))$$

for $i = 1, 2, 3$.

We can transform every term of this sum according to the rules of the calculus of probability and to the Axioms 5,6 and 8, as follows:

$$P(G_i \cdot FG_j \cdot MG_k/P^1(Z)) = P(FG_j \cdot MG_k/P^1(Z)) \cdot P(G_i/FG_j \cdot$$

$$\cdot MG_k \cdot P^1(Z)) = P(FG_j/P^1(Z)) \cdot P(MG_k/P^1(Z)) \cdot P(G_i/FG_j \cdot MG_k)$$

$$= P(G_j/F \cdot Z) \cdot P(G_k/M \cdot Z) \cdot P(G_i/FG_j \cdot MG_k).$$

The first two terms in each of the above products are determined by the distribution of genotypes in the set Z, while the third term can be calculated on the basis of the Theorem T.3.a and b. We thus obtain, for e.g. the homozygous genotype G_1, the following equality:

$$P(G_1/P^1(Z)) = p'_1 \cdot p''_1 \cdot 1 + p'_1 \cdot q'' \cdot \tfrac{1}{2} + p'_1 \cdot p''_2 \cdot 0 + q' \cdot p''_1 \cdot \tfrac{1}{2} +$$

$$+ q' \cdot q'' \cdot \tfrac{1}{4} + q' \cdot p''_2 \cdot 0 + p'_2 \cdot p''_1 \cdot 0 + p'_2 \cdot q'' \cdot 0 + p'_2 \cdot p''_2 \cdot 0.$$

After simplification we obtain therefrom the Theorem T.4.a for $i=1$. We derive similarly equalities for the homozygous genotype G_2, and for the heterozygous genotype G_3, in the form T.4.b.

It follows from the Axiom 7 that, for $n \geqslant 1$: the probabilities of particular genotypes in the set $P^n(Z)$ which we shall denote by $p_1^{(n)}$, $p_2^{(n)}$ and $q^{(n)}$, are equal in the subsets of male and female organisms (heredity is independent of sex). Therefore the knowledge of the distribution of probabilities in Z allows to calculate, according to the equations T.4.a and b, the whole distribution of probabilities in $P^1(Z)$ necessary to calculate the probabilities in $P^2(Z)$.

For we have $p_i^{(1)} = P(G_i/F \cdot P^1(Z)) = P(G_i/M \cdot P^1(Z))$ for $i=1$, 2, and $q^{(1)} = P(G_3/F \cdot P^1(Z)) = P(G_3/M \cdot P^1(Z))$.

Thus the Theorem T.4, after the substitution of numbers $p_i^{(1)}$ for p'_i and p''_i and $q^{(1)}$ for q' and q'', will allow us to calculate in turn the probabilities of particular genotypes in $P^2(Z)$, etc., for any n. However, these equations can be simplified already for $n=2$. As in the set $P^1(Z)$ the probabilities of genotypes are equal in male and female subsets, we obtain, as it can be easily seen:

T.5.a. $\quad P(G_i/P^n(Z)) = (p_i^{(n-1)} + \tfrac{1}{2} q^{(n-1)})^2 \quad$ for $i=1, 2$ and $n \geqslant 2$,

T.5.b. $\quad P(G_3/P^n(Z)) = 2 \cdot (p_1^{(n-1)} + \tfrac{1}{2} q^{(n-1)}) \cdot (p_2^{(n-1)} + \tfrac{1}{2} q^{(n-1)})$

$\quad$ for $n \geqslant 2$.

Further consequence of the Axiom 7 is one of the most important theorems of genetics, that of stabilization of distribution:

T.6. $\qquad P(G_i/P^n(Z)) = P(G_i/P^2(Z)) \qquad$ for $i=1, 2, 3$ and $n \geqslant 2$.

The proof of this theorem consists in simple arithmetic transformations of the right sides of the equality T.5 for $P^n(Z)$ and for $P^{n+1}(Z)$. This proof has been presented in J. Neyman's *First Course in Probability and Statistics* (cf. Note 5).

We shall now present two theorems which are immediate consequences of the Axiom 9, which determines the relations between genotypes and phenotypes:

T.7. $\qquad$ a. $P(C_1/P^n(Z)) = P(G^1/P^n(Z))$,

$\qquad$ b. $P(C_2/P^n(Z)) = P(G_2/P^n(Z)) + P(G_3/P^n(Z))$.

The distribution of genotypes in Z thus determines the distributions of phenotypes in any generations of the offspring of Z.

3. EMPIRICAL DEFINITIONS OF GENOTYPES

It has been said in the introductory remarks that the attempt at defining genotype in this paper will basically proceed in the same direction as that carried out by Woodger; we shall thus use such terms as names of observable properties of organisms (C_1, C_2) and names of sets of offspring for a given set Z, defined in terms of the relation of being direct offspring of a pair of organisms. We have assumed here that these terms are empirical ones.

We shall regard as strictly empirical such definitions which do not contain in the definienses any extralogical non-empirical terms. Thus, for example, Woodger's definitions of genotypes which determine genotypes of organisms in terms of observable characteristics of the offspring, are empirical definitions in the above sense. We shall, however, try to demonstrate that the adequacy of definitions of genotypes referring to observable properties of the offspring requires that the definiens also contain the term 'probability', and such will be the definitions of genotypes formulated here.

The problem whether and in what sense such definitions can be regarded as empirical will be dealt with in the next chapter.

Let us first note that the Axiom 9a states the equality:

$$G_1 = C_1,$$

and so the empirical definition of one of the homozygous (for the recessive property) genotypes is simply one of the axioms of the theory.
With respect to the second homozygous genotype G_2, however, as well
as to the heterozygous genotype G_3, the theory does not determine
such a correlation. The Axiom 9.b assigns to both genotypes the same
observable property C_2. If we then want to find the empirical definitions
of the genotypes G_2 and G_3, we must solve the problem of finding the
appropriate empirical criterion, permitting to differentiate organisms
having these genotypes.

Although organisms having the genotypes G_2 and G_3 do not differ
by phenotypes, differences due to their genotypes can nevertheless be
observed: indeed, the distribution of phenotypes varies among the
offspring of such organisms (cf. Section 1, Example B). The search
for empirical criteria for genotypes should then be oriented towards
the theorems of the theory concerning the properties of the offspring.

According to the Theorem T.3.a and to the Axioms 2 and 9, we obtain
the equality:

$$P(C_2/FG_2 \cdot MG_2) = 1.$$

Therefore, on the grounds of the Theorems T.4.a and T.5.a, we obtain
the following equalities for the set of organisms Z in which all male
and female organisms have the homozygous genotype G_2:

$$P(C_2/P^1(Z)) = 1 \quad \text{and} \quad P(C_2/P^2(Z)) = 1.$$

Further generations can be left out, as starting from the second
generation the stabilization of distribution begins (Theorem 6). It is
also easy to determine that in every case when Z contains some organisms
having the genotype G_1 or G_3, the probability of the property C_2 in the
second generation will be smaller than one.

This can make us suppose that the following definition might be
adopted for the genotype G_2:

D_1. $Z \subset G_2 \leftrightarrow Z \subset C_2 \wedge P^1(Z) \subset C_2 \wedge P^2(Z) \subset C_2 \wedge P^2(Z) \neq \emptyset.$

D_1 is a simplification of the definition of homozygous genotype presented by Woodger (cf. Note 1). M. Przełęcki has pointed out in *On the Concept of Genotype* (cf. Note 2) some paradoxical consequences of such a definition. The latter allows to asign the genotype G_2 to every organisms x having the property C_2 if only x has no progeny with any organism belonging to Z; but such an organism x can have the heterozygous genotype G_3. In view of this shortcoming of Woodger's definition Przełęcki further considers the possibility of a modification of this definition, which we present here in a slightly modified form as a reduction sentence:

D_2. $\quad \bigwedge_x [x \in Z \wedge \bigvee_y (y \in Z \wedge P^2\{x, y\} \neq)\emptyset]$

$\rightarrow [Z \subset G_2 \leftrightarrow Z \subset C_2 \wedge P^1(Z) \subset C_2 \wedge P^2(Z) \subset C_2].$

D_2 might be considered as the consequence of the above-presented axioms if we assumed that the probability P satisfies the condition:

C. $\qquad A \neq \emptyset \rightarrow (P(B/A) = 1 \leftrightarrow A \subset B).$

Przełęcki has shown in his work that the definition D_2 is also inadequate. We shall now show that on the grounds of the theory presented here this definition leads to a contradiction.

Let us suppose that a set Z satisfies the initial condition of D_2 and that it is composed of 200 organisms, among which one half are male and the other half–female; let us further suppose that 198 of these organisms have the genotype G_2, and that the remaining two have the heterozygous genotype G_3, one of them being male and the other female. In this situation we calculate on the grounds of T.5 and T.7 that:

$$P(G_1/P^1(Z)) = P(C_1/P^1(Z)) = 0.000025, \quad \text{and}$$

$$P(G_1/P^2(Z)) = P(C_1/P^2(Z)) = 0.000025.$$

With so small a probability of the property C_1 in the sets of the offspring of Z this property may appear neither in the first generation nor in the second (and not even in the further ones), should even these generations be relatively numerous. We then shall have as a result of observation:

$$Z \subset C_2 \quad \text{and} \quad P^1(Z) \subset C_2 \quad \text{and} \quad P^2(Z) \subset C_2,$$

which entails on the grounds of D_2:

(1) $Z \subset G_2$.

It may occur, however, that one of these two heterozygous organisms in Z, say x, has a mother or a father y with the characteristic C_1; this entails that y has the genotype G_1, and therefore x cannot have inherited the genotype G_2 from its parents; it must have the genotype G_3. On the grounds of these experimental data and of the theorems of the theory we thus obtain the consequence:

(2) $\sim Z \subset G_2$,

which contradicts the consequence (1).

As we can see, in genetic theory the probability P cannot satisfy the condition C. What is more, this probability cannot be simply identified with relative frequency. For it is easy to notice that by such an interpretation the theory would be false: incompatible with the results of experience. The theorems of the theory about the probabilities of genotypes, and therefore of observable properties in sets of offspring, can be considered as true only on an interpretation which would not require a precise realization of these probabilities in the form of actual proportions in any sets of offspring, but only an approximate realization in cases when these sets would be very numerous.

It seems to be fully obvious that it is assumed in genetics that actual proportions of genotypes and of the related observable properties among the offspring differ as a rule from the values of probabilities determined by the assumptions of the theory.

These considerations lead to the conclusion that if for a given sort of properties there is no unique assignment of genotypes of these properties (i.e. if one of them dominates), it is impossible to formulate an empirical citerion for genotypes in the form of an equivalence between the possessing of a definite genotype by the organisms of a set Z and the possessing of a definite observable property by these organisms and by their offspring. The condition requiring the second (or further) generation to be non-empty or even very numerous will not eliminate the main shortcoming of the definition; it consists in stating too close a relation between the genotypes of the ancestors and the observable properties of the offspring, a relation which is much looser according to genetic theory,

for it has a probabilistic character. If we deal with a set of organisms Z having a dominant property, genetic theory will never allow to state in an infallible way that all the organisms in Z have a homozygous genotype on the grounds that all the offspring of the second and possibly of further generations also have this property. For it will always be theoretically possible that the recessive property will not appear in any of the existing sets P^nZ, in spite of the existence of heterozygous organisms in the set Z. It seems that the assertion can be safely accepted saying that genetics does not know any empirical facts which would constitute infallible criteria of homozygousness with respect to dominant properties. In order to avoid the contradictions to which the above-discussed definitions have led it would be necessary, in defining genotype, to take into consideration the fact that the relation between the genotypes of the ancestors and the corresponding phenotypes of the progeny has in genetics a probabilistic character.

The most natural way of formulating such a definition seems to be the formulation of its definiens in the form of a probabilistic condition. In the case of the situation described in the Section 1, Example B, to which the above-presented theory is restricted, the following equivalence can constitute the definition of a homozygous genotype with respect to any of the two properties C_i:

$$\text{D.Hom.}\ x \in G_i \leftrightarrow x \in C_i \wedge \bigvee_y [y \in C_i \wedge P(C_i/P^1\{x, y\})$$

$$= 1 \wedge P(C_i/P^2\{x, y\}) = 1] \quad \text{for} \quad i = 1, 2.$$

The organism x has a homozygous genotype G_i if and only if it has the property C_i and, if there is an organism y also having the property C_i, the probability of C_i in the first and in the second generation of the offspring of the pair x, y equals one. This definition is a consequence of the axioms presented above.

Let us first of all note that if x satisfies the condition:

$$\text{C*.} \quad \bigvee_y P(C_i/P^2\{x, y\}) = 1,$$

then the other conditions of the definition also will be satisfied. Indeed, the couple of organisms x, y satisfies the condition:

$$P(C_i/P^2\{x, y\}) = 1$$

only—which will be shown below—if both x and y have the genotype G_i, from which it follows that these both organisms have the property C_i and that in the first generation of their offspring the probability of C_i also equals one (T.7.a and T.4.a).

The definition of homozygous genotype can therefore also be presented in a shorter form, leaving out the conditions resulting from C^*:

D.Hom.* $x \in G_i \leftrightarrow \bigvee_y P(C_i/P^2\{x,y\}) = 1$.

We shall now present the proof of D.Hom.* on the grounds of the axioms of the above-presented theory; it will at the same time be the proof of **D.Hom.**

PROOF. Let $Z = \{x, y\}$ and x be of different sex than y, say $x \in F, y \in M$. If x and y have a homozygous genotype G_i, then T.3.a, A.8 and T.5.a allow to compute that the probability of the genotype G_i in the second generation of $\{x, y\}$ equals one, from which it follows that the probability of the property C_i will also equal one, on the grounds of T.7:

$$(1) \qquad x \in G_i \wedge y \in G_i \rightarrow P(C_i/P^2\{x,y\}) = 1 \qquad \text{for } i = 1, 2.$$

It results from the Axiom 1 that if it is not true that x and y have the same homozygous genotype, then there are 8 possibilities of combinations of their genotypes. We calculate for ach of them the probability of C_i in $P^2\{x, y\}$; the calculations show that in every case the probability of C_i differs from one. Therefore:

$$(2) \qquad x \in G_i \wedge y \in G_i \leftrightarrow P(C_i/P^2\{x,y\}) = 1 \qquad \text{for } i = 1, 2.$$

(Calculations show that this equivalence holds for the dominating C_i as well as for the recessive C_i.)

From the equivalence (2) it follows that:

$$(3) \qquad x \in G_i \rightarrow (y \in G_i \leftrightarrow P(C_i/P^2\{x,y\}) = 1).$$

Now, we can derive the following consequence from the Axiom 1.c, which assumes that every genotype is non-empty in both the set of male and female organisms:

$$(4) \qquad x \in G_i \rightarrow \bigvee_y (y \in G_i \text{ and } y \text{ is of different sex than } x).$$

The implications (3) and (4) yield a consequence:

(5) $x \in G_i \to \bigvee_y P(C_i/P^2\{x, y\}) = 1$ for $i = 1, 2$.

From the equivalence (2), in turn, if follows that:

(6) $\sim x \in G_i \to \sim \bigvee_y P(C_i/P^2\{x, y\}) = 1$ for $i = 1, 2$.

The implications (5) and (6) provide the proof of the definition D.Hom.*, and thus also of D.Hom.

Besides, the axioms also have as consequence the equivalence:

$$x \in G_i \leftrightarrow \bigvee_y P(C_i/P^n\{x, y\}) = 1 \quad \text{for any } n \geqslant 2.$$

The equivalence D.Hom.* is a scheme of the definition of the genotypes G_1 and G_2. From the Axiom 1 the definition of the genotype G_3 follows:

D.Het. $x \in G_3 \leftrightarrow \sim x \in G_1 \wedge \sim x \in G_2$.

However, the definition of heterozygous genotype can also be formulated independently of homozygous genotype:

D.Het.* $x \in G_3 \leftrightarrow \bigvee_y P(C_1/P^1\{x, y\}) = \frac{1}{4}$.

A proof analogical to that of D.Hom.* can be carried out for the latter definition. One should also refer here to the fact that there are at least two organisms of different sexes having the same heterozygous genotype, and calculate that in the case of such a pair the probability of the recessive property C_1 equals 1/4 in the first generation, while this probability is different for any pair with another combination of genotypes.

4. THE EMPIRICAL CHARACTER OF DEFINITIONS
OF GENOTYPES

The Definitions D.Hom.* and D.Het.* are sentences in which on the right side of the equivalence there are, apart from logical terms, certain empirical ones and the mathematical term 'probability'. The definienses are sentences stating that there is an organism y which together with a

given organism x satisfies the equality:

$$P(C_i/P^2\{x, y\}) = 1,$$

resp.

$$P(C_1/P^1\{x, y\}) = \tfrac{1}{4}.$$

As it has already been said these equalities should not be interpreted as sentences about actual relative frequencies of phenotypes in sets of offspring of $\{x, y\}$: generally — equalities of the form $P(C_i/P^n(Z)) = p$ cannot be considered as sentences about actually observed relative frequencies in actually existing sets of offspring, for in such sets there will be as a rule deviations from probabilities determined by the theory.

It seems that in genetics the term 'probability' does not have a full empirical interpretation which would be expressible by means of a definition. Therefore the definitions D.Hom.* and D.Het.* are not empirical in the strict sense we mentioned at the beginning of Section 3. They can be considered empirical, though, in some weaker sense, provided that the term 'probability' has some, at least partial, empirical interpretation.

It might be worth quoting here J. Neyman's opinion about the interpretation of probability, as the genetic theory presented here is based on his axiomatics. In *Mathematical Statistics and Probability*[6] (Section 1) the author says that probability can be understood, among others, as relative frequency in some 'hypothetical' sets. Thus, e.g., the sentence saying that the probability of obtaining a 6 when throwing a die equals 1/6 can be understood either as speaking about the relative frequency of throwing 6 in the set of actual throws of the die or as speaking about the relative frequency in a hypothetical set of throws of the die (in the set of possible throws?); the latter interpretation is considered to be more fruitful than the former. In our case we would then be concerned with the relative frequencies of phenotypes in some hypothetical (possible) sets of offspring $P^n\{x, y\}$, and not in real sets accessible to observation. Of course, we could hardly consider such an interpretation empirical.

A somewhat different view is presented by Popper in his paper *The Propensity Interpretation of Probability*.[7] As I understand him — Popper considers probabilities of events as their dispositions to appear with

certain relative frequencies: these are, however, 'merely' dispositions, and not realizations. Therefore — Popper says — "the probability of an event is its property to be measured by a potential statistical frequency rather than by an actual one".

As a matter of fact, this view seems to be close to that of Neyman's. Anyway, the 'potential' relative frequency, as well as frequency in a 'hypothetical' set, cannot be considered empirical concepts. It seems, however, that in many cases of using the term 'probability' in empirical sciences, although its empirical interpretation encounters difficulties as in genetics, this term receives at least partial empirical interpretation determined by more or less precise rules of accepting sentences about probabilities on the basis of results of experiences.

This is precisely the situation in Mendelian genetics which — we can say — has been formulated on the basis of some statistical observations. Statistical observations are here as a rule the basis of adopting hypotheses about theoretical probabilities. With respect to hypotheses concerning the genotypes of particular organisms this procedure could be presented as follows:

Let us suppose that the theory assumes probabilities of genotypes conforming to the axiomatic presented here; let us further suppose that a pair of organisms $\{x, y\}$ has very numerous offspring in the first generation and that it has been observed that the relative frequencies of the phenotypes C_1 and C_2 in the set $P^1\{x, y\}$ are respectively: 0.22 and 0.78.

The theory allows 6 possible hypotheses concerning the genotypes x and y: $G_1 G_1$, $G_1 G_2$, $G_1 G_3$, $G_2 G_2$, $G_2 G_3$ and $G_3 G_3$. For these hypotheses the theory determines the probabilities of the property C_1 (recessive) in the set $P^1\{x, y\}$ as follows (respectively): $p_1 = 1$, $p_2 = 0$, $p_3 = 1/2$, $p_4 = 0$, $p_5 = 0$, $p_6 = 1/4$. As we can see, the sixth hypothesis saying that the organisms x and y both have heterozygous genotypes G_3 determines a probability p_6 of C_1 in the set $P^1\{x, y\}$ which is the closest to the observed relative frequency C_1; in the other hypotheses these probabilities considerably differ from that frequency. The sixth hypothesis therefore provides the greatest probability of the observed result of experience. The sixth hypothesis can then be accepted, for example, on the grounds of the rule requiring the acceptance of the hypotheses for which the probability of a given result of observation is

the greatest. Of course, such an approach is considerably simplifying because of the extreme simplification of reality made in the present theory. The problem of the choice of the hypothesis is in reality much more complicated. But this is roughly the situation.

Therefore one is inclined to 'explain' this case by the fact that the observable (actually existing) sets $P^n(Z)$ are considered here as random samples (they would be somehow perfect random samples) from—as Neyman says—some hypothetical sets of offspring, to which the theoretical probabilities refer. Independently, however, from how this case will be explained (I do not know of any satisfactory explanation) it is a fact that a relation is assumed in genetics between theoretical probabilities and actual relative frequencies by using statistical rules of accepting hypotheses. That is how the term 'probability' contained in the theory obtains to a degree an empirical interpretation.

Therefore, although we cannot say that the definitions D.Hom.* and D. Het.* are empirical in the strict sense, we can nevertheless consider that these definitions have an empirical character in a weaker sense, as they give to the defined concepts a partial empirical interpretation through statistical rules of accepting the sentences contained in the definienses.

Let us make a final remark concerning the difference between the 'probabilistic' definitions formulated here and the sort of postulates called by H. Mehlberg 'statistical definitions'.

In his paper *Positivisme et Science*, [8] devoted to the relations between various ways of defining theoretical terms and the empirical testability of sentences containing these terms, H. Mehlberg calls 'statistical definition' a certain set of statistical postulates usually called in methodology statistical (or probabilistic) indices, as for example:

$$P(T/Q_1)=p \quad \text{and} \quad P(T/Q_2)=q,$$

where T is a theoretical term, Q_1 and Q_2 are empirical terms and P is the relative frequency, and both such sentences about frequencies are analytic. Such definitions are further discussed by Mehlberg in his work *The Reach of Science*. [9]

The utility of such 'definitions' is now widely recognized in the methodology of certain sciences, as among others sociology or psychology. They are not, however, definitions in the strict sense of the word.

They can be considered as a weakening of postulates determining some separate necessary and sufficient conditions for theoretical terms of the form:

$$\bigwedge_x (x \in Q_1 \rightarrow x \in T) \quad \text{and} \quad \bigwedge_x (x \in Q_2 \rightarrow \sim x \in T).$$

The definitions of genotypes formulated in this work are normal equivalence definitions. They have been called 'probabilistic' simply because the term 'probability' is contained in the definiens.

It can be said, however, that they have something in common with Mehlberg's so-called 'statistical definitions'; namely, that neither give infallible criteria of accepting, on the grounds of observations, sentences containing defined terms, but only statistical criteria.

REFERENCES

* A shortened version of an article first published in *Studia Logica* **15** (1964). Translated by S. Wojnicki.

[1] J. H. Woodger, *Biology and Language*, Univ. Press, Cambridge 1952,

[2] M. Przełęcki, 'On the Concept of Genotype', in: *Form and Strategy in Science*, (edited by John R. Gregg and F. T. C. Harris), D. Reidel, Dordrecht 1964.

[3] The example has been taken from S. Skowron's handbook *Zarys nauki o dziedziczności* (*Outline of the Science of Heredity*), Wiedza — Zawód — Kultura, Kraków 1947, p. 20.

[4] The example is taken from the same source, p. 104.

[5] J. Neyman, *First Course in Probability and Statistics*, H. Holt and Co., New York 1950. In the 3rd chapter of this book, entitled *Probabilistic Problems of Genetics* (pp. 96–163), Neyman presents the axioms and some theorems of simplified Mendelian genetics, as an example of the application of probabilistic calculus to a concrete science. The theorems presented in this paper, and denoted as T2, 3, 4, 5, 6, correspond to the theorems in Neyman's book. On the other hand, the axiomatics adopted here differs considerably from that formulated by Neyman. Some of Neyman's axioms have been left out here because of the restriction to problems of inheriting genotypes for only one sort of properties. The modification of Neyman's Axioms 1, 4 and 5 consists in that Neyman's axioms concern some regularities in the division and joining of pairs of alleles in connection with the division of cells into gametes and their joinign into zygotes; therefore the probabilities of inheriting definite genotypes with respect to the genotypes of the pair of parents are calculated in the sets of gametes of the parents organisms; further, considering questions of heredity in populations of organisms Neyman utilizes probabilities calculated in these populations, i.e. in classes of organisms. In the theory presented

here all probabilities are calculated in classes of organisms. Moreover we take here as axioms some assumptions tacitly adopted by Neyman in the course of deriving the theorems, as well as those adopted by him only for some situations (panmixia and the independence of genotypes from sex). Finally, we adopt here an axiom establishing certain relations between genotypes and observable properties of organisms which is indispensable for deriving empirical definitions of genotypes.

[6] J. Neyman, *Mathematical Statistics and Probability*, Washington 1952.

[7] Karl R. Popper, 'The Propensity Interpretation of Probability', *British Journal for the Philosophy of Science* X (1959).

[8] J. Mehlberg, 'Positivisme et science', *Studia Philosophica* III (1948).

[9] H. Mehlberg, *The Reach of Science*, Toronto Press, 1958.

ADAM NOWACZYK

ANALYTIC SENTENCES IN THE SEMANTIC
SYSTEM*

I

One might look for connections between the notion of analyticity and
the problems of semantics or — in broader terms — semiotics in the
definition of analytic judgements deriving from Kant, which, as it is
known, involves the inclusion of the meaning of the predicate in the
meaning of the subject. The above mentioned connections manifest
themselves at present in numerous current definitions of the term 'analytic
sentence', which, as a rule, involve such semantic terms as: 'meaning',
'definition', 'synonymy' and others. Many authors express the opinion
that the concept of analyticity is indispensable in order to define certain
semantic concepts. The concept of meaning comes to the foreground
here. Rudolf Carnap, the author of one of the most important attemps
to explicate this concept has used the concept of logical truth which — according to his intentions — is supposed to correspond to the old notion of
analyticity.

Not infrequently the tendency to identify analytic sentences with
so called definitional theorems is met in the literature on the logical
theory of language. Not all theorems thus called are definitions in the
exact meaning of this word, though they all play a role similar to definitions. This role consists in constituting semantic relations. The above
mentioned tendency to identify analytic sentences with the totality of
definitional theorems is often caused by the conviction of the particular
methodological or epistemological status of these theorems. It is often
the conviction of the conventional, i.e. — in some sense — *a priori* character
of these theorems. This conviction cannot be maintained in view of
the argumentation presented by Kazimierz Ajdukiewicz in his work
[2] (at least without imposing some quite serious restrictions upon the

difinitional theorems). The realization of this state of affairs has caused a
tendency opposite to the one mentioned above: not all, but only some
definitional theorems are considered to be analytic sentences. This new
tendency has been manifested in the works of Rudolf Carnap [7], Ryszard Wójcicki [30] and Marian Przełęcki [25]. It seems that both the
broad concept of analyticity comprising all the definitional propositions,
as well as some concepts narrower in scope, may find applications in
the field of semantics. However, in all atempts at reconstructing the
concept of analytic sentence for semantic (or semiotic) use the division
between problems of semantics and the epistemological question of *a
priori* justification should be observed. We believe that there is no need
for analytic sentences, understood as a category in the field of semantics,
to be characterized in terms of a peculiar type of justification, distinguished in the theory of knowledge. In particular, they do not have
to be *a priori* theorems in the philosophical sense of the word. Indeed,
attempts to distinguish sentences that are, in a sense, independent of
experience are noteworthy also from the methodological point of view.
Their success, however, clearly depends on how this independence is
understood. It seems that a certain result may be obtained only if we
do not have in mind the independence of sensory experience in general,
but the independence of the definite results formulated in a definite
language, obtained in the course of empirical examination of a definite
model of this language.

    In constructing an explicatum of the concept of analyticity that may
be useful in semantics we have considerable freedom of choice among
various concepts, which is connected with the fact that the field within
which this explicatum is to function is, in part, constructed simultaneously
with it. The choice of a definite concept of language to which the explicatum of the concept of analyticity is relativised plays an important
role in constructions of this type. This is not a natural language, as a
rule, but rather an entity that may be treated as the result of making
it more precise. For instance, according to Carnap, a semantic system
is an entity of this sort. In contrast with the natural language, the syntactic and semantic properties of the expressions are precisely determined
in the semantic system. Carnap characterizes the semantic system with
the help of sentences of metalanguage called respectively syntactic and
semantic rules. The former characterize a certain formalized language *L*

specifying its vocabulary and the rules of formulation of complex expressions, and the latter may be considered to be the definition of a certain model $\mathfrak{M}$ of this language. Hence the semantic system may be treated as an ordered pair:

(S) $\langle L, \mathfrak{M} \rangle$.

The main idea that Carnap uses in order to distinguish the analytical sentences from the semantic system S is the following principle:

(I.1)   *A sentence α is an analytic sentence in S if and only if the sentence 'α is a true sentence in the model $\mathfrak{M}$', is a consequence of the semantic rules of system S.*[1]

However, this principle is a deceptive one. It is clear that semantic systems S_1 and S_2 may be identical, while rules R_1 and R_2 subsequently making up its characteristics may be logically non-equivalent. Then — according to principle (I.1) — the same sentence α may be analytic in S_1 and synthetic in S_2, which is a paradox. This paradox may be avoided by identifying the semantic system with a more complex structure, namely:

(S′) $\langle L, \mathfrak{M}, A \rangle$

in which A is just the set of analytic sentences in S'. We assume at the same time that the sentences A are included in the set of true sentences in model $\mathfrak{M}$. The phrase 'analytic sentence in S'' is then a primitive, undefined expression. Only in an informal way the explanation is given that A is a deductive system based on the set of sentences adopted as meaning postulates.[2]

However, the concept of the semantic system in harmony with the conceptual apparatus that is often used in the investigations in the field of semantics of languages of empirical theories seems to be the most adequate one for the discussion of analyticity.[3] We have in mind a semantic system here understood as the structure:

(S″) $\langle L, \mathfrak{M}_0, \sigma \rangle$.

L is an elementary formalized language here; its descriptive terms have been devided into O-terms (observational) and T-terms (theoretical terms). It is assumed that O-terms have been interpreted in a direct

way. Thus a certain model, namely $\mathfrak{M}_0$, has been ascribed to language L_0, which is a fragment of language L constructed by the omission of the T-terms. σ is a set of sentences adopted as meaning postulates, which, as a rule, contain both types of terms. Since intuition makes us consider all the definitional theorems analytic we may assume that the set of analytic sentences in the semantic system S'' is a deductive system based on the set of postulates σ. The adopted characteristics of the proper model $\mathfrak{M}^*$ of the semantic system S'' determines the definitional character of these sentence. For, among others, the following condition is included in the characteristics:

(I.2)   *The set of sentences σ is included in the set of sentences true in the model $\mathfrak{M}^*$.*

This condition is equivalent to the following one:

(I.3)   *The set of sentences $\mathrm{Cn}\,(\sigma)$ is included in the set of sentences true in the model $\mathfrak{M}^*$.*

So condition (I.2) demands that descriptive terms of the language L be interpreted in such a way that the analytic sentences are true.

II

The concept of a semantic system that has been outlined above seems to be an satisfactory frame of reference for the investigations concerning the concept of analyticity. Even the fact that the concept of a meaning postulate, to which this concept refers in the sphere of intuition, remains undefined, is not an obstacle here. For the semantic role of the meaning postulates has been made clear in the characteristics of the proper model of the semantic system. We shall, howewer, use a different concept of language than the concepts of the semantic system that have been mentioned above. We shall use mainly the concepts of syntax and pragmatics in making the concept precise. The motif that we might — using the term suggested by Roman Suszko — call the conviction of the *priority* of pragmatics over semantics, speaks in favour of such an approach.

While precising the concept of language in some way we should like, however, to refer clearly to the current meaning of the word 'lan-

guage'. We believe that one may speak of language in the ordinary meaning of the word only if a certain system of signs functions in some human community, and its use, i.e. verbal behaviour of the members of this social group, is adequately uniform. As is known—this idea lay at the basis of the concept of language that has been presented by Kazimierz Ajdukiewicz in his works [4] and [5]. We shall often refer to this concept in further discussion.

Some patterns of behaviour are connected with every language and are observed by its users. The basic difference between a natural and an artificial language consists in that in the former case these patterns are formed by reciprocal interactions between members of the language community, and in the latter they are established by way of decisions (called linguistic conventions) that are communicated to those interested in them in the form sentences of metalanguage.

Linguistic conventions are a kind of decisions, resolutions. Obviously, decisions are not sentences. Only a message about making a decision may be a sentence. It is equally obvious that only human behaviour may be subject to decision. Thus, if we sometimes speak about *semantic* conventions that are manifested in a sentence of the type: 'We decide that the term λ will denote such and such an object', then the meaning of this sentence is approximately the following: we have decided to *use* the term λ as the name of such and such an object. Thus, speaking of *semantic* conventions we probably express the conviction that by making appropriate decisions concerning verbal behaviour and by realizing them consistently we are capable of causing that (at least when some additional conditions objective in character are also fulfilled) the semantic relations we have postulated will hold. Thus, the existence of certain dependencies between the pragmatic properties and the semantic properties of language expressions is stated. These dependencies are of the type that the system of behaviours (or—rather—dispositions for behaviour) connected with language determines—together with some objective situations—the semantic properties of expressions.

As it has been mentioned above, the conviction about the priority of pragmatics over semantics has been expressed in the concept of language we intend to present. This priority manifests itself in the process of constructing the language, as well as in the investigation of it. We establish, by way of convention, certain patterns of verbal behaviour

in constructing a language. On the other hand, the desired semantic relationship – for instance that a certain term will be, for some person, the name of specified objects (especially of unobservable objects) – is achieved indirectly, by inculcating the established system of behaviour on a person. Similarly, observations of behaviour provide initial material in an empirical examination of language. The semantic relations are – in contrast with verbal behaviour – principally unobservable. Thus we reach the hypotheses concerning the semantic properties of the expressions of a language under investigation in an indirect way. We then take advantage of some, usually undefined, assumptions concerning the relations between the pragmatic and semantic properties of expressions.

The purpose of the above remarks was to justify the use of the concept of language that will be characterized more precisely presently. From now on we shall call language the structure of the type:

(J) $\langle L, C_0, E, A \rangle$.

The members of this structure are, subsequently:

L – formalized language, given by means of a list of simplex expressions and the rules of formation complex expressions,

C_0 – a relation called the *direct consequence in language J* (its domain consists of small sets of sentences of the formalized language L, its counterdomain of sentences in this language),

E – a function defined on a certain established subset of terms of the formalized language L, possessing values that are extralinguistic objects, namely, observable individuals, sets, or sets of ordered pairs of such individuals,

A – distinguished set of sentences (finite or quasifinite) in the formalized language L that we shall call the *axioms of language J*.

In constructing a concrete language we may determine the function E by enumeration. The set A – also by enumeration (if it is finite) or by specifying the propositional schemata that belong to it. When establishing the relation C_0 we use the patterns in terms of which rules of inference are usually described.

The intuitive sense of the terminology that has been introduced will become visible when we formulate the necessary and sufficient condition for belonging to a language group speaking the language J.

One might also say, that we have in mind here the necessary and sufficient condition to *understand appropriately* the expressions of language J, though under the condition that we do not identify the understanding of expressions with a psychological process, but with a disposition rather to adequate use of these expressions. The formulation of the above mentioned condition requires the use of certain concepts from the field of pragmatics. The concept of *assertive attitude* is one of them. The assertive attitude is a relation possessing the character of a disposition that holds between a person and an expression that has the grammatical structure of a sentence. We distinguish three types of this attitude (to put it simply) that we symbolize respectively with the abbreviations: assert$^+$ (acceptance), assert$^-$ (rejection) and assert^0 (the attitude of indifference, being a complementary relation to the sum of the first two). We do not define the expressions: assert$^+$, assert$^-$, and assert^0, treating them as the primitive terms of pragmatics. We believe that finding certain criteria of empirical applicability for them would not be a difficult task. Similarly to what Ajdukiewicz says in the above quoted works, we believe that assuming an appropriate assertive attitude as regards some sentences (always, or under specified circumstances) is a necessary condition for proper understanding of expressions. Another necessary condition is the adequate use of terms that are usually called observational. It seems that a person using some intersubjective language is compelled to predicate a given observational term — let it be a one-place predicate O_i — about objects belonging to an established set p_i at the moment of perceiving them (under the circumstances of good observation); he refuses, on the other hand, to predicate O_i (under analogous circumstances) about observable objects that do not belong to the set p_i. In order to characterize the way observational terms are used we shall employ the phrase 'person X has the disposition to predicate a term λ about an object a', which we shall further note as: pred (X, λ, a). The object a should be an observable individual (when λ is an individual constant or a one-place predicate), or an ordered pair of such individuals (when λ is a two-place predicate). The following explanation in the form of a conditional definition provides a closer characteristics of the meaning of the expression pred (X, λ, a); we formulate it here in a loose way, treating the expression pred (X, λ, a) as a primitive term of pragmatics.

Explanation: If a person X has been presented an object a under the circumstances of good observation together with a question 'is this λ?' and the person X has decided this question, then: pred (X, λ, a) if and only if the person X has given a positive answer.

Let us establish that the variables α and β represent sentences, and the variable Z sets of sentences of the formalized language L. We assume that the following equivalence holds:

(D1)   *A person X belongs to the language group using the language J (in other words a person X understands the expressions of the language J appropriately) if and only if he fulfils the following three conditions:*

    (i) *for any α (if $\alpha \in A$, then X assert$^+$ α),*

    (ii) *for any Z and α {if $\langle Z, \alpha \rangle \in C_0$, then [if (for any $\beta \in Z$, X assert$^+$ β), then (X assert$^-$ α does not occur)]},*

    (iii) *for any λ and p {if $E(\lambda)=p$, then for any a [pred(X, λ, a) if and only if $a \in p$]}.*

(Condition (iii) has been formulated—as an example—only for the case, when λ is a one-place predicate. Hence the variable p represents sets of individuals and the variable a—the individuals). We consider the above equivalence to be the definition of the phrase appearing on its left side.

We shall now define the concepts of the *consequence in language J*, the *thesis of language J* and the *expanded consequence in language J*.

(D2)   *A sentence α is a consequence in the language J of the set of sentences Z (symbolically: $\alpha \in C_J(Z)$) if and only if there exists a finite sequence of sentences $\langle \beta_1, ..., \beta_n \rangle$ such that $\beta_n = \alpha$ and for any $k = 1, ..., n$, sentence β_k belongs to the set Z, or there exists the set K of sentences that form the sequence $\langle \beta_1, ..., \beta_n \rangle$ with indicators smaller than k and $\langle K, \beta_k \rangle \in C_0$.*

(D3)   *A sentence α is the thesis of the language J (symbolically: $\alpha \in T_J$) if and only if $\alpha \in C_J(A)$.*

(D4)   *A sentence α is an expanded consequence in the language J of the set of sentences Z (symbolically: $\alpha \in C_J^A(Z)$) if and only if $\alpha \in C_J(A \cup Z)$.*

The syntactic structure of the formalized language L belonging to the structure (J) may be, in principle, arbitrary. Since, however, we intend to speak about the models of the language L, it will be convenient to assume that L is a language of the first order logic. We also assume that the characteristics of the formalized language L contains the division of its expressions into logical and descriptive, the definition of tautology and of logical consequence.

We also impose some restrictions upon the language J treated as a whole. Namely, we require that the logical constants be understood in this language in the way adopted in classical logic. We assume that this takes place, when the language J fulfils the following two conditions:

(II.1)   *All the logical tautologies in the formalized language L are theses in the language J.*

(II.2)   *If a formal schema of some sentence in the formalized language L is not a tautological schema, then not all its substitution instances are theses of the language J.*

By adopting condition (II.2) we express our conviction that if—for instance—all substitution instances of the schema $(p \to q) \to (q \to p)$ were theses, then we wouldn't be able to say that the connective $\to$ is understood as the material implication, but differently (e.g. as material equivalence).

The next restriction is supposed to guarantee that the set of theses of language J is a system, i.e. fulfils the condition:

(II.3) $\mathrm{Cn}(T_J) \subset T_J$.[5]

We are going to demonstrate that to obtain this it is enough to adopt the condition:

(II.4) $\mathrm{Cn}(X) \subset C_J^A(X)$.

The proof that (II.4) implies (II.3) is as follows: It follows from the condition (II.4) and the definition (D4) that: $\alpha \in \mathrm{Cn}(X) \to \alpha \in C_J(A \cup X)$.

By substituting T_J for the variable X we obtain: $\alpha \in \mathrm{Cn}(T_J) \to \alpha$ $\in C_J(A \cup T_J)$. Since $A \cup T_J = T_J$ and $T_J = C_J(A)$, the implication holds: $\alpha \in \mathrm{Cn}(T_J) \to \alpha \in C_J(C_J(A))$. Since the consequence C_J fulfils the condition: $C_J(C_J(X)) \subset C_J(X)$ (which is implied by definition (D2)), and thus: $\alpha \in \mathrm{Cn}(T_J) \to \alpha \in T_J$. And hence (II.3) is implied.

III

We now turn to the discussion of the problems connected with the semantic characteristics of language understood, as in the previous section, as a structure of the type:

(J) $\langle L, C_0, E, A \rangle$.

This characteristics is secondary with respect to the pragmatic characteristics in the sense that the semantic properties of the language J are determined by the previously established pragmatic properties as well as by some general assumptions concerning the relations between those two types of properties. Since from now on we are going to speak about the models of the formalized language L we shall have to characterize its syntactic properties in more detail. We have already agreed that L is the language of the first order logic. The syntax of this language is described in many textbooks of logic and hence the definition of a well formed formula is not necessary here. Let us, then, merely establish that the only descriptive terms of the formalized language L are (one- or many-place) predicates

$$P_1, \ldots, P_n.$$

Let us additionally assume — as it is usual in the discussions of languages of empirical theories — that the predicates constitute two separate sequences

$$\langle O_1, \ldots, O_l \rangle \quad \text{and} \quad \langle T_1, \ldots, T_m \rangle.$$

We shall call the elements of the first sequence O-terms, and of the second sequence T-terms. Let L_0 denote the formalized language that we obtain from the language L after all T-terms, and none others are deleted from its vocabulary. Let us establish that variables $\mathfrak{M}, \mathfrak{M}', \mathfrak{M}'', \ldots$

range over the class of models of language L, whereas variables $\mathfrak{M}_0$, $\mathfrak{M}_0'$, $\mathfrak{M}_0''$, ... represent models of language L_0.

Models of language L are structures of the type:

$$(\mathfrak{M}) \qquad \langle U, x_1, \ldots, x_l; y_1, \ldots, y_m \rangle,$$

whereas models of language L_0 are structures of the type:

$$(\mathfrak{M}_0) \qquad \langle U, x_1, \ldots, x_l \rangle.$$

Object U is a non-empty set, called the *universe of a model*. Objects $x_1, \ldots, x_l$ as well as $y_1, \ldots, y_m$ are relations defined in the set U and they form the so-called characteristics of a model. Exactly one relation corresponds to each descriptive term of the formalized language L (or L_0) in its model. If the term is an n-place predicate, then the relation corresponding to it is an n-ary one. The relation that is correlated to the descriptive term in a given model is called the *value* (or *denotation*) of this term in this model. We shall use the functional symbol Val with an index indicating model to symbolize the value of the term in an appropriate model, for instance: $\mathrm{Val}_{\mathfrak{M}}(T_i)$, $\mathrm{Val}_{\mathfrak{M}_0}(O_i)$, etc. The set of sentences of the formalized language L that are true in model $\mathfrak{M}$ will be denoted briefly by the symbol Ver $(\mathfrak{M})$. Similarly, Ver $(\mathfrak{M}_0)$ denotes the set of sentences of language L_0 true in model $\mathfrak{M}_0$.

Let us now assume that the function E, which is a member of the structure (J), is defined on O-terms, i.e. predicates

$$O_1, \ldots, O_l$$

and it univocally correlates to them relations

$$p_1, \ldots, p_l$$

according to the syntactic category of the respective predicates. As has been said above, these relations hold exclusively among observable objects and a person competent in language J is compelled — under specified circumstances — to apply the predicate O_i to perceived objects, if the relation p_i holds among them. Thus the O-terms may really be treated as observational in a certain meaning of the word. On the other hand, the T-terms, which do not have such a direct connection with experience, may be ascribed the character of theoretical terms.

We may assume that function E distinguishes one of the models of the formalized language L_0. It is the model

$(\mathfrak{M}_0^E)$ $\langle \varDelta, p_1, \ldots, p_l \rangle$.

The elements of the characteristics of the model $\mathfrak{M}_0^E$ are identical with values of function E for appropriate predicates. On the other hand, its universe $\varDelta$ may be considered to be the class of all observable objects, or its subclass which has been established in some way (e.g. the sum of the fields of relations $p_1, \ldots, p_l$).

Our approach to language is founded on the conviction that while establishing the members of the system

(J) $\langle L, C_0, E, A \rangle$

and postulating the principles of using the expressions of the formalized language L that have been indicated in definition (D1) we indirectly establish interpretations of these expressions, i.e. we assign the objects of which people using the language J speak to the expressions. Since the choice of interpretation of the expressions of a formalized language may be reduced to distinguishing certain model of this language, the following question may be posed: which of the models of the formalized language L is the *proper model* of the language J, i.e. the model, about which the users of the language J speak? We shall, from now on, denote the proper model of the language J by the symbol M^*.

We believe that this model should fulfil the following conditions: The theses of the language J should be — if it is possible — sentences true in the model M^*, i.e. inclusion should hold:

(III.1) $T_J \subset \mathrm{Ver}(\mathfrak{M}^*)$.

Moreover — since function E determines in some way the problem of interpreting the O-terms by establishing about which observable objects these terms are to be predicated — this fact should be taken into consideration in the choice of the proper model $\mathfrak{M}^*$. This is expressed in the condition:

(III.2)   *for every* $i = 1, \ldots, l$, *if* $x_i = \mathrm{Val}_{M^*}(O_i)$, *then* $x_i \mid \varDelta = E(O_i)$. [6]

Postulates (III.1) and (III.2) exhaust the list of conditions characterizing the proper model of language J, which — to my mind — are indis-

pensable from the intuitive point of view. It is clear that they do not characterize the model $\mathfrak{M}^*$ univocally. In particular, they do not even characterize univocally its universe. It will be advantageous for our discussion if we make them stronger in some way. Before we deal with this, however, we shall introduce an auxiliary concept of prolongation of a model. It will be defined thus:

(D5)   The model $\mathfrak{M} = \langle U, x_1, ..., x_l; y_1, ..., y_m \rangle$ is a *prolongation of the model* $\mathfrak{M}_0 = \langle U'_1, x'_1, ..., x'_l \rangle$ (symbolically $\mathfrak{M}$ prol $\mathfrak{M}_0$) if and only if

(i) $U = U'$,

(ii) for every $i = 1, ..., l$, $x_i = x'_i$.

We shall characterize the proper model of the language J by the axiom:

(A1) $\mathfrak{M}^*$ prol $\mathfrak{M}_0^E$ and $T_J \subset \text{Ver}(\mathfrak{M}^*)$.

We treat this axiom as one of the possible general assumptions establishing relations between the pragmatic and semantic properties of language. It is easy to note that it implies postulates (III.1) and (III.2), but at the same time it does not characterize the model $\mathfrak{M}^*$ univocally, except in the case when the set of sentences T_J is consistent and the T-terms are semantically definable with the help of O-terms on the grounds of the sentences T_J. Otherwise language J is semantically open, since the interpretations (denotations) of T-terms are established in it ambiguously.

We must, however, make the reservation here that by adopting axiom (A1) as the characteristics of the proper model we have made a simplification, as a result of which the language J cannot be treated as an adequate reconstruction of languages of such empirical theories, in which the properties of the observable objects are explained in terms of hypotheses concerning the unobservable constituents of these objects (atoms, elementary particles, etc.). Since — according to the axiom (A1) — the universe of the model $\mathfrak{M}^*$ is identical with the set $\varDelta$, which has been assumed to be a set of observable objects, the above mentioned hypotheses could not be formulated in the language J.

Using the concept of the proper model of the language J we may define the concept of a sentence true in this language.

(D6)   The sentence α is a *true sentence in the language* J if and
only if $\alpha \in \mathrm{Ver}\,(\mathfrak{M}^*)$.

The above way of defining the 'absolute' concept of truth with the help
of the 'relative' concept of truth in the model is one of the five possible
solutions of this problem that have been presented in the work of Przełęcki [23]. In contrast to the others, the above solution allows the preservation of the two-valued logic together with the classical metalogical principles. The following circumstance is noteworthy here, however. The model $\mathfrak{M}^*$ characterized by the axiom (A1) may not exist. It
does not exist not only when the set of theses of the language J is contradictory, but also—for instance—when among the theses of the language J there are sentences formulated exclusively with the help of
logical constants and O-terms, false in the model $\mathfrak{M}_0^E$. In this situation
the concept of the proper model of the language J, as well as of the
sentence true in this language and other semantic concepts become
pointless. We would then tend to say that the language J is not a *semantic
system*. We use the term 'semantic system' here in a different sense than
that known from Carnap's works. The following definition explains this
sense:

(D7)   The language J is a *semantic system* if and only if such a
model $\mathfrak{M}$ exists that $\mathfrak{M}$ prol $\mathfrak{M}_0^E$ and $T_J \subset \mathrm{Ver}\,(\mathfrak{M})$.

IV

The concept of language presented above suggests that sentences which
are theses of language should be considered analytic. The concept of an
analytic sentence thus understood would coincide with the one Ajdukiewicz uses in his work [3]. However, since it is generally believed that
analytic sentences in a language are always sentences true in this language, and the concept of truth (as understood according to the definition (D6)) has no application to languages which are not semantic
systems, the definition of the concept of analyticity should be as follows:

(DA1)   Sentence α is an *analytic sentence in the language* J if and
only if language J is a semantic system and α is a thesis in
the language J.

According to the above definition there are no analytic sentences in languages that are not semantic systems. But the implication holds:

(IV.1)   If α is an analytic sentence in the language J, then α is a true sentence in this language.

Since—according to the adopted assumption (II.1) (which is supposed to guarantee that the logical constants are understood in the classical way in the language J)—all the logical tautologies are theses in language J, then:

(IV.2)   If language $J = \langle L, C_0, E, A \rangle$ is a semantic system, then all the logical tautologies in the formalized language L are analytical sentences in language J.

Moreover, for languages that fulfil the condition

(II.4) $Cn(X) \subset C_J^A(X)$,

from which—as has been proved—follows that

(II.3) $Cn(T_J) \subset T_J$,

the following proposition is valid:

(IV.3)   The logical consequences of any class of analytic sentences are analytic sentences.

Definition (DA1) expresses the intuition that makes us identify analytic sentences with definitional theorems. Here the definitional character of the theses of language consists in the fact that together with function E they establish (although, as a rule, not univocally) the proper model of a language, and thus—similarly to definitions—they establish the semantic relations between the expressions of language and objects this language refers to. On the other hand, the definition (DA1) does not agree with the intuitions involved in the current definition of analytic sentences as 'sentences true on the basis of meaning of expressions', if the above quoted expression is supposed to denote sentences that are true exclusively on the basis of decisions establishing the meanings of expressions. Having in mind language as understood in this work, i.e. a structure of the type

(J) $\langle L, C_0, E, A \rangle$

it might be said that the meanings of expressions were determined with the establishment of the language elements and the principles of using the expressions to which definition (D1) applies. However, not every decision establishing the contents of the relations E, C_0 and the set A guarantees truth of the theses of language, for not every such decision constitutes a language which is a semantic system. As it is known, whether a certain language is a semantic system or not depends in some cases on — speaking figuratively — the states of affairs in the model $\mathfrak{M}_0^E$. Then the efficacy of our decisions, if their aim is to guarantee truth of the theses of language, depends on our knowledge (empirical!) about the relations that hold in the model $\mathfrak{M}_0^E$.

V

The semantic characteristics of the language J that has been presented in the Section III may — from some point of view — seem to be unsatisfactory. This brings forth certain suggestions for its modification. These, in turn, allow to define the concept of analytic sentence in the way different from above.

The consequence of the previously adopted statements consisting in that in some cases the proper model of a language may not exist and then the concepts of true sentence, relation of denoting and other semantic concepts relativized to this language are pointless, seems to be undesirable from the logical point of view.

The semantic characteristics of language J that has been outlined in Section III allows to define — it seems — the relation of denoting in language J only in the following way:

(D8)   *The predicate P_i denotes in language J an object x if and only if $x = \mathrm{Val}_{\mathfrak{M}^*}(P_i)$.*

(Note that the constant symbol $\mathfrak{M}^*$ denotes here a proper model of the language J.) But — as it has been stated before — a proper model of the language J may not exist. Thus there is an alternative: either all descriptive terms in language J denote certain objects (namely, those which are their values in a proper model), or — when a proper model does not exist — the concept of denoting in language J is pointless and

then it is impossible to state about any of the descriptive terms of this language that it denotes a specified object.

This alternative is bound to arise objection. Having natural language while others denote none. The distinction between the denoting and the while others denote none. The distinction between the denoting and the non-denoting terms is based on the fact that some sentences of the language under discussion are considered to be the definitions of respective terms. The aim of these definitions is to establish the denotation of the defined terms by imposing upon them certain conditions. In the case when no object fulfils conditions contained in a definition of a term we say that this term does not denote anything. The 'definitions' we have in mind here do not have to be definitions in the strict meaning of the word. They do not have to fulfil the conditions of non-creativity, translatability or uniqueness. They may also jointly define several terms. Such sentences are usually called meaning postulates, or shortly, *postulates*, if we treat the concept of postulate as a generalization of the concept of definition. The typical situation, in which we usually speak of postulates, is connected with the passing from a certain language J to a language richer in terms. A postulate is then a sentence containing terms that are new with regard to J. We say that it *is the postulate for* these new terms that it contains, or that it *concerns* those terms. Naturally, such postulate may also contain certain 'old' terms and — what is more — if we do not require of it non-creativity it may impose some conditions upon their denotations, i.e. restrict the range of their possible interpretations. From the pragmatic point of view postulates are *axioms* of language, and thus, while speaking a given language, we are compelled to have the positive assertive attitude towards them.

In presenting the first modification of the previous approach to the semantic problems connected with the adopted concept of language we shall use the concept of *semantic meaningfulness*. The intuitions that we bind with this concept tell us not to include a descriptive term among the semantically meaningful ones if the postulate which concerns this term imposes such conditions upon its possible denotation that are not fulfilled by any object. Moreover, the terms introduced ('defined') with the help of such terms cannot also be considered semantically meaningful. It seems that thus understood concept of semantic meaningfulness may be useful in the study of the semantics of languages approximating

natural ones as well as the concept of syntactic meaningfulness (i.e. the concept of a well formed formula) and the concept of pragmatic meaningfulness (i.e. the concept of an expression whose use is adequately regulated in a given language). For the three above mentioned concepts may come to be different in scope.

We shall formulate the precise definition of semantic meaningfulness with regard to language understood, as in provious sections, as a structure

(J) $\langle L, C_0, E, A \rangle$.

We shall, however, introduce certain modifications into its characteristics. Namely, we shall assume—for simplicity's sake—that the relation of consequence in the language J is identical with the classical logical consequence, i.e. that for any set X of sentences in the formalized language L the equation holds:

(V.1) $Cn(X) = C_J(X)$.

In this state of affairs it may be assumed that the set of axioms A of the language J is finite. We shall from now on treat this set as an *ordered* one, and its particular axioms as the postulates for respective terms. We establish: that: 1) for each term there exists only one postulate which concerns this term, 2) the postulates concern only T-terms, although they may also contain O-terms, 3) each postulate concerns exactly one descriptive term. We should like to emphasise that the last assumption does not play any important role in our discussion, but it promotes the simplification of certain formulations. As it has already been mentioned, we assume a certain order among the postulates. This order should observe the principle: if a postulate with an index k contains a term T_i, then either this postulate concerns the term T_i, or the term T_i occurs in postulates with indexes smaller than k. This principle does not always determine only one ordering. Thus certain factors historical or utilitarian in nature may gain influence on the choice of a defined ordering of postulates.

The symbol A, previously denoting the (unordered) set of axioms of language will now denote a sequence of postulates:

(A) $\langle \pi_1, \ldots, \pi_m \rangle$.

Let us assume that these postulates concern subsequently the terms:

$$T_1, \ldots, T_m.$$

We may now give the definition of a semantically meaningful term the following inductive form:

(D9) 1. O-terms are *semantically meaningful terms in the language J*.

2. If K is a set of all and only those sentences from among the elements of the sequence A with indexes smaller than i, which contain exclusively semantically meaningful terms in the language J, then T_i is a *semantically meaningful term in the language J* if and only if

(i) for any $k < i$: if T_k occurs in π_i, then T_k is a semantically meaningful term in the language J,

(ii) there exists such model $\mathfrak{M}$ that $\mathfrak{M}$ prol $\mathfrak{M}_0^E$ and $K \cup \{\pi_i\} \subset \mathrm{Ver}(\mathfrak{M})$.

The notion of a semantically meaningful sentence in language J is defined as follows:

(D10) Sentence α is a *semantically meaningful sentence in the language J* if and only if it contains (apart from the logical constants) merely semantically meaningful descriptive terms in the language J.

It may be proved that:

(V.2) *Language J is a semantic system (according to the definition (D7)) if and only if all the descriptive terms in the language J are semantically meaningful terms in this language.*

This theorem is a consequence of the definitions (D7), (D9), (D10) and the equation (V.1).

We may define the concepts of truth, denotation and analytic sentence in the way presented above (cf. the definitions (D6), (D8) and (DA1)) for a language which is a semantic system. However, the characteristics of language that we have now adopted allows the formulation of the

definitions of these concepts also in the case of a language, which is not a semantic system. In a language like this — according to the theorem (V.2) — not all descriptive terms are semantically meaningful. The intuitive arguments that we have applied in constructing the definition of semantic meaningfulness make us treat the semantically meaningless terms as lacking denotation (not denoting any object). The definitions of the concept of denoting as well as truth and felsehood that we shall formulate below are a manifestation of such an attitude.

Let us first note, however, that for each language it is possible to indicate univocally such its fragment that is a semantic system. We shall call the *fragment of a given language*, which is a structure

(J) $\langle L, C_0, E, A \rangle$

the structure

(J*) $\langle L^*, C_0^*, E^*, A^* \rangle$

if: 1) the formalized language L^* can be obtained from the langauge L by deleting from its vocabulary at most some of its descriptive terms, 2) E^* is the relation E of the domain restricted to the terms of the formalized language L^*, 3) C_0^* is the relation C_0 bilaterally restricted to the set of sentences of the formalized language L^*, 4) A^* is the sequence A restricted to the set of sentences of the formalized language L^*.

Let $S(J)$ be this fragment of the language J that is formed by deleting from its vocabulary all those terms, and only those, which do not fulfil the condition of semantic meaningfulness. In view of the theorem (V.2) it is clear that

(V.3) $S(J)$ *is a semantic system.*

Thus the concepts of truth and denotation may be defined for language $S(J)$ in the way that has been presented in the previous sections. The above mentioned concepts may be transferred onto the language J, with the reservation however, that no denotations will be ascribed to the semantically meaningless terms in language J and neither of the two classical logical values will be ascribed to the semantically meaningless sentences in language J. This may be obtained by adopting the following definitions:

(D11)   *Predicate P_i denotes in the language J an object x* if and only if

    (i) P_i is a semantically meaningful term in the language J,

    (ii) P_i denotes in the language $S(J)$ an object x.

(D12)   Sentence α is a *true sentence in the language J* if and only if

    (i) α is a semantically meaningful sentence in the language J,

    (ii) α is a sentence true in the language $S(J)$.

(D13)   Sentence α is a *false sentence in the language J* if and only if

    (i) α is a semantically meaningful sentence in the language J,

    (ii) α is a false sentence in the language $S(J)$.

It is easy to notice that in the case when not all the terms of the language J are semantically meaningful and, hence, languages J and $S(J)$ are not identical, there exist sentences in the language J which, according to the above definitions, are neither true nor false in this language. Thus the principles of two-valued logic is not fulfilled in such a language. This consequence probably reflects the state of affairs in certain natural languages. On the other hand, another consequence of the above mentioned definitions seems to be undesirable: in the case, when $J \neq S(J)$, not all logical consequences of sentences true in language J are true sentences in this language (they may be sentences devoid of logical value).

On the basis of the same intuitions that accompanied the definition (DA1) of the concept of analytic sentence it would have to be now defined as follows:

(DA2)   The sentence α is an *analytic sentence of the language J* if and only if α is a thesis of the language $S(J)$.

Since theses of the language $S(J)$ are true sentences in this language and – according to the definition (D12) – they are also true sentences in language J, then if the definition (DA2) is assumed the theorem (IV.1) is also satisfied. On the other hand the theorem (IV.3) holds only for those languages which are semantic systems, i.e. only when $J = S(J)$.

In the remaining cases not all logical consequences of analytic sentences are analytic sentences. As a result the set of analytic sentences, similarly to the set of true sentences, is not always a system.

VI

Both of the above mentioned consequences may be avoided if we establish the relation between language and its proper model in a different way. So far, a proper model of the language J has been characterized as a certain distinguished member of the family of models that included models which, in place of the symbol $\mathfrak{M}^*$, satisfied the axiom

(A1) $\mathfrak{M}^* \operatorname{prol} \mathfrak{M}_0^E$ and $T_J \subset \operatorname{Ver}(\mathfrak{M}^*)$.

This family could, in some cases, turn out empty and then the concept of the proper model of a language would become pointless. The modification that we now introduce consists in characterizing the proper model of language J as an element of such a family of models, whose non-emptiness would be guaranteed independently of the relations existing in the model $\mathfrak{M}_0^E$ and the properties of the set of theses of language J.

The axiom (A1) required — among other things — that all the theses of the language J be sentences true in its proper model $\mathfrak{M}^*$. This condition will now be substituted with a weaker one, namely

(VI.1) $\mathfrak{S} \subset \operatorname{Ver}(\mathfrak{M}^*)$.

The symbol $\mathfrak{S}$ denotes here the set of all those sentences, and only those, from among the postulates $\pi_1, \ldots, \pi_m$, which — according to the definition (D10) — are semantically meaningful sentences in the language J. The set of sentences $\mathfrak{S}$ may also be defined inductively, with no reference to the concept of semantic meaningfulness, in the following way:

(D14)   If X is a set of all those sentences, and only those, from among the expressions of the sequence $A = \langle \pi_1, \ldots, \pi_m \rangle$ with indexes smaller than i, which belong to $\mathfrak{S}$, then π_i *belongs to* $\mathfrak{S}$ if and only if:

(i) for any $k < i$: if T_k occurs in π_i, then π_k belongs to $\mathfrak{S}$,

(ii) there exists such a model $\mathfrak{M}$ that $\mathfrak{M} \operatorname{prol} \mathfrak{M}_0^E$ and $K \cup \{\pi_i\} \subset \operatorname{Ver}(\mathfrak{M})$.

The equivalence of both the above mentioned definitions of the set $\mathfrak{S}$ may be proved inductively on this basis of the definitions (D9), (D10) and the assumptions concerning the sequence of postulates.

According to the remarks made above we shall now characterize the proper model $\mathfrak{M}^*$ of language J by means of the axiom:

(A2) $\mathfrak{M}^* \operatorname{prol} \mathfrak{M}_0^E$ and $\mathfrak{S} \subset \operatorname{Ver}(\mathfrak{M}^*)$.

The existence of a model fulfilling the above conditions results directly from the definition (D14). Thus, it is guaranteed for any language J. As a result, definitions of the concepts of truth and denotation equiform to the definitions (D6) and (D8) can be applied to any language J. In this way we avoid the consequence consisting in that the set of sentences true in the language J is not a system. It is also possible to demonstrate that all the classical metalogical principles are now valid for any language J. In contrast to the approach outlined before it is now possible to state about each descriptive term of the language J that it denotes something. In the light of these remarks it seems that it is now possible to say of every language J that it is—in a certain sense of the word—a semantic system. This does not mean, however, that every language J is a rational construction. Languages in which the sets of sentences $\mathfrak{S}$ are not coextensive with the sets of all the postulates cannot be considered rational for two reasons. First: If a certain postulate π_i does not belong to the set $\mathfrak{S}$, then, according to the definition (D14), the term T_i does not occur in the sentences $\mathfrak{S}$. The denotation of such terms remains nearly completely undefined. We only know about it that it is a relation (of a known arity) holding between elements of the set $\varDelta$, which is the universe of the model $\mathfrak{M}_0^E$. Second: if not all the postulates belong to the set $\mathfrak{S}$, then—as can be proved—there is among them such a postulate that is false in the proper model $\mathfrak{M}^*$. Since—as has been already mentioned—the postulates are at the same time axioms of the language, we are, using language J, compelled to accept them. The result is that—in the situation when not all the postulates belong to the set $\mathfrak{S}$—we are compelled to accept certain false sentences in this language, when we speak the language J.

We should like to preserve now the name of analytic sentences also for theorems of definitional character. It is clear that—according to the axiom (A2)—this character is possessed by the sentences of $\mathfrak{S}$,

or the sentences of $Cn(\mathfrak{S})$, because the condition $\mathfrak{S} \in Ver(\mathfrak{M}^*)$ is equivalent to the condition $Cn(\mathfrak{S}) \subset Ver(\mathfrak{M}^*)$. Since we have adopted the assumption:

(V.1) $Cn(X) = C_J(X),$

the concept of analytic sentence may be defined as follows:

(DA3)   The sentence α is an *analytic sentence in the language J* if and only if α is a consequence in the language J of the set of sentences $\mathfrak{S}$ (symbolically: $\alpha \in C_J(\mathfrak{S})$).

Thus characterized set of analytic sentences possesses the following properties: 1) it consists exclusively of sentences true in the language J, 2) it contains all the logical tautologies of the language J, 3) it is closed under the relation of logical consequence.

VII

If we assume that between the language J and its proper model $\mathfrak{M}^*$ the relation determined by the axiom

(A2) $\mathfrak{M}^* \operatorname{prol} \mathfrak{M}_0^E$ and $\mathfrak{S} \in Ver(\mathfrak{M}^*)$

occurs, then it is possible to define the concept of analyticity narrower in scope than that established by the definition (DA3). It is possible to maintain about analytic sentences in this narrower sense that they are, in a sense, independent of the data from experience. Before we explain the nature of this independency note that whether a certain postulate π_i belongs to the set of sentences $\mathfrak{S}$ is determined — in some cases — by experience. Let us consider the following situation. Let K be the set of those postulates with indices lower than i, which belong to $\mathfrak{S}$. Moreover, let the postulate π_i fulfil the condition (i) of the definition (D14), i.e. it will contain no term $T_j (j < i)$, which does not occur in sentences K. In this situation — according to the condition (ii) of the definition (D14) — the postulate π_i does not belong to the set $\mathfrak{S}$ if and only if there does not exist such a model $\mathfrak{M}$ that $\mathfrak{M} \operatorname{prol} \mathfrak{M}_0^E$ and $K \cup \{\pi_i\} \subset Ver(\mathfrak{M})$. Such a model does not exist, among others cases, if certain sentences in the formalized language L_0, which are the consequences of the sentences $K \cup \{\pi_i\}$, are false in the model $\mathfrak{M}_0^E$, and thus also in its every prolongation. But only experience may solve the problem whether

certain non-tuatological sentences in language L_0 are true or false in the model $\mathfrak{M}_0^E$. Thus experience sometimes decides whether certain postulates belong to the set $\mathfrak{S}$, and also — indirectly — whether certain sentences are analytic in the language J. For it is easy do demonstrate that in the above presented situation the postulate π_i belongs neither to the set $\mathfrak{S}$, nor to the set $\mathrm{Cn}(\mathfrak{S})$. Moreover, it is a false sentence in the language J, because it is a false sentence in the model $\mathfrak{M}^*$.[8]

Thus it is clear that though — generally speaking — the analytic sentences according to the definition (DA3) are always true sentences, the properties of the model $\mathfrak{M}_0^E$ discernible by experience in some cases determine whether a given sentence is an analytic one and, indirectly, whether it is a true sentence. Hence only those sentences, which are analytic and thus true irrespective of *what the model $\mathfrak{M}_0^E$ is like*, may be considered as independent of experience. No results obtained in the course of empirical investigations of this model could question such sentences.

In order to express this idea precisely let us consider the infinite sequence of models of the formalized language L_0:

$$\mathfrak{M}_{0_1}, \ldots, \mathfrak{M}_{0_n}, \ldots$$

and the sequence of classes of sentences:

$$\mathfrak{S}_{\mathfrak{M}_{0_1}}, \mathfrak{S}_{\mathfrak{M}_{0_n}}, \ldots$$

which are connected with appropriate models in the same way as the set of sentences $\mathfrak{S}$ is — according to the definition (D14) — connected with the model $\mathfrak{M}_0^E$. Instead of speaking, as above, of 'sentences, which are analytic irrespective of *what model $\mathfrak{M}_0^E$ is like*', we may now use a more precise formulation: sentences, which remain analytic irrespective of *which* element of the sequence of models under discussion is identical with $\mathfrak{M}_0^E$. It is clear that a set of sentences that can be defined as below corresponds to such characteristics:

$$\mathfrak{S}^* = \bigcap_{i=1}^{\infty} \mathrm{Cn}(\mathfrak{S}_{\mathfrak{M}_{0_i}}),$$

or simply:

(DA4) $\alpha \in \mathfrak{S}^*$ *if and only if for any model* $\mathfrak{M}_0$, $\alpha \in \mathrm{Cn}(\mathfrak{S}_{\mathfrak{M}_0})$.

Sentences which belong to the set $\mathfrak{S}^*$ will be called *strictly analytic* sentences of language J.

Since for some i, $\mathfrak{M}_{0_i} = \mathfrak{M}_0^E$ and $\mathfrak{S}_{\mathfrak{M}_{0_i}} = \mathfrak{S}$, and the set $\mathrm{Cn}(\mathfrak{S})$ is a set of analytic sentences according to the definition (DA3), an obvious inclusion holds:

(VII.1) $\mathfrak{S}^* \subset \mathrm{Cn}(\mathfrak{S})$.

Hence strictly analytic sentences are a subset of the set of analytic sentences.

We now prove that the set of strictly analytic sentences is closed under the relation of logical consequence, i.e. that the inclusion holds:

(VII.2) $\mathrm{Cn}(\mathfrak{S}^*) \subset \mathfrak{S}^*$.

It follows from the definition (DA4) that: for every $\mathfrak{M}_0$, $\mathfrak{S}^* \subset \mathrm{Cn}(\mathfrak{S}_{\mathfrak{M}_0})$. Utilizing the theorem: If $Y \subset \mathrm{Cn}(X)$, then $\mathrm{Cn}(Y) \subset \mathrm{Cn}(X)$,[9] we obtain: For any $\mathfrak{M}_0$, $\mathrm{Cn}(\mathfrak{S}^*) \subset \mathrm{Cn}(\mathfrak{S}'_{\mathfrak{M}_0})$, and hence the theorem: If $\alpha \in \mathrm{Cn}(\mathfrak{S}^*)$, then for any $\mathfrak{M}_0$, $\alpha \in \mathrm{Cn}(\mathfrak{S}_{\mathfrak{M}_0})$, which—according to the definition (DA4)—is equivalent to the theorem (VII.2).

The direct conclusion from the proposition (VII.2) is the theorem:

(VII.3)   *All logical tautologies in language J are strictly analytic sentences in language J.*

VIII

We intend to present in this section the solution of a certain formal problem, connected with the concept of strict analyticity. We have called *strictly analytic* the sentences which are consequences of any set of postulates Z that fulfil the conditions:

(VIII.1)   *There exists a model $\mathfrak{M}_0$ such that $Z = \mathfrak{S}_{\mathfrak{M}_0}$.*

In order to make it possible to determine whether any sentence in the formalized language L is strictly analytic in this language it is necessary that the family of sets fulfilling the condition (VIII.1) is effectively given.[10] We should like to present here a certain method of constructing such a family of sets. The method consists in utilizing the logical dependencies between *Ramsey's counterparts* of the sentences of the formalized

language L. According to our previous assumption L is the language of the first order logic that does not contain variables of predicate type. Ramsey's counterparts of the sentences in the language L are not sentences in this language, but in the formalized language L_R. Language L_R will be obtained by deleting from the vocabulary of language L all the T-terms

$$T_1, ..., T_m$$

and introducing in their place the predicate variables

$$X_1, ..., X_m.$$

We assume that for any $i = 1, ..., m$, the variable X_i is of the same syntactic category as the predicate T_i. The variables $X_1, ..., X_m$ may be bound by quantifiers in the language L_R.

Ramsey's counterpart of sentence α of the formalized language L containing the T-terms $T_{k_1}, ..., T_{k_r}$ will be the expression

$$\bigvee_{X_{k_1}} ... \bigvee_{X_{k_r}} \alpha(T_{k_1}/X_{k_1}, ..., T_{k_r}/X_{k_r}),$$

i.e. the expression obtained from sentence α by substituting the variables $X_{k_1}, ..., X_{k_r}$ for the terms $T_{k_1}, ..., T_{k_r}$, respectively, and by binding those variables by existential quantifiers. The counterpart of sentence α will be denoted by symbol $R(\alpha)$. In the case when Z is a finite set of sentences we shall write $R(Z)$ assuming that this symbol denotes Ramsey's counterpart of the *conjuntion* of sentences of which the set Z consists. We shall consider the conjunction of members of an empty set of sentences identical with some arbitrarily chosen tautology in language L.

We shall call *models* of the formalized language L_R structures of the type

$$(\mathfrak{M}_0) \qquad \langle U, x_1, ..., x_1 \rangle,$$

i.e. structures identical with models of the formalized language L_0. We shall denote the set of sentences in language L_R true in the model $\mathfrak{M}_0$ by $\mathrm{Ver}_{L_R}(\mathfrak{M}_0)$. We shall not give the full definition of this set except the remark that only the relations *defined in the set* U, which is the universe of the model $\mathfrak{M}_0$, are values of the predicate variables $X_1, ..., X_m$

in this model.[11] We also omit proofs of the following two lemmas, which are consequences of the above mentioned definition.

LEMMA 1. $R(Z) \in \mathrm{Ver}_{L_R}(\mathfrak{M}_0)$ *if and only if there exists a model* $\mathfrak{M}$ *such that* $\mathfrak{M}$ prol $\mathfrak{M}_0$ *and* $Z \subset \mathrm{Ver}(\mathfrak{M})$.

LEMMA 2. *If* $Z \subset Z'$, *then for any model* $\mathfrak{M}_0$, *if* $R(Z') \in \mathrm{Ver}_{L_R}(\mathfrak{M}_0)$, *then* $R(Z) \in \mathrm{Ver}_{L_R}(\mathfrak{M}_0)$.

The immediate consequence of the above adopted definition of the symbol $R(Z)$ is the following lemma:

LEMMA 3. *If* $Z = Z'$, *then the expresion* $R(Z)$ *is identical with the expression* $R(Z')$.

(In the lemmas given above the variables Z and Z' represent sets of the formalized language L, the variable $\mathfrak{M}$ — as previously — the models of this language, and the variable $\mathfrak{M}_0$ the models of the formalized language L_R.)

We are now going to introduce several auxiliary symbols and to formulate the definition of the set of sentences $\mathfrak{S}_{\mathfrak{M}_0}$ determined by the variable parameter $\mathfrak{M}_0$.

(Def. 1) Symbol $*i$ $(i = 1, \ldots, m, m+1)$ denotes the set of all the postulates in the language J with indices smaller than i.

(According to this definition *1 denotes an empty set of postulates, and *$m+1$ the set of all the postulates in the language J.)

(Def. 2) Let Z be an arbitrary set of postulates in the language J and let the variable i range over the set of positive integers from 1 to $m+1$. The symbol $\mathfrak{B}(Z, i)$ then denotes the set of all and only such postulates π_k, which fulfil the following three conditions:
(a) $k < i$,
(b) $\pi_k \bar{\in} Z$,
(c) for every $s < k$, if T_s occurs in π_k, then $\pi_s \in Z$.[12]

(Def. 3) Symbol $\tilde{\mathfrak{R}}_{Z,i}$ denotes the expression of the formalized language L_R which is the conjunction of all and only such sentences α that fulfil the condition: There is a natural

number $k = 1, \ldots, m$ such that $\pi_k \in \mathfrak{B}(Z, i)$ and α is identical with the expression $\sim R(*k \cap Z \cup \{\pi_k\})$. If the set of sentences α fulfilling the above condition, is an empty one, then the symbol $\widetilde{\mathfrak{R}}_{Z,i}$ denotes an arbitrary chosen tautology in the language L_R.

The inductive definition of the set $\mathfrak{S}_{\mathfrak{M}_0}$ may now be given the following form:

(D15) $\pi_i \in \mathfrak{S}_{\mathfrak{M}_0}$ if and only if:

(i) for every $k < i$, if T_k occurs in π_i, then $\pi_k \in \mathfrak{S}_{\mathfrak{M}_0}$,

(ii) there exists a model $\mathfrak{M}$ such that $\mathfrak{M}$ prol $\mathfrak{M}_0$ and $*i \cap \mathfrak{S}_{\mathfrak{M}_0} \cup \{\pi_i\} \subset \mathrm{Ver}(\mathfrak{M})$.

The aim of the present section is, as mentioned earlier, to present a certain method of constructing the family of sets Z that fulfil the condition:

(VIII.1) *There exists a model $\mathfrak{M}_0$ such that $Z = \mathfrak{S}_{\mathfrak{M}_0}$.*

Let us denote this family of sets by the symbol $\mathfrak{X}$. In order to obtain this family we construct subsequently elements of the following sequence:

$$\mathfrak{X}_1, \ldots, \mathfrak{X}_m, \mathfrak{X}_{m+1}.$$

The elements of this sequence are families of sets defined as follows: $K \in \mathfrak{X}_i$ if and only if there exists a set Z of postulates such that $Z \in \mathfrak{X}$ and $K = *i \cap Z$; or in an equivalent way:

(Def. 4) $K \in \mathfrak{X}_i$ if and only if there exists a model $\mathfrak{M}_0$ such that $K = *i \cap \mathfrak{S}_{\mathfrak{M}_0}$.

(According to the definition $\mathfrak{X}_1 = \{\Lambda\}$ but $\mathfrak{X}_{m+1} = \mathfrak{X}$.)

It is easy to prove that the following theorems are consequences of (Def. 4):

THEOREM 1. *If $K \in \mathfrak{X}_i$, then $K \in \mathfrak{X}_{i+1}$ or $K \cup \{\pi_i\} \in \mathfrak{X}_{i+1}$.*

THEOREM 2. *If $K' \in \mathfrak{X}_{i+1}$, then there exists a set K such that $K \in \mathfrak{X}_i$ and $K' = K \cup \{\pi_i\}$ or $K' = K$.*

The above theorems indicate that when the family of sets $\mathfrak{X}_i$ is given, then it is possible to establish the content of the family of sets $\mathfrak{X}_{i+1}$ by determining whether each set K belonging to $\mathfrak{X}_i$ also belongs to $\mathfrak{X}_{i+1}$

and whether $K \cup \{\pi_i\}$ belongs to $\mathfrak{X}_{i+1}$. At the same time it is known that one of these sets certainly belongs to $\mathfrak{X}_{i+1}$ and hence the negative answer to one of the above questions implies the positive answer to the other. The solution of the above mentioned questions may be based on the following two theorems:

THEOREM 3. *If $K \in \mathfrak{X}_i$, then $K \cup \{\pi_i\} \in X_{i+1}$ if and only if the following two conditions are fulfilled*:

  (i) *for any $s < i$, if T_s occurs in π_i, then $\pi_s \in K$,*

  (ii) *there exists a model $\mathfrak{M}_0$ such that the expression $R(K \cup \{\pi_i\}) \wedge \tilde{\mathfrak{R}}_{K,i}$ belongs to $\mathrm{Ver}_{LR}(\mathfrak{M}_0)$.*

THEOREM 4. *If $K \in \mathfrak{X}_i$, then $K \in \mathfrak{X}_{i+1}$ if and only if at least one of the following conditions is fulfilled*:

  (i) *there exists $s < i$ such that T_s occurs in π_i, and $\pi \bar{\in} K$,*

  (ii) *there exists a model $\mathfrak{M}_0$ such that the expression $R(K) \wedge \tilde{\mathfrak{R}}_{K,i}$ belongs to $\mathrm{Ver}_{LR}(\mathfrak{M}_0)$, and the expression $R(K \cup \{\pi_i\})$ does not belong to $\mathrm{Ver}_{LR}(\mathfrak{M}_0)$.*

Both the above theorems are consequences of the lemmas and definitions given in this section. The Appendix that follows this section contains the proofs of the above theorems.

  It is convenient to use a kind of graph called a *tree*[13] in constructing the family of sets $\mathfrak{X}$. The figure below presents a fragment of such a tree.

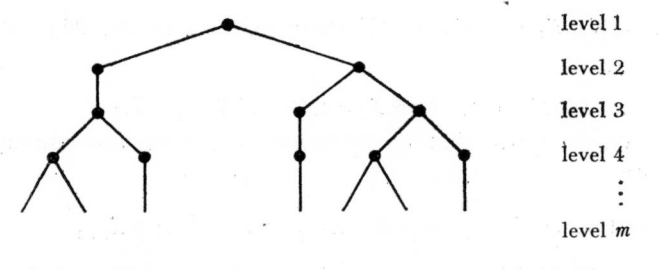

level 1

level 2

level 3

level 4

⋮

level m

The sets of postulates belonging to $\mathfrak{X}_i$ correspond to the vertices on the level i. The only vertex on the level 1 (called the root of the tree) corresponds to the empty set of postulates. The vertices on the level $m+1$ correspond to the searched sets belonging to the family $\mathfrak{X}$. The construction of the graph consists in hanging at every vertex on the level

$i \leqslant m$, corresponding to the set of postulates K belonging to $\mathfrak{X}_i$, of one or two edges depending on whether only one of the sets K and $K \cup \{\pi_i\}$ belongs to the family $\mathfrak{X}_{i+1}$, or both.

In constructing the family of sets according to the method that has been outlined here we are compelled to answer questions of two types. The first type of questions concerns the purely syntactic relations between postulates. These may be easily resolved. The questions of the second type concern the semantic properties of the sentences in language L_R. It is easy to note that they can all be reduced to the following form:

(∗) Does there exist a model $\mathfrak{M}_0$ such that the sentence α in language L_R is true in $\mathfrak{M}_0$?

To prove the positive answer to the question of the type (∗) we construct an example of an appropriate model. To prove the negative answer we make use of the fact that the formalized language L_R is the language of the second order logic. Since every thesis of this logic is a tautology (i.e. a sentence true in every model $\mathfrak{M}_0$), then proving that $\sim\alpha$ is a thesis we thus prove that there does not exist a model $\mathfrak{M}_0$ in which the sentence α would be true. However, the second order logic is not a complete system (not every tautology is a thesis).[14] Hence it may happen that the sentence $\sim\alpha$ is not a thesis even though it is true in every model $\mathfrak{M}_0$. In this case the problem (∗) is in principle undecidable. The above consideration indicates that the family of sets $\mathfrak{X}$ cannot always be constructed. If, on the other hand, for a certain language J the family of sets $\mathfrak{X}$ cannot be constructed, then there are sentences such that we cannot decide whether they are strictly analytic in this language.

APPENDIX

We employ the lemmas and definitions included in Section VIII in the proofs presented here.

LEMMA 4. *If a set of postulates K fulfils the condition*:
    (t) *if $\pi_k \in K$, then: for any $s < k$ if T_s occurs in π_k, then $\pi_s \in K$, then for every $i = 1, \ldots, m+1$, if the expression $R(*i \cap K) \wedge \wedge \tilde{\mathfrak{R}}_{K,i}$ belongs to $\mathrm{Ver}_{L_R}(\mathfrak{M}_0)$, then $*i \cap K = *i \cap \mathfrak{S}_{\mathfrak{M}_0}$.*

PROOF. We assume that the antecedent of the Lemma 4 holds and we inductively prove the general implication, which is its consequent. We easily find out that this implication holds for $i=1$, because the set $*1$ is empty, hence $*1 \cap K = *1 \cap \mathfrak{S}_{\mathfrak{M}_0} = *1$.

We subsequently prove that if the above mentioned implication holds for $i=k$, then it also holds for $i=k+1$. This part of the proof is carried out indirectly. We assume that

(1)   If the expression $R(*k \cap K) \wedge \tilde{\mathfrak{R}}_{K,k}$ belongs to $\mathrm{Ver}_{LR}(\mathfrak{M}_0)$, then $*k \cap K = *k \cap \mathfrak{S}_{\mathfrak{M}_0}$,

and that for some model $\mathfrak{M}_0'$ holds:

(2)   The expression $R(*k+1 \cap K) \wedge \tilde{\mathfrak{R}}_{K,k+1}$ belongs to $\mathrm{Ver}_{LR}(\mathfrak{M}_0')$,

but

(3) $*k+1 \cap K \neq *k+1 \cap \mathfrak{S}_{\mathfrak{M}_0'}$,

(4) $*k \cap K \subset *k+1 \cap K$ [(Def. 1)],

(5) $\mathfrak{B}(K, k) \subset \mathfrak{B}(K, k+1)$ [(Def. 2)],

(6)   Conjunction $\tilde{\mathfrak{R}}_{K,k}$ is the component of the conjunction $\tilde{\mathfrak{R}}_{K,k+1}$ [(5), (Def. 3)],

(7)   The expression $R(*k \cap K) \wedge \tilde{\mathfrak{R}}_{K,k}$ belongs to $\mathrm{Ver}_{LR}(\mathfrak{M}_0')$ [(2), (4), Lemma 2, (6)],

(8) $*k \cap K = *k \cap \mathfrak{S}_{\mathfrak{M}_0'}$ [(1), (7)],

(9) $\pi_k \in K$ and $\pi_k \,\bar{\in}\, \mathfrak{S}_{\mathfrak{M}_0'}$ or $\pi_k \,\bar{\in}\, K$ and $\pi_k \in \mathfrak{S}_{\mathfrak{M}_0'}$ [(3), (8)].

Here the demonstration branches. Let us first assume in addition:

(1.1) $\pi_k \in K$ and $\pi_k \,\bar{\in}\, \mathfrak{S}_{\mathfrak{M}_0'}$,

(1.2) $*k \cap K \cup \{\pi_k\} = *k+1 \cap K$ [(1.1), (Def. 1)],

(1.3)   The expression $R(*k \cap \mathfrak{S}_{\mathfrak{M}_0'} \cup \{\pi_k\})$ belongs to $\mathrm{Ver}_{LR}(\mathfrak{M}_0')$ [(8), (1.2), Lemma 3, (2)].

Hence, since $\pi_k \in K$, and K fulfils — according to the assumption — the condition (t), we conclude that the condition (i) of the Definition (D15)

is fulfilled. It follows from (1.3) and the Lemma 1 that also the condition (ii) of this definition is fulfilled. Hence:

(1.4) $\pi_k \in \mathfrak{S}_{\mathfrak{M}_0'}$,

which leads to the contradiction in the first branch of the proof [(1.1), (1.4)].

We subsequently assume that:

(2.1) $\pi_k \bar{\in} K$ and $\pi_k \in \mathfrak{S}_{\mathfrak{M}_0'}$,

(2.2)   For every $s<k$, if T_s occurs in π_k, then $\pi_s \in \mathfrak{S}_{\mathfrak{M}_0'}$ [(2.1), (D15)],

(2.3)   The expression $R(*k \cap \mathfrak{S}_{\mathfrak{M}_0'} \cup \{\pi_k\})$ belongs to $\text{Ver}_{L_R}(\mathfrak{M}_0')$ [(2.1), (D15), Lemma 1],

(2.4)   For every $s<k$, if T_s occurs in π_k, then $\pi_s \in K$ [(2.2), (8)],

(2.5) $\pi_k \in \mathfrak{B}(K, k+1)$ [(2.1), (2.4), (Def. 2)],

(2.6)   The expression $\sim R(*k \cap K \cup \{\pi_k\})$ is the component of the conjunction $\tilde{\mathfrak{R}}_{K,k+1}$ [(2.5), (Def. 3)],

(2.7)   The expression $\sim R(*k \cap K \cup \{\pi_k\})$ belongs to $\text{Ver}_{L_R}(\mathfrak{M}_0')$ [(2.6), (2)],

(2.8)   The expression $\sim R(*k \cap \mathfrak{S}_{\mathfrak{M}_0'} \cup \{\pi_k\})$ belongs to $\text{Ver}_{L_R}(\mathfrak{M}_0')$ [(8), Lemma 3, (2.7)].

The contradiction is implied by assumptions (2.3) and (2.8). This contradiction ends the demonstration of Lemma 4.

LEMMA 5. *For every $i=1, \ldots, m, m+1$, if $*i \cap K = *i \cap \mathfrak{S}_{\mathfrak{M}_0}$, then the expression $R(*i \cap K) \wedge \tilde{\mathfrak{R}}_{K,i}$ belongs to $\text{Ver}_{L_R}(\mathfrak{M}_0)$.*

PROOF. Let us assume that for some $k=1, \ldots, m, m+1$ and for some model $\mathfrak{M}_0'$

(1) $*k \cap K = *k \cap \mathfrak{S}_{\mathfrak{M}_0'}$

holds, but

(2)   The expression $R(*k \cap K) \wedge \tilde{\mathfrak{R}}_{K,k}$ does not belong to $\text{Ver}_{L_R}(\mathfrak{M}_0')$.

It follows from the definition (D15) and Lemma 1 that

(3)   The expression $R(\mathfrak{S}_{\mathfrak{M}_0'})$ belongs to $\mathrm{Ver}_{L_R}(\mathfrak{M}_0')$.

Since $*k \cap \mathfrak{S}_{\mathfrak{M}_0'} \subset \mathfrak{S}_{\mathfrak{M}_0'}$, then — according to Lemma 2 — also:

(4)   The expression $R(*k \cap \mathfrak{S}_{\mathfrak{M}_0'})$ belongs to $\mathrm{Ver}_{L_R}(\mathfrak{M}_0')$,

(5)   The expression $R(*k \cap K)$ belongs to $\mathrm{Ver}_{L_R}(\mathfrak{M}_0')$ [(1), (4), Lemma 3],

(6)   The expression $\tilde{\mathfrak{R}}_{K,k}$ does not belong to $\mathrm{Ver}_{L_R}(\mathfrak{M}_0')$ [(2), (5)]

(7) $\mathfrak{B}(K, k) = \mathfrak{B}(\mathfrak{S}_{\mathfrak{M}_0'}, k)$ [(1), (Def. 2)],

(8)   The expression $\tilde{\mathfrak{R}}_{\mathfrak{S}_{\mathfrak{M}_0'},k}$ does not belong to $\mathrm{Ver}_{L_R}(\mathfrak{M}_0')$ [(7), (Def. 3), (6)].

In view of (8) and (Def. 3) for some $s < k$ we obtain:

(9) $\pi_s \in \mathfrak{B}(\mathfrak{S}_{\mathfrak{M}_0'}, k)$ and the expression $R(*s \cap \mathfrak{S}_{\mathfrak{M}_0'} \cup \{\pi_s\})$ belongs to $\mathrm{Ver}_{L_R}(\mathfrak{M}_0')$,

(10) $\pi_s \bar{\in} \mathfrak{S}_{\mathfrak{M}_0'}$ [(9), (Def. 2)],

(11)   For any $r < s$, if T_r occurs in π_s, then $\pi_r \in \mathfrak{S}_{\mathfrak{M}_0'}$ [(9), (Def. 9)].

Note that the conditions (i) and (ii) of the definition (D15) are fulfilled, which follows from (11), (9) and Lemma 1. Hence:

(12) $\pi_s \in \mathfrak{S}_{\mathfrak{M}_0'}$.

We have obtained a contradiction between (10) and (12).

THEOREM 3. *If $K \in \mathfrak{X}_i$, then $K \cup \{\pi_i\} \in \mathfrak{X}_{i+1}$ if and only if the following two conditions are fulfilled:*

(i) *For every $s < i$, if T_s occurs in π_i, then $\pi_s \in K$,*

(ii) *There exists a model $\mathfrak{M}_0$ such that the expression $R(K \cup \{\pi_i\}) \wedge \tilde{\mathfrak{R}}_{K,i}$ belongs to $\mathrm{Ver}_{L_R}(\mathfrak{M}_0)$.*

PROOF.

(1) $K \in \mathfrak{X}_i$ (assumption),

(2) $K \subset *i$ [(Def. 4), (1)],

(3) $\quad K \cup \{\pi_i\} \subset {}^*i+1$ [(2)],

(4) $\quad {}^*i+1 \cap (K \cup \{\pi_i\}) = K \cup \{\pi_i\}$ [(3)],

(5) $\quad \pi_i \in K \cup \{\pi_i\},$

(6) $\quad \pi_i \bar{\in} \mathfrak{B}(K \cup \{\pi_i\}, i+1)$ [(5), (Def. 2)],

(7) $\quad \mathfrak{B}(K, i) = \mathfrak{B}(K \cup \{\pi_i\}, i+1)$ [(6), (Def. 2)],

(1.1) $\quad K \cup \{\pi_i\} \in \mathfrak{X}_{i+1}$ (additional assumption).

It follows from the assumption (1.1) and (Def. 4) that for some model $\mathfrak{M}_0'$

(1.2) $\quad K \cup \{\pi_i\} = {}^*i+1 \cap \mathfrak{S}_{\mathfrak{M}_0'}$ holds,

(1.3) $\quad$ The expression $R(K \cup \{\pi_i\}) \wedge \tilde{\mathfrak{R}}_{K,i}$ belongs to $\text{Ver}_{LR}(\mathfrak{M}_0')$ [(4), (7), (1.2), Lemma 3, (Def. 3), Lemma 5].

From existential generalization of sentence (1.3) we obtain:

(1.4) $\quad$ The condition (ii),

(1.5) $\quad \pi_i \in \mathfrak{S}_{\mathfrak{M}_0'}$ [(1.2)],

(1.6) $\quad K = {}^*i \cap \mathfrak{S}_{\mathfrak{M}_0'}$ [(2), (1.2)],

(1.7) $\quad$ The condition (i) [(1.5), (1.6), (D15)].

Applying the rule subjoining implication to the proof we obtain:

(8) $\quad$ If $K \cup \{\pi_i\} \in \mathfrak{X}_{i+1}$, then the conditions (i) and (ii) are fulfilled,

(2.1) $\quad$ Both the condition (i) and the condition (ii) occur (additional assumption),

(2.2) $\quad$ The set K fulfils the condition (t) of the Lemma 4 [(1), (Def. 4), (D15)],

(2.3) $\quad$ The set $K \cup \{\pi_i\}$ fulfils the condition (t) of the Lemma 4 [(2.2), (2.1)],

(2.4) $\quad$ If the expression $R(K \cup \{\pi_i\}) \wedge \tilde{\mathfrak{R}}_{K,i}$ belongs to $\text{Ver}_{LR}(\mathfrak{M}_0)$, then $K \cup \{\pi_i\} = {}^*i+1 \cap \mathfrak{S}_{\mathfrak{M}_0}$ [(4), Lemma 3, (7), (Def. 3), Lemma 4, (2.3)],

(2.5)   There exists such a model $\mathfrak{M}_0$ that $K \cup \{\pi_i\} = *i+1 \cap \mathfrak{S}_{\mathfrak{M}_0}$
             [(2.1), (2.4)],

(2.6) $K \cup \{\pi_i\} \in \mathfrak{X}_{i+1}$ [(2.5), (Def. 4)].

By subjoining the implication to the proof we obtain:

(9)   If the conditions (i) and (ii) are fulfilled, then $K \cup \{\pi_i\} \in \mathfrak{X}_{i+1}$.

The conjunction of (8) and (9) is equivalent to the consequent of the Theorem 3.

THEOREM 4. *If $K \in \mathfrak{X}_i$, then $K \in \mathfrak{X}_{i+1}$ if and only if at least one of the following conditions is fulfilled:*

(i) *There exists such $s < i$ that T_s occurs in π_i, and $\pi_s \bar{\in} K$.*

(ii) *There exists such a model $\mathfrak{M}_0$ that the expression $R(K) \wedge \wedge \tilde{\mathfrak{R}}_{K,i}$ belongs to $\mathrm{Ver}_{L_R}(\mathfrak{M}_0)$, and the expression $R(K \cup \{\pi_i\})$ does not belong to $\mathrm{Ver}_{L_R}(\mathfrak{M}_0)$.*

PROOF.

(1) $K \in \mathfrak{X}_i$ (assumption),

(2) $K \subset *i$ [(1), (Def. 4)],

(3) $*i \cap K = K$ [(2)],

(4) $\pi_k \bar{\in} K$ [(2), (Def. 1)].

As the first additional assumption we take

(1.1)   The condition (i).

According to the assumption (1) and (Def. 4), for some model $\mathfrak{M}'_0$, we obtain:

(1.2) $K = *i \cap \mathfrak{S}_{\mathfrak{M}'_0}$,

(1.3) $\pi_i \bar{\in} \mathfrak{S}_{\mathfrak{M}'_0}$ [(1.1), (1.2), (D15)],

(1.4) $*i+1 \cap \mathfrak{S}_{\mathfrak{M}'_0} = *i \cap \mathfrak{S}_{\mathfrak{M}'_0} = K$ [(1.3), (1.2)],

(1.5)   There exists such a model $\mathfrak{M}_0$ that $K = *i+1 \cap \mathfrak{S}_{\mathfrak{M}_0}$ [(1.4)],

(1.6) $K \in \mathfrak{X}_{i+1}$ [(1.5), (Def. 4)].

Applying the rule of subjoining the implication we obtain:

(5)   If condition (i) holds, then $K \in \mathfrak{X}_{i+1}$.

We shall subsequently adopt as a second assumption:

(2.1)   The condition (ii).

According to this assumption, for some model $\mathfrak{M}_0''$ we obtain:

(2.2)   The expression $R(K) \wedge \tilde{\mathfrak{R}}_{K,i}$ belongs to $\mathrm{Ver}_{L_R}(\mathfrak{M}_0'')$, and the expression $R(K \cup \{\pi_i\})$ does not belong to $\mathrm{Ver}_{L_R}(\mathfrak{M}_0'')$,

(2.3)   The set K fulfils condition (t) of the Lemma 4 [(1), (Def. 4), (D15)],

(2.4) $K = {}^*i \cap \mathfrak{S}_{\mathfrak{M}_0''}$ [(2.2), (3), Lemma 3, (2.3), Lemma 4],

(2.5) $K \cup \{\pi_i\} = {}^*i \cap \mathfrak{S}_{\mathfrak{M}_0''} \cup \{\pi_i\}$ [(2.4)],

(2.6)   The expression $R({}^*i \cap \mathfrak{S}_{\mathfrak{M}_0''} \cup \{\pi_i\})$ does not belong to $\mathrm{Ver}_{L_R}(\mathfrak{M}_0'')$ [(2.2), (2.5), Lemma 3],

(2.7) $\pi_i \bar{\in} \mathfrak{S}_{\mathfrak{M}_0''}$ [(2.6), Lemma 1, (D15)],

(2.8) ${}^*i+1 \cap \mathfrak{S}_{\mathfrak{M}_0''} = {}^*i \cap \mathfrak{S}_{\mathfrak{M}_0''} = K$ [(2.7), (2.4)],

(2.9)   There exists a model $\mathfrak{M}_0$ such that $K = {}^*i+1 \cap \mathfrak{S}_{\mathfrak{M}_0}$ [(2.8)],

(2.10) $K \in \mathfrak{X}_{i+1}$ [(2.9), (Def. 4)].

By subjoining the implication to the proof we obtain:

(6)   If the condition (ii) holds, then $K \in \mathfrak{X}_{i+1}$,

(7)   If either the condition (i) or the condition (ii) holds, then $K \in \mathfrak{X}_{i+1}$ [(5), (6)].

We now assume that:

(3.1) $K \in \mathfrak{X}_{i+1}$.

Then – according to (Def. 4) – for some model $\mathfrak{M}_0'''$, we have:

(3.2) $K = {}^*i+1 \cap \mathfrak{S}_{\mathfrak{M}_0'''}$,

(3.3) $\pi_i \in \mathfrak{S}_{\mathfrak{M}_0'''}$ [(4), (3.2)],

(3.4) $*i+1 \cap \mathfrak{S}_{\mathfrak{M}_0'''} = *i \cap \mathfrak{S}_{\mathfrak{M}_0'''} = K$ [(3.3), (3.2)].

(3.5)   For model $\mathfrak{M}_0'''$ one of the conditions (i) or (ii) of the definition (D15) does not hold [(3.3)].

(3.6)   If the condition (i) of the definition (D15) does not hold for the model $\mathfrak{M}_0'''$, then the condition (i) of the theorem under proof is fulfilled [(3.4), (D15)],

(3.7)   The expression $R(K) \wedge \tilde{\mathfrak{R}}_{K,i}$ belongs to $\mathrm{Ver}_{L_R}(\mathfrak{M}_0''')$ [(3), (3.4), Lemma 3, Lemma 5],

(3.8)   If the condition (ii) of the definition (D15) does not hold for the model $\mathfrak{M}_0'''$, then the expression $R(K) \wedge \tilde{\mathfrak{R}}_{K,i}$ belongs to $\mathrm{Ver}_{L_R}(\mathfrak{M}_0''')$, and the expression $R(K \cup \{\pi_i\})$ does not belong to $\mathrm{Ver}_{L_R}(\mathfrak{M}_0''')$ [(D15), (3.7), (3.4), Lemma 3, Lemma 1],

(3.9)   Either the condition (i) or the condition (ii) of Theorem 4 holds.

By subjoining the implication we obtain:

(8)   If $K \in \mathfrak{X}_{i+1}$, then either the condition (i) or the condition (ii) holds.

The conjunction of (7) and (8) is equivalent to the consequent of the theorem under proof.

BIBLIOGRAPHY

[1] K. Ajdukiewicz, *Język i poznanie* (*Language and Knowledge*), PWN, Warszawa, vol. I 1960, vol. II 1965.

[2] K. Ajdukiewicz, 'Le problème du fondement des propositions analytiques', *Studia Logica* **VIII** (1958), included in the present volume as: 'The Problem of Justifying Analytic Sentences'.

[3] K. Ajdukiewicz, 'Logika a doświadczenie' ('Logic and Experience'), *Przegląd Filozoficzny* **XLIII** (1947) (also in [1]).

[4] K. Ajdukiewicz, 'O znaczeniu wyrażeń' ('On the Meaning of Expressions'), in: *Księga Pamiątkowa Polskiego Towarzystwa Filozoficznego we Lwowie*, Lwów 1931.

[5] K. Ajdukiewicz, 'Sprache und Sinn', *Erkenntnis* **IV** (1934) (also in [1]).

[5] K. Ajdukiewicz, 'Axiomatic Systems From the Methodological Point of View', *Studia Logica* **IX** (1960) (also in [1]), included in the present volume.

[7] R. Carnap, 'Beobachtungssprache und theoretische Sprache', *Dialectica* 12 (1958).

[8] R. Carnap, *Meaning and Necessity*, Univ. of Chicago Press, 1947.

[9] G. Frege, *The Foundation of Arithmetic*, Oxford 1950.

[10] H. P. Grice and P. F. Strawson, 'In Defence of Dogma', in: *Classics of Analytic Philosophy* (ed. by R. R. Ammerman), McGraw-Hill 1965.

[11] A. Grzegorczyk, 'Próba ugruntowania semantyki języka opisowego' ('An Attempt to Establish the Semantics of Descriptive Language'), *Przegląd Filozoficzny* **XLIV** (1948).

[12] T. Hobbes, 'Elementa Philosophiae', in: *Latin Works* (ed. by M. Molesworth), vols. 1–5 (1893–45).

[13] I. Kant, *Kritik der reinen Vernunft*, Philosophische Bibliothek Bd. 73a, Meiner, Leipzig 1926.

[14] W. Kneale, 'Are Necessary Truths True by Convention?', *Aristotelian Society Proceedings* (1947) (supplement).

[15] W. Kneale and M. Kneale, *The Development of Logic*, Oxford 1962.

[16] M. Kokoszyńska, 'O dwojakim rozumieniu uzasadnienia dedukcyjnego' ('On the Double Meaning of Deductive Justification'), *Studia Logica* **XIII** (1962).

[17] M. Kokoszyńska, 'O dyrektywach inferencji' ('On the Rules of Inference'), *Rozprawy logiczne*, PWN 1964.

[18] C. I. Lewis, *An Analysis of Knowledge and Valuation*, La Salle, Illinois, 1946.

[19] H. Mehlberg, *The Reach of Science*, Toronto 1958.

[20] O. Ore, *An Introduction to the Theory of Graphs.*

[21] A. Pap, 'Necessary Propositions and Linguistic Rules', *Archivio di Filosofia*, Roma 1955.

[22] A. Pap, *Semantics and Necessary Truth*, Yale Univ. Press, 1958.

[23] M. Przełęcki, 'Z semantyki pojęć otwartych' ('From Semantics of Open Concepts'), *Studia Logica* **XV** (1964).

[24] M. Przełęcki, 'W sprawie istnienia przedmiotów teoretycznych' ('On the Existence of Theoretical Objects'), *Teoria i doświadczenie*, PWN, 1966.

[25] M. Przełęcki, 'O pojęciu zdania analitycznego' ('On the Concept of Analytic Sentence'), *Studia Logica* **XIV** (1963).

[26] W. V. O. Quine, 'Two Dogmas of Empiricism', in: *From a Logical Point of View*, Harvard Univ. Press, 1953.

[27] B. Stanosz, 'Formalne teorie zakresu i treści wyrażeń' ('Formal Theories of the Scope and Content of Expressions'), *Studia Logica* **XV** (1964).

[28] R. Suszko, 'Logika formalna a niektóre zagadnienia teorii poznania', (Formal Logic and some Problems of Epistemology'), *Myśl Filozoficzna* (1956).

[29] M. G. White, 'The Analytic and the Synthetic: An Untenable Dualism', in: *Semantics and the Philosophy of Language* (ed. by W. L. Linsky), Univ. of Illinois Press, 1952.

[30] R. Wójcicki, Analityczne komponenty definicji arbitralnych' ('Analytic Components of Arbitrary Definitions'), *Studia Logica* **XIV** (1963).

[31] R. Wójcicki, 'Analityczność, syntetyczność, empiryczna sensowność zdań', *Studia Filozoficzne* 3 (1966); appeared as: 'Analyticity, Syntheticity, Empirical Meaningfulness of Sentences', *Studia Filozoficzne* 4 (1970), supplementary volume in English.

[32] R. Wójcicki, 'O warunkach empirycznej sensowności terminów' ('On the Conditions of Empirical Meaningfulness of Terms'), *Teoria i doświadczenie*, PWN, 1966.

[33] R. Wójcicki, 'Semantical Criteria of Empirical Meaningfulness', *Studia Logica* **XIX** (1966).

REFERENCES

* A fragment of Ph.D. dissertation (1967). Translated by E. Ronowicz.

[1] This principle may be considered to be the result of stating more precisely the condition of adequacy for the definition of logical truth adopted by Carnap. In the original this condition runs as follows: "A sentence $\mathfrak{S}_i$ is *L-true* in a semantic system S if and only if $\mathfrak{S}_i$ is true in S in such a way that its truth can be established on the basis of the semantical rules of the system S alone, without any reference to (extra-linguistic) facts." Carnap [8], p. 10.

[2] We find this approach in the work by Stanosz [27].

[3] We have in mind here mainly the works by Przełęcki [23] and [24] as well as Wójcicki [32] and [33].

[4] For simplicity's sake we have disregarded the phenomenon of vagueness of observational terms here.

[5] The symbol Cn denotes here the classical logical consequence. The relation Cn may in certain languages be different from C_J (consequence in language J), and it is certainly different from C_0 (immediate consequence), which canot be a transitive relation, since in that case the fulfilment of the condition (ii) of the definition (D1) by a person X would surpass human abilities.

[6] The symbol $x_i|\Delta$ denotes the relation x_i restricted to the set Δ.

[7] The concept of prolongation of models was introduced into discussions in the field of semantics of languages of empirical theories by M. Przełęcki [23], who used the term 'enrichment'. R. Suszko suggested the use of the term 'prolongation'.

[8] In the situation under discussion there is no such prolongation of model $\mathfrak{M}_0^K$ in which sentences $K \cup \{\pi_i\}$ would be true. There would exist such a model — against the assumption — and it would be $\mathfrak{M}^*$, if $K \cup \{\pi_i\} \subset \mathrm{Ver}(\mathfrak{M}^*)$. Since $K \subset \mathfrak{S}$ and hence $K \subset \mathrm{Ver}(\mathfrak{M}^*)$, then $\pi_i \bar{\in} \mathrm{Ver}(\mathfrak{M}^*)$.

[9] This theorem is the immediate consequence of the known theorems on logical consequence: If $Y \subset X$, then $\mathrm{Cn}(Y) \subset \mathrm{Cn}(X)$ and $\mathrm{Cn}(\mathrm{Cn}(X)) \subset \mathrm{Cn}(X)$.

[10] We would obtain the simplest criterion of strict analyticity by presenting the set of strictly analytical sentences, i.e. the set of sentences $\bigcap_{i=1}^{\infty} \mathrm{Cn}(\mathfrak{S}_{\mathfrak{M}_{0_i}})$ as the set of consequences of some definite set of sentences Z. Though the set of strictly analytical sentences is — as we have demonstrated — a system, we do not know

whether it is always finitely axiomatizable. It is known, on the other hand, that it is not always axiomatizable in terms of a certain set of postulates. In particular, the equation given below does not always hold:

$$\text{Cn}\left(\bigcap_{i=1}^{\infty} \mathfrak{S}_{\mathfrak{M}_{0_i}} \right) = \bigcap_{i=1}^{\infty} \text{Cn}\left(\mathfrak{S}_{\mathfrak{M}_{0_i}} \right) \ .$$

[11] To be precise, the models of the formalized language L_R possess several universes. One of them corresponds to the range of variability of the individual variables, the others — occurring in the number equal to the number of types of predicate variables — correspond to the sets of values of these variables. Since the latter ones can be determined with the help of U (e.g.: 2^U, $2^{U \times U}$, $2^{U \times U \times U}$, etc.) there is no need to mention them.

[12] The intuitive sense of this definition becomes clear when we assume that $Z = \mathfrak{S}_{\mathfrak{M}_0}$. The symbol $\mathfrak{B}(Z, i)$ then denotes the set of those postulates with indices smaller than i, which do not belong to $\mathfrak{S}_{\mathfrak{M}_0}$ although they fulfil condition (i) of the definition (D15) formulated below.

[13] The terminology concerning the graphs has been taken from Ore's book [20].

[14] This follows from the known theorem of Gödel on the incompleteness of richer logical systems.

LESZEK NOWAK

THE MODEL OF EMPIRICAL SCIENCES IN THE CONCEPTS OF THE CREATORS OF MARXISM*

1. THE AIMS OF THE ARTICLE

It is a widely accepted view in Marxist philosophy that the creators of Marxism held methodological views different from those expressed in various trends of contemporary philosophy of the sciences. The aim of this article is the systematization of these concepts and their presentation in the conceptual apparatus that will make possible their testability. Anticipating the general conclusion of the article we may say that, indeed, a certain model of the empirical sciences is assumed in the works of K. Marx and F. Engels, which differs from the models presented by contemporary trends in the philosophy of science, and in particular from the positivistic model. This is especially worth noting since the epistemological concepts of Marx and Engels, although they were not presented by them systematically anywhere, were formed more or less at the same time as the positivistic trend dominated in the philosophy of science.

In order to reconstruct the model of empirical sciences that is *implicite* contained in the works of Marx and Engels I shall look in these works for answers to the following questions:

I. What is the criterion for the propositions to belong to the empirical sciences?

II. What types of sentences are basic theorems of the empirical sciences?

III. What types of sentences are extra-basic theorems of the empirical sciences?

IV. What types of relations of justification hold between the basic and extra-basic theorems in the empirical sciences?

V. How are the propositions belonging to the empirical sciences explained?

VI. Do extra-basic theorems refer to reality?

VII. What is the course of development of the empirical sciences?

VIII. Are all the empirical sciences homogeneous with respect to their methodological characteristics?

By basic theorems I understand sentences for the justification of which reference to other sentences is not necessary, and by extra-basic theorems – sentences for the justification of which reference to other sentences is inevitable, and, eventually, to basic theorems.[1]

The answers to the above eight questions will constitute the model of the empirical sciences in the concepts of the creators of Marxism. The word 'model' is particularly convenient here since we are not going to simply report the concepts of the classics of Marxism, but to reconstruct them, for a number of these questions (e.g. the first three) were not asked in the works of Marx and Engels at all. Thus we shall have to infer indirectly their putative answers to these questions from what they wrote on other, related topics, or from their methodological comments to their own works in the fields of economics, sociology or history. This, naturally, increases the risk of the inadequacy of interpretation. There is, however, no other way of reconstructing the concepts of the creators of Marxism in this field than putting forth interpretative hypotheses and then testing whether they agree with all their statements, their scientific practice, whether they make possible an explanation of the sense of statements difficult to comprehend without these hypotheses, whether they allow to arrange the concepts of the classics in a consistent system. If our hypotheses fulfil all these conditions, we shall be able to accept them as justified to a considerable extent. This does not mean, naturally, that they aspire at absolute certainty, that the possibility of some other reconstruction of the epistemological concepts of the classics of Marxism is excluded. For we should demand of all the interpretative hypotheses merely appropriate justification in the light of the analysis of the texts and scientific practice of the creators of Marxism, and not – as it sometimes happens – promoting of one's suppositions by unjustified forwardness and categoric statements.

2. THE CRITERION OF SCIENTIFIC CHARACTER
(FOR THE EMPIRICAL SCIENCES)

The problem (I) was not stated clearly in the works of Marx or Engels. They did not even use uniformly the term 'empirical sciences'. When Engels states, e.g. that Marx used in *Capital* his own modification of

Hegel's dialectics "diese Methode auf die Tatsachen einer empirischen Wissenschaft, der politischen Ökonomie, angewandt zu haben",[2] he uses the term in the sense approximating its present use, also in this work; namely—to denote the non-formal disciplines. When, however, he states that in the 19th century the natural science "aus einer empirischen in eine theoretische Wissenschaft, und bei der Zusammenfassung des Gewonnenen in ein System der materialistischen Naturerkenntnis sich verwandelte",[3] he is using the term 'empirical science' in a much narrower sense. However, although the problem (I) was not stated clearly by the creators of Marxism, it is possible—by analysing the contexts, in which they spoke of the scientific or non-scientific character of some concepts—to find in their works a consistently maintained answer to this question.

In most general terms the answer is as follows: concepts that are scientific in character are testable in confrontation with facts, they are empirically decidable. Thus, e.g. Marx gives the following contrast between Ricardo's and Proudhon's theories of value in *The Poverty of Philosophy*: "Ricardo shows us the real movement of the bourgeois production establishing value. Mr. Proudhon abstracts from this real movement and he 'goes quite beside himself' to find some new ways of acting and to arrange the world according to supposed new formula (...) Ricardo takes as his starting point the present community in order to show us how it establishes the value; Mr. Proudhon takes as his starting point a value that has already been constituted in order to establish with the help of this value a new social world. (...) Ricardo's theory of value is a scientific interpretation of contemporary economic life; the theory of value of Mr. Proudhon is a utopian interpretation of Ricardo's theory. Ricardo asserts about the truth of his formula deriving it from all the economic processes and explaining with its help all the phenomena, even those, which at the first glance seem to contradict it like, e.g., pension, accumulation of capitals and the relation of pays to profits; *and this makes his doctrine a scientific system*; Mr. Proudhon, who has discovered this formula of Ricardo's anew with the help of entirely arbitrary hypotheses is then forced to look for isolated economic facts, which he tortures and falsifies, so that they could pretend to be examples, already existing applications, the beginnings of the realization of his idea made alive"[4] (emphasis—L. N.). That is—Ricardo's

theory is scientific in character, since it explains reality and may be
confronted with facts and Proudhon's concepts do not possess this
character, because they cannot be confronted with facts, since Proudhon
postulates rather than explains and he uses facts (and false ones, too)
in order to demonstrate that his utopian postulates are realistic.

An analogous idea: that economy should provide theoretical explanations of facts, which are to be used as a criterion for selecting the true
theories was expressed – it seems – in the following thesis of Marx's:
"Political economy may be transformed into a real science only in such a
way that in place of conflicting dogmas conflicting facts are substituted
as well as real contradictions which form their hidden background".[5]

F. Engels expressed analogous views in the essay *The Natural Science
in the World of Ghosts*, in which he objected that well known biologists
took advantage of their scientific authority to promote spiritualistic
experiments. Engels wrote: "Es zeigt hier handgreiflich, welches der
sicherste Weg von der Naturwissenschaft zum Mystizismus ist. Nicht
die überwuchernde Theorie der Naturphilosophie, sondern die aller-
platteste, alle Theorie verachtende, gegen alles Denken mißtrauische
Empirie. Es ist nicht die aprioristische Notwendigkeit, die die Existenz
der Geister beweist, sondern die erfahrungsmäßige Beobachtung der
Herren Wallace, Crookes & Co. Wenn wir den spektralanalytischen
Beobachtungen von Crookes glauben, die zur Entdeckung des Metalls
Thallium führten, oder den reichen zoologischen Entdeckungen von
Wallace im Malayischen Archipel, so verlangt man von uns denselben
Glauben für die spiritistischen Erfahrungen und Entdeckungen dieser
beiden Forscher. Und wenn wir meinen, daß hier doch ein kleiner
Unterschied stattfinde, nämlich der, daß wir die einen *verifizieren können*
und die andern nicht, so entgegnen uns die Geisterseher, daß dies
nicht der Fall"[6] (emphasis – *L.N.*). The same criterion: empirical decidability manifests itself in Engel's considerations when he contrasts
his contemporary natural science to some of the systems of speculative
philosophy. For he maintains that 'modern natural science' is "Die
moderne Naturwissenschaft – die einzige, von der qua (als) Wissenschaft
die Rede sein kann gegenüber den genialen Intuitionen der Griechen
und den sporadisch zusammenhangslosen Untersuchungen der Ara-
ber".[7] And the difference between excellent, though speculative suppositions of Greek philosophy of nature and modern natural science con-

sists in that "was bei den Griechen geniale Intuition war, bei uns Resultat streng wissenschaftlicher, erfahrungsmäßiger Forschung ist und daher auch in viel bestimmterer und klarerer Form auftritt".[8] This does not mean, however, that Engels considers philosophy to be a discipline devoid of cognitive value. On the contrary. For certain philosophical systems possess scientific character; moreover, philosophy as a whole, i.e. together with the doctrines speculative in nature, is indispensable for the development of theories in the empirical sciences. Thus Engels states that for the 'empirical natural science' of his times "unabweisbar wird es, die einzelnen Erkenntnisgebiete unter sich in den richtigen Zusammenhang zu bringen. Damit aber begibt sich die Naturwissenschaft auf das theoretische Gebiet, und hier versagen die Methoden der Empirie, hier kann nur das theoretische Denken helfen. Das theoretische Denken ist aber nur der Anlage nach eine angeborne Eigenschaft. Diese Anlage muß entwickelt, ausgebildet werden, und für diese Ausbildung gibt es bis jetzt kein andres Mittel als das Studium der bisherigen Philosophie".[9] It is just the needs of the natural science itself which cause that "Die Naturforscher mögen sich stellen, wie sie wollen, sie werden von der Philosophie beherrscht. Es fragt sich nur, ob sie von einer schlechten Modephilosophie beherrscht werden wollen oder von einer Form des theoretischen Denkens, die auf der Bekanntschaft mit der Geschichte des Denkens und mit deren Errungenschaften beruht".[10] "Der Mangel an Bekanntschaft mit der Geschichte der Philosophie tritt hier aber oft und grell genug hervor. Sätze, die in der Philosophie seit Jahrhunderten aufgestellt, die oft genug längst philosophisch abgetan sind, treten oft genug bei theoretisierenden Naturforschern als funkelneue Weisheit auf und werden sogar eine Zeitlang Mode".[11] Since they ignore the philosophical problems, the natural scientists sometimes unconsciously adopt very speculative philosophical assumptions against the method of the natural sciences which cannot be empirically tested. Thus, e.g. the widely spread among the natural scientists themselves version of mechanicism leads to the 'theory of the absolute qualitative identity of matter', and "Die Theorie von der absoluten qualitativen Identität der Materie —sie ist empirisch ebensowenig widerlegbar wie beweisbar".[12]

Thus, summing up, it should be stated that for the creators of Marxism the concepts scientific in character are concepts which are empirically decidable; this is also the difference between science and mysticism, and

certain philosophical systems—those that Engels called speculative (this—as we have seen—does not determine the negative evaluation of their cognitive value at all) whose end he connected with the future development of the natural and social theories.[13]

3. THE CHARACTERISTICS OF BASIC THEOREMS

The distinction between basic and extra-basic sentences presupposes the view that our senses provide us with information about the world; it seems obvious that thus understood empirical thesis may be ascribed to the creators of Marxism. They did not, however, use the concept of basic theorem—they did not distinguish the theorems of this type from extra-basic theorems, since they did not undertake the problems of scientific language at all. Problems of semiotics entered the philosophy of science much later. It is possible, however, to use this distinction to present the concepts of the creators of Marxism (mostly Engels) on the subjects we are interested in here. And if we aim at presenting these concepts in the form comparable to the contemporary methodological concepts, i.e., if we want to make possible the confrontation of Marx's–Engels's model of the empirical sciences with other models of these sciences, then this distinction is indispensable. Moreover, it is worth noting that the conceptual distinction corresponding to the contemporary distinction between the theoretical and observational concepts appears several times in F. Engels's *Dialectics of Nature*. Namely, Engels uses Hegel's term 'mental category' speaking of "Atom, Molekül als *Gedanken*bestimmungen, worüber das Denken zu entscheiden hat",[14] since "Atom und Molekül usw. kann man nicht mit dem Mikroskop beobachten, sondern nur mit Denken".[15] In order to obtain the distinction we need, i.e. between the basic and extra-basic theorems it is enough to transfer this distinction of 'mental categories' denoting unobservable objects and the remaining 'categories' denoting observable objects into the field of sentences.

It ought to be stated at once that it is not possible to solve in the light of the texts of the creators of Marxism a certain important problem concerning the status of the basic sentences, namely, whether only extraspective sentences are acceptable as basic theorems or the intro-

spective sentences as well. For it seems they did not concern themselves with the problem of the cognitive validity of introspection.

However one may find in the works of the classics the answers to a number of other problems connected with the status of basic theorems which are important for the contemporary methodology of the sciences. Thus, e.g. Engels states that each observation or experiment is directed by a theory adopted by the researcher, which establishes the aim of his experiments and divides the observable facts into relevant and irrelevant ones. And thus e.g. Engels states after Hegel that "Bei der Erfahrung kommt es darauf an, mit welchem Sinn man an die Wirklichkeit geht. Ein großer Sinn macht große Erfahrungen und erblickt in dem bunten Spiel der Erscheinung das, worauf es ankommt". [16] This dependence of experiment upon the adopted theories is so important that without the 'theoretical natural science', which "theoretischen Naturwissenschaft, ihre Naturanschauung möglichst zu einem harmonischen Ganzen verarbeitet heutzutage selbst der gedankenloseste Empiriker nicht vom Fleck kommt". [17]

However, the relationship between basic sentences and theory go much further. According to Engels' concepts 'pure' experiment, independent of theory, which is acceptable at present, does not exist at all. If somebody maintains that in accepting some definite results of observations he does not adopt any non-observational assumptions, then it is merely the sign that he does not realize the assumptions that he tacitly adopts; moreover, these theoretical assumptions are often derived from outdated theories. As Engels says "Die exklusive Empirie bildet sich ein, nur mit unleugbaren Tatsachen zu hantieren. In Wirklichkeit aber hantiert sie vorzugsweise mit überkommenen Vorstellungen, mit größtenteils veralteten Produkten des Denkens ihrer Vorgänger. So skeptisch diese Art Empirie sich verhält gegen die Resultate des gleichzeitigen Denkens, so gläubig steht sie da vor jenen des Denkens ihrer Vorgänger. Sogar die experimentell festgestellten Tatsachen sind ihr allmählich untrennbar geworden von den zugehörigen überlieferten Deutungen". [18] And as all the observable facts are interpreted—consciously or unconsciously—in the light of the accepted theories, the problem at stake is only to interpret them in the light of the best theories. It is the lack of sufficiently general and reliable theories in some field that leads to the interpretation of observable facts in this theory on the basis of outdated

or even common knowledge, i.e. to the interpretation that is not adequate. Thus, e.g. "die Feststellung einer umfassenden Theorie einstweilen unmöglich machende, zerfahrene Stand der Elektrizitätslehre, der es bedingt, daß auf diesem Gebiet die einseitige Empirie vorherrscht, jene Empirie, die sich das Denken möglichst selbst verbietet, und die eben deshalb nicht nur falsch denkt, sondern auch nicht imstande ist, den Tatsachen treu zu folgen oder nur sie treu zu berichten, die also in das Gegenteil von wirklicher Empirie umschlägt".[19] This dependency of the results of observation, i.e. observational sentences, upon the assumed theories is common in all the empirical sciences, not only in physics or chemistry. As Marx says: "similarly to the geologists, even the best, like Cuvier, absolutely wrongly interpreting certain facts, the philologists of such force as Grimm wrongly translated the simplest Latin sentences because they were under the influence of Möser etc.".[20]

That we interpret observational sentences in the light of accepted theories does not mean, however, that these theories create real facts in some mystical way — we only conceptualize these facts differently on the basis of various theories. Engels formulates it as follows: "Unrichtige Vorstellungen in jeder Wissenschaft sind schließlich, wenn wir von Beobachtungsfehlern absehn, unrichtige Vorstellungen von richtigen Tatsachen. Die letzteren bleiben, wenn wir auch die ersten als falsch nachgewiesen. Haben wir die alte Kontakttheorie abgeschüttelt, so bestehn noch die festgestellten Tatsachen, denen sie zur Erklärung dienen sollte".[21]

Since, by adopting basic sentences we accept at the same time the theory that allows to interpret these sentences, it is clear that a basic sentence may be rejected not only if we have accepted a different one, but also if we have rejected the theory that this sentence assumed. Engels formulates the thesis about the hypothetical character of observational sentences in his polemics with the physicists who were proponents of the positivist empiricism, which said that such sentences were unretractable, certain, that they formed the ultimate premises of science: "Die exklusive Empirie, die sich das Denken höchstens in der Form des mathematischen Rechnens erlaubt, bildet sich ein, nur mit unleugbaren Tatsachen zu hantieren. In Wirklichkeit aber hantiert sie vorzugsweise mit überkommenen Vorstellungen, mit größtenteils veralteten Produkten des Denkens ihrer Vorgänger, als da sind positive und negative Elektri-

zität, Kontakttheorie. Diese dienen ihr zur Grundlage endloser mathematischer Rechnungen, in denen sich die hypothetische Natur der Voraussetzungen über der Strenge der mtahematischen Formulierung angenehm vergessen läßt. Sogar trennbar geworden von den zugehörigen überlieferten Deutungen; die einfachste elektrische Erscheinung wird in der Darstellung verfälscht, z.B. durch Einschmuggelung der beiden Elektrizitäten; diese Empirie kann die Tatsachen nicht mehr richtig schildern, weil die überkommene Deutung mit in die Schilderung unterläuft".[22]

As can be seen from the above, the creators of Marxism, or at least Engels, who devoted more thought to this question, adopted the following characteristics of basic theorems: (1) they are sentences describing the results of observations, where, what should be observed, and with what aim in view, is decided by the theory adopted by the researcher; (2) basic sentences are interpreted on the basis of the adopted theory and thus the sense we give to these sentences depends not only upon their relation to the sensory experiences, but also upon their relation to a certain theory; (3) basic sentences are hypothetical in character – they are sometimes also retracted when the theory is rejected that provides appropriate interpretation to them. The last conclusion will gain additional confirmation if we analyse the regularities of development of empirical sciences as perceived by Engels.

4. THE CHARACTERISTICS OF EXTRA-BASIC THEOREMS

Let us start with the question of the logical structure of theorems which the classics accepted as extra-basic theorems. Naturally, one cannot expect of them, e.g. an explicit drawn distinction between a strictly general sentence and a historical generalization. Nevertheless, as we shall soon see, the concept of a strictly general sentence itself is clearly present in the works of Engels. When, e.g. Engels discussed the question of the general character of the laws formulated by the natural sciences, we can see clearly from his statements that he had in mind what we call today the strictly general character of scientific laws: "Wir wissen, daß Chlor und Wasserstoff innerhalb gewisser Druck- und Temperaturgrenzen und unter Einwirkung des Lichts sich unter Explosion zu Chlorwasserstoffgas verbinden, und sobald wir dies wissen, wissen wir auch,

daß dies *überall* und *immer* geschieht, wo obige Bedingungen vorhanden, und es kann gleichgültig sein, ob sich dies einmal oder millionenmal wiederholt und auf wieviel Weltkörpern. Die Form der Allgemeinheit in der Natur ist *Gesetz*".[23] This is connected with the interpretation of Engels' thesis on the 'eternal character of natural laws'. Namely, Engels maintains that the laws of the natural sciences are, at the same time, eternal (one should add — if we assume their truth) and historical. Here is Engels's statement, in which he draws the distinction between the eternal and historical character of these laws: "*Die ewigen Naturgesetze* verwandeln sich auch immer mehr in historische. Daß Wasser von 0–100°C flüssig ist, ist ein ewiges Naturgesetz, aber damit es Geltung haben kann, muß 1. Wasser, 2. die gegebne Temperatur und 3. Normaldruck da sein. Auf dem Mond ist kein Wasser, auf der Sonne nur seine Elemente, und für diese Weltkörper existiert das Gesetz nicht. — Die Gesetze der Meteorologie sind auch ewig, aber nur für die Erde oder für einen Körper von der Größe, Dichtigkeit, Achsenneigung und Temperatur der Erde, und vorausgesetzt, daß er eine Atmosphäre von gleicher Mischung Sauerstoff und Stickstoff und gleiche Mengen aufsteigenden und sich niederschlagenden Wasserdampfs hat".[24] The idea Engels had in mind might be explained in the following way. The theorem he discusses has the following form:

(1)   if x is a portion of water and x finds itself in the temperature from 0 to 100°C, and x finds itself under the pressure of 1 Atm, then x is liquid.

Thus, the above theorem is represented by the general scheme:

(2) $F(x) \rightarrow G(x)$.

Now, that theorem (1) is an eternal law (if we make the simplifying assumption that it is true, and thus does not require any changes) means simply that whenever the conditions mentioned in its antecedent are satisfied, its consequent will also be satisfied. That it is at the same time historical means that it is non-vacuously satisfied in some spatio-temporal area, i.e. — in other words — the antecedent of this law is satisfied only by bodies that are in some specified place and time. Generally speaking — that the law of the form (2) is eternal (if we assume it is true, i.e. it is not a relative truth — cf. below) means that whenever a

condition $F(x)$ is satisfied, also a condition $G(x)$ will be satisfied; that the law is a historical one means that the condition $F(x)$ is not satisfied for all the objects of the universe under discussion, but only for some of them.[25] And Engels' introductory thesis that the eternal laws of nature (more precisely—the natural science) become more and more historical means that together with the development of our knowledge we find that the antecedents of laws of the type (2) have to be more and more expanded; which —as we shall see—is in full agreement with Engels' concept of relative truth. The adequacy of the above interpretation of the concepts of the eternal and historical character of natural laws is further reinforced by the fact that Engels additionally distinguishes between historical and universal laws: "Wenn wir also von allgemeinen Naturgesetzen sprechen wollen, die auf *alle* Körper — vom Nebelfleck bis zum Menschen — gleichmäßig passen, so bleibt uns nur die Schwere und etwa die allgemeinste Fassung der Theorie von der Umwandlung der Energie".[26] Thus the universal natural law is a sentence of the form:

(3) $G(x)$,

i.e. a sentence which (if it is true) can refer to all the objects of the universe under discussion. So it does not have the historical character in the Engels' sense.

It follows from the above that Engels allowed in the natural sciences strictly general theorems in the form of extra-basic theorems. The question arises whether he allowed such theorems in the humanities as well. For the creators of Marxism undoubtedly allowed theorems in the humanities that we call today historical generalizations. Many theorems of this type can be found in the economics and sociological works of both the creators of Marxism. They even called certain historical generalizations laws; thus, e.g. F. Engels wrote: "It seems to be the law of historical development that in no European country the bourgoises can—at least for a longer period of time—acquire so exclusive power as had been preserved by the feudal aristocracy in the Middle Ages".[27] This does not mean, however—as some Marxists would have it—that according to the classics all the theses of the humanities possess the character of historical sentences (i.e. either singular sentences, or historical generalizations). In order to find arguments supporting this

view the following characteristics of political economy, done by Engels, should suffice: "Thus political economy is a historical science in its essence. It speaks of a historical subject, i.e. of one constantly changing; it investigates mainly the specific laws of each particular phase of development of production and exchange—and only by the end of this investigation it will be able to establish the not numerous, absolutely general laws obligatory in production and exchange at all times. It is understood, at the same time, that the laws obligatory for some defined ways of production and forms of exchange are also obligatory for all the historical periods for which these ways of production and forms of exchange are common. Thus, e.g., the moment metallic currency has been introduced a number of laws begins to act obligatorily in all countries and in all historical periods in which metallic currency takes part in the exchange".[28] Formulating this idea in the terms that Engels used in his *Dialectics of Nature* we might say the following: all economic laws (assuming their truth) are eternal; few of them are universal and the remaining ones—most of them—are historical. It follows that—according to Engels—strictly general sentences are acceptable in the humanities, for the historical character of some of them is not connected with the fact that in formulating them individual names occur or their derivatives, but with the fact that the actual area of their non-vacuous satisfiability is limited in time and space.

We have so far spoken of a certain kind of strictly general sentences, namely of factual theorems, i.e. of conditional theorems of the type that their non-vacuous satisfaction is not excluded on the basis of the empirical knowledge we possess. However, as I tried to demonstrate elsewhere, according to Marx, the idealizing laws, i.e.—roughly speaking—conditional theorems of the type that their non-vacuous satisfiability is excluded on the basis of the empirical knowledge we possess, form the basic corpus of the laws of economy. Marx also tended to see in laws of this type the basic type of laws in all the empirical sciences. Since I have presented the argumentation for the accuracy of these theses in another works,[29] I shall now only attempt to demonstrate that also Engels distinguished idealizing laws in the natural sciences, although in comparing Engels' and Marx's concepts it is difficult to avoid the impression that Marx thought the method of idealization more important than Engels.

It may be assumed that Engels saw that at least certain laws of the natural sciences possess idealizing character, because he mentioned whole fields of physics in which definite idealizing assumptions were adopted, and because in analysing concrete natural laws he emphasised that they were true only under the condition that certain assumptions, idealizing (non-realistic) in character are adopted. He wrote, e.g. that "Die Mechanik betrachtet die Wirkung des Stoßes als *rein vorgehend*. Aber in der Wirklichkeit geht's anders zu. Bei jedem Stoß wird ein Teil der mechanischen Bewegung in Wärme umgesetzt, und Reibung ist gar weiter nichts als eine Form des Stoßes, die fortdauernd mechanische Bewegung in Wärme umsetzt".[30] Analysing the research practice of physicists he repeatedly stated that they are right in applying the method of idealization. He writes, e.g. about the investigations of Wiedemann on the electric pile: "(...) hat derselbe Wiedemann ganz recht, wenn er in seinen theoretischen Ableitungen auf Nebenumstände, die die Reinheit des Prozesses fälschen".[31] Analysing Galileo's law of free fall Engels writes that "(...) das Fallgesetz schon bei mehreren Minuten Fallzeit unrichtig wird, weil der Erdhalbmesser dann nicht mehr ohne Fehler $= \infty$ gesetzt werden kann und die Attraktion der Erde zunimmt, statt sich gleich zu bleiben, wie Galileis Fallgesetz voraussetzt. Trotzdem wird dies Gesetz noch fortwährend gelehrt, die Reserve aber weggelassen!"[32] In other words, Engels states that in the light of Newton's theory the law of free fall cannot be formulated in the way Galileo had done:

$$(4) \qquad h = g\,\frac{t^2}{2},$$

where h is the path of the falling object, $t-$time of falling, and g is acceleration treated as a constant value. For in the light of later findings it is known that g is not constant, but it changes depending on the distance of the falling object from the centre of the earth. Thus (4) is false, and the idealizing law is true, which possesses in its antecedent the adequate idealizing assumptions, e.g. that the radius of the earth is infinite. Engels repeatedly drew attention to the necessity of introducing such 'idealizing corrections' to the circulatory formulations of laws in physical theories.[33]

Thus the conclusion from the above discussion on the form of extra-basic theorems in the concept of the creators of Marxism is that they

accept as extra-basic theorems of the empirical sciences: (1) historical generalizations, (2) strictly general factual sentences (factual laws), (3) strictly general idealizing sentences (idealizing laws). The sentences of the categories (1) and (2) are factual theorems.

The fact that idealizing laws are present among the extra-basic theorems automatically presupposes the negative answer to the question, whether all the extra-basic theorems are equivalent to the open (or closed) sets of basic theorems. But we do not have to turn to idealizing laws in order to find that, according to the concepts of the creators of Marxism, the answer to this question is negative. For, as we shall see further, they adopted the realistic interpretation of all the extra-basic theorems, i.e. among others of the strictly general factual sentences; so they treated them as describing certain regularities inaccessible to observation. In that case, however, it follows that — being consistent — they could not adopt at the same time the positivistic point of view saying that the content of the laws of science can be brought down to what definite sets of observational sentences state. Anyway, Engels expressed his opinion on the point of view which did not accept the fact that theories possess their own cognitive values quite clearly: "Erst macht man Abstraktionen von den sinnlichen Dingen und dann will man sie sinnlich erkennen, die Zeit sehn und den Raum riechen. Der Empiriker vertieft sich so sehr in die Gewohnheit des empirischen Erfahrens, daß er sich noch auf dem Gebiet des sinnlichen Erfahrens glaubt, wenn er mit Abstraktionen hantiert".[34]

Let us now take up the question of the hypothetical character of extra-basic theorems in Engels. It is closely connected with the problem of the relativity of truth. Let us then begin with an attempt at defining what Engels understood by this concept. The contexts in which Engels spoke about relative truths in general terms (e.g. "(...) what is accepted as truth today has its hidden false side that will come out openly later, and what we think to be false today, has its true side, thanks to which it might have been considered truth before"[35]) can be explained, when we pass to Engels' analyses of concrete examples of relative truths. Boyle's law is, according to Engels, a law possessing the character of relative truth: "Let us take as an example the well known Boyle's law, stating that in the same temperature the volume of gas is inversely proportionate to the pressure it is under. Regnault has discovered

that the law is not confirmed in certain cases. And if he were 'the philosopher of reality', he would have to say: Boyle's law is variable, i.e. it is not a real truth, i.e. it is not truth at all, i.e. it is a mistake. But he would thus comit a much bigger mistake than that in Bolye's law; the grain of truth he has discovered would be overshadowed by the greatness of mistake: thus he would make a mistake of his initially correct discovery in comparison with which Boyle's law with small mistake in it would be truth. But Regnault, as a man of science, did not let himself be carried away like that, he continued his investigations and has found out that Boyle's law is only approximately correct, and in particular, it is not valid for gases that undergo condensation under the influence of pressure — when the pressure reaches the point at which condensation occurs. Thus Boyle's law appeared to be correct only within certain limits. Is it, however, absolutely and finally true within these limits? No physicist would hold this view. He would say that the law is valid within certain limits of temperature and pressure and only with respect to certain gases, and within these narrower limits further narrowing or modification on the basis of further investigations cannot be excluded".[36] We might, on the basis of the above, assume that, according Engels' intentions, the general theorem of the form:

$$(5) \qquad F_1(x) \wedge F_2(x) \wedge \ldots \wedge F_k(x) \rightarrow G(x)$$

is relatively true, if it is false, and the product of the set of objects satisfying the antecedent of law (5) and the set of objects satisfying the consequent of this law is non-empty; in other words — if not all objects possessing properties $F_1, F_2, \ldots, F_k$ possess also a property G, but at the same time certain objects possessing these properties possess also the property G. The evolution of Boyle's law that Engels discussed in the fragment quoted above consisted in adding new conditions to the antecedent. At the same time, it is clear that Engels did not exclude the possibility that future research would demonstrate the inadequacy of the original formulation of Boyle's law. So Engels did not hold the view that narrowing the range of the laws of science is the only aspect of progress in the empirical sciences. It is noteworthy, however, that the scope of the concept of relative truth may embrace both the factual theorems, and the idealizing laws; in the latter case the ideal types of real objects are the objects non-vacuously satisfying the law (5). And in view of the fact

that all the scientific theorems, except singular sentences and certain simple general sentences,[37] are relative truths, we may say that Engels adopted the hypothetical character of the extra-basic theorems of the empirical sciences. As relative truths "(...) they always contain, with no exception, more data requiring qualifications than data not requiring qualifications, i.e. correct data",[38] and in that case they may undergo further modifications in the development of science. Because of this very hypothetical character of the extra-basic theorems "(...) the most valuable result would be making us as doubtful as possible towards our present cognition since it is most probable that we find ourselves at the beginnings of the history of mankind and generations that will correct us rather; there will, it seems, be far more than those whose cognition we shall have to correct – how often with considerable contempt".[39]

The thesis that every extra-basic theorem is hypothetical in character, i.e. that it is retractable, does not mean the same, however, as the thesis that every extra-basic theorem is a hypothesis in the sense of the term 'hypothesis' that was used by Engels himself. The first of these theses Engels accepted – as I have tried to demonstrate above. He rejected the other. For by hypothesis Engels understood a sentence which fulfils all the requirements of the law of science apart from one thing: it has not been empirically justified yet[40] (we shall say more on this after we pass to question (IV)).

5. THE PROBLEM OF JUSTIFYING EXTRA-BASIC THEOREMS

The most general formulation of the criterion for justifying the extra-basic theorems in the empirical sciences is contained in the following statement of Engels: "Darüber sind wir alle einig, daß auf jedem wissenschaftlichen Gebiet in Natur wie Geschichte von den gegebenen *Tatsachen* auszugehn ist, in der Naturwissenschaft also von den verschiednen sachlichen und Bewegungsformen der Materie; daß also auch in der theoretischen Naturwissenschaft die Zusammenhänge nicht in die Tatsachen hineinzukonstruieren, sondern aus ihnen zu entdecken und, wenn entdeckt, erfahrungsmäßig soweit dies möglich nachzuweisen sind".[41] Note a very characteristic thing – that according to Engels, the discovery of a regularity, and the justification that it takes place

(i.e. justification of the law describing it), are two different matters. And especially the fact that the regularity has been discovered 'in facts' (possibly through observation) is not the justification of the law describing it at all; in order to justify the law it should be tested 'in an empirical way'. The following statement formulates it clearly: "Die Entwicklungsform der Naturwissenschaft, soweit sie denkt, ist die *Hypothese*. Eine neue Tatsache wird beobachtet, die die bisherige Erklärungsweise der zu derselben Gruppe gehörenden Tatsachen unmöglich macht. Von diesem Augenblick an werden neue Erklärungsweisen Bedürfnis — zunächst gegründet auf nur beschränkte Anzahl von Tatsachen und Beobachtungen. Ferneres Beobachtungsmaterial epuriert diese Hypothesen, beseitigt die einen, korrigiert die andren, bis endlich das Gesetz rein hergestellt. Wollte man warten, bis das Material fürs Gesetz *rein* sei, so hieße das, die denkende Forschung bis dahin suspendieren, und das Gesetz käme schon deswegen nie zustande".[42] Engels substantiates this statement by analysing concrete examples from the history of the natural science — and from this analysis he draws the conclusion that a theorem in the natural science will be considered justified if the observational predictions derived from it are confirmed. Thus, e.g., Engels commented on the justification of the heliocentric system in the following way: "The Copernican solar system was a hypothesis for 300 years for which one could bet 100, 1000, 10,000 to 1, however, only a hypothesis; when, however, Leverrier, on the basis of the data provided by this system, not only derived the necessity of existence of an unknown planet, but calculated the place in the sky where it should be found, and when, next, Galle really found this planet, then the Copernican system was proved".[43] Naturally, Engels uses the term 'proof' in a different sense from that common today — a proved proposition in the sense of Engels (i.e. tested — in contemporary terminology) is also a relative truth, since it cannot be guaranteed that further investigations will show no need of its modification. Thus neither the hypothesis, nor the law (i.e. a proposition that has been empirically proved) are absolute truths in the sense of Engels; they are at the most relatively true. We might therefore, say that, according to Engels, a hypothesis is such an extra-basic theorem which, at that given time had not undergone empirical testing. Law, on the other hand, is such an extra-basic theorem which

has already undergone empirical testing, and which has been confirmed as a result of this testing; thus, for the time being, we cannot see what further modifications that theorem might undergo, which does not mean that such modifications are excluded in future—it is enough for this theorem to loose the status of a law, and to assume the status of a hypothesis that may undergo further changes, if observations contradicting it come to light.

Empirical testability of the idealizing laws shows certain methodological peculiarities in comparison with the testability of ordinary factual laws—peculiarities to which the creators of Marxism themselves drew our attention. Since I have discussed the matter extensively elsewhere I shall omit it here, merely stating that, according to Marx and Engels, the justification of the idealizing laws as well as the justification of the factual laws consists in testing them on the basis of experiment—only in the case of the idealizing laws the procedure is more complex and marked by certain peculiarities.[44]

I have said above that, according to Engels, observation alone does not determine the theoretical theorems (as we have seen above—it can merely serve the purpose of testing the predictions derived from them). The following thesis formulates it emphatically: "Man mag noch so viel Geringschätzung hegen für alles theoretische Denken, so kann man doch nicht zwei Naturtatsachen in Zusammenhang bringen oder ihren bestehenden Zusammenhang einsehn ohne theoretisches Denken. Es fragt sich dabei nur, ob man dabei richtig denkt oder nicht, und die Geringschätzung der Theorie ist selbstredend der sicherste Weg, naturalistisch und damit falsch zu denken".[45] Naturally, not a word is said about deriving the idealizing laws from observation, since they concern directly the ideal types, and not the really existing objects. It is for this reason that Marx was such a radical oponent of positivist phenomenalism. Marx radically distinguished two ways of investigating: the method of idealization as distinct from the positivist method of collecting observations and generalization; and this according to the principle that "(...) One of these conceptions fathoms the inner connection, the physiology, so to speak, of the burgeois system, whereas the other takes the external phenomena of life, as they seem and appear and merely describes, catalogues, recounts and arranges them under formal definitions. (...) the one expresses the intrinsic connections more or less

correctly, the other, with the same justification – and without any connection to the first method of approach – expresses the apparent connections without any internal relation".[46] Marx explains why the method of description of observations and generalization does not give epistemologically important results. We can observe only phenomena, which undergo so numerous and diversified influences that, without a theory, which explains the dependencies between them, they are entirely incomprehensible, and the observation of the coincidence of these phenomena may lead to false results. For this reason it is necessary to apply the method of idealization (abstraction in Marx's terminology): take into consideration only some of those influences and establish simple dependencies that will provide the starting point for applying the method of concretization: taking into consideration more and more of the previously disregarded factors and modifying the primary dependencies until the real course of phenomena in all their complexity is reconstructed. Speaking of the method of his main work, Marx described it as follows: "In Book I we analysed the phenomena which constitute the *process of capitalist production* as such (...) with no regard for any of the secondary effects of outside influences. But this immediate process of production does not exhaust the life span of capital. It is supplemented in the actual world by the *process of circulation*, which was the object of study in Book II. (In this Book) we must locate and describe the concrete forms which grow out of the movements of capital as a whole. In their actual movement capitals confront each other in such concrete shape, for which the form of capital in the immediate process of production, just as its form in the process of circulation, appear only as special instances. The various forms of capital, as evolved in this book, thus approach step by step the form which they assume on the surface of society, in the action of different capitals upon one another, in competition, and in the ordinary consciousness of the agents of production themselves".[47] "And it is these ready relations and forms – says Marx – that occur in real production as premises, for capitalist production carries out its process in the forms created by itself and these forms, its result, occur to the same extent in the process of reproduction as ready premises. As such premises they define in practice the way of acting of the particular capitalists etc., they provide motifs for them in the form in which they are reflected in their minds. Vulgar economy does only

one thing: it expresses as doctrines the minds of the capitalists which have, according to their motifs and images the power of ruling over the phenomena of the capitalist way of production. The less deeply does vulgar economy reach under the surface of phenomena, the more it is merely a systematized echo of these phenomena, the more it is convinced that it is 'in agreement with the laws of nature' and is far from any abstract speculations".[48]

Connected with the above presented opposition of the creators of Marxism to the positivistic phenomenalism is their opposition to the tendency in reasoning, that might be defined as linguistic phenomenalism. This tendency may briefly be characterized as the belief that natural language is the optimal language for cultivating the sciences (at least the humanities), and even if the conceptual apparatus of the natural language is modified, the common intuitions are taken into consideration as a matter of programme. However, scientific reasoning differs from common thinking in that, in the first place, it applies the method of idealization: "It should not astonish us, then, that vulgar economy feels particularly at home in the estranged outward appearances of economic relations in which these *prima facie* absurd and perfect contradictions appear and that these relations seem the more self-evident the more their internal relationships are concealed from it, although they are understandable to the popular mind".[49] Thus—thanks to the application of the method of idealization, which allows to establish the 'heart of the matter', and concretization, which allows the derivation of the 'form in which the thing demonstrates itself'—the science may merely explain the things, which 'common beliefs' concern. It cannot take them as a starting point, however, if it is not to be merely a simple description of what the 'common sense' sees. It is in this connection —states Engels—that the conceptual apparatus of the Marxist political economy does not take natural language as its starting point, while pre-Marxist "Political Economy has generally been content to take just as they were the terms of commercial and industrial life, and to operate with them, entirely failing to see that by so doing, it confined itself within the narrow circle of ideas expressed by those terms".[50] It is only in the light of the idealized reconstruction of the real phenomena, i.e. after the method of idealization has been applied to explain these phenomena, that it comes out that the commonly used concepts make

it impossible, in fact, to grasp the basic regularities that govern these phenomena. Thus, e.g. the common concept of profit not only gives no explanation as to where the capitalist's income comes from, but it even hinders the establishment that this income is the result of unpaid labour of the worker.[51] The Marxist concept of surplus value, which has no common counterpart, was necessary in order to find out that "The profit (...) is thus the same as surplus value, only in a mistified form that is nonetheless a necessary outgrowth of the capitalist mode of production".[52] In order to state that commonly understood profit is the surplus value — only a little 'modified' first, the conceptual apparatus of natural language had to be put aside. For those who "made the common notion of profit their starting point, i.e. they gave way to the suggestion inbuilt in it that profit is a form (...) assigned to capital as a separate tool of production in the same way as pension is assigned to land" [53] — for all those the law of surplus value is "a deeply hidden secret".[54]

On the basis of the outlined above system of concepts, which rather uniquely rejects the idea of deriving the extra-basic theorems from observation Engels' remarks concerning the role of induction in the empirical sciences do not seem very clear. E.g., Engels writes "Induktion und Deduktion gehören so notwendig zusammen wie Synthese und Analyse. Statt die eine auf Kosten der andern einseitig in den Himmel zu erheben, soll man suchen, sie jede an ihrem Platz anzuwenden, und das kann man nur dann, wenn man ihre Zusammengehörigkeit, ihr wechselseitiges Sichergänzen im Auge behält".[55] If we assume — which is not self-evident at all — that Engels uses in this fragment the term 'induction' to mean, what is today understood by ampliative induction, then at least three interpretations are possible that could be in agreement with the above presented image of the methodological concepts of the creators of Marxism.

a. Among the extra-basic theorems of the empirical sciences (more precisely — among the factual theorems) a subclass of phenomenological generalizations (formulated with the help of observational terms) should be distinguished, which are justified with the help of ampliative induction.

b. In the early phases of development of the empirical sciences the scheme of ampliative induction is used in order to justify certain (and

possibly all?) extra-basic theorems; with the development of the empirical sciences conclusions of this type lose their importance, and in highly developed sciences they entirely give way to the procedure of testing (among others — testing in terms of the concretization of idealized laws).

c. The scheme of ampliative induction is used in the empirical sciences, however, not to justify extra-basic theorems, but to discover them; thus it is a heuristic scheme, and not an evidencing one — the justifications are provided for extra-basic theorems only by successful attemps to test them.

The fact that Engels rejects the basic principle of using ampliative induction which says that the number of singular premises increases the justification of the general conclusion speaks against the first two interpretations. For, in the course of his discussion of the mechanical equivalent of heat he stated: "Die Dampfmaschine gab den schlagendsten Beweis, daß man Wärme einsetzen und mechanische Bewegung erzielen kann. 100 000 Dampfmaschinen bewiesen das nicht mehr als *eine*, drängten nur mehr und mehr die Physiker zur Notwendigkeit, dies zu erklären".[56] Besides, interpretation (a) would lead to a not very clear image of the empirical sciences: it is not known what role should these phenomenological generalizations play in the set of extra-basic theorems. It would be clear if, e.g., we assumed that Engels accepted Mill's approach: first the phenomenological generalizations should be established inductively, and then further theorems should be deduced from them, which should undergo testing. But it is clear that idealized laws can be neither deduced from the phenomenological generalizations, nor justified inductively. And Engels was well aware of the latter. He wrote: "Die Dampfmaschine gab den schlagendsten Beweis, daß man Wärme einsetzen und mechanische Bewegung erzielen kann. 100 000 Dampfmaschinen bewiesen das nicht mehr als *eine*, drängten nur mehr und mehr die Physiker zur Notwendigkeit, dies zu erklären. Sadi Carnot war der erste, der sich ernstlich dranmachte. Aber nicht per Induktion. Er studierte die Dampfmaschine, analysierte sie, fand, daß bei ihr der Prozeß, auf den es ankam, nicht *rein* erscheint, von allerhand Nebenprozessen verdeckt wird, beseitigte diese für den wesentlichen Prozeß gleichgültigen Nebenumstände und konstruierte eine ideale Dampfmaschine (oder Gasmaschine), die zwar ebensowenig herstellbar ist wie z.B. eine geometrische Linie oder Fläche, aber in ihrer Weise den-

selben Dienst tut wie diese mathematischen Abstraktionen: Sie stellt den Prozeß rein, unabhängig, unverfälscht dar".[57]

A certain arbitrariness of interpretation (b) speaks against it — as well as the above mentioned violation by Engels of the basic principle of reasoning through ampliative induction. Why should a conclusion that was once legitimate lose legitimacy simply as the time goes by? On the other hand, the fact that Engels actually put forth the thesis that the empirical sciences pass from the phase of collecting facts to the phase of looking for a theory, speaks in favour of this interpretation. Thus, e.g. Engels wrote about the modern biological sciences: "Zoologie und Botanik bleiben zunächst Sammelwissenschaften, bis die Paläontologie hinzutritt — Cuvier — und bald darauf die Entdeckung der Zelle und die Entwicklung der organischen Chemie. Damit vergleichende Morphologie und Physiologie möglich, und von da an beide wahre Wissenschaften".[58]

The fact speaking in favour of interpretation (c) is that it is in agreement with all the statements of Engels that have been quoted. In the case of this interpretation also the thesis that during the first phases of their development the empirical sciences use ampliative induction, which they give up later, can easily be explained — and this avoiding the objection of arbitrariness. For, since induction is both earlier and later merely a method of discovery, and not justification, this thesis is obvious — because in the course of development of science the scientific theories become more and more abstract and thus it is more difficult to find ideas for new theories through observation. It has to be stated, however, that Engels did not make a clear distinction between the 'context of discovery' and the 'context of justification' and that we may accept interpretation (c) only according to the principle that after joining it to the other epistemological concepts of the creators of Marxism we shall obtain the most consistent system.

6. MARX'S MODEL OF EXPLANATION

I have put this section here merely for the sake of orderly presentation since I have dealt with the model of explanation in Marx exhaustively elsewhere.[59] I shall present here only the general conclusion in order to obtain a possibly full image of the epistemological concepts of the

creators of Marxism: Marx's model of explanation has the form

(6) $T^k \dashv T^{k-1} \dashv \ldots \dashv T^i \leadsto| T^0 \wedge P \rightarrow E$,

where $T^k, \ldots, T^i$ are idealized laws (with, respectively, $k, \ldots, i$ idealizing assumptions), T^0 is a factual law, P—initial conditions of law T^0, and E—a sentence being explained, $\dashv$ is the relation of strict concretization, and $\leadsto|$ is the relation of approximate concretization. In other words, explanation according to the scheme (6) consists in successive derivation from the initial idealized laws (and certain additional premises) of idealized laws with fewer idealizing assumptions and which, at the same time, are somehow modified as regards the initial laws; in this way an idealized law is reached which is based on a relatively small number of idealizing assumptions and then, possibly, approximate concretization is used: from this law the factual law is derived, which says that for cases slightly differing from the idealizing assumptions of the law that is undergoing concretization the consequent of this law will also be satisfied with sufficient approximation; from this approximative law we may derive sentences about real facts that have been observed. It is noteworthy that Marx rejected explanations which referred exclusively to factual laws which were not based in the adopted idealizing theory, i.e. they were not concretizations (strict or approximate) of any idealized laws.

7. THE PROBLEM OF REFERENCE OF THE EXTRA-BASIC THEOREMS

The question (VI): do extra-basic theorems refer to reality, should be discussed separately for factual theorems and for idealizing laws.

As far as the first problem is concerned the positive answer to this question should not rise any doubts. The creators of Marxism undoubtedly believed that factual scientific laws concerned objective regularities—that they were their true or false descriptions. After all the natural sciences are "(...) *ordering* sciences, sciences about processes about the origins and development (...) of things and about the relation that combines these natural processes into one great whole".[60] Also concrete natural laws factual in character were interpreted by Engels in a realistic, and not instrumental way. Thus, e.g. characterizing the law of conservation of energy (which Engels treated as factual) the author of *Dialectics of Nature* writes: the law of conservation of energy

demonstrates that "(...) all the so-called forces, mainly active in the inorganic matter, the mechanical force and its supplement known as potential energy, heat, radiation (light or radiating heat), electricity, magnetism, chemical energy, are various forms expressing the universal movement, which pass from one form into another under defined quantitative conditions, so that in place of the disappearing defined quantity of one of them a defined quantity of another appears and thus the whole movement in nature is reduced to this constant process of transformation of one form into another".[61] As can be seen, Engels was far from interpreting this law as 'convenient fiction', whose only function is connecting observational sentences; he treated the above mentioned law as a description of real dependencies between various types of energy; so, he treated it realistically, similarly to spontaneous physicists' attitude. Incidentally, Engels often opposed the instrumental tendencies he observed among the physicists-theoreticians. For instance, analysing the development of Helmholtz's views on the relation of the concepts of 'a live force' (kinetic energy) and labour Engels writes: "So wenig klar war sich Helmholtz 1847 über die gegenseitige Beziehung von lebendiger Kraft und Arbeit, daß er gar nicht einmal merkt, wie er das frühere proportionelle Maß der lebendigen Kraft in ihr absolutes verwandelt; daß ihm ganz unbewußt bleibt, welche bedeutende Entdeckung er mit seinem kühnen Griff gemacht, und er sein $mv^2/2$ nur aus Bequemlichkeitsrücksichten empfiehlt gegenüber dem mv^2! Erst allmählich hat man das $mv^2/2$ auch mathematisch bewiesen (...)".[62] Engels' view concerning the realistic interpretation of these concepts is demonstrated in the following statement: "Für uns, die wir gesehn haben, daß lebendige Kraft nichts andres ist als das Vermögen einer gegebnen mechanischen Bewegungsmenge, Arbeit zu leisten, für uns ist es selbstverständlich, daß der mechanische Maßausdruck dieses Arbeitsvermögens und der der von ihm wirklich geleisteten Arbeit einander gleich sein müssen; daß also, wenn $mv^2/2$ die Arbeit mißt, die lebendige Kraft ebenfalls $mv^2/2$ zum Maß haben muß. Aber so geht's in der Wissenschaft. Die theoretische Mechanik kommt auf den Begriff der lebendigen Kraft, die praktische der Ingenieurs auf den der Arbeit, und zwingt ihn den Theoretikern auf. Und so sehr hat man sich über dem Rechnen des Denkens entwöhnt, daß man jahrelang den Zusammenhang beider nicht erkennt, die eine nach mv^2, die andere nach $mv^2/2$ mißt, und endlich für beide $mv^2/2$

akzeptiert, nicht aus Einsicht, sondern der Einfachheit der Rechnung halber!".[63]

The question of realistic interpretation of the idealizing laws is a little different. The creators of Marxism assumed a realistic attitude towards them as well, but the argumentation they used to justify this attitude was slightly different. Let us first discuss Engels' views on this matter. Answering K. Schmidt's objection that the law of value is only a cognitive fiction, although a profitable one, Engels wrote: "What you, Sir, have against the law of value concerns *all* concepts, if we are to consider them from the point of view of reality. (...) the concept of a thing and its reality run alongside like two asymptotes which come nearer and nearer to each other and never meet. (...) Although a concept has the basic nature of a concept, i.e. it is not coextensive *prima facie* with reality from which it had to be abstracted, it is nevertheless something more than fiction, unless you consider fiction all the results of reasoning, for reality corresponds to them only indirectly and only in an asymptotic approximation". [64] The same applies to other scientific concepts idealizing in character — Engels asks rhetorically: "Did feudalism ever correspond to its concepts? (...) Was this system a fiction because it had existed in its absolutely classical form only for a short time in Palestine, and even there usually only on paper?".[65] Engels' views on the realistic character of the idealizing laws may be formulated as follows: an idealizing law possesses realistic character to the extent, to which its corresponding approximate law is true (i.e., if the deviations of real cases from the ideal cases described in the idealizing laws do not go beyond a certain acceptable measure of deviations). This attitude was, however, liable to the objection that the law of value has no realistic character, since its corresponding approximate law ("the evaluations of real goods approximate the values of these goods") is not fulfilled in contemporary free competitive capitalism. Engels realized this and he found the following solution in the appendix to the 3rd volume of the *Capital*. Although the law of value cannot be treated even as an approximate description of the free competition economy (for the deviations of the prices from values are too large), it may, however, be treated as an approximate description of the relations of goods exchange in historically earlier epochs. On the basis of rich historical material Engels documents the thesis that "The Marxian law of value holds generally, as far as economic

laws are valid at all, for the whole period of simple commodity production, that is, up to the time when the latter suffers a modification through the appearance of the capitalist form of production".[66] Formulating it in terms of the conceptual apparatus used above: the law of value may be interpreted realistically for, although its corresponding approximate law is not satisfied for developed capitalist production, it is satisfied for simple commodity economy.

Engels' interpretation of the idealizing laws justified their realistic character by the fact that they can be 'roughly' applied to the description of real phenomena (more precisely: that their correspondent approximate laws are true).

Marx understood the realistic character of the idealizing laws a little differently. He treated them as descriptions of really active regularities (only isolated from the disturbing factors thanks to the adoption of adequate idealising assumptions). He often stated that it is the adoption of idealizing assumptions that makes possible the establishment of actual regularities. Thus, e.g. justifying the adoption of the idealizing assumption that the demand is equal to supply Marx writes: "It is evident that the real inner laws of capitalist production cannot be explained by the interaction of supply and demand (...), because these laws cannot be observed in their pure state until supply and demand cease to act, i.e. are equated".[67]

Marx's views are certainly more difficult to explicate although they are, at the same time, more radical, i.e. in less agreement with instrumentalism. They are also, it seems, more adequate, for surely the idealizing laws are so abstract (i.e. based on so many idealizing assumptions) that their correspondent approximate laws are false. Anyway, there is no reason why the existence of such laws should be excluded. The laws of this type might be treated as realistic in the sense of Marx, although they would not be realistic in the sense of Engels. On the other hand, it is extremely difficult to explicate Marx's concept, for it would require the explication of the meaning of the statement that a law which is vacuously satisfied by real phenomena describes the regularities that occur in them. There is no doubt, however, that idealizing laws are interpreted in Marx's sense by physicists, chemists, economists, etc. — at least as long as they cultivate their own field and do not deal with reflection on their own proceedings.

8. THE PRINCIPLES OF DEVELOPMENT OF THE
EMPIRICAL SCIENCES

We saw in Section 3 that, according to Engels, one of the aspects of progress in science was that additional assumptions (realistic or idealizing) are
attached to the law of science (factual or idealizing) which is a relative
truth, in this way providing a law narrower in range, i.e. with a more
developed antecedent and the same consequent. Let us assume that we
are dealing with an idealising law of the form:

$$(7) \qquad f_1(x) \wedge f_2(x) \wedge \ldots \wedge f_k(x) \rightarrow G(x),$$

which is relatively true. Now, the process of selecting the additional
idealizing assumptions to the initial law is carried out with the help of
the procedure of concretization of this law. For we may learn about
the need of a new assumption only on the basis of failure of the empirical
predictions derived from the law (7), i.e. – on the basis of this law being
rejected; and this, as has been mentioned, is possible only through its
concretization. In other words, the process of narrowing the range of
law (7) is carried out as follows. After the law (7) has been formulated,
it undergoes concretization: its idealizing assumptions are removed and
definite corrections (strict or only approximate) are introduced to its
consequent. As a result a factual law is obtained, which is a concretization (strict or approximate) of the idealizing law (7). Let us assume that
observations speak for rejecting it. Unless the subsidiary premises assumed in the concretization procedure are removed, we may conclude
on the basis of the falsity of this factual law that the idealizing law (7)
is false. However, not all the assumptions of the thesis (7) are given
up at once, but an attempt is made at narrowing its range in such a way
that it would not speak of the ideal types of the phenomenon falsifying
the law. Thus a subsequent hypothesis is put forth of the form:

$$(8) \qquad f_1(x) \wedge f_2(x) \wedge \ldots \wedge f_k(x) \wedge f_{k+1}(x) \rightarrow G(x).$$

The relation between the law (8) and the law (7) may be described with
the phrase: the law (8) is relatively more true than the law (7). Testing
through concretization of law (8) follows and, if it appears false again a
subsequent assumption is attached to it. However, as has been mentioned
above, the procedure described here does not exhaust all the aspects
of empirical sciences development.

For there are cases in the natural sciences when the theories require not range modification, but entire change, when they are substituted by new theories that disagree with them. Engels writes on the problem as follows: "In der Naturwissenschaft selbst aber begegnen uns oft genug Theorien, in denen das wirkliche Verhältnis auf den Kopf gestellt, das Spiegelbild für die Urform genommen ist, und die daher einer solchen Umstülpung bedürfen. Solche Theorien herrschen oft genug für längere Zeit. Wenn die Wärme während fast zwei Jahrhunderten als eine besondre geheimnisvolle Materie galt, statt als eine Bewegungsform der gewöhnlichen Materie, so war das ganz derselbe Fall, und die mechanische Wärmetheorie vollzog die Umstülpung. Nichtsdestoweniger hat die von der Wärmestofftheorie beherrschte Physik eine Reihe höchst wichtiger Gesetze der Wärme entdeckt, und besonders durch Fourier und Sadi Carnot die Bahn freigemacht für die richtige Auffassung, die nun ihrerseits die von ihrer Vorgängerin entdeckten Gesetze umzustülpen, in ihre eigne Sprache zu übersetzen hatte. Ebenso hat in der Chemie die phlogistische Theorie durch hundertjährige experimentelle Arbeit erst das Material geliefert, mit Hilfe dessen Lavoisier in dem von Priestley dargestellten Sauerstoff den reellen Gegenpol des phantastischen Phlogiston entdecken und damit die ganze phlogistische Theorie über den Haufen werfen konnte. Damit aber waren die Versuchsresultate der Phlogistik durchaus nicht beseitigt. Im Gegenteil. Sie bleiben bestehn, nur ihre Formulierung wurde umgestülpt, aus der phlogistischen Sprache in die nunmehr gültige chemische Sprache übersetzt, und behielten soweit ihre Gültigkeit".[68] The following conclusions may be drawn from the above: (1) subsequent theories (at least some of them) are as a whole in disagreement with the earlier theories (which does not mean—as we have seen—that each constituent theorem of the earlier theory must disagree with some theorem of the later theory); (2) observational sentences that confirm the earlier theory and were explained by it are also explained by the later theory, although they are newly interpreted—on the basis of the later theory; (3) the sense of the theoretical terms of the latter theory is at least partially different from the sense of the theoretical terms of the earlier theory; thus basic theoretical terms do not preserve their meaning in case of change of theory.

Let us illustrate the above conclusions with some more statements

of Engels. In order to support the conclusion (1) we might quote, his analyses of concrete examples of overthrowing physical or chemical theories, as well as the general statements he made. Since Engels is frequently thought to hold the concept of cummulative science development, we shall quote his lengthy statement concerning Wiedemann's concepts in the field of electrostatics. Engels says that the author operates three theories "(...) Theorie der Strombildung bei ihm: zuerst die alt-überkommene, vermittelst des reinen Kontakts; zweitens die vermittelst der schon abstrakter gefaßten elektrischen Scheidungskraft (...). Wie er nicht merkt, daß die zweite Erklärung die erste umstößt, ebensowenig ahnt er, daß die dritte ihrerseits die zweite über den Haufen wirft. Im Gegenteil, der Satz von der Erhaltung der Energie wird ganz äußerlich an die alte, von der Routine überkommene Theorie angefügt, wie man einen neuen geometrischen Lehrsatz an die früheren anhängt. Keine Ahnung davon, daß dieser Satz eine Revision der ganzen traditionellen Anschauungsweise auf diesem wie auf allen andern Gebieten der Natur-wissenschaft nötig macht".[69] Anyway, Engels states the thesis of anti-cummulativism in quite a general form. After the description of the revision of subsequent hypotheses of the natural science that has been quoted on p. 515 Engels writes: "Die Anzahl und der Wechsel der sich verdrängenden Hypothesen—bei (...) der Naturforscher—bringt dann leicht die Vorstellung hervor, daß wir das *Wesen* der Dinge nicht erkennen können (...). Dies ist der Naturwissenschaft nicht eigentümlich, da alle menschliche Erkenntnis in einer vielfach verschlungnen Kurve sich entwickelt, und die Theorien auch in den geschichtlichen Disziplinen inklusive Philosophie sich ebenso verdrängen (...)".[70]

The conclusion (2) consists of two parts—the statement that basic sentences explained by the old theory are explained also by the later theory and the statement that these sentences obtain a new interpretation from the point of view of the later theory. As for the first of these statements we might turn to Engels' statement that he made while discussing the contact supra-theory in the science of electricity: "Haben wir die alte Kontakttheorie abgeschüttelt, so bestehn noch die festgestellten Tatsachen, denen sie zur Erklärung dienen sollte".[71] As for the second one, we might turn to the statement in which Engels says that the introduction of a new theory is connected with reinterpretation of the previously established observational facts: "Und in der Tat ist nicht abzusehn,

wodurch anders der Lehre vom Galvanismus und damit in zweiter Linie derjenigen vom Magnetismus und von der Spannungselektrizität eine feste Grundlage gegeben werden kann als durch eine chemisch-exakte Generalrevision aller überkommenen, unkontrollierten, auf einem überwundnen wissenschaftlichen Standpunkt angestellten Versuche (...)".[72]

The thesis about the changeability of the meanings of theoretical terms as the theory changes is formulated by Engels in a general way: "Every new aspect of a science involves a revolution in the technical terms of that science. This is best shown by chemistry, where the whole of the terminology is radically changed about once in twenty years, and where you will hardly find a single organic compound that has not gone through a whole series of different names".[73] But we have the same situation in economy. Especially in the Marxist theory are used "(...) certain terms in a sense different from what they have, not only in common life, but in ordinary Political Economy".[74] "It is, however, self-evident — states Engels — that a theory which views modern capitalist production as a mere passing stage in the economic history of mankind, must make use of terms different from those habitual to writers who look upon that form of production as imperishable and final".[75]

It remains to coordinate the conclusions (1)–(3) with the thesis given at the beginning of this section that at least some of the scientific laws are substituted by laws which are relatively more true. Note that the conclusions (1)–(3) are connected with whole theories, whereas the above thesis concerns only the particular laws. Thus it cannot be excluded that, according to Engels, the development of the natural sciences goes as follows.

a. A hypothesis in the form of an idealizing theorem is put forth (let us restrict ourselves only to hypotheses of this type); it undergoes concretization and from the factual theorem obtained in this way observational predictions are derived; in case they are false the hypothesis is rejected.

b. In place of this hypothesis another one is adopted — that possesses additional conditions specially chosen in the antecedent, i.e. a hypothesis which is relatively more true; this procedure is repeated until one of the subsequent hypothesis is considered justified; it then obtains the status

of law; naturally, this law is also merely a relative truth—no scientist "will exclude the possibility of further narrowing them or modification of the formulation on the basis of future investigations" [76] within the established range of the law.

c. This 'modification of the formulation' may occur when the theory to which the law belongs is substituted by a new theory which disagrees with it; then this law may be invalidated by a theorem (initially adopted hypothesis) belonging to the new theory that is in disagreement with it; this may happen, but it does not have to happen, for not all the theorems of the old theory are in disagreement with some theorems of the new theory.

d. All the observational sentences explained by the old theory are also explained by the new one; however, the interpretation of these sentences undergoes a change—they are reinterpreted according to the principles of the new theory.

e. The sense of the basic theoretical terms of the old theory undergoes at least partial change according to the new theory.

f. The hypotheses included in the new theory undergo testing through their concretization until laws are established etc., etc.

Naturally, a question arises, whether the changes of theory may be described in exclusively sociological terms (such as acceptance of the theory by a group of scientists) or in epistemological terms (such as approximating the absolute truth) as well. It is well known that Engels held the latter view—human knowledge, according to him, goes in the direction of full cognition of reality. I shall not deal with these matters here though, since they enter the field of the epistemological concepts of the creators of Marxism, while the aim of this article is the analysis of their methodological concepts.

9. THE PROBLEM OF THE PECULIARITY OF THE HUMANITIES

The question (VIII) that we approach now is connected with the controversy between the naturalist and the anti-naturalist understanding of the humanities; according to the naturalists the empirical sciences possess an approximately uniform methodological characteristics, according to the anti-naturalists—it is on the contrary: it is possible to distinguish two basically different, from the point of view of methodolog-

ical characteristics, groups of sciences—the natural sciences and the humanities. Naturally, the above statements are merely slogans; in particular, the naturalist's thesis is clearly elliptical—the thesis about the methodological uniformity of the empirical sciences assumes a different meaning each time depending on how these sciences are characterized as regards their methodology. Thus, the thesis of methodologist naturalism is not a statement that shows uniformity of theoretical views; it is a slogan demonstrating uniformity of reasoning tendencies aiming at approaching the humanities similarly to the characteristics of the natural sciences, which is considered to be adequate. Thus, for methodologists that possess different theoretical concepts the naturalist thesis will have a different sense. From the point of view of the concepts of the creators of Marxism it might be formulated in the following way:

(9)   the methodological characteristics of a science that consists of answers to questions (I)–(VII) concerns both the natural sciences and the humanities equally.

The following theses, which are often confused with the thesis of naturalism thus understood, should be treated apart from it:

(10)   all the presently formulated laws of the humanities are deducible from the presently formulated laws of the natural sciences,

(11)   all the presently formulated laws of the humanities are deducible from laws of the natural science that can be formulated,

(12)   the conceptual apparatus of the humanities may be defined in terms of the concepts of the natural science.

I shall now try to present some arguments that speak for the supposition that the creators of Marxism rejected the theses (10) and (12), accepted the thesis (9) and did not decide the thesis (11).

A number of statements may be found in the works of the creators of Marxism that speak in favour of the supposition that they rejected the thesis (10). Engels, e.g., says: "The important difference between the human and the animal communities consists in that animals at the most *collect*, while people *produce*. This one, but very important dif-

ference itself makes impossible the transfer with no qualifications of the laws governing animal communities to human communities".[77] We may, it seems, assume that Engels had a fortiori the same views as regards the laws of chemistry, physics, etc., especially as he criticized the mechanistic materialists, among other things for "(...) die Anmaßung, die Naturtheorien auf die Gesellschaft anzuwenden".[78] In other words: laws of the social sciences cannot be explained in terms of laws of the natural science that we know at the present moment. This does not, however, determine the negative answer to the question whether, in future, the explanation of laws concerning the society will be possible in terms of the laws of the natural science. It is also difficult to find a positive answer to this question in the works of the creators of Marxism. Thus we have to leave open the question whether they accepted the thesis (11).

On the other hand, we may presume that they rejected also the thesis (12). Thus, e.g. Marx repeatedly stated that the basic concepts of political economy cannot be reduced to concepts of the natural science. For instance, the concept of commodity is a concept peculiar to economy. Commodity is not the same as an object possessing defined physical properties, thanks to which it fulfils human needs. As Marx says: "(...) the bodies of commodities, are combinations of two elements — matter and labour. If we take away the useful labour expended upon them, a material substratum is always left, which is furnished by Nature without the help of man".[79] But "As values commodities are social magnitudes, i.e. something absolutely different from their properties as things. As values they reflect not only the relations between people in their productive activity. Value really implies exchanges, but exchanges are exchanges of things between men, exchanges which do not concern the things as such".[80] That is why the concept of commodity is not a concept from the natural science and it cannot be reduced to concepts of this type. One must not say that "(...) exchange value of things is defined by their properties as things, that it is in general a natural property of these things", for, "No student of nature has discovered so far thanks to what natural properties tobacco and paintings are equivalents of each other in some defined proportion".[81]

That basic concepts of political economy cannot be defined in terms of the concepts of the natural science is based on the Marxist assumption

that "Everything that is not the result of human work is nature and as such is not a social value".[82] And in order to understand properly the sense of this assumption one has to realize that for Marx work is a purposeful activity. "We presuppose labour in a form that stamps it as exclusively human. A spider conducts operations that resemble those of a weaves, and a bee puts to shame many an architect in the construction of the cells. But what distingushes the worst architect from the best of bees is this, that the architect raises his structure in imagination before he erects it in reality. At the end of every labour-process, we get a result that already existed in the imagination of the labourer at its commencement. He not only effects a change of form in the material on which he works, but he also realizes a purpose".[83] The conclusion we might draw from the above is that the reason for which the economical conceptual apparatus cannot be reduced to the concepts of the natural science is that basic economical concepts call upon, among other things, the concept of purposeful activity which, as a result, should be considered as one of the initial concepts of Marxist economy, and may be of all the humanities.[84] Independently of whether there are some other reasons for the fact that the humanistic concepts cannot be reduced to those of the natural sciences we have to state that thesis (12) was decidedly rejected by Marx.

We may presume, on the other hand, that the creators of Marxism accepted the thesis of methodological naturalism (9), i.e. they referred in their methodological theses, which were answers to the questions (I)–(VII), not only to the natural science, but to the humanities as well. I have tried to provide arguments for it by reconstructing these particular methodological theorems. And thus we saw that Marx applied the same criterion of scientific character, which Engels applied to the natural science, to his economical discussions; that Engels attributed to some of the natural laws the same idealizing character that is the feature of the theorems of Marxist economy; that Engels used the same concept of general law (i.e. – a strictly general sentence) both to the propositions of the natural science and the humanities; that both, Marx in the field of economy, and Engels in the field of the natural science protested against positivist phenomenalism; that Engels criticized the point of view that we have called linguistic phenomenalism both with regards to chemistry and economy; that Marxist model of explana-

tion, based on the principle of abstraction and concretization can equally well be used in the humanities and the natural science; that Engels protested against instrumentalism and argued for realism in the fields of both physics and historical materialism; that he formulated the thesis that subsequent theories are in disagreement with each other, clearly with a general intention in view—so that it concerned both the natural and the historical sciences, as he called the humanities; that Engels illustrated the thesis about the variability of meaning of the basic theoretical terms both on the example of the history of chemistry and on the example of the history of political economy; that, finally, he considered the concept of the relativity of truth as a general epistemological theory, and not as a concept related only to the natural sciences. The only point in which it cannot be stated without any doubt that the creators of Marxism had the generalizing intention treating both the natural sciences and the humanities alike, is the problem of the characteristics of basic theorems. For it is difficult to find statements which would confirm that basic theorems of the humanities possess the same methodological character as base theorems of the natural sciences. Nevertheless, there is nothing that speaks for a contrary conclusion. The creators of Marxism were far from looking for some separate cognitive faculty that would justify basic sentences of the humanities, like ordinary observation justifies basic sentences of the natural science. And as they did not adopt this type of concept—it is difficult to find any reasons for which basic sentences of the humanities should possess a different methodological status than those of the natural sciences. Thus, it seems, that we shall obtain a more consistent image of the epistemological concepts of the creators of Marxism, if we assume that the characteristics of the basic propositions concerns the natural sciences and the humanities alike, than if we adopted a contrary assumption.

10. SUMMARY

Thus, the model of the empirical sciences that may, it seems, be reconstructed on the basis of the works of the creators of Marxism is as follows:

I. A theorem possesses scientific character (may be a thesis of the empirical sciences) inasmuch as it is decidable on the basis of observation.

II. 1. Every observation or experiment is directed by theoretical assumptions accepted by the researcher;

2. basic sentences of the empirical sciences describing the results of observations are always interpreted in agreement with the theory accepted at a given time;

3. basic sentences possess hypothetical character; they may be rejected also as a result of rejecting the theory that is the foundation for their interpretation.

III. 1. Extra-basic theorems of the empirical sciences are: (1) historical generalizations, (2) factual laws, (3) idealizing laws;

2. neither the factual laws, nor the idealising laws are equivalent to any sets of basic theorems;

3. extra-basic theorems of the empirical sciences are relative truths and, in this connection, they possess hypothetical character (although they do not have to be hypotheses in the sense of Engels).

IV. 1. Extra-basic theorems of the empirical sciences are considered justified after they have been empirically tested; this concerns, among other things, the idealizing laws, which are tested through concretization;

2. induction plays a heuristic role in the sciences — it is a method of discovery, and not of justification of extra-basic theorems; this role of induction diminishes in the course of development of the empirical sciences;

3. extra-basic theorems should not be formulated exclusively in terms of intuitive notions, but rather new concepts should be introduced that disregard the common intuitions.

V. Explanation in the empirical sciences proceeds according to the scheme (6).

VI. 1. Extra-basic factual theorems are realistic in character;

2. idealizing laws are realistic in character, since their corresponding approximate laws are either true (Engels) or they describe the relevant factors (Marx).

VII. The development of the empirical sciences proceeds according to the scheme described in the points (a)–(f) on pp. 529–30.

VIII. The methodological characteristics of the sciences that has been presented above in the points (I)–(VII) applies to both the natural sciences and the humanities.

Due to lack of space it would be difficult to compare here the above model of the empirical sciences with the models functioning in contemporary methodological literature. Nevertheless, it should be said that it is different from all three contemporary models of the empirical sciences: the positivist, the instrumentalist and Popper's. Marx's and Engels' model differs from all the other models in that it indicates the role of idealizing laws in the sciences, it gives a different model of explanation, a different—for taking into consideration idealization, concretization and only relative truth of the extra-basic theorems—concept of the empirical sciences development.

Naturally, this comparison is not equivalent to the demonstration of the superiority of Marx's and Engels' model over the other models of the empirical sciences [85]; it is only an indication of its different character and originality, which is worth noting considering that the works that, it seems, contain it, were written in the period of exclusive domination of the positivist model. This superiority may be demonstrated in one way only—by carrying out possibly accurate investigations on the cognitive procedures that are actually used in the empirical sciences, which might provide arguments that would allow the solution between the competitive models of these sciences.

REFERENCES

* First published in *Studia Filozoficzne* 2 (75) (1972). Translated by E. Ronowicz.
[1] J. Kmita and L. Nowak, *Studia nad teoretycznymi podstawami humanistyki* (*Studies on the Theoretical Foundations of the Humanities*), Poznań 1968, p. 20.
[2] F. Engels, *Dialektik der Natur*, Berlin 1952, p. 38.
[3] *Ibid.*, p. 208.
[4] K Marx, *Das Elend der Philosophie*, in: K. Marx and F. Engels, *Werke*, Vol. 4, Dietz Verlag, Berlin 1974, p. 82.
[5] K Marx, A letter to Engels of 10th Oct. 1868, in: K. Marx and F. Engels, *Letters on 'Capital'* (in Polish), Warszawa 1957, p. 191.
[6] F. Engels, *Dialektik der Nature*, ed. cit., p. 51.
[7] *ibid.*, p. 205.
[8] *ibid.*, pp. 18–19.
[9] *ibid.*, p. 32.
[10] *ibid.*, p. 223.
[11] *ibid.*, pp. 32–33.
[12] *ibid.*, p. 270.
[13] *ibid.*, p. 223.

[14] *ibid.*, p. 219.
[15] *ibid.*, p. 217.
[16] *ibid.*, p. 218.
[17] *ibid.*, p. 25.
[18] *ibid.*, p. 142.
[19] *ibid.*, p. 114.
[20] K. Marks. Letter to Engels of 25th March 1868, in: K. Marx and F. Engels, *Selected Correspondence* (in Polish), Warszawa 1951, p. 256.
[21] F. Engels, *Dialektik der Natur, ed. cit.*, p. 167.
[22] *ibid.*, p. 142.
[23] *ibid.*, p. 249.
[24] *ibid.*, pp. 253–254.
[25] This interpretaton goes in the same direction as the interpretation done by W. Krajewski, although it differs from it in the details because of a different conceptual apparatus used there. Cf. W. Krajewski, 'Engels o prawach przyrody' ('Engels on the Laws of Nature'), in: W. Krajewski, *Pojęcie prawa nauki w XIX wieku* (*The Concept of a Scientific Law in the 19th Century*), Warszawa 1967, pp. 155–157.
[26] F. Engels, *Dialektik der Natur, ed. cit.*, p. 254.
[27] F. Engels, *Socialism Utopian and Scientific*, International Publishers, New York 1945.
[28] F. Engels, *Herr Eugen Dühring's Revolution in Science*, London 1934.
[29] Cf. L. Nowak, *Essays*, Grüner, Amsterdam 1975, p. 8ff; L. Nowak, 'Idealization: A Reconstruction of Marx's Ideas', *Poznań Studies in the Philosophy of the Sciences and the Humanities* I (1) (1975), as well as L. Nowak, 'The Problem of Explanation in Marx's Capital', *Quality and Quality* V (1971).
[30] F. Engels, *Dialektik der Natur, ed, cit.*, p. 298.
[31] *ibid.*, p. 163, cf. also *ibid.*, pp. 289, 82 ff.
[32] *ibid.*, p. 292.
[33] *ibid.*, pp. 243, 285, 106 ff.
[34] *ibid.*, pp. 250–251.
[35] F. Engels, *Ludwig Feuerbach and the End of the Classical German Philosophy*, Progress, Moscow 1969.
[36] F. Engels, *Herr Eugen Dühring's Revolution in Science, ed. cit.*, p. 90.
[37] *ibid.*, p. 88.
[38] *ibid.*, p. 85.
[39] *ibid.*, p. 85.
[40] F. Engels, *Dialektik der Natur*, p. 256.
[41] *ibid.*, p. 37.
[42] *ibid.*, p. 256.
[43] F. Engels, *Ludwig Feuerbach...*, *ed. cit.*, pp. 309–310.
[44] L. Nowak, *Essays, ed. cit.*, p. 67; W. Patryas, 'An Analysis of the Caeteris Paribus Clause', *Poznań Studies...* I (1) (1975), p. 60.
[45] F. Engels, *Dialektik...*, p. 51.

[46] K. Marx, *Theories of Surplus Value*, Part II, Progress Publishers, Moscow 1968, p. 165.

[47] K. Marx, *Capital*, vol. III, Part I, Foreign Languages Publishing House, Moscow 1962, p. 25.

[48] K. Marx, *Theories of Surplus Value*, Part III, Warszawa 1966, pp. 565–566.

[49] K. Marx, *Capital*, *ed. cit.*, p. 797.

[50] F. Engels, Preface to the English Edition of the 1st volume of *Capital*, in: K. Marx, *Capital*, Volume 1, *ed. cit.*, p. 4.

[51] Cf. e.g., K. Marx, *Theories...*, Part III, *ed. cit.*, p. 566.

[52] K. Marx, *Capital*, *ed. cit.*, Vol. III, p. 36.

[53] K. Marx, *Theories...*, Part III, *ed. cit.*, p. 566.

[54] *ibid.*, p. 565.

[55] F. Engels, *Dialektik...*, *ed. cit.*, p. 242.

[56] *ibid.*, p. 243.

[57] *ibid.*, p. 243.

[58] *ibid.*, p. 196.

[59] L. Nowak, *Essays*, *ed. cit.*, p. 59, and L. Nowak, *The Problem of Explanation...*, *op. cit.* Cf. also L. Nowak, 'Idealizational Laws and Explanation', *Logique et Analyse* (1972) 59–60.

[60] F. Engels, *Ludwig Feuerbach...*, *op. cit.*, p. 331.

[61] *ibid.*, p. 331.

[62] F. Engels, *Dialektik...*, *ed. cit.*, p. 98.

[63] *ibid.*, p. 99.

[64] F. Engels, Letter to Schmidt of 12th March, 1895, in: K. Marx and F. Engels, *Letters on 'Capital'*, *ed. cit.*, p. 351.

[65] *ibid.*, p. 354.

[66] F. Engels, Supplement to *Capital*, Volume Three, in: K. Marx, *Capital*, *ed. cit.*, Vol. III, p. 876.

[67] K. Marx, *Capital*, *ed. cit.*, Vol. III, p. 186, More on Marx's views on the subject cf. J. Kmita, Marxism and the Controversy: Realism – Instrumentalis', *Studia Metodologiczne* 10 (foreign issue); cf. also L. Nowak, *Essays*, *ed. cit.*, p. 58; and L. Nowak, 'Relative Truth, the Correspondence Principle and Absolute Truth', *Philosophy of Science* 42 (2) (1975).

[68] F. Engels, *Dialektik...*, pp. 38–39.

[69] *ibid.*, pp. 150–151.

[70] *ibid.*, pp. 256–257.

[71] *ibid.*, p. 167.

[72] *ibid.*, p. 177.

[73] F. Engels, Preface to the English Edition of *Capital*, in: K. Marx, *Capital*, *ed. cit.*, Vol. 1, p. 4.

[74] *ibid.*, p. 4.

[75] *ibid.*, p. 5.

[76] F. Engels, *Herr Eugen Dühring's Revolution in Science*, *ed. cit.*, p. 90.

[77] F. Engels, Letter to P. L. Lavrov.

[78] F. Engels, *Dialektik...*, *op. cit.*, p. 215.

[79] K. Marx, *Capital, ed. cit.*, vol. 1, p. 43.

[80] K. Marx, *Theories of Surplus Value*, Part III, Warszawa 1966, p. 149.

[81] *ibid.*, p. 150.

[82] *ibid.*, p. 499.

[83] K. Marx, *Capital*, Vol. 1, *ed. cit.*, p. 178.

[84] That basic concepts of Marxist economy have a 'mixed' character, i.e. that besides subjective concepts, concerning the aim of activity, there occur objective concepts, concerning the unintended consequences of activity, is indicated by J. Kmita,' Marx's Way of Explaining of Social Processes', *Poznań Studies in the Philosophy of the Sciences and the Humanities* I (1) (1975), p. 86ff.

[85] The superiority of this model in questions connected with idealization has been dealt with in the works: L. Nowak, 'Laws of Science, Theories and Measurement', *Philosophy of Science* **39** (1972); L. Nowak, 'On an Interpretation of Marxist Methodology', *Poznań Studies in the Philosophy of the Sciences and the Humanities* I (1) (1975), p. 86ff.

TADEUSZ PAWŁOWSKI

ON THE EMPIRICAL MEANINGFULNESS OF SENTENCES *

The criteria of empirical meaningfulness of sentences have been subject to numerous changes, taking a more and more liberal character. The last known version of the criterion [1] formulated by Carnap gives a very wide extension to the concept of empirical meaningfulness. It is so wide that it covers expressions which should rather be excluded, and recognized as meaningless. I mean here certain sentences stating quantitative relations between magnitudes. Further on we shall give arguments for considering these sentences as nonsensical expressions, and we shall demonstrate that indeed these sentences fall under the concept of meaningfulness formulated by Carnap. We shall also sdicuss an attempt to formulate empirical criteria of meaningfulness for sentences stating quantitative relations. It is the definition of meaningfulness given by P. Suppes in *Measurement, Empirical Meaningfulness and Three-Valued Logic*. [2] Although his definition cannot be considered as an equivalent condition of meaningfulness and although it does not cover all forms of quantitative sentences, [3] it can nevertheless constitute a basis for formulating a general condition of meaningfulness which, added to Carnap's definition, will eliminate the doubtful quantitative sentences from the set of meaningful sentences determined by this definition.

Let us consider the following examples of quantitative sentences:

(1)   The distance between Łódź and Warsaw is 133.

(2)   The distance between Łódź and Warsaw is two times shorter than the distance between Świdnica and Bielsko-Biała.

(3)   The quotient of a day's temperature by pressure is never bigger then the quotient of humidity by temperature:

$$\frac{t}{p} \leqslant \frac{w}{t}.$$

It is obvious that the sentence (1) does not state anything definite, its content will become clear only if we specify the unit of length with which the distance has been measured. On the contrary, the sentence (2) is fully comprehensible, and even true without any additional information about the measuring unit used, or more exactly: the sense and the true of (2) will remain unchanged independently of the unit used to measure the distance.

And how is the case with sentence (3)? We must say that this sentence, similarly to (1) will not have any definite sense as long as it is not indicated what units have been used to measure the magnitudes involved in the stated relationship. Let us suppose that the relationship (3) has been stated on the basis of numerous observations, in the course of which temperature was always measured by using Celsius scale. If we now were to describe the results of the same observations in terms of Fahrenheit scale, then on the grounds of the known definitional relationship between the two scales the sentence (3) would assume the form:

$$(4) \qquad \frac{\frac{9}{5}t+32}{p} \leqslant \frac{w}{\frac{9}{5}t+32}.$$

Sentences (3) and (4) are not equivalent, which is shown by the following substitution:

$$(5) \qquad \frac{30}{75} \leqslant \frac{22}{30}, \quad \frac{86}{75} \leqslant \frac{22}{86}.$$

If we then were to use Fahrenheit scale in measuring temperature, the relationship between temperature, pressure and humidity would assume a totally different form from the relationship described in (3).

The situation is then the following: sentence (3) does not state any definite empirical relation until it is specified what units have been used for measurement; on the other hand, if we supply such an information it appears that we are not dealing with a single empirical relationship among pressure, temperature and humidity, but with numerous various relations, the form of which depends on the measuring units used. As the determining of units is a matter of free choice, we can say that the sentence (3), completed with information about the units used,

states something more than only the results of observation. All these reasons speak in favour of excluding expressions of the type (3) from the set of meaningful sentences.[4]

The sentence (1) is in another situation, as it does not state any general relation, but an individual fact. In its present form it is, of course, deprived of any sense. But completing this sentence with information about unit of measure used does not lead to the above-mentioned undesirable consequence; nothing then opposes the inclusion of the sentence (1) thus completed in the set of meaningful sentences. All further remarks about the meaningfulness of quantitative sentences apply only to sentences stating general relations between magnitudes. Of course, when giving the definition of a meaningful sentence it should be exactly determined what is understood by a 'quantitative sentence stating a general relationship'. Such a general and, at the same time, exact definition has not yet been formulated; further on we shall present its particular form, constructed on the basis of a concrete scientific theory the axioms of which are expressed in a syntactically determined language.

The way of measuring a given magnitude, i.e. the kind of scale used, can leave to free choice not only the measuring unit, but also other elements of structure. In interval scale, they are: the unit and zero; in ordinal scale, the assigning of all numbers to the measured magnitudes is arbitrary, with the reservation that all the possible assignments preserve the relations of equality and precedence holding between the magnitudes. As we know, the type of scale is determined by a set of transformations of numerical values, isomorphically assigned to the measured magnitudes, transformations which leave the scale structure unchanged. Every set of such numerical values is isomorphic to the measured magnitude. This supplies us with a criterion for meaningfulness of quantitative statements. According to the criterion a sentence stating relations among quantities measured on a given type of scale is an empirically meaningful sentence if its truth value remains invariant under transformations admissible relative to this scale. In other words, a sentence S_1 is a meaningful sentence if admissible transformations carry S_1 into an equivalent sentence S_2.[5] The above-mentioned sentence (3) is an example of an expression which does not fulfil this condition. And here is an example of a sentence stating in a correct way the relation between the degree of intelligence and age, with the assumption that intelligence

is measured on at least an ordinal scale:

(5) $I(x) < I(y) \rightarrow W(x) < W(y)$.

We shall now demonstrate that the definition of meaningful sentence formulated by Carnap covers also sentences which do not fulfil the above presented condition of invariance, that is, sentences which according to the discussed conception should be considered as meaningless expressions. Following Carnap, the expression A is a meaningful sentence in the language L_T if and only if A is admitted by the formation rules of the language L_T, and if, moreover, every descriptive constant in A is a meaningful term. The definition of 'meaningful term', in turn, consists of two parts; we shall here adduce only the first one. The term M is significant relative to the class K, with respect to L_T, L_O, T, C if and only if the terms K belong to V_T, M belongs to V_T, but not to K; there are sentences $S(M)$ and $S(K)$ in L_T and sentence $S(O)$ in L_O such that the following conditions are fulfilled:

a) $S(M)$ contains M as the only descriptive term.
b) The descriptive terms in $S(K)$ belong to K.
c) The conjunction $S(M) \cdot S(K) \cdot T \cdot C$ is consistent.
d) $S(O)$ is logically implied by the conjunction $S(M) \cdot S(K) \cdot T \cdot C$.
e) $S(O)$ is not logically implied by the conjunction $S(K) \cdot T \cdot C$.[6]

In constructing the proof we shall use the language and the axioms of the theory of decision making, formulated in *Decision Making. An Experimental Approach*.[7] The language L_T of this theory consists of the constant Z, denoting a small set of observable objects, of the constants P and $\approx E$, denoting the relations between these objects, of the functional term φ denoting the function defined on the set Z, and assuming values from the set of real numbers; theses are the specific constants of the theory. Moreover, the language of the theory contains the names of real numbers and the signs $+$, $\cdot$, $<$. The language L_O, in turn, contains the same constant Z, the constants W and JE denoting observable relations between objects in Z, and individual constants and variables. Coordinating definitions C have the following form:

$x P y \equiv x W y$ (from among two objects x and y the subject chooses the object y).

$x, y \approx Eu, v \equiv x, y JEu, v$ (the alternative of getting the objects x and y when the event E happens is by the subject valued

equally highly as the alternative of getting the objects u and v when the event **E** does not happen).

The class of terms K is here empty, there is therefore also no sentence $S(K)$. The sentences $S(M)$ and $S(O)$ have the form $\exists(x)\,\exists(y)$ $[\varphi(x)<\varphi(y)]$, $a_1\,W a_2$, where a_1 and a_2 are names of objects in the set Z, and φ is precisely the term M. The theory T, which we shall use here, allows to explain and to predict the behaviour of a person who is to make a choice from among a set of objects. The axioms of the theory are given in the Annex. It is possible to prove, on the grounds of these axioms the existence of function φ, which assigns to objects numbers representing the subjective value of these objects for the person choosing. The function φ meets the following condition:

$$(6) \qquad \prod_{x}\prod_{y}[\varphi(x)<\varphi(y)\equiv x\,P\,y].$$

It can also be demonstrated that the function φ determines the interval values scale, as it is unique up to linear transformation; i.e. if φ_1 is such a function, then also is $\varphi_2 = a\varphi_1 + b$, where $a>0$, and b is any real number.

To prove that the conjunction $S(M)\cdot T\cdot C$ is consistent we find an interpretation under which the conjunction is true. This interpretation is determined by the following equivalences:

$$(7) \qquad Z=\{1,2\}, \quad x\,P\,y\equiv x<y, \quad xy\approx Eu, \quad v\equiv x+y=u+v,$$

$$x\,W\,y\equiv x<y, \quad x,\,yJEu, \quad v\equiv x+y=u+v.$$

As we can see, under this interpretation the axioms of the theory, the sentence $S(M)$, and the coordinating definitions are all true sentences. It is also easy to prove that the sentence $S(O)$, here $-a_1\,W a_2$, follows from the conjunction $S(M)\cdot T\cdot C$. In order to do that we omit quantifiers in the equivalence (6), substituting for the variables x and y the constants a_1, a_2, and we obtain:

$$(8) \qquad \varphi(a_1)<\varphi(a_2)\equiv a_1\,P\,a_2.$$

We carry out a similar operation on the sentence $S(M)$ and on the coordinating definition $x\,P\,y\equiv x\,W\,y$:

(9) $\varphi(a_1) < \varphi(a_2)$,

(10) $a_1 P a_2 \equiv a_1 W a_2$.

From the sentences (8), (9) and (10) we obtain the sentence $S(O)$.

In order to demonstrate that the sentence $S(O)$ is not implied by the conjunction $T \cdot C$, we choose an interpretation differing from the previous one only in the set Z consists now of a unique element $\{1\}$. Under this interpretation the conjunction $T \cdot C$ is true, while the sentence $S(O)$ is false, as it then reduces to the sentence $1 < 1$.

We have demonstrated that the functional term φ is significant on the grounds of Carnap's definition; then every well-formed sentence containing the term φ (and possibly other significant terms) is also meaningful. For instance, the sentence:

(11) $\exists(x)\exists(y)\exists(z)[\varphi(x) + \varphi(y) < \varphi(z)]$

fulfils these conditions and is a meaningful sentence under Carnap's definition. But it can be demonstrated that the truth value of this sentence is not invariant under transformations admissible relative to the interval scale. Such transformations carry (11) into a non-equivalent sentence:

(12) $\exists(x)\exists(y)\exists(z)[a\varphi(x) + b + a\varphi(y) + b < a\varphi(z) + b]$.

The non-equivalence of these two sentences is shown by the interpretation in which the universe is a three-element set $\{1, 2, 3\}$, and the constants a and b as well as the function φ are defined by the equations $a = b = 2$; $\varphi(x) = x$. Under this interpretation the sentence (11) is true, while the sentence (12) − false.

We have demonstrated that the definition of meaningful sentence formulated by Carnap covers also quantitative sentences, which for reasons presented at the beginning of this paper should be excluded from that extension. Such exlusion can be obtained by adding to Carnap's definition a supplementary condition, postulating for quantitative sentences the invariance of their truth value under transformations admissible relative to the used scales. As we have said before, a general and, in the same time, fully precise definition of that condition of invariance has not yet been formulated. In his already quoted work [8] P. Suppes discusses its particular case, which we shall present here. The definition is relativized to the formalized language L_m. It is the

language of the theory formulated by A. Tarski in *A Decision Method for Elementary Algebra and Geometry*, enriched by individual variables, individual constants $O_1, ..., O_{10}$, denoting ten physical objects, and by the term m, where $m(a)$ is a real number denoting the mass of the object a. The individual constants and the constant m are then the only descriptive terms of the language L_m. The following definitions are given for this language: of a term, of an atomic formula and of a sentence. The function m is supposed to determine the ratio scale of mass. The definition of meaningful sentence, formulated on the basis of this language, consists of the following two parts:

A. The atomic sentence S of the language L_m is empirically meaningful if and only if the closure of the formula

$$\alpha > 0 \rightarrow [S \equiv S(a)]$$

is arithmetically true for every α, while $S(\alpha)$ results from S by replacing the term m by the quotient αm in every place where the term m occurs in the sentence S.

B. The sentence S of the language L_m is empirically meaningful if and only if every atomic sentence in S is empirically meaningful.[9]

Suppes considers two more definitions of empirical meaningfulness, which are modifications of the above-presented definition, formulated in the language of model theory. We do not intend to quote them here, we shall mention only two aspects pointed out by the author in relation to these definitions. First, the carrying out logical operations on sentences requires – if these definitions are assumed – to introduce a three-valued logic, including besides the values of truth and falsehood also a third one – meaninglessness. The example of the following two sentences indicates the need of introducing such logic:

$$\exists(x)[m(x)=1] \lor \sim \exists(x)[m(x)=1],$$

$$\exists(x)[m(x)=1] \lor \exists(y)[m(y)=2].$$

The first alternative is – according to one of the author's definitions – a meaningful and true sentence, although both of its components are meaningless sentences. In the second alternative both, the components and the whole are meaningless sentences. As we can see, the alternative has not here its usual properties of truth-functional connective. The situation is the same in the case of other connectives.

Another important point to which the author pays attention is that, on the ground of his definitions, the set of meaningful sentences does not constitute a deductive system, which means that logical consequences of such a set may not be meaningful sentences. For example, a logical consequence of a sentence meaningful on the grounds of the previously quoted definition of meaningfulness

$$\prod_x [x > 2 \rightarrow x > 1]$$

is the meaningless sentence

$$m(o_1) > 2 \rightarrow m(o_1) > 1 .$$

To take care of such cases it is necessary to introduce, besides the concept of logical consequence — that of meaningful logical consequence, which is defined by the author. The use of the concept of meaningful logical consequence requires, in turn, the change of formal logical rules. The author proposes such a change for the systems of natural deduction. It would consist in the additional restriction that every line of the proof be a meaningful sentence.

Finally, let us go back for a moment to the previously given definition of meaningfulness. The author formulates its both parts in the form of equivalences. It can be seen, however, that the fulfilment by a sentence of the conditions set by the definition does not yet suffice to assure empirical meaningfulness of the sentence. For those conditions do not take into consideration the relation of the sentence to empirical data. Suppes' definitions can therefore be considered only as formulations stating a necessary condition of meaningfulness with respect to quantitative sentences. Only by adding this condition to Carnap's definition (or else to some other alternative definition) we will obtain a complete definition of the empirical meaningfulness of sentences.

ANNEX

The domain $\langle Z, P, \approx E \rangle$ is an equal-interval preference structure if and only if for every x, y, u, v, z in Z the following axioms are satisfied:

(A₁)   The set Z is finite, and the relation P is antisymmetric, transitive and connected in Z,

(A_2) $x, y \approx \mathbf{E}\, y, x,$

(A_3) $x, y \approx \mathbf{E}\, u, v \to u, v \approx \mathbf{E}\ x, y,$

(A_4) $x, y \approx \mathbf{E}\, u, v \cdot u, v \approx \mathbf{E}\, z, w \to x, y \approx \mathbf{E}\, z, w,$

(A_5) $x, y \approx \mathbf{E}\, u, v \cdot x\, P\, u \to v\, P\, y,$

(A_6) $x\, N\, y \cdot u\, N\, v \to x, v \approx \mathbf{E}\, u, y.$

N is the relation which holds between x and y when y is the only immediate successor of x. The definition of this relation has the following form:

xNy if and only if xPy and for every w in Z: if xPw, then $y = w$ or yPw.

REFERENCES

* First published in *Fragmenty Filozoficzne* **III**, Warszawa 1967, Translated by S. Wojnicki.
[1] Cf. 'The Methodological Character of Theoretical Concepts', in: *Minnesota Studies in the Philosophy of Science* I, Minneapolis 1965.
[2] Published in: *Measurement; Definitions and Theories*, New York 1959.
[3] Further on I shall use this abbreviation to denote sentences stating quantitative relations among magnitudes.
[4] P. Suppes suggests such an exclusion in his work mentioned above; a similar view seems to result from D. Luce's analyses, presented in his article 'On the Possible Psychophysical Laws', *Psychological Review* **66** (1959).
[5] A wider account of the problems of measurement and types of scales is given in my book: *Methodologische Probleme in den Geistes- und Sozialwissenschaften*, Vieveg Verlag, Braunschweig – Polnischer Verlag der Wissenschaften, Warszawa 1975. Cf. also: S. S. Stevens, 'On the Theory of Scales of Measurement', *Science* (1946).
[6] *Op. cit.*, pp. 51 and 60.
[7] By D. Davidson, P. Suppes and S. Siegel, Stanford University Press, Stanfrod 1957. The axioms are given in the Annex.
[8] See note 2.
[9] *Op. cit.*, p. 136 ff.

MARIAN PRZEŁĘCKI

A MODEL-THEORETIC APPROACH TO THE
PROBLEM OF INTERPRETATION
OF EMPIRICAL LANGUAGES *

One of the basic problems of the logical methodology of empirical sciences is that of the interpretation of the language of these sciences – the problem of the way in which the terms of an empirical language are endowed with meaning and objective reference. We assume here that the language of every empirical theory is a meaningful one, and that it refers to a definite domain of reality (or to a definite class of such domains), and we then ask what is the procedure which gives it such a character. The point here – as in other problems of logical methodology – is not the description of the actual interpretation procedures, but their logical reconstruction – such characterization of the interpretation procedure which would provide reasons for assigning to a given language its appropriate interpretation. It becomes particularly important when, leaving out the empirical sciences using colloquial language, or language which only slightly differs from it, we shift to empirical theories the language of which includes specific terms not belonging to colloquial language. Their sense and the way to determine it are far from being obvious. The terms of these theories are to refer to objects very remote from those which our everyday experience deals with, so that also the way of assigning the objects to them must have here a specific character.

To present it we must distinguish two different components within the whole constituted by an interpreted empirical language. The uninterpreted language, being a purely formal entity, and its interpretation, understood as the domain of reality to which the language refers. Both of these concepts have their exact counterparts in the formal apparatus of contemporary logical semantics; they are the concepts of formalized language and of its model. Model theory is this part of contemporary

551

logic which investigates semantic relations between formalized languages and their models. For this reason we shall use the logical means of that theory to analyze the problem of interpretation of empirical language. The basic aim of the analysis is to give an example of the applicability of model-theoretic concepts to the semantics of empirical languages. It is therefore a continuation of similar analyses presented both in my earlier works (in particular, in the monograph *The Logic of Empirical Theories*, London 1969) and in the works of other authors (Suszko, Wójcicki, Montague). The formal apparatus used here does not go beyond the limits of the elementary concepts of model theory, and it is nowadays known not only to logicians, but also to methodologists; moreover, it belongs to the set of standard investigation means of contemporary methodology. There is no need, therefore, to explain it again in a detailed and systematic way. I shall confine myself to a short review.[1]

I

1. We shall consider the problem of interpretation of empirical languages on the example of the simplest possible languages. They are so-called standard formalized languages, and they contain predicates as the only extra-logical terms. (As we shall see, this simplifying assumption will not restrict undesirably the generality of our considerations.) The vocabulary of such a language L contains, besides individual variables, sentential connectives, quantifiers and an identity predicate — as logical constants, and predicates $r_1, ..., r_n$ — as extra-logical constants. We define the set of sentential formulas of the language L and its set of sentences (sentential formulas not containing free variables) in a common, purely syntactic way. We also characterize the operation of logical consequence in the language L, Cn, and hence the set of logical theorems of the language L as the set of logical consequences of an empty set, $Cn(0)$ (i.e. the theorems of the first-order predicate calculus with identity) in a purely syntactic way.

The so-characterized language L remains, of course, an uninterpreted language. Without referring to any definite domain of reality, it is suitable to refer to various such domains. Every domain which can be referred to in the language L constitutes a model of this language.

Thus by a model of the language L we shall understand every relational system

$$\mathfrak{M} = \langle U; R_1, \ldots, R_n \rangle,$$

consisting of a non-empty set U, called the universe of the model $\mathfrak{M}$, and of the relations $R_1, \ldots, R_n$ holding between its members, each of which has as many arguments as the corresponding predicate of the language L. Every model $\mathfrak{M}$ of the language L determines one of the possible interpretations of this language. The universe U is assigned to its variables as their set of values, and the relations $R_1, \ldots, R_n$ — to the predicates $r_1, \ldots, r_n$ as their denotations.

The concept of model $\mathfrak{M}$ of the language L allows us to introduce, in a well-known way, elementary semantic concepts relativized to the model $\mathfrak{M}$. The fundamental concept of this construction is that of satisfiability. It is not possible to present here its rather complicated definition, but it would be unnecessary, as the concept is well-known and intuitively clear. I shall therefore restrict myself to a short informal explanation. Let $\alpha(x_1, \ldots, x_k)$ be a formula in the language L with free variables $x_1, \ldots, x_k$, $\mathfrak{M} = \langle U; R_1, \ldots, R_n \rangle$ — a model of this language, and $\langle a_1, \ldots, a_k \rangle$ — a sequence of members of the universe U.

> The formula $\alpha(x_1, \ldots, x_k)$ is satisfied in the model $\mathfrak{M}$ by the sequence $\langle a_1, \ldots, a_k \rangle$ if what the formula $\alpha \langle x_1, \ldots, x_k \rangle$ states is the case under the following interpretation: its bound variables range over the set U, the free variables $x_1, \ldots, x_k$ assume the function of names of the members $a_1, \ldots, a_k$, the predicates $r_1, \ldots, r_n$ denote the relations $R_1, \ldots, R_n$, and the logical constants have their usual, classical interpretation.

The concept of satisfiability allows, in turn, to define the concept of truth in the model $\mathfrak{M}$.

> The formula $\alpha(x_1, \ldots, x_k)$ is true in the model $\mathfrak{M}$ if the formula $\alpha(x_1, \ldots, x_k)$ is satisfied in that model by every sequence $\langle a_1, \ldots, a_k \rangle$ of members of the universe U.

Thus, for formulas not containing free variables we obtain the concept of sentence true in the model $\mathfrak{M}$. We shall symbolize the set of

sentences in the language L true in the model $\mathfrak{M}$ by Ver $(\mathfrak{M})$, and the set of their negations—i.e. sentences in the language L false in the model $\mathfrak{M}$—by Fls $(\mathfrak{M})$. If all the sentences of the set X are true in $\mathfrak{M}$, we say that $\mathfrak{M}$ is the model of the set of sentences X, in short $\mathfrak{M} \in M(X)$. By using this concept we can give a semantic characteristic of the concepts of consequence and of logical theorem introduced above in a syntactic way:

$\alpha \in Cn(X)$ if every model of the set of sentences X is a model of the sentence α;

$\alpha \in Cn(0)$ if every model of the language L is a model of the sentence α.

It is generally accepted that the language L becomes an interpreted language when from among all the domains which can be referred to in that language one is distinguished as that to which this language actually refers, in other words, when from among all the models of the language L one is distinguished as the so-called proper or intended model. The interpreted language is therefore most often identified with the pair $\langle L, \mathfrak{M}^* \rangle$ consisting of the formalized language L and its proper model $\mathfrak{M}^*$. We can introduce 'absolute' (and not 'relative', as before) semantic concepts, with respect to the interpreted language, and particularly that of a sentence true *tout court*, identifying 'truth' with 'truth in the proper model'. The sentence α is true if it is true in $\mathfrak{M}^*$; it is false, when false in $\mathfrak{M}^*$. Denoting by Ver the set of true sentences of language L, and by Fls—that of false sentences, we thus assume that:

$$\text{Ver} = \text{Ver}(\mathfrak{M}^*), \qquad \text{Fls} = \text{Fls}(\mathfrak{M}^*).$$

Such a concept of interpreted language, however, seems to be too restrictive, especially with respect to empirical languages. What an empirical language refers to is almost never uniquely determined. The pragmatic factors which decide what the terms of such a language refer to determine its proper model in an ambiguous way. This is clearly manifested by the notorious vagueness of all empirical terms. These factors, then, do not determine one proper model $\mathfrak{M}^*$, but a class of such models M^*, containing as a rule more than one member. Further

analysis of the interpretation of empirical languages clearly confirms that assumption. We shall therefore accept here a more liberal concept of an interpreted language, according to which we can speak of such a language as soon as we have distinguished a class of its proper models M^* – non-empty and not equal to the totality of its models. We then identify every interpreted language with the pair $\langle L, M^* \rangle$, where L is the formalized language and M^* – the class of its models fulfilling the mentioned conditions. This concept includes the previous one as a boundary case. If M^* is a unit class, we get a uniquely interpreted language. The problem of defining the 'absolute' semantic concepts, in particular the 'absolute' concept of truth, for ambiguously interpreted languages meets some difficulties and remains still an open question. Several proposals of its solution have been offered and discussed. [2] Although they differ in many respects, they all have in common the following assumption:

> If the sentence α is true in every model of the class M^*, this sentence is true; if the sentence α is false in every model of the class M^*, this sentence is false.

The differences between various solutions concern merely the qualification of so-called indeterminate sentences, i.e. sentences which are true in some models of the class M^* and false in others. There is no need to decide this issue in our considerations.

2. Let us now try to characterize in a most general way the interpretation of any empirical language. What can we assume about the properties of the class M^* and the ways to determine it? First of all, we shall distinguish two basic ways of interpreting a formalized language: verbal and non-verbal. The verbal interpretation of the language L consists in characterizing the class of its proper models M^* as the class of models of a definite set of sentences of the language L, that is, as the class of models in which all the sentences of that set are true. We shall call that set of sentences symbolized by P the set of postulates for the language L (or for its terms). The verbal interpretation of the language L can then be reduced to the definition:

$$M^* = M(P),$$

identifying the class M^* with a class definable in the language L. We shall regard as non-verbal any other way of determining the class M^*. Now, it should be clearly stated that if the language L is to be empirical, its interpretation can not be purely verbal. This fact is quite obvious and often mentioned. I shall therefore not give any wider account of it (I have done so among others in the quoted monograph), restricting myself to only a few remarks. Whatever consistent set of sentences in the language L should we choose as the set of postulates P, the class of its models $M(P)$ will be so inclusive and will contain so different models that it can not be considered the interpretation of an empirical language. Among its models there will always be such models that their universe consists of natural numbers, as well as models having universes consisting of linguistic expressions. Therefore, if the class of proper models of a language includes all the models of the class $M(P)$ the language cannot have an empirical character, no matter how widely should this term be understood. The fact that in a so interpreted language every extralogical term remains completely vague reflects this feature.

We then accept that the interpretation of our language L is not purely verbal. The class of its proper models M^* is not identical with any class definable in the language L. How is it determined? What non-verbal methods are used? Our assumption in this case might be considered as characteristic of semantical empiricism. It says that there are only two ways of assigning denotations to empirical terms; one is direct, in fact amounting to indicating in some way objects which are to belong to the denotation of a given term (it is called ostensive definition of the interpreted term), the other is indirect, and consists in characterizing the denotation of the term by means of linguistic postulates, which refer to already interpreted empirical terms (in some cases these postulates assume the form of an explicit definition). The direct interpretation is applied only in the case of so-called observational terms, i.e., loosely speaking, terms denoting observable properties of observable objects (or observable relations between such objects). All other empirical terms, qualified as so-called theoretical terms, can be interpreted only in an indirect way. Thus, in conformity with these assumptions, we shall always understand the class of proper models M^* of the empirical language L as a non-empty proper subclass

of the class of all the models of the set of postulates P, $M(P)$:

$$0 \neq M^* \subset M(P),$$

this subclass being distinguished from the latter by means of direct interpretation procedures, which have the character of ostensive definitions of some terms of the language L. The set P stands here for the set of all the postulates for the terms of the language L. These postulates – and their logical consequences, $Cn(P)$ – are the only sentences in the language L whose truth in models of the class M^* is guaranteed by the very way this class is determined. The direct way of interpreting empirical terms is a procedure the analysis of which escapes methods of formal logic.[3]

We shall devote our further considerations to the analysis of indirect interpretation. We shall be interested in the procedure of extending the empirical language by adding theoretical terms, i.e. terms permitting exclusively indirect interpretation. Our present problem is how to account – on the basis of a definite interpretation of the primary language – for the interpretation of the language extended. As we shall see, there are several answers to that question – depending on the situation in which the extension takes place. To formulate these answers precisely we have to introduce some formal concepts and symbols.

3. Let L_1 be our primary empirical language, containing as only extralogical terms predicates $r_1, ..., r_n$. This language is extended by adding new predicates $q_1, ..., q_m$, and thus becomes a language L_2, containing L_1. We adopt here the convention that all syntactic and semantic concepts concerning the language L_i ($i = 1, 2$) will have the subscript i. In particular, we shall symbolize by $\mathfrak{M}_1$ the models of the language L_1, i.e. the relational systems $\langle U; R_1, ..., R_n \rangle$, and by $\mathfrak{M}_2$ – the models of the language L_2, i.e. the relational systems $\langle U; R_1, ..., R_n, Q_1, ..., Q_m \rangle$. The symbol $\mathfrak{M}_{2/1}$ will denote the fragment of the model $\mathfrak{M}_2$ corresponding to the language L_1, i.e. the model of the language L_1 obtained from $\mathfrak{M}_2$ by eliminating the denotations of the predicates $q_1, ..., q_m$. We assume that P_1 is the set of postulates for the predicates $r_1, ..., r_n$ in the language L_1, and P_2 – the set of postulates for the newly introduced predicates $q_1, ..., q_m$ in the language L_2. According to what has been established above, we assume that L_1 is an interpreted language,

and that the class of models M_1^*, which determines that interpretation, is a non-empty proper subclass of the class of models $M(P_1)$, distinguished from the latter by direct interpretation procedures:

$$0 \neq M_1^* \subset M(P_1).$$

How are we to account for the interpretation of L_2? How are we to characterize the class M_2^* of its proper models? The answer to these questions depends on some pragmatic factors, in particular on the intentions of the users of L_2, who decide about the character of its interpretation. We shall restrict our considerations to situations — undoubtedly the most frequent — in which the extension of L_1 is to have, according to the intentions of its users, a 'conservative' character, preserving its previous interpretation. In these situations the language L_2 is to be interpreted so that, first, all the terms belonging to L_1, i.e. the predicates $r_1, \ldots, r_n$, preserve their previous interpretation determined by the class M_1^*; second, all the introduced terms, i.e. the predicates $q_1, \ldots, q_m$, be interpreted according to the postulates P_2, that is, in such a way that the latter are true. Any definition of the class M_2^*, which determines the interpretation of language L_2, must meet those requirements. It seems that the second condition allows only one explication. It simply requires that every model $\mathfrak{M}_2$, belonging to the class M_2^*, should be a model of the set of postulates P_2: $\mathfrak{M}_2 \in M(P_2)$. The first condition, on the other hand, allows various explications. The preservation of their previous interpretation by the terms of L_1 may be understood in a more or less rigorous way, and it seems that in different actual situations it is, indeed, differently understood. To account for these differences, we must refer to some model-theoretic concepts — which seem to play a particularly important role in the applications of this theory to the semantics of empirical languages. I mean here such model-theoretic concepts as those of extension and elementary extension of models. We shall explain their meaning in application to the language L discussed above.

Let $\mathfrak{M} = \langle U; R_1, \ldots, R_n \rangle$ and $\mathfrak{M}' = \langle U'; R_1', \ldots, R_n' \rangle$ be two models of the language L.

$\mathfrak{M}'$ is an extension of $\mathfrak{M}$, symbolically $\mathfrak{M} \subset \mathfrak{M}'$, if (i) $U \subset U'$ and (ii) $R_i'|U = R_i$ $(i = 1, \ldots, n)$; that is, when the universe of the model $\mathfrak{M}'$ contains the universe of the model $\mathfrak{M}$,

and the relations of model $\mathfrak{M}'$ restricted to the universe of model $\mathfrak{M}$ coincide with the corresponding r lations of model $\mathfrak{M}$.

$\mathfrak{M}'$ is an elementary extension of $\mathfrak{M}$, symbolically $\mathfrak{M} < \mathfrak{M}'$, if (i) $\mathfrak{M}'$ is an extension of $\mathfrak{M}$ and (ii) for any formula $\alpha(x_1, \ldots, x_k)$ in the language L and for any sequence $\langle a_1, \ldots, a_k \rangle$ of members of the universe of the model $\mathfrak{M}$: the formula $\alpha(x_1, \ldots, x_k)$ is satisfied by the sequence $\langle a_1, \ldots, a_k \rangle$ in the model $\mathfrak{M}'$ if and only if the formula $\alpha(x_1, \ldots, x_k)$ is satisfied by the sequence $\langle a_1, \ldots, a_k \rangle$ in the model $\mathfrak{M}$.

Note some consequences of these definitions. If $\mathfrak{M}'$ is an extension or elementary extension of $\mathfrak{M}$, the universe of the model $\mathfrak{M}'$ may contain new members, not belonging to the universe of the model $\mathfrak{M}$. In both cases the interpretation of the predicates $r_1, \ldots, r_n$ in the old part of the universe of the model $\mathfrak{M}'$, i.e. in the set U, must be identical with their interpretation in the model $\mathfrak{M}$. And what about their interpretation in the new part of the universe of the model $\mathfrak{M}'$, i.e. in the set $U' - U$? If $\mathfrak{M}'$ is a simple extension of $\mathfrak{M}$, the interpretation is quite arbitrary. But if $\mathfrak{M}'$ is an elementary extension of $\mathfrak{M}$, it must fulfil the condition (ii). What is its intuitive sense? Note that it entails the condition of the so-called elementary equivalence of the models $\mathfrak{M}'$ and $\mathfrak{M}$:

$$\text{Ver}(\mathfrak{M}') = \text{Ver}(\mathfrak{M}).$$

Every sentence in the language L true in the model $\mathfrak{M}'$ must be true in the model $\mathfrak{M}$, and vice versa. The predicates $r_1, \ldots, r_n$ must then be interpreted in the model $\mathfrak{M}'$ so as not to affect the logical value of any sentence of L which contains these predicates. Anything (expressible in the language L) which was true of them in the model $\mathfrak{M}$ must remain true in $\mathfrak{M}'$. Actually, this relation between the interpretation of the predicates $r_1, \ldots, r_n$ in the models $\mathfrak{M}$ and $\mathfrak{M}'$ is even closer. It is worth noting that in the case of a model $\mathfrak{M}$ with a finite universe every elementary extension of $\mathfrak{M}$ must be identical with $\mathfrak{M}$.

The concepts introduced so far allow to explicate the condition requiring the preservation of the previous interpretation of the language L_1 when shifting to L_2. By means of these concepts we can distinguish at least three versions of the condition, each of them being less rigorous

than the previous one. As we remember, the interpretation of the language L_1 is determined by the class of models M_1^*, and that of L_2 — by the class of models M_2^*. There are now three possibilities: the fragment of any model $\mathfrak{M}_2 \in M_2^*$, corresponding to the language L_1 may be identical with a certain model $\mathfrak{M}_1 \in M_1^*$: $\mathfrak{M}_1 = \mathfrak{M}_{2/1}$; it may be its elementary extension: $\mathfrak{M}_1 < \mathfrak{M}_{2/1}$; and, finally, it may be its simple extension: $\mathfrak{M}_1 \subset \mathfrak{M}_{2/1}$. Each of these possibilities leads to a different definition of the class M_2^*. We shall first present their symbolic formulations,[4] and then consider their intuitive sense.

D1. $M_2^* = \{\mathfrak{M}_2 : \mathfrak{M}_2 \in M(P_2) \text{ and } \mathfrak{M}_1 = \mathfrak{M}_{2/1}, \text{ for some } \mathfrak{M}_1 \in M_1^*\}$,

D2. $M_2^* = \{\mathfrak{M}_2 : \mathfrak{M}_2 \in M(P_2) \text{ and } \mathfrak{M}_1 < \mathfrak{M}_{2/1}, \text{ for some } \mathfrak{M}_1 \in M_1^*\}$,

D3. $M_2^* = \{\mathfrak{M}_2 : \mathfrak{M}_2 \in M(P_2) \text{ and } \mathfrak{M}_1 \subset \mathfrak{M}_{2/1}, \text{ for some } \mathfrak{M}_1 \in M_1^*\}$.

As we can see, each of these definitions includes in the class M_2^*, determining the interpretation of the language L_2, only such models of L_2 which are models of the set of postulates P_2. Thus each of the definitions fulfils the requirement of interpreting the newly introduced predicates $q_1, \ldots, q_m$ according to the postulates P_2. Can we also maintain that each of the given definitions fulfils the requirement of interpreting the previous predicates $r_1, \ldots, r_n$ in the same way as they were interpreted by the class M_1^*? Certainly, each of them does to some degree. According to each of the definitions, the universe of any model $\mathfrak{M}_2$ of the class M_2^* includes the universe of a certain model $\mathfrak{M}_1$ of the class M_1^*. According to each of these definitions the former predicates $r_1, \ldots, r_n$ are interpreted in the model $\mathfrak{M}_2$ within the old part of its universe (i.e. in the universe of the model $\mathfrak{M}_1$) in exactly the same manner as in model $\mathfrak{M}_1$. In this sense each of the definitions preserves the former interpretation of the terms of L_1. What are then the differences? The definition D1 preserves that interpretation in the strictest possible sense. As, according to it, the universe of any model $\mathfrak{M}_2$ of the class $\mathfrak{M}_2^*$ is exactly identical with the universe of some model $\mathfrak{M}_1$ of the class M_1^*, the interpretation of the predicates $r_1, \ldots, r_n$ in the model $\mathfrak{M}_2$ is simply identical with their interpretation in the model $\mathfrak{M}_1$: the predicates in both models denote the same relations. According to the other definitions, the universes of the models of the class M_2^* may contain objects which do not belong to any universe of the models of the class M_1^*.

The interpretations of the predicates of L_1 in models of the class M_2^* may then differ from their interpretations in models of the class M_1^*; the denotations of these predicates in a model of the class M_2^* may include objects from the new part of the universe, that is, objects which did not belong to their denotations in any model of the class M_1^*. The definition D2, however, considerably restricts such modifications of interpretation. By requiring that the fragment of any model $\mathfrak{M}_2$ of the class M_2^*, corresponding to the language L_1, be an elementary extension of some model $\mathfrak{M}_1$ of the class M_1^*, the definition D2 requires, as we have seen, that these changes be inexpressible in the language L_1 (even in certain extensions of this language), i.e. that the extension of the interpretation of the language L_1 be in some sense unnoticeable for the users of this language. The definition D3 does not impose such restrictions. The interpretation of L_1 may be here arbitrarily extended (as long as it conforms with the postulates P_2).

But the primary interpretation of the language L_1 is also characterized by the fact that it conforms with the postulates P_1: every model of the class M_1^* is the model of the set of postulates P_1. Is this property preserved in the shift to the language L_2? There is an important difference in this respect between the definition D3 and the others. If the class M_2^* is determined according to the definition D1 or D2, every model belonging to it must be the model of the set of postulates P_1, as its fragment corresponding to the language L_1 is identical with some model of the class M_1^*, or is its elementary extension. The situation is different in the case of the definition D3. The fact that the fragment of a model $\mathfrak{M}_2$ of the class M_2^* which corresponds to L_1 is an extension of some model $\mathfrak{M}_1$ of the class M_1^* guarantees the truth in the model $\mathfrak{M}_2$ of only such sentences true in the model $\mathfrak{M}_1$ which are purely existential (i.e. logically equivalent to sentences in normal form which do not contain universal quantifiers). There can thus exist, among the postulates P_1, sentences which turn out to be false in models of the class M_2^*. But even then the postulates P_1 are preserved in a restricted form. Let us assume that there is in L_2 a one-place predicate q_1, which in every model $\mathfrak{M}_2$ of the class M_2^* denotes the old universe, that is the universe of the model $\mathfrak{M}_1$ of the class M_1^* which has model $\mathfrak{M}_2$ as its extension. Let us now relativize in every sentence of the set P_1 all the bound variables to the predicate q_1. (This operation consists in replacing normal quantifiers by cor-

responding quantifiers restricted to the predicate q_1.) Let us symbolize the set of the transformed postulates by $P_1(q_1)$. The definition D3 of the class M_2^* guarantees that every model of that class is the model of the set of sentences $P_1(q_1)$. In each of them the postulates P_1 restricted to the old universe remain true.

4. We have considered three possible ways of interpreting the language L_2. Of course, the question arises whether these possibilities correspond to some actual situations, especially to situations typical for the development of empirical sciences. It is clear that the schemes presented here can at the most constitute a considerable simplification and idealization of the actual state of affairs. But, in my opinion, the question can be answered positively provided that this reservation is made. We may, I think, assume that there are actual scientific procedures of extending empirical languages with new theoretical terms which conform to the patterns just described. It would be a separate task to justify this assumption. I shall only outline the situations involved.

The situation corresponding to the definition D1 — in which extension of a given empirical language preserves its interpretation, and particularly its universe — is probably the most frequent. Such situation seems to be characteristic of all definitional extensions. When introducing a new term into the language L_1 by means of an explicit definition there is no need to change its former interpretation, as we have the guarantee that with the existing one we shall always find the required (i.e. conforming with that definition) interpretation for the introduced term.

However, we may not have such a guarantee if the postulates for the introduced term assume a form which differs from the explicit definition. It may then turn out that we have to extend our universe in order to find an interpretation for the introduced term which would satisfy these postulates. In some cases — corresponding to the definition D2 — it can be an elementary extension. This is possible when we extend our universe by adding objects of the same kind as those which already belong to it, i.e. objects having the same properties (expressible in the previous language L_1) as the former objects. As we have seen, such an extension does not change the existing interpretation of the language L_1 in a noticeable way. For someone who does not go beyond that language the modification remains unnoticeable.

There are, however, also situations in which, in the course of extending our language, we consciously change its former interpretation, and particularly its former universe, in a noticeable way. We do so when we introduce new terms referring to objects which fundamentally differ from those of the former universe. Such is for instance the case when theoretical terms referring to unobservable objects are introduced into an observational language, the universe of which consists only of observable objects. It is obvious that such an extension of the former universe cannot remain unnoticeable, as the new objects have, as a rule, properties different from those of the old objects — and these properties are expressible in the old language L_1. Thus not all that was true in the former interpretation of the language L_1 will remain true also in the present interpretation. This concerns also the postulates P_1; they will remain true only if restricted to the old universe. To illustrate the point let us assume that one of the postulates of the observational language says that 'every object has a colour'. This postulate, although true in models with universes consisting of observable objects, ceases to be true if the universe is extended to include objects in principle unobservable (such as atoms or electrons). In these models the postulate remains true if restricted to the old universe, as for instance the statement 'every observable object has a colour'. Thus the extension involved cannot be an elementary one. These situations seem to fit the scheme corresponding to the definition D3.[5]

II

1. Whether the above given definitions of the class of proper models of the language L_2 — D1, D2, D3 — characterize the interpretation of an empirical language adequately, that is, in conformity with the actual interpretations of such a language — depends, among others, on what kind of set the set P_2 is. This set, representing the totality of postulates for the terms newly introduced, cannot be arbitrary if the definitions are to characterize properly the interpretation of the language L_2. Let us try to formulate the conditions imposed by each of the definitions on the set P_2. This problem has already been considered with regard to some of these definitions. I shall therefore restrict myself to brief and general remarks, presenting the results previously obtained.

The answer to the above question depends on some general assumptions concerning the semantic properties of empirical languages. We shall also make such assumptions. Namely, we assume that an empirical language – the language of an empirical theory – is always an interpreted one, and that this fact is independent of experience. This meets the tendency – predominant in contemporary semantics – to consider the meaningfulness of linguistic expressions as independent of experience. Experience is to decide about the truth or falsity of an empirical statement, and not about its meaningfulness; the latter is to be guaranteed *a priori*, by the very construction of the given empirical language.

We have previously identified the interpreted language with the pair $\langle L, M^* \rangle$, where L is the formalized language and M^* – a non-empty proper subclass of the class of its models. We then can consider a given language as interpreted only if the class of its proper models is non-empty. What is more, according to the assumption just adopted, its non-emptiness should be guaranteed *a priori*. We must treat our language L_2 similarly. The statement of the non-emptiness of the class of its proper models:

$$M_2^* \neq 0$$

is to be true, and true independently of experience. When can we consider such a condition as fulfilled? Let us examine it in application to one of the presented definitions of the class M_2^*, for instance the definition D1. On the basis of it, the statement that the class M_2^* is non-empty is equivalent to the following one:

(*)   For some model $\mathfrak{M}_1 \in M_1^*$ there is a model $\mathfrak{M}_2 \in M(P_2)$ such that $\mathfrak{M}_1 = \mathfrak{M}_{2/1}$.

When can we consider this statement as true independently of experience? I think that only if the following general statement is true:

C1.   For each model $\mathfrak{M}_1 \in M(P_1)$ there is a model $\mathfrak{M}_2 \in M(P_2)$ such that $\mathfrak{M}_1 = \mathfrak{M}_{2/1}$.

Let us notice that we have assumed what follows about the class of proper models of the language L_1:

$$0 \neq M_1^* \subset M(P_1).$$

According to this assumption, the truth of the statement C1 guarantees that of the statement (∗), and obviously is itself independent of experience. The statement C1 formulates a certain model-theoretic condition and−for given sets P_1 and P_2−can be settled by model theory itself. On the other hand, the truth of the statement (∗) may be guaranteed *a priori* only if the statement C1 is true. As we know, the class M_1^* has been distinguished from the class $M(P_1)$ in terms of direct interpretation procedures of the type of ostensive definition. The only *a priori* property of its models is that they are models of the postulates P_1. If we then want to be *a priori* sure that among the models of the class M_1^* there is such a model which is the fragment of a model of postulates P_2, we must *a priori* know that each model of the class $M(P_1)$ fulfils that condition. And this is what the statement C1 says.[6]

We can consider this statement as a certain condition imposed upon the set of the postulates P_2. Only such set P_2 can be considered as set of postulates for the predicates $q_1, ..., q_m$ of the language L_2 which fulfils, for a given P_1, the condition formulated in this statement. We can analogously express the conditions for the set of postulates P_2 in the case of other definitions of the class M_2^*, i.e. the definitions D2 and D3. As before, their realization guarantees the non-emptiness of the class M_2^*. Here are the conditions corresponding to the definitions D1, D2, D3, respectively:

C1.   For every model $\mathfrak{M}_1 \in M(P_1)$ there is a model $\mathfrak{M}_2 \in M(P_2)$ such that $\mathfrak{M}_1 = \mathfrak{M}_{2/1}$,

C2.   For every model $\mathfrak{M}_1 \in M(P_1)$ there is a model $\mathfrak{M}_2 \in M(P_2)$ such that $\mathfrak{M}_1 < \mathfrak{M}_{2/1}$,

C3.   For every model $\mathfrak{M}_1 \in M(P_1)$ there is a model $\mathfrak{M}_2 \in M(P_2)$ such that $\mathfrak{M}_1 \subset \mathfrak{M}_{2/1}$.

Formulated as above the conditions have a clearly semantic character. Therefore the question arises whether−as in the case of many other semantic concepts−syntactic counterparts may be found for them. There appears to be a substantial difference in this respect between the condition C1 and the others. In contrast to C1, the conditions C2 and C3 may be formulated in a purely syntactic way. Thus C2 is equiv-

alent to the condition:

C2'. $Cn(P_2) \cap Z_1 \subseteq Cn(P_1)$,

where Z_1 symbolizes the set of sentences of the language L_1. C3 is equivalent to the condition:

C3'. $Cn(P_2) \cap A_1 \subseteq Cn(P_1)$,

where A_1 stands for the set of purely universal sentences in the language L_1 (i.e. sentences in the language L_1 logically equivalent to sentences of normal form which do not contain any existential quantifiers).[7] As we see, the conditions C2' and C3' are of a purely syntactic character. The first requires that all logical consequences of the postulates P_2, belonging to the language L_1, follow from the postulates P_1, the second requires the same only in case of these logical consequences of the postulates P_2 which are purely universal sentences in the language L_1.

As I have said, the condition C1 does not have any syntactic counterpart; more precisely, it has no such counterpart corresponding to the language L_2. In particular the condition C2' is not such a counterpart—it is a standard condition of the non-creativity of the set P_2 with respect to the set P_1. C1 entails C2', but not conversely. It can be demonstrated that there are sets of sentences P_1 and P_2, which satisfy the condition C2' and do not satisfy the condition C1.[8] This point is both interesting and important. If the set P_2 is non-creative with regard to the set P_1, it does not impose on the interpretation of the terms of the language L_1 any other conditions expressible in the language L_1 than those imposed by the set P_1. It appears, however, that it may impose on the interpretation of these terms some conditions different from P_1 and inexpressible in the language L_1! There may, therefore, exist models of the set P_1 which are not fragments of any model of the set P_2. The condition C2' entails C1 only in some special cases. Here are some of them:

1) C2' implies C1 if P_2 is a set of purely universal sentences.[9]

2) C2' implies C1 if every model of the set P_1 is a model of finite universe (i.e. if P_1 entails a consequence stating the existence of at most n objects).

In cases when the fulfilment of the condition of non-creativity does not guarantee the fulfilment of the condition C1, the users of the language L_2 do not possess any criterion of checking whether C1 has been satisfied. They cannot decide it while remaining 'inside' their language; in order to do that, they must go beyond its limits. This is the reason why the existence of syntactic criteria is considered very important for conditions of the type we are speaking of, and why the usefulness in scientific practice of conditions which do not have such criteria seems somewhat doubtful.

2. The semantic analysis of a given empirical language implies the necessity of determining a set of postulates for its extralogical terms. While analysing the language L_2 we must, in particular, define the set of postulates P_2 for the newly introduced predicates $q_1, \ldots, q_m$. How can this be done? The procedure of extending a given empirical language by adding new theoretical terms always involves the act of accepting by the users a definite set of sentences which characterize the introduced terms. Can this set of sentences be considered a set of meaning postulates? Let T_2 be the set of sentences accepted in the procedure of extending the language L_1 by adding the predicates $q_1, \ldots, q_n$. Can this set be identified with P_2 — the set of postulates determining in the above-mentioned way the class of proper models of the language L_2? As we have seen P_2 cannot be an arbitrary set. If the interpretation of the language L_2 is to accord with the accepted semantic assumptions, the set P_2 must fulfil certain conditions. Generally speaking, it must be a set in some sense non-creative (with regard to the set P_1). Depending on how we define the class M_2^* of proper models of the language L_2 — i.e. according to the definition D1, D2, or D3 — the condition of non-creativity assumes a corresponding form — C1, C2, or C3. It is not excluded that the set T_2 of actually accepted sentences does not fulfil such a condition. I think that this situation is not only possible, but also common in actual scientific practice. Without justifying this presumption I should only like to point out that it often happens in empirical sciences that theoretical terms are characterized by the totality of axioms of an empirical theory, within which it is impossible to distinguish, on pragmatic grounds, definitional theorems and factual hypotheses. It is clear that the totality of axioms of such a theory cannot fulfil the condition of non-creativity, as this would conflict with the

theory's empirical character. Therefore, when proceeding to the logical reconstruction of the language of such a theory, we cannot consider the totality of its axioms as the set of postulates for the introduced terms. It is the task of the logician who attempts the reconstruction to distinguish this set. How can this be done? Let T_2 stand for the set of sentences accepted by the users of the language L_2 while introducing the predicates $q_1, \ldots, q_m$. Let us assume that the type of interpretation of the language L_2 is represented by the scheme corresponding to one of the definitions of the class M_2^* just considered—D1, D2, or D3. The set of sentences of the language L_2, which can be considered the set of postulates P_2 in these definitions, must fulfil the appropriate condition of non-creativity—C1, C2, or C3. If the set T_2 fulfils such a condition, we shall take T_2 as the set P_2. In the opposite case we cannot do that. We must then consider the set T_2 as one including not only the postulates for the predicates $q_1, \ldots, q_m$, but also some factual statements, and we have to accept as P_2 a correspondingly weaker set. What conditions should it be determined by? First, it must be a set that fulfils the appropriate condition of non-creativity—C1, C2, or C3. Moreover, the set must 'correspond' to T_2, i.e. it must include the postulates included in T_2. This is a rather vague requirement, allowing various explications. I shall give one of them here, first explaining its sense in application to the scheme corresponding to the definition D2.

In this situation the condition of non-creativity to be fulfilled by P_2 assumes the form of the condition C2, or else that of the equivalent condition C2'. The set P_2 cannot entail any sentences of the language L_1 which are not consequences of the set P_1:

$$Cn(P_2) \cap Z_1 - Cn(P_1) = 0.$$

At the same time, if the set T_2 does not fulfil such a condition, we must state that:

$$Cn(T_2) \cap Z_1 - Cn(P_1) \neq 0.$$

This set represents the totality of those conditions supplied by the set T_2 for the interpretation of the language L_1, which cannot be supplied by the set P_2. It then seems that P_2 should be strong enough to yield—together with the set just mentioned —a set logically equivalent

to the set T_2. This is just the requirement of the condition:

A. $\quad Cn\big((Cn(T_2) \cap Z_1 - Cn(P_1)) \cup P_2\big) = Cn(T_2).$

The set P_2 may therefore be characterized as a set fulfilling the conditions C2′ and A.

This same condition A may be used to determine the set P_2 in the situation corresponding to the definition D1. It is true that in this case the set P_2 must fulfil the condition of non-creativity C1, which is stronger than C2′. Speaking loosely, the set P_2 cannot impose on the interpretation of the language L_1 any conditions which would differ from those imposed by the set P_1 — neither those expressible in the language L_1 nor those which are not. But for this reason the set representing all these conditions — expressible in the language L_1 — imposed on the interpretation of this language by the set T_2, which cannot be imposed by P_2, is the same set as before:

$$Cn(T_2) \cap Z_1 - Cn(P_1).$$

Nothing else can be added to the set P_2 which would not go beyond our language L_1. As a result, the set P_2 can be here defined as one fulfilling the conditions C1 and A.

On the other hand, we must formulate otherwise the condition discussed for the scheme corresponding to the definition D3. In this case the set P_2 must fulfil merely partial condition of non-creativity, C3, which demands in its syntactic version C3′ that:

$$Cn(P_2) \cap A_1 - Cn(P_1) = 0,$$

i.e. that the set P_2 does not entail any purely universal sentences of the language L_1, which would not be consequences of the set P_1. Thus all the restrictions imposed on the interpretation of the language L_1 by the set T_2, which cannot be imposed by P_2, are represented by the set:

$$Cn(T_2) \cap A_1 - Cn(P_1).$$

This set, together with the set P_2, should then constitute a set logically equivalent to T_2:

B. $\quad Cn\big((Cn(T_2) \cap A_1 - Cn(P_1)) \cup P_2\big) = Cn(T_2).$

In this situation we characterize P_2 as a set satisfying the conditions C3′ and B.

Let us put together the results of our considerations, giving the conditions characterizing the set P_2 for each of the interpretations of the language L_2 just distinguished. The set P_2 is to be the set of sentences in the language L_2 which (for the given sets P_1 and T_2) fulfils:

(1) in the case of D1 — the conditions C1 and A,

(2) in the case of D2 — the conditions C2′ and A,

(3) in the case of D3 — the conditions C3′ and B.

Note that when the set T_2 itself fulfils the appropriate condition of non-creativity, the set P_2 becomes — on the basis of the definitions presented above — logically equivalent to T_2. In such cases the latter may then be considered the set of postulates P_2. In the other cases it must be a correspondingly weaker set. It is worth stressing that in these cases the definitions accepted by us do not uniquely determine the set P_2. Examples can be easily given of such logically non-equivalent sets P_2' and P_2'', each of which fulfils (for the same sets P_1 and T_2) the conditions mentioned in (1), and thereby also the weaker conditions of (2).[10] From the semantic point of view the difference between such sets seems irrelevant. Each of them may be considered the set of postulates coresponding to the set T_2. The choice must be carried out on the pragmatic basis.

The most important problem resulting from the accepted definitions is that of the existence of sets fulfilling the conditions they contain. Is there always a set P_2 fulfilling these conditions for any sets P_1 and T_2? The answer to this question depends on the conditions involved. The answer is negative in the cases of (1) and (2), and positive in the case of (3). Detailed investigations have been carried out concerning the problem of existence of sets satisfying the conditions C1 and A, and C2′ and A; the following results have been obtained:

1) For some sets P_1 and T_2 there is neither such a set P_2 which would fulfil the conditions C1 and A, nor such a set P_2 which would fulfil the conditions C2′ and A.

2) For some sets P_1 and T_2 there is no set P_2 which would fulfil the conditions C1 and A, but there is a set P_2 fulfilling the conditions C2′ and A.[11]

As I have said, the problem of existence of sets fulfilling the conditions C3′ and B has a positive solution:

For any sets P_1 and T_2 there is always a set P_2 fulfilling the conditions C3′ and B.

As this problem, unlike the others, has not yet been considered, I shall outline the proof of this theorem. In order to do it I repeat the conditions imposed on the set P_2:

C3'. $Cn(P_2) \cap A_1 \subseteq Cn(P_1)$;

B. $Cn((Cn(T_2) \cap A_1 - Cn(P_1)) \cup P_2) = Cn(T_2)$.

If $Cn(T_2) \cap A_1 - Cn(P_1) = 0$, we assume that $P_2 = T_2$. Let us consider the case when $Cn(T_2) \cap A_1 - Cn(P_1) \neq 0$, and assume that the sentence $\alpha \in Cn(T_2) \cap A_1 - Cn(P_1)$. We take as set P_2 the set of all the conditional sentences, obtained from the sentences of the set T_2 by preceding each of them with a common antecedent α:

$$P_2 = \{\alpha \rightarrow \beta_i\}_{\beta_i \in T_2}.$$

We are to demonstrate that the so defined set P_2 fulfils the conditions C3' and B. The fact of fulfilling B is obvious. In order to demonstrate that P_2 fulfils the condition C3' let us assume that $\gamma \in Cn(P_2) \cap A_1$. As $\gamma \in Cn(P_2)$, γ is the consequence of a finite subset of the set of sentences P_2, i.e.:

$$\gamma \in Cn(\{\alpha \rightarrow \beta_1, \ldots, \alpha \rightarrow \beta_k\}),$$

for some k. This is equivalent to saying that $\gamma \in Cn(\{\alpha \rightarrow \beta_1 \wedge \ldots \wedge \beta_k\})$. It is easy to notice that this is possible only if $\gamma \in Cn(\{\sim\alpha\})$. Since α is a purely universal sentence, $\sim\alpha$ is a purely existential one. As γ is also a purely universal sentence, it is easy to demonstrate (with the help of the interpolation theorem) that it is entailed by the sentence $\sim\alpha$ only if either $\alpha \in Cn(0)$, or $\gamma \in Cn(0)$. But the former case is impossible, as by definition $\alpha \notin Cn(P_1)$. The latter alternative, therefore, must hold. This means that $\gamma \in Cn(P_1)$, which was to be proved.

Thus, if the interpretation of language L_2 corresponds to the scheme of the definition D3, and, in consequence, the set of its postulates P_2 is determined by the conditions (3) intended for this situation, we have the guarantee that the set so determined exists, and we know how it can be constructed. In other situations we do not have any such guarantee. In some cases it may then appear that the set P_2 determined by the conditions (1) or (2) intended for the given situation does not exist at all. If we want to distinguish a set of postulates for the language

L_2, we must characterize it otherwise. This can be done by weakening in an appropriate way the condition A characterizing the set.[12]

3. Finally—some remarks about the degree of generality of the constructions presented here. The languages to which we have restricted our considerations are very simple: they are based on the first order predicate calculus (with identity) and they do not contain any extralogical terms besides predicates. The doubt then arises whether our considerations may be applied to real empirical languages, and particularly to the languages of typical empirical theories, such as physical theories. These theories are characterized by the fact that they use a powerful mathematical apparatus. Can it be accounted for in the framework of the constructions we have accepted? Does the character of the language L considered here and of its interpretation M^* allow such a possibility? The answer to this question depends to a considerable extent on the strength of that apparatus. If it goes beyond elementary devices and involves such concepts as the general concept of set, or of function, it cannot be accounted for without substantial extension of the constructions accepted here, as the latter are by definition limited to elementary languages. Undoubtedly, many of the existing physical theories use such a non-elementary apparatus. It seems, however, that there are also numerous interesting physical theories which do not go beyond the elementary apparatus; what is more, some—seemingly non-elementary physical theories—can be formulated in elementary language without any substantial loss.[13] The application of our constructions to elementary languages of physical theories does not seem to present any basic difficulties. However, some explanations and qualifications are necessary, as both these languages and their interpretations considerably differ at the first glance from those considered before.

The language of such a physical theory—let us call it L_f—is as a rule conceived of as a so-called many-sorted language and containing functional symbols. In the simplest case two types of variables are distinguished: $x_1, x_2, ...,$ and $y_1, y_2, ...,$ as well as two kinds of functional symbols as extralogical constants: $g_1, ..., g_k$ and $f_1, ..., f_l$. Models of such a language L_f assume the form of so-called two-range models:

$$\langle U_1, U_2 ; G_1, ..., G_k, F_1, ..., F_l \rangle,$$

where U_1 is the set of values of the variables $x_1, x_2, ...,$ U_2—the set

of values of the variables $y_1, y_2, ..., G_1, ..., G_k$—are functions with arguments and values from the set U_1, and $F_1, ..., F_l$—functions with arguments from the set U_2 and values from U_1. According to the intended interpretation of the language L_f, U_1 is to be a set of numbers (most often a set of real numbers), U_2—a set of physical objects, $G_1, ...,$ G_k—some mathematical operations and $F_1, ..., F_l$—some physical magnitudes. The set U_1 and the functions $G_1, ..., G_k$ thus represent the mathematical apparatus assumed in the given theory.

The differences between this language and its models and our language L and its models $\mathfrak{M}$—although undoubtedly existing—have a merely technical character. As we know, everything that can be expressed in a many-sorted language with functions is also expressible, loosely speaking, in an appropriate one-sorted language without functions, i.e. in a language such as L. I cannot formulate or justify here this—as a matter of fact, very well known—theorem (this is done in any advanced handbook of logic). I shall then restrict myself to some explanatory remarks.

The shift from a language with functions to the corresponding language without functions can be reduced from the syntactic point of view to replacing each k-place functional symbol by a $(k+1)$-place predicate. We thus obtain, instead of a language with functional symbols $g_1, ..., g_k, f_1, ..., f_l$, a language containing $k+l$ corresponding predicates $r_1, ..., r_k, q_1, ..., q_l$. Then, the shift from a two-sorted language to a one-sorted one consists in adopting one type of variables and introducing instead of the two precedent types of variables two additional one-place predicates s_1, s_2. As a result, we obtain a language of the same type as those considered previously—having only one type of variables and $n = 2 + k + l$ predicates, $s_1, s_2, r_1, ..., r_k, q_1, ..., q_l$. Let us call it, as before, the language L. We assume that its models, i.e. structures of the type:

$$\langle U; S_1, S_2, R_1, ..., R_k, Q_1, ..., Q_l \rangle$$

are connected with the models of L_f in the following way:

1) the universe U is the sum of the two previous ranges:

$$U = U_1 \cup U_2,$$

2) the subsets S_1, S_2 coincide respectively with the ranges U_1, U_2:

$$S_1 = U_1, \quad S_2 = U_2,$$

3) each relation R_i $(i=1, ..., k)$ corresponds to a function G_i according to the schema:

$$R_i(x_1, ..., x_{k_i}, y) \equiv G_i(x_1, ..., x_{k_i}) = y\,;$$

in the same way each relation Q_i $(i=1, ..., l)$ corresponds to the function F_i.

These assumptions make it intuitively understandable how everything the language L_f refers to is expressible in the language L. (Speaking loosely, it is enough to replace in a given sentence in L_f terms with functional symbols by appropriate predicates, and to unify the bound variables and relativize them to the predicate s_1 or s_2.)

We have already said that, according to the intended interpretation of the language L_f, some of its terms are given a mathematical interpretation. After the shift to the language L their role is taken over by the predicates $s_1, r_1, ..., r_k$. It is possible to maintain that a given theory is endowed with an appropriate mathematical apparatus only when the predicates denote definite mathematical entities, fixed in advance. Thus the predicate s_1 is to denote the set of real numbers — let us symbolize it by S_1, and the predicates $r_1, ..., r_k$ — definite relations between these numbers (corresponding to the operations of addition, multiplication, etc.) — let us symbolize them by $R_1, ..., R_k$. Such an assumption is of course expressible in the terminology we have accepted. It demands that the class M^* of proper models of the considered language L contains only such models $\mathfrak{M}$:

$$\langle U\,;\, S_1, S_2, R_1, ..., R_k, Q_1, ..., Q_l\rangle$$

in which the predicates $s_1, r_1, ..., r_k$ are given that fixed mathematical interpretation: $S_1, R_1, ..., R_k$. Let us call them standard models of the language L. However, if such a condition is imposed on the classes of proper models of the languages considered in this paper, some considerations presented in its second part should undergo certain modifications. These considerations are based on the assumption that the class of proper models M^* has been distinguished from the class of models of postulates $M(P)$ in terms of direct interpretational procedures of the type of ostensive definition. This was the only kind of direct interpretational procedure which we have referred to in this paper, according to its empirical assumptions. In the case now considered this kind of

direct interpretation is of course out of question. The distinguishing of the class of standard models of the language L must refer to another kind of direct interpretation which, therefore, entails different consequences.

I think, though, that the fact of providing an appropriate mathematical apparatus for a physical theory may also be understood in a more liberal way, fully harmonious with the constructions outlined in the present paper and without making any modification necessary. Namely, the mathematical terms can be considered as admitting — like all theoretical terms — only of an indirect interpretation, i.e. given by some set of postulates. This set would be simply a part of the set of postulates P, and thus would determine the interpretation of the mathematical terms of the language L. If we want this interpretation to be determined possibly precisely, we must take as the set of postulates for the predicates s_1, $r_1, \ldots, r_k$ a certain maximal set. It can simply be the totality of mathematical truths expressible in the language L. We denote it by Mt and define as follows:

> $\alpha \in Mt$ if α is a sentence true in every standard model of the language L.

Of course, the condition demanding that every proper model of the language L be a model of the set of sentences Mt is weaker than the condition requiring that this model be a standard one. It is worth pointing out, however, that in many applications — important from the logical point of view — this difference does not play any essential role. This can be illustrated by the following theorem.

> If the set of sentences X is a finite set, the sentence α is true in every standard model of the set X if and only if $\alpha \in Cn\,(Mt \cup X)$.[14]

This shows that a more liberal way of interpreting mathematical terms also allows to account for the mathematical apparatus, assumed by some empirical languages. What follows, they also come within range of our considerations.

REFERENCES

* First published in *Studia Filozoficzne* I (1972). Translated by S. Wojnicki.
[1] All the formal concepts employed in this paper are explained in any advanced handbook of logic. The most important of them are presented in a simple way

by R. Wójcicki, 'Analityczność, syntetyczność, empiryczna sensowność zdań', *Studia Filozoficzne* **3** (1966); appeared as: 'Analyticity, Syntheticity, Empirical Meaningfulness of Sentences', *Studia Filozoficzne* **4** (1970) supplementary volume in English.

[2] I discuss them among others in the paper 'Z semantyki pojęć otwartych' ('On the Semantics of Open Concepts'), *Studia Logica* **15** (1964).

[3] I have considered this problem in the paper 'O definiowaniu terminów spostrzeżeniowych' ('On Defining Observational Terms'), *Rozprawy Logiczne* (1964).

[4] The symbol $\{x: W(x)\}$ stands for the set of x's fulfilling the condition W.

[5] This kind of extension of a language and its interpretation has been distinguished by R. Suszko in his paper 'Logika formalna a rozwój poznania' ('Formal Logic and the Development of Knowledge'), *Studia Filozoficzne* **1** (**44**) (1966). The principle of preservation of postulates in a restricted form has also been formulated there.

[6] This argumentation is of a sketchy and intuitive character. It might be presented in a full and precise way if we developed further our considerations. It would then be necessary to characterize precisely the metalanguage of the language L_2, which would enable us to explicate a somehow puzzling concept of 'truth independent of experience'. A statement true independently of experience is, roughly speaking, one which may be demonstrated on the basis of the metalanguage. It might then be easily shown that it is possible to prove the statement (∗) on the basis of the metalanguage only if it is possible to prove the statement C1.

[7] The equivalence of the conditions C2 and C2′ is stated, among others, in J. Shoenfield's *Mathematical Logic* (1967); its proof is easy. The equivalence of C3 and C3′ has been first shown in the paper 'On the Extending of Models', by J. Łoś, *Fundamenta Mathematicae* **42** (1955).

[8] Cf., among others, J. Shoenfield, *op. cit.*

[9] Cf. J. Łoś, *op. cit.*

[10] Examples of such sets have been constructed for some special cases, in which the set T_2 consisted of a pair of so-called reduction sentences. Cf., among others, 'The Logic of Empirical Theories', *op. cit.*

[11] This problem is dealt with in the paper 'Inessential Parts of Extensions of First-Order Theories', by M. Przełęcki and R. Wójcicki, *Studia Logica* **28** (1971). The concepts discussed there differ slightly from the definitions (1) and (2) considered here. However, the proofs of the theorems given above may be obtained from those presented in that paper by a minor modification of the latter.

[12] Some suggestions in this matter can be found in the already quoted paper 'Inessential Parts of Extensions of First-Order Theories' by M. Przełęcki and R. Wójcicki, and in the paper 'The Problem of Analyticity' by the same authors, *Synthese* **19** (1969).

[13] In his paper 'Deterministic Theories', published in *Decisions, Values and Groups* (1962), R. Montague shows in detail how it is possible to formulate in such a way classical particle mechanics.

[14] This theorem is quoted by R. Montague in the work mentioned before.

MARIAN PRZEŁĘCKI

EMPIRICAL MEANINGFULNESS
OF QUANTITATIVE STATEMENTS*

I

The main object of our analysis is Suppes's well-known criterion of empirical meaningfulness for quantitative statements proposed in Suppes (1959), and our main purpose is to show this criterion to be a particular instance of some general criterion of empirical meaningfulness applicable to arbitrary statements. Such a criterion has been put forward in Przełęcki (1969). It is based on the same fundamental idea that underlies Suppes's criterion: a connection between meaningfulness and invariance in truth value of a given statement. The criterion is applicable to all empirical languages that can be formalized within first-order predicate logic.

Let L_0 and L_1 be two such languages and let L_1 be an extension of L_0 obtained from the latter by adding to its extralogical constants — called O-terms — new extralogical constants — called T-terms. In what follows, the models of language L_0 (in other terminology: its semi-models, realizations) will be symbolized by $\mathfrak{M}_0$, those of L_1 — by $\mathfrak{M}_1$. By $\mathfrak{M}_1|_0$ we shall denote the reduct of model $\mathfrak{M}_1$ to language L_0, i.e., the model of L_0 obtained from $\mathfrak{M}_1$ by eliminating from it the interpretations of all T-terms. When $\mathfrak{M}_1|_0 = \mathfrak{M}_0$, $\mathfrak{M}_1$ will be called an expansion of $\mathfrak{M}_0$. It is assumed known under what conditions a sentence α of language L_i, for $i=0,1$, is true in a model $\mathfrak{M}_i$. If all sentences of a set X_i are true in a model $\mathfrak{M}_i$, $\mathfrak{M}_i$ is said to be a model of the set X_i, in symbols $\mathfrak{M}_i \in M(X_i)$. When $X_i = \{\alpha\}$, we write $\mathfrak{M}_i \in M(\alpha)$ for short.

We shall accept the usual semantical assumption to the effect that the meanings of extralogical constants in the languages considered are governed by certain sets of postulates: set P_0 for O-terms, and set P_1 for T-terms. What this assumption amounts to may be stated in terms of the intended (proper) interpretation of both kinds of terms: both of

them should be interpreted in such a way that the corresponding meaning postulates are true. Moreover, the interpretation of T-terms should preserve the intended interpretation of O-terms: T-terms are to be understood in such a way that the sentences of P_1 are true under the intended interpretation of O-terms. And, in contrast to O-terms, there is assumed to be no other way in which the intended interpretation of T-terms is determined. We arrive thus at the following assumptions: (1) the intended model of L_0 is chosen from the class $M(P_0)$ of all possible models for L_0 by some non-verbal means (i.e. without stipulating the truth of any other sentences of L_0); (2) the intended model of L_1 is determined only by two conditions: (i) it is an expansion of the intended model of L_0, and (ii) a model of set $P_1{}^1$.

Since postulates P_1 are not meant to stipulate meanings for O-terms but only for T-terms, they should not impose any restrictions upon interpretations of O-terms that are not imposed by postulates P_0. One way of making precise this intuitive idea is to require the set P_1 to be non-creative (in the model-theoretic sense) with respect to the set P_0. P_1 is non-creative with respect to P_0 if and only if every model of P_0 can be expanded to a model of P_1, i.e., if and only if the following condition is fulfilled:

(C1)   For every $\mathfrak{M}_0 \in M(P_0)$ there is an $\mathfrak{M}_1 \in M(P_1)$ such that $\mathfrak{M}_1|_0 = \mathfrak{M}_0$.

In what follows, P_1 is always assumed to satisfy the condition C1.

Now, the semantical characterization of language L_1 outlined above makes it possible to formulate some criteria of empirical meaningfulness with regard to all statements expressible in this language. Any such criterion has necessarily a relative character: empirical meaningfulness of a given sentence being relativized to some class of sentences (or terms) whose empirical meaningfulness is regarded by us as unquestionable. According to a usual assumption, these are the so-called observational sentences (or terms) – in some of the various meanings of this highly ambiguous notion. In our case the role of such expressions is played by O-terms. It is just a kind of connection between the interpretations of T-terms and O-terms that proves to be decisive in classifying a given statement of language L_1 as empirically meaningful or meaningless. All the connections between these two kinds of terms are established by

postulates P_1. If, for any interpretation of O-terms, the postulates determined the interpretation of T-terms in a unique way, all statements containing T-terms would be empirically meaningful. But this is hardly an acceptable assumption: with regard to most empirical languages it is obviously false. And when the interpretation of T-terms is determined ambiguously, there always might appear statements which under one interpretation turn out to be true and under another false. These are just the empirically meaningless sentences of L_1.

This roughly stated criterion can be made precise in more than one way. I shall here give two of its explications: a stronger and a weaker one. Let us call them EM_1 and EM_2. Their definitions read as follows:

(D1)   A sentence α of L_1 is EM_1 if and only if for every $\mathfrak{M}_0 \in M(P_0)$ and for any $\mathfrak{M}_1$, $\mathfrak{M}_1'$: if $\mathfrak{M}_1|_0 = \mathfrak{M}_1'|_0 = \mathfrak{M}_0$, $\mathfrak{M}_1 \in M(P_1)$ and $\mathfrak{M}_1' \in M(P_1)$, then $\mathfrak{M}_1 \in M(\alpha)$ if and only if $\mathfrak{M}_1' \in M(\alpha)$.

(D2)   A sentence α of L_1 is EM_2 if and only if there is an $\mathfrak{M}_0 \in M(P_0)$ such that for any $\mathfrak{M}_1$, $\mathfrak{M}_1'$: if $\mathfrak{M}_1|_0 = \mathfrak{M}_1'|_0 = \mathfrak{M}_0$, $\mathfrak{M}_1 \in M(P_1)$ and $\mathfrak{M}_1' \in M(P_1)$, then $\mathfrak{M}_1 \in M(\alpha)$ if and only if $\mathfrak{M}_1' \in M(\alpha)$.[2]

The intuitive meaning of the two notions of empirical meaningfulness seems clear enough. Two models of set P_1 which are expansions of the same model of language L_0 may differ in the interpretation of T-terms. If these differences, however, do not affect the truth value of a given statement, the statement may be classified as an empirically meaningful one. Its truth value is invariant over the class of all those expansions of model $\mathfrak{M}_0$ which satisfy the postulates P_1. If that is the case with respect to all possible models $\mathfrak{M}_0$ of language L_0 (i.e., all models of postulates P_0), the statement is meaningful in sense EM_1, if with respect to at least one of such models, it is meaningful in sense EM_2.

Let us compare these two notions. Evidently $EM_1 \subseteq EM_2$. All sentences of the sublanguage L_0 are empirically meaningfull in both senses. And so are all the analytic and contradictory sentences of L_1 (i.e. logical consequences of postulates P_0 and P_1, and their negations). The difference between EM_1 and EM_2 may best be illustrated with a simple example. Let O-terms comprise two one-place predicates O_1, O_2 and an individual constant o_1, and T-terms a one-place predicate T_1. Let P_0

consist of mere tautologies and P_1 of one reduction sentence for T_1:

(1) $\forall x [O_1(x) \rightarrow (T_1(x) \leftrightarrow O_2(x))]$.

It is easily seen that, according to definition D1, a sentence of L_1 will belong to EM_1 if and only if it is translatable, by virtue of postulate (1), into a sentence of L_0. Thus, e.g., $O_1(o_1) \wedge T_1(o_1)$ (equivalent on the basis of (1) to $O_1(o_1) \wedge O_2(o_1)$) will have to be qualified as EM_1, while $T_1(o_1)$ as non-EM_1. According to definition D2, all sentences of L_1 will belong to EM_2: $O_1(o_1) \wedge T_1(o_1)$ as well as $T_1(o_1)$. The example seems to speak in favour of the weaker criterion of empirical meaningfulness, EM_2. A language whose only theoretical predicate has been defined conditionally by means of observation terms hardly contains any empirically meaningless statements.

It seems worth noticing that the notion of EM_2 may be characterized in an alternative, especially perspicuous way. It may be easily checked that a sentence α of L_1 is EM_2 if and only if either the set $P_1 \cup \{\alpha\}$ or the set $P_1 \cup \{\sim\alpha\}$ is creative with respect to P_0. Thus every empirically meaningful statement (or its negation) asserts, in conjunction with P_1, something more about the interpretation of O-terms than has been stipulated by the meaning postulates P_0 alone. Every such statement bears certain amount of empirical import.

II

Suppes's criterion of empirical meaningfulness for quantitative statements has been expounded in detail for one specific (but typical and important) case: statements about results of mass measurements. In order to reconstruct that criterion within the conceptual framework outlined above, we have to provide a syntactical and semantical characterization of a language adequate to express the results of mass measurements as it differs in some respects from the language considered thus far. The only formal apparatus contained in the latter has been that of the first-order predicate logic. The language needed now must include some mathematical apparatus as well. The one presupposed in our further considerations is, roughly speaking, that of the arithmetic of real numbers.

We shall thus assume that the vocabulary of languages L_0 and L_1 as conceived now includes, besides a logical, a mathematical part: numerical

variables x, y, ... and a set of arithmetical constants including a two-place predicate $\leqslant$, a two-place function symbol $+$, and possibly some other terms (e.g., a stock of numerical constants). All mathematical symbols are assumed to have the same, uniquely fixed, interpretation in all models of language L_0 and L_1; viz. their standard interpretation in terms of the usual system of real numbers. In all models of language L_0 and L_1 the variables x, y, ... range over the set of real numbers, and the constants $\leqslant$, $+$ are interpreted as the relation less than or equal and the operation of addition on the set of real numbers. Only models satisfying these conditions, the so-called standard models of L_0 and L_1, are from now on being referred to by symbols $\mathfrak{M}_0$ and $\mathfrak{M}_1$. The extralogical and extra-mathematical part of the vocabulary of language L_0 comprises individual variables a, b, ... and a set of O-terms including a two-place predicate R, a two-place function symbol $\circ$, and possibly some other terms (e.g., a number of individual constants). The language L_1 is obtained from L_0 by adding to it one T-term: a one-place function symbol m. In any model $\mathfrak{M}_1$ of L_1 the variables a, b, ... range over some arbitrary non-empty set A, R is interpreted as a binary relation on A, $\circ$ as a binary function from $A \times A$ into A, and m as a unary function from A into the positive real numbers.[3]

Now, the intended interpretation of language L_1 as a language expressing results of mass measurements (by means of an equal-arm balance) might be roughly stated as follows. Set A is taken to be a set of physical objects, R is interpreted as the relation 'weighs less than or just as much as', and the interpretation of $\circ$ is physical combination ($a \circ b$ is the object obtained by combining the objects a and b). The only T-term m is to denote the mass function: $m(a)$ designating a real number, the mass of a physical object a.

With this interpretation in mind we shall define two sets of meaning postulates: P_0 for O-terms R and $\circ$, and P_1 for T-term m. P_0 is taken to be a set of axioms for extensive quantities (Suppes, 1951), consisting of the closures of following formulas:

(1) $a R b \wedge b R c \rightarrow a R c$,

(2) $(a \circ b) \circ c R a \circ (b \circ c)$,

(3) $a R b \rightarrow a \circ c R c \circ b$,

(4) $\sim a\,R\,b \rightarrow \exists\,c\,(a\,R\,b \circ c \wedge b \circ c\,R\,a)$,

(5) $\sim a \circ b\,R\,a$,

(6) $a\,R\,b \rightarrow \exists\,n\,(b\,R\,na)$,

where variable n takes as values the natural numbers, and na is defined in the usual recursive way:

(a) $1a = a$,

(b) $na = (n-1)\,a \circ a$.

The set P_1 contains the closures of two formulas:

(i) $a\,R\,b \leftrightarrow m(a) \leqslant m(b)$,

(ii) $m(a \circ b) = m(a) + m(b)$.

Sets P_0 and P_1 defined in this way satisfy the condition C1 of non-creativity, which has been imposed on any sets of meaning postulates for language L_0 and L_1: every model of P_0 can be expanded to a model of P_1. The fact is guaranteed by the Theorem of Representation, which states that every extensive system is homomorphic to some subsystem of real numbers with the relation less than or equal and the operation of addition. The homomorphism in question provides an interpretation for the function symbol m satisfying the postulates P_1.

The assumptions made above allow us to apply the general criteria of empirical meaningfulness defined in Section I to statements of language L_1 characterized in the way indicated. Definitions D1 and D2 provide two such criteria, EM_1 and EM_2. Now, my claim is that EM_1, the stronger one, is equivalent to one of Suppes's criteria defined in Suppes (1959), viz. to the criterion of empirical meaningfulness in sense B, or rather to a reconstruction of it adapted to our present purposes.

Let $\mathfrak{M}_1$ and $\mathfrak{M}_1'$ be two models of language L_1 such that $\mathfrak{M}_1|_0 = \mathfrak{M}_1'|_0$, and let m and m' be the interpretations of the function symbol m in $\mathfrak{M}_1$ and $\mathfrak{M}_1'$, respectively. If there is a positive number k such that $\mathrm{m}(a) = k \cdot \mathrm{m}'(a)$, for all arguments of m, $\mathfrak{M}_1$ and $\mathfrak{M}_1'$ are said to be related by a similarity transformation. Now Suppes's criterion of empirical

meaningfulnes, symbolized here by EM_3, may be rendered as follows:

(D3)   A sentence α of L_1 is EM_3 if and only if for every $\mathfrak{M}_0 \in M(P_0)$ and for any $\mathfrak{M}_1$, $\mathfrak{M}'_1$: if $\mathfrak{M}_1|_0 = \mathfrak{M}'_1|_0 = \mathfrak{M}_0$, $\mathfrak{M}_1 \in M(P_1)$ and $\mathfrak{M}'_1$ is related to $\mathfrak{M}_1$ by a similarity transformation, then $\mathfrak{M}_1 \in M(\alpha)$ if and only if $\mathfrak{M}'_1 \in M(\alpha)$.[4]

To show the criterion EM_3 to coincide with EM_1 it suffices to adduce the following theorem, based on some well-known facts:[5]

(T1)   For every $\mathfrak{M}_1$ and $\mathfrak{M}'_1$: if $\mathfrak{M}_1|_0 = \mathfrak{M}'_1|_0$ and $\mathfrak{M}_1 \in M(P_1)$, then $\mathfrak{M}'_1 \in M(P_1)$ if and only if $\mathfrak{M}'_1$ is related to $\mathfrak{M}_1$ by a similarity transformation.

On the basis of T1, the definition D3 turns out to be equivalent to definition D1.

The intuitive meaning of the criterion EM_3 seems quite plausible. Two models of the postulates (i)–(ii) being expansions of the same possible model of L_0 may differ in the interpretation of the mass function m. As theorem T1 clearly shows, that interpretation is unique up to a similarity transformation. And so, if a statement of L_1 has a truth value invariant with respect to all such transformations it is empirically meaningful; otherwise — empirically meaningless. To illustrate this idea with some simple examples let us assume that among O-terms there are three individual constants: o_1, o_2, o_3. Then it may easily be seen that statements like

$$m(o_1) \leqslant m(o_2),$$

or

$$m(o_1) + m(o_2) = m(o_3)$$

are EM_3, while those like

$$m(o_1) \leqslant 4,$$

or

$$m(o_1) = 4$$

are non-EM_3.

EM_1 is the stronger of our two notions of empirical meaningfulness defined previously. Does the weaker one also provide a useful criterion for the particular case being considered now? To see what the difference between them in this case amounts to let us formulate the criterion

EM_2 for our language of mass measurements in a way analogous to that of EM_3:

(D4) A sentence α of L_1 is EM_4 if and only if there is an $\mathfrak{M}_0$ $\in M(P_0)$ such that for any $\mathfrak{M}_1$, $\mathfrak{M}_1'$: if $\mathfrak{M}_1'|_0 = \mathfrak{M}_1|_0 = \mathfrak{M}_0$, $\mathfrak{M}_1 \in M(P_1)$ and $\mathfrak{M}_1'$ is related to $\mathfrak{M}_1$ by a similarity transformation, then $\mathfrak{M}_1 \in M(\alpha)$ if and only if $\mathfrak{M}_1' \in M(\alpha)$.

Classification of statements mentioned above into EM_4 and non-EM_4 turns out to be consistent with that given with regard to EM_3. The difference between the two notions appears in the case of certain compound sentences of L_1. The statement

$$m(o_1) \leqslant m(o_2) \wedge m(o_3) = 4$$

may serve as an example. It is empirically meaningful in sense EM_4, and meaningless in sense EM_3. This will easily be seen if we realize that there always are such possible models of L_0 that in all their expansions being models of P_1 the first component of the above statement is false. Then the whole statement is false too, and hence meaningful in sense EM_4. It would be meaningful in sense EM_3 if that were the case with respect to all possible models of L_0. But among them there always exist such that in their expansions being models of P_1 the first component turns out to be true. And since within that class of expansions the truth value of the second component varies, the truth value of the whole statement varies with it, showing the statement to be meaningless in sense EM_3. A choice between the two notions of empirical meaningfulness depends on whether or not we want to include statements of that kind into the class of empirically meaningful expressions.

III

What raises the question of empirical meaningfulness (in the sense considered) of a given statement is a kind of ambiguity in the interpretation of some of its terms. It is just because the interpretation of T-terms, under a given interpretation of O-terms, is determined in an ambiguous way that some sentences containing the former appear to be empirically meaningless. If that interpretation were unique any statement would become empirically meaningful. This, in particular, is true of our language

for mass measurements. The mass function is fixed, for a given interpretation of terms R and $\circ$, to a certain extent only. This is the reason
why some meaningless statements are bound to appear. A unique determination of the mass function would turn all of them into meaningful
ones.

Such a unique determination may easily be accomplished. And often
it is accomplished in actual scientific theories. The ambiguity in the
determination of mass function is due to lack of a unit specification.
Once the unit of mass is specified the mass function is fixed completely. A
choice of unit is expressed in our conceptual framework by adding to the
meaning postulates (i) and (ii) in set P_1 a third one to the effect that:

(iii) $m(o_1) = 1.$

If in definitions D3 and D4 set P_1 is understood as made up of sentences
(i), (ii) and (iii) the definitions will classify all statements of language
L_1 as empirically meaningful, i.e. as EM_3 and EM_4. And so, for language
like that, our criteria will seem to be devoid of any real importance.

I shall not enter into the question which, if any, languages of actual
scientific theories belong to the unitless languages of the kind described
in Section II. Rather I shall try to argue that even for languages with a
unit specification our general criteria of empirical meaningfulness do not
lose their importance. There might appear, within such languages,
statements which, according to those criteria, should be reckoned
among empirically meaningless expressions.

As mentioned before, every criterion of empirical meaningfulness
has to be based on a class of expressions endowed with an unquestionable
empirical meaning. Their role in our constructions has been played by
the class of O-terms, including among others terms R and $\circ$. But might
the latter rightly be regarded as terms whose empirical meaning is so
straightforward and evident that it does not raise any questions? Might
they, in particular, be treated as observation terms in any plausible
meaning? Their usual explications suggest the positive answer. The relation 'weighs less than or just as much as', denoted by R, is normally
explicated in terms of such simple physical operations as placing two
weights on the opposite pans of an equal-arm balance and comparing
their relative positions. The operation of physical combination, denoted
by $\circ$, is often identified with placing two weights on the same pan.

But it is evident that these rough and cursory explications cannot be taken too literally. The set of postulates P_0 for R and $\circ$ entails that A is an infinite set, containing objects of an arbitrarily great mass. Since the relation of comparison is defined on the whole set A it cannot be identified with operations which, by their very nature, are restricted to certain medium size physical objects. The same is true of the interpretation proposed for $\circ$. Moreover, the operation of combination, which according to postulates P_0 does not satisfy the idempotence law: $a \circ a = a$, cannot be understood as an idempotent operation of placing together.

  All these difficulties can be avoided if the terms R and $\circ$ are treated, not as observational, but as theoretical expressions (as they, in fact, have been treated by some writers, cf. Sneed (1971)). Then their connections with observational terms may be accounted for in the same way as in the case of other theoretical expressions. The connections are of an indirect and incomplete nature. They are expressible by meaning postulates which typically assume forms looser than that of a full-fledged definition. Those which express some operational criteria for terms R and $\circ$ may be formulated as a kind of reduction sentences. One of the meaning postulates for R may roughly be rendered as follows:

> If a and b are placed on the opposite pans of an equal-arm balance then $a R b$ if and only if the pan with a does not outweigh that with b.

The postulate has a form of a reduction sentence:

(I) $a R_1 b \rightarrow (a R b \leftrightarrow a R_2 b)$,

and provides only a partial operational criterion for the term R. Other such criteria may be supplied by additional reduction sentences, based on different operations.

  The interpretative procedures for the term $\circ$ seem more complicated. There has been suggested (in Sneed (1971)) the following interpretation: Let $a \circ b$ be the physical object that you get by pouring enough sand on one pan to balance a and b placed on the other pan. But the procedure is not applicable when $a = b$. In that case it might be modified as follows: Let us first balance a with sand on the other pan, then place a on the pan with sand and balance the pan loaded in this way by pouring enough sand on the other pan; the amount of sand on this pan may be identified

with the physical object $a \circ a$. On a closer scruting these interpretative procedures also appear to be expressible by means of some reduction sentences such as:

(II) $S_1(a, b, c) \rightarrow (a \circ b = c \leftrightarrow S_2(a, b, c))$,

or the like, in which predicates S_1, S_2 refer to certain observable relations.

This, of course, is a highly oversimplified and incomplete account, which cannot be elaborated in this paper. It has been given only to exhibit one peculiar feature of the interpretation of terms R and $\circ$, a feature characteristic of the interpretation of all typical theoretical terms: its ambiguity. The postulates connecting terms R and $\circ$ with terms such as R_1, R_2, S_1, or S_2 do not determine the interpretation of the former, for a given interpretation of the latter, in a unique way. This fact proves to be decisive in our question.

If we insist on basing our criteria of empirical meaningfulness upon a class of expressions whose empirical meaning is guaranteed by their observational status then our construction should start from terms R_1, R_2, S_1 and S_2 rather than from R and $\circ$. The former only will be classified as O-terms. R and $\circ$ will, together with m, be reckoned among T-terms. Accordingly, L_0 will now symbolize a language whose (extra-logical and extramathematical) vocabulary consists of terms R_1, R_2, S_1, S_2, and L_1 will be its extension obtained by the addition of terms R, $\circ$ and m. The set formerly symbolized by P_0 and consisting of the axioms for extensive quantities (1)–(6) will, together with the reduction sentences (I)–(II) (and possibly some others) constitute the set of meaning postulates for R and $\circ$, while the set P_1 consisting of sentences (i)–(iii) will, as before, represent the set of meaning postulates for m. The sum of these two sets will now be symbolized by P_1. Terms R_1, R_2, S_1, S_2 also will be governed by certain, suitably chosen, set of postulates, now symbolized by P_0.

It is easily seen that under the present interpretation of all symbols occurring in definitions D1 and D2 the definitions will classify some statements of L_1 as empirically meaningless. Although the interpretation of function symbol m is fixed, for a given interpretation of terms R and $\circ$, uniquely, the interpretation of the latter is determined, for a given interpretation of O-terms R_1, R_2, S_1, S_2, only ambiguously.

And the ambiguity is, in effect, transferred to the interpretation of m. As a result, there might appear some statements expressing the results of mass measurements which under one such interpretation of m turn out to be true and under another false. A specification of the unit mass does not exclude this possibility, and thus does not eliminate the problem of empirical meaningfulness.

BIBLIOGRAPHY

M. Przełęcki, *The Logic of Empirical Theories*, Routledge and Kegan Paul, London 1969.

J. D. Sneed, *The Logical Structure of Mathematical Physics*, D. Reidel, Dordrecht 1971.

P. Suppes, 'A Set of Independent Axioms for Extensive Quantities', *Portugaliae Mathematica* **10** (1951), 163–172.

P. Suppes, 'Measurement, Empirical Meaningfulness, and Three-Valued Logic', in *Measurement: Definitions and Theories* (ed. by C. W. Churchman and P. Ratoosh), Wiley, New York 1959, pp. 129–143.

REFERENCES

* Reprinted from *Synthese* **26** (1974), D. Reidel Publishing Company, Dordrecht-Holland.

[1] For a closer examination of the kind of interpretation characteristic of empirical languages see Przełęcki (1969).

[2] The notions of EM_1 and EM_2 defined above coincide with those of DT_1 and DT_2 as defined in Przełęcki (1969) provided the set P_0 contains nothing but tautologies.

[3] Although two kinds of variables are used in L_0 and L_1 as characterized above (and, consequently, two domains appear in their models), the languages may easily be modified to conform to the general pattern described in Section I: that of a one-sorted language formalized within first-order predicate logic.

[4] The original definition in Suppes (1959) abstracts from any connections between the mass function m and the empirical relations corresponding to terms R and $\circ$. Therefore it lacks the clauses requiring that $\mathfrak{M}_0$ should be a model of P_0 and $\mathfrak{M}_1$ of P_1.

[5] See e.g. Suppes (1951), (1959), and Sneed (1971), p. 86.

MARIAN PRZEŁĘCKI and RYSZARD WÓJCICKI

THE PROBLEM OF ANALYTICITY*

Our purpose in the present paper is to give an account of certain results that have been obtained within the field of logical methodology by the Polish logicians in recent years. The problem of analyticity constitutes one of the main points on which these investigations have concentrated. The studies on the problem have brought about a series of papers written by several authors and containing results which seem interesting enough to be presented in this place. The work in this field has, to a great extent, been initiated by Ajdukiewicz. His interesting paper [1] poses the problem concerning the dependence of the analytic sentences on experience in a way which stimulates further inquiries. Almost all the recent papers concerned with this problem might be thought of as a kind of response to that stimulus. It should be noticed, however, that their authors make use of very different conceptual and formal frameworks. Any attempt to present their results in each author's original terminology would thus lead to a very complicated and incoherent exposition. Therefore, we have decided to present all the results which have been taken into account within one uniform conceptual framework. It has been possible to arrive, in this way, at a comparatively simple and coherent exposition. But not all of the existing results could be included into it. Some papers must have been left out completely, some others — partly; and those which are included have undergone a more or less essential reformulation. For this reason, the present exposition can by no means be considered as an exhaustive survey of all results obtained in the field in question. The more so that it often goes beyond the mere statement of the existing results, filling some gaps in the picture to be presented.

The formal conceptual apparatus employed in the paper is that of the model theory. The languages considered belong to formalized languages. The fundamental formal concepts are explained in a general way in Section I.

I

The languages discussed in this paper are so-called applied first-order languages and their second-order extensions, i.e. languages based on the first- or second-order logic, respectively. They thus all contain: (a) a denumerable set of individual variables (for a given language $\mathscr{L}$ all individual variables are treated as ranging over the same set of elements), (b) certain basic logical constants: connectives 'not', 'and', 'or', 'if ... then', 'if and only if', and quantifiers 'for all', 'for some'; these constants will be symbolized by $\sim$, $\wedge$, $\vee$, $\rightarrow$, $\equiv$, $\bigwedge$, $\bigvee$. Those languages which are based on the second-order logic contain, in addition, predicate variables. Besides being first or second order, the languages may also differ in their supply of non-logical constants which in general are of three kinds: (1) the predicates or relational symbols, (2) the function symbols, (3) the individual constants. For reasons of simplicity we confine ourselves to the languages not containing constants from categories (2) and (3). The set of the non-logical constants of a language $\mathscr{L}$ will be called the *vocabulary* of $\mathscr{L}$ and denoted by $V(\mathscr{L})$.

Let $V(\mathscr{L}) = \{r_1, ..., r_k\}$. By a *model* of $\mathscr{L}$ we shall understand a relational system $\mathfrak{M} = \langle U, R_1, ..., R_k \rangle$ suitable as an interpretation for the non-logical constants of $\mathscr{L}$ in the sense that each R_i takes the same number of arguments as the corresponding r_i. The set of models of $\mathscr{L}$ will be denoted by $\mathrm{Mod}\,(\mathscr{L})$. We assume it to be clear under what conditions a sentence φ of a language $\mathscr{L}$ is said to *be true* (to *hold*) in a relational system $\mathfrak{M} \in \mathrm{Mod}\,(\mathscr{L})$, in symbols $\varphi \in \mathrm{Ver}\,(\mathfrak{M})$. Any couple $L = \langle \mathscr{L}, \mathfrak{M} \rangle$, where $\mathscr{L}$ is a language, $\mathfrak{M} \in \mathrm{Mod}\,(\mathscr{L})$, is said to be an *interpreted language* (or *semantical system*). The model $\mathfrak{M}$ will be called the *proper model* of L. A sentence φ of an interpreted language $L = \langle \mathscr{L}, \mathfrak{M} \rangle$ is said to be true if, and only if, $\varphi \in \mathrm{Ver}\,(\mathfrak{M})$, i.e. it is true in the proper model of L.

The languages considered here may be regarded as formalized counterparts of languages of various branches of science. The methods of formal logic (formal syntax and formal semantics) are far from sufficient for analyzing the exact nature of the very complicated process of endowing the expressions of such languages with their meaning and thus establishing their interpretation. Nevertheless these methods have proved their usefulness in dealing with the case when non-logical constants are

characterized by a purely verbal way, i.e. by postulates or definitions. In our further discussion we shall concern ourselves with such cases only.

Some notational conventions will simplify the ensuing discussion. If $\mathscr{L}$ is a first-order language, we shall denote by $\mathscr{L}^*$ its second-order extension, i.e. the language which possesses the same vocabulary ($V(\mathscr{L}) = V(\mathscr{L}^*)$) but is based on the second-order logic. In general, if a symbol stands for a notion related to $\mathscr{L}$, the same symbol provided with a star will denote this notion related to $\mathscr{L}^*$. The subscripts 'm' and 'p' will be used to discern between the model theoretic (or semantical) notions anp their proof theoretic (or syntactical) counterparts. In particular, we shall discern the model theoretic notion of consequence, which will be denoted by Cn_m, from the proof theoretic one, to be denoted by Cn_p. Thus, e.g. the formula $\varphi \in Cn_m(A)$ reads: φ is a model theoretic sonsequence of A. (We need not explain that this formula is meaningless if φ and A do not belong to the same language). Let Sent $(\mathscr{L})$ denote the set of sentences of $\mathscr{L}$. If Cn_m and Cn_p are both consequences defined on the subsets of Sent $(\mathscr{L})$, Cn_m^* and Cn_p^* are understood to be their counterparts defined on the subsets of Sent $(\mathscr{L}^*)$. Sometimes we shall use the notion of consequence in an ambiguous way as a notion which may be as well a semantical or syntactical one. In that case we shall apply the symbol Cn. We shall write A equiv$_m$ B instead of $Cn_m(A) = Cn_m(B)$. The formula A equiv$_m$ B reads: the set A is equivalent in the model theoretic sense to B. The symbols equiv$_p$ and equiv are understood in an analogous way.

We shall say that $\mathscr{L}'$ is an *enlargement* of $\mathscr{L}$, in symbols $\mathscr{L} \in \text{Enlarg}(\mathscr{L})$, if and only if $V(\mathscr{L}) \subseteq V(\mathscr{L}')$. The languages $\mathscr{L}$ and $\mathscr{L}'$ may differ as to their order. Let $\mathscr{L}' \in \text{Enlarg}(\mathscr{L})$, $\mathfrak{M} \in \text{Mod}(\mathscr{L})$, $\mathfrak{M}' \in \text{Mod}(\mathscr{L}')$. By $\mathfrak{M}'|_{\mathscr{L}}$ we shall denote this part of the model $\mathfrak{M}'$ which provides the interpretations for the non-logical constants of $\mathscr{L}$. If $\mathfrak{M}'|_{\mathscr{L}} = \mathfrak{M}$, the relational system $\mathfrak{M}'$ will be said to be a *prolongation* of $\mathfrak{M}$. Assume again that $\mathscr{L}' \in \text{Enlarg}(\mathscr{L})$. Let $\mathfrak{M} \in \text{Mod}(\mathscr{L})$ and $A \subseteq \text{Sent}(\mathscr{L}')$. The relational system $\mathfrak{M}$ will be said to be *admitted* by A, in symbols $\mathfrak{M} \in \text{Adm}(A)$, if and only if there is $\mathfrak{M}' \in \text{Mod}(\mathscr{L}')$ such that $\mathfrak{M}'|_{\mathscr{L}} = \mathfrak{M}$ and $A \subseteq \text{Ver}(\mathfrak{M}')$.

The following two notions will play an important role in our further considerations. Suppose that $\mathscr{L}' \in \text{Enlarg}(\mathscr{L})$, $A \subseteq \text{Sent}(\mathscr{L}')$. The set A will be said to be *non-creative with respect to $\mathscr{L}$ in the model theoretic*

sense, in symbols $A \in \text{Noncreat}_m(\mathscr{L})$, if and only if for each $\mathfrak{M} \in \text{Mod}(\mathscr{L})$, $\mathfrak{M} \in \text{Adm}(A)$. The set A is called *non-creative in the proof-theoretic sense with respect to $\mathscr{L}$*, in symbols $A \in \text{Noncreat}_p(\mathscr{L})$ if and only if for every sentence $\varphi \in \text{Sent}(\mathscr{L})$, if $\varphi \in \text{Cn}_p(A)$, then $\varphi \in \text{Cn}_p(\emptyset)$. These notions do not coincide. If both languages $\mathscr{L}$ and $\mathscr{L}'$ are first-order languages, it can be proved that the semantical notion of non-creativity is stronger than the syntactical one: if $A \in \text{Noncreat}_m(\mathscr{L})$ then $A \in \text{Noncreat}_p(\mathscr{L})$, but in general not conversely.

All the notions defined above for sets of sentences will be in an obvious way applied to single sentences. Thus we shall sometimes write $\text{Cn}_m(\varphi)$, $\varphi \in \text{Noncreat}_p(\mathscr{L})$ etc., φ being a sentence. As a rule we shall not distinguish between a finite set of postulates A and the conjunction of the sentences belonging to A (in particular, we shall denote the latter by the same symbol A). Similarly, we shall not distinguish between a sentence φ and the unit set which contains this sentence as its single element.

II

A typical situation in which analytic sentences are bound to appear is that of enlarging a semantical system by means of postulates. It may roughly be described as follows. Let $L_0 = \langle \mathscr{L}_0, \mathfrak{M}_0 \rangle$ be a first-order semantical system and let $V(\mathscr{L}_0) = \{a_1, \ldots, a_m\}$. In order to simplify our further analyses we shall assume that there are no postulates (in particular, no definitions) for the non-logical constants $a_1, \ldots, a_m$. These terms are assumed to be characterized in a non-verbal way, e.g. by the so-called ostensive definitions. No analysis of this procedure will be given here, as no such analysis can be carried out by the methods of formal logic. There is only one point which will be presupposed in all our further considerations. Whatever this interpretative procedure might be, it will be assumed to result in the following fact: there are no sentences in L_0 which are true in virtue of the meaning of terms $a_1, \ldots, a_m$ alone.

Suppose now that a semantical system $L_1 = \langle \mathscr{L}_1, \mathfrak{M}_1 \rangle$ results from L_0 by adding to the latter some new non-logical constants, say $b_1, \ldots, b_n$, characterized in a verbal way, namely, by a set of postulates P. P is taken to be a set of sentences of $\mathscr{L}_1$ which, according to a decision of the

users of language are to stipulate the meaning of terms $b_1, ..., b_n$: the latter are to be understood in such a way that the sentences of P are true. (Notice that definitions stated in metalanguage can be replaced by definitions which belong to the corresponding object language, and those are postulates of a particular kind. Therefore the case of enlarging a semantical system by postulates covers that of enlarging the system by definitions.)

Clearly, in the situation described some sentences of $\mathscr{L}_1$ will turn out to be true in virtue of the meaning of terms $b_1, ..., b_n$ as stipulated by the set of postulates P. According to philosophical tradition those are the sentences which deserve the name of analytic sentences in L_1. In what follows we shall present some attempts at constructing a precise and adequate definition of that class of sentences. The solution of this problem depends largely on what the set of postulates P is like; in particular, on whether or not is a creative set of sentences, and in which sense, if so. Let us notice that if the set of postulates P is non-creative (in the semantical, and hence also in the syntactical, sense) with respect to $\mathscr{L}_0$, the sentences which belong to it do not assert anything concerning the denotations of terms $a_1, ..., a_m$. Their only function is to stipulate the meaning of terms $b_1, ..., b_n$. They may thus safely be classified as analytic sentences in L_1. Now assume that P is creative (in the semantical sense) with respect to $\mathscr{L}_0$. In this case the sentences which belong to P, besides stipulating the meaning of terms $b_1, ..., b_n$, do assert something about the denotations of terms $a_1, ..., a_m$. If, in addition, P is creative in the syntactical sense, what is asserted by the sentences of P with regard to the denotations of $a_1, ..., a_m$ can be expressed in the language $\mathscr{L}_0$ itself. Otherwise, this could be done only within an extension of language $\mathscr{L}_0$. In either case, P may be said to possess a cognitive content. The truth of sentences which belong to P is not merely a matter of deciding to interpret the terms $b_1, ..., b_n$ in an appropriate way. The question then arises which sentences of $\mathscr{L}_1$ should here be qualified as analytic sentences in L_1. There seems to be no straightforward answer to this question. Different solutions to it have been proposed by different authors who have been dealing with the problem. The main of them will be presented in this paper, in which the concept of postulate will be taken in its most general sense. In particular, no requirement of non-creativity will be imposed on it. Any set of sentences of language $\mathscr{L}_1$,

whether creative or not, will be regarded as a possible set of postulates for its non-logical constants. The use of creative definitions in actual scientific practice seems to justify such an approach.

The solution to the problem of analyticity with regard to a given language clearly depends on how this language is characterized from the semantical point of view. Different definitions of the notion of an anylytic sentence in L_1 result from different sematical characteristics of the system L_1. As this system has been identified with a couple $\langle \mathscr{L}_1, \mathfrak{M}_1 \rangle$, its semantical characteristic amounts to a characteristic of the proper model $\mathfrak{M}_1$. How then is $\mathfrak{M}_1$ to be defined?

The language L_1 constitutes, as assumed, an enlargement of the already interpreted language L_0. This fact seems to impose a requirement to the effect that the terms $a_1, \ldots, a_m$ of language L_0 retain in L_1 the interpretation they have possessed in L_0. As that intepretation has been given by model $\mathfrak{M}_0$, the requirement may be expressed — in the most natural and strict way — by the following semantical rule:

$$(\mathbf{R}_0) \qquad \mathfrak{M}_1|_0 = \mathfrak{M}_0$$

(here we write $\mathfrak{M}_1|_0$ instead of $\mathfrak{M}_1|_{\mathscr{L}_0}$, for short), which demands that the fragment of model $\mathfrak{M}_1$ corresponding to language L_0 coincide with the model $\mathfrak{M}_0$. The rule $\mathbf{R}_0$ clearly guarantees that in lagnuage L_1 the terms $a_1, \ldots, a_m$ will be interpreted exactly as before. It has been suggested (cf. Przełęcki [14]) that the requirement in question might be understood in a more liberal way: the rule $\mathbf{R}_0$ might be replaced by a weaker one, which demands that the fragment $\mathfrak{M}_1|_0$ be an extension of the model $\mathfrak{M}_0$.[1] We shall not examine this possibility in the present paper.

Observe that the rule $\mathbf{R}_0$ cannot generate any analytic sentences. This follows from our assumption stating that the model $\mathfrak{M}_0$ is determined in such a way that the truth of any sentences of L_0 describing this model but tautologies cannot be established in advance without examining the properties of $\mathfrak{M}_0$. Since $\mathbf{R}_0$ does not change the characteristic of $\mathfrak{M}_0$ and at the same time brings no characteristic of the new terms, $\mathbf{R}_0$ does not guarantee the truth of any sentence of L_1 by purely verbal reasoning. Thus, apart from the tautologies, the analytic sentences may appear in L_1 only as a result of accepting the semantical conditions (rules) which are to determine the denotations of the terms added to L_0, i.e. the terms $b_1, \ldots, b_n$. Let $\Omega(\mathfrak{M}_1)$ be the conjunction of all such conditions (hence

we assume that there are finitely many of them). Suppose now that α is a sentence of $\mathscr{L}_1$ which is true in every model $\mathfrak{M} \in \mathrm{Mod}(\mathscr{L}_1)$ satisfying the conditions $\Omega(\mathfrak{M}_1)$ imposed on $\mathfrak{M}_1$. If it is the case, we may say that α is true by virtue of the semantical rules of L_1 alone. In view of the well-known Carnap's Convention (cf. R. Carnap, *Meaning and Necessity*, Chicago 1958, p. 10), α should be considered as an analytic sentence in L_1. This seems to justify the following definitional schema (in which α runs over Sent $(\mathscr{L}_1)$):

$$(*) \qquad \alpha \in \mathrm{Anal}(L_1) \equiv \bigwedge \mathfrak{M} \in \mathrm{Mod}(\mathscr{L}_1)\left[\Omega(\mathfrak{M}) \to \alpha \in \mathrm{Ver}(\mathfrak{M})\right].$$

As we shall see, the schema yields different definitions of the set $\mathrm{Anal}(L_1)$ according to different ways in which the semantical rules $\Omega(\mathfrak{M})$ might be specified. In what follows, we shall examine, in general terms, the main interpretations of these semantical conditions.

A. Since, under our assumption, the terms $b_1, \ldots, b_n$ are to be understood in such a way that the postulates P become true, the simplest explication of the conditions $\Omega(\mathfrak{M}_1)$, which characterize the introduction of these terms, seems to be the following one:

$$(\mathrm{R_I}) \qquad P \subseteq \mathrm{Ver}(\mathfrak{M}_1).$$

The rule $\mathrm{R_I}$ surely guarantees that in language L_1 the terms $b_1, \ldots, b_n$ will be interpreted in accordance with postulates P. By applying the schema $(*)$ to the conditions $\mathrm{R_I}$ (i.e. by identifying $\Omega(\mathfrak{M}_1)$ with $\mathrm{R_I}$), we obtain the definition:

$$\alpha \in \mathrm{Anal}(L_1)$$
$$\equiv \bigwedge \mathfrak{M} \in \mathrm{Mod}(\mathscr{L}_1)\left[P \subseteq \mathrm{Ver}(\mathfrak{M}) \to \alpha \in \mathrm{Ver}(\mathfrak{M})\right],$$

apparently equivalent to the following one:

$$(\mathrm{I}) \qquad \alpha \in \mathrm{Anal}(L_1) \equiv \alpha \in \mathrm{Cn_m}(P).$$

According to I, the analytic sentences in L_1 turn out to be logical consequences of the postulates P.

What are the properties of the set $\mathrm{Anal}(L_1)$ so defined? Now, if only a semantical system characterized by the rules $\mathrm{R_0}$ and $\mathrm{R_I}$ exists, all sentences $\mathrm{Anal}(L_1)$, and only those, are true in it by virtue of the semantical rules alone. However, the existence of a system satisfying

the above conditions is guaranteed only if P is, with respect to $\mathscr{L}_0$, a non-creative set of sentences. Otherwise P will imply such assertions about the model $\mathfrak{M}_0$ which may well turn out to be false. And then, clearly, there will not exist any model of $\mathscr{L}_1$ which would satisfy both of the above conditions.

This situation may be dealt with in different ways. Let us, first, mention an approach which has not actually been taken by any of the authors but which presents a possible way of avoiding the above situation. It consists in modifying the condition R_0, and might be supported by the following argumentation. The set P is here regarded as the set of postulates for all the terms of language $\mathscr{L}_1$: not only $b_1, ..., b_n$, but $a_1, ..., a_m$ as well; the latter also should be understood in such a way that P be true. Now, if this new semantical rule comes into conflict with the old ones, which determine the proper model of L_0, it is here the latter that must give way. The requirement to the effect that $\mathfrak{M}_1|_0 = \mathfrak{M}_0$ is here imposed on $\mathfrak{M}_1$, not categorically, but conditionally: if it is realizable, that is, if $\mathfrak{M}_0$ is admitted by P. Thus, the semantical rule R_0 is replaced by a looser one:

$$(R'_0) \qquad \mathfrak{M}_0 \in \mathrm{Adm}(P) \to \mathfrak{M}_1|_0 = \mathfrak{M}_0 .$$

The existence of the semantical system L_1 as characterized by the rules R'_0 and R_I is obviously guaranteed, if only P is a consistent set of sentences. The price we have to pay for it is a possible change in meaning of the terms of language $\mathscr{L}_0$. With regard to typical cases of enlarging a semantical system, this seems to be an unacceptable consequence. All the remaining solutions presented in this paper respect the categorical requirement expressed in the rule R_0.

Now, if we insist on preserving the meanings of the terms $a_1, ..., a_m$ unchanged, we may, of course, retain the rules R_0 and R_I as characterizing the semantical system L_1, and simply accept the fact that in some cases a system so characterized will not exist. This is a standpoint which has been taken by some of the authors, in the first place, by Kokoszyńska [5]. Under this approach, however, the definition of $\mathrm{Anal}(L_1)$ implies certain consequences, which to some other authors sound paradoxical.

(i) The analytic sentences, defined as above, may be — in a sense — dependent on experience. If the semantical system L_1 does not exist,

all sentences of $\mathscr{L}_1$ (and so, all analytic sentences) are devoid of any interpretation; as sentences of an uninterpreted language they are neither true nor false. But the semantical system L_1 exists if, and only if, there is a model of $\mathscr{L}_1$ which satisfies the conditions R_0 and R_I. In the case of a creative set of postulates P, this, as we have seen, depends on what the model $\mathfrak{M}_0$ proves to be like. Since $\mathfrak{M}_0$ is assumed to have been determined in a non-verbal (ostensive) way, what it is like may be said to be a matter of experience: the state of model $\mathfrak{M}_0$ is by no means presupposed by the manner in which it has been determined. And so, experience may decide whether the analytic sentences are meaningful statements; they cannot thus be said to be true 'come what may'.

Being, in this sense, dependent on experience, they differ, however, in that respect from the synthetic sentences. Experience can decide only on the meaningfulness of an analytic sentence; it may show the sentence to be meaningless, but it cannot show it to be false: once meaningful, it must be true. In the case of the synthetic sentences, the role of experience is not confined to this. Experience decides, not only on the meaningfulness, but also on the truth-value of a synthetic sentence; it may show the sentence to be meaningful but false.

This characteristic of analytic sentences has been widely discussed and elaborated by Kokoszyńska [5]. She has emphasized that so understood the analytic sentences may still be said to be true "by virtue of the meaning of the terms" alone: if only all terms which occur in an analytic sentence do acquire the stipulated meaning, the sentence must be true.

(ii) The set of analytic sentences in L_1, defined as above, may include sentences of $\mathscr{L}_0$ which belong to the synthetic sentences in L_0. As, according to our assumption, there are no postulates in L_0, all non-tautological (and non-contradictory) sentences of $\mathscr{L}_0$ are synthetic in L_0. Now, if P is a creative (in the syntactical sense) set of postulates, some of those sentences will belong to its logical consequences, and will, thus, turn into analytic sentences in L_1. This seems hardly possible without a change in the meaning of terms $a_1, \ldots, a_m$; while retaining their former denotations, they must have changed their meanings. But, under the present approach, the postulates P are regarded as postulates for terms $b_1, \ldots, b_n$ only, and not for terms $a_1, \ldots, a_m$; it seems thus counterintuitive that the acceptance of those postulates might affect the meaning of terms $a_1, \ldots, a_m$.

The situation described above appears to be quite common in science. Let us point out the following fact. Postulates introducing new predicates into a given language often assume the form of so-called reduction sentences (cf. Section III). Let a_1, a_2 be predicates of language $\mathscr{L}_0$, b_1 a predicate of language $\mathscr{L}_1$, and let the set of postulates P consist of a reduction pair for b_1:

$$\bigwedge x[a_1(x) \to b_1(x)],$$

$$\bigwedge x[a_2(x) \to \sim b_1(x)].$$

Set P is clearly a creative one; it entails a non-tautological (and non-contradictory) sentence of $\mathscr{L}_0$:

$$\bigwedge x[a_1(x) \to \sim a_2(x)],$$

which may thus be taken to be a synthetic sentence in L_0. Being a consequence of P, it will be reckoned among the analytic sentences in L_1. In consequence, the predicates a_1, a_2 must have changed their meanings.

To some writers on the subject (e.g. Kokoszyńska) this seems to be an acceptable conclusion. But to some others, both the consequences, (i) and (ii), appear paradoxical. Let us notice that they can be avoided only through a suitable modification of the semantical rule R_I. This has been done in different ways, which will be presented briefly in what follows.

B. One of these modifications has been suggested by following considerations. The rule R_I requires that P be true in $\mathfrak{M}_1$. But this requirement is realizable only if $\mathfrak{M}_1|_0 \in \mathrm{Adm}(P)$. The last condition, in view of the rule R_0, is equivalent to the condition $\mathfrak{M}_0 \in \mathrm{Adm}(P)$, which, in the case of a creative set of postulates P, may well prove false. If this is the case, i.e. if the model $\mathfrak{M}_0$ is not admitted by P, the requirement to the effect that $P \subseteq \mathrm{Ver}(\mathfrak{M}_1)$ cannot be realized, and hence cannot govern the meanings of terms $b_1, \ldots, b_n$. We may thus conclude that the requirement $P \subseteq \mathrm{Ver}(\mathfrak{M}_1)$ is to be treated as obligatory only if $\mathfrak{M}_1|_0 \in \mathrm{Adm}(P)$. This seems to justify the following interpretation of the semantical rule characterizing the enlargement of L_0 to L_1:

(R$_{II}$) $\mathfrak{M}_1|_0 \in \mathrm{Adm}(P) \to P \subseteq \mathrm{Ver}(\mathfrak{M}_1)$.

(The matter has been discussed in a more detailed way by Wójcicki [17]. A condition similar to R$_{II}$ has been proposed by Przełęcki [13].)

By applying our definitional schema (∗) to the condition R_{II}, we obtain the following definition of the set Anal (L_1):

(II) $\alpha \in \text{Anal}(L_1)$

$\equiv \bigwedge \mathfrak{M} \in \text{Mod}(\mathscr{L}_1) \left[(\mathfrak{M}|_0 \in \text{Adm}(P) \rightarrow P \subseteq \text{Ver}(\mathfrak{M})) \right.$

$\left. \rightarrow \alpha \in \text{Ver}(\mathfrak{M}) \right].$

It is easily seen that the definition II avoids all the troublesome consequences of the definition I, discussed above. Some of the theorems quoted later seem to confirm its adequacy.

C. All other ways of modifying the semantical rule R_I may, generally, be characterized as follows. Any creative set of postulates P might be thought of as consisting of sentences which fulfil a double function. First, they state some empirical facts (express some factual knowledge); second, they determine an interpretation of the terms being introduced (stipulate their meaning). There will then be a problem of dividing the whole set P into two parts: one factual, fulfilling only the first function, and the other definitional, fulfilling only the second. They have been called a synthetic and an analytic component of P—an S-component and an A-component, for short.

Let P_A represent an A-component of the set of postulates P. The modification of the semantical rule R_I, being considered now, consists in imposing on $\mathfrak{M}_1$ a condition to the effect that the A-component P_A, and not the whole set P, be true in $\mathfrak{M}_1$. Thus, as a rule characterizing the enlargement of L_0 to L_1, we accept here:

(R_{III}) $P_A \subseteq \text{Ver}(\mathfrak{M}_1).$

If we apply the definitional schema (∗) to condition R_{III}, we get the definition:

$$\alpha \in \text{Anal}(L_1) \equiv \bigwedge \mathfrak{M} \in \text{Mod}(\mathscr{L}_1) [P_A \subseteq \text{Ver}(\mathfrak{M}) \rightarrow \alpha \in \text{Ver}(\mathfrak{M})],$$

logically equivalent to:

(III) $\alpha \in \text{Anal}(L_1) \equiv \alpha \in \text{Cn}_m(P_A).$

Thus, the analytic sentences in L_1 are here identified, not with logical

consequences of the whole set of postulates, but only with consequences of its analytic component.

The task of splitting up a set of postulates into an S- and an A-component has, at first, been accomplished for some special sets, e.g. for a set consisting of two reduction sentences. The solutions proposed for this case will be presented in the next section. A general solution, applicable to any finite set of postulates P, has been given by Carnap. We shall present briefly his main ideas. Let P represent a conjunction of the elements of P, and let $\Sigma_0(P)$ (the so-called Ramsey-sentence) be defined as follows:

$$\Sigma_0(P) = \ulcorner \bigvee x_1...x_n P(x_1, ..., x_n) \urcorner,$$

where $P(x_1, ..., x_n)$ is the formula obtained from P by a proper simultaneous substitution of predicate variables $x_1, ..., x_n$ for the occurrences of $b_1, ..., b_n$, respectively. Carnap has proposed to regard $\Sigma_0(P)$ as an S-component, and $\Sigma_0(P) \to P$ as an A-component of P. The proposal has been motivated by the following properties of the two components (cf. also Section III):

(a) $\Sigma_0(P)$ axiomatizes the set of all those consequences of P which belong to $\mathscr{L}_0^*$;

(b) $\Sigma_0(P) \to P$ does not entail any non-tautological sentences of $\mathscr{L}_0^*$;

(c)   the conjunction of $\Sigma_0(P)$ and $\Sigma_0(P) \to P$ is logically equivalent to P.

Now, if in the rule R_{III} and definition III we take as P_A the Carnap's A-component $\Sigma_0(P) \to P$, we obtain the rule:

(R_{III}^*) $\Sigma_0(P) \to P \in \text{Ver}(\mathfrak{M}_1)$,

and the definition:

(III*) $\alpha \in \text{Anal}(L_1) \equiv \alpha \in \text{Cn}_m^*(\Sigma_0(P) \to P) \cap \text{Sent}(\mathscr{L}_1)$.

It can easily be shown that the present definition avoids the shortcomings of the definition I, mentioned under (i) and (ii). The existence of a model of $\mathscr{L}_1$ as characterized by conditions R_0 and R_{III}^* does not depend on experience, in particular—on what the model $\mathfrak{M}_0$ may prove to be

like; its existence is guaranteed in advance. If $\Sigma_0(P)$ is false in $\mathfrak{M}_0$, these conditions are satisfied vacuously. Suppose then that $\Sigma_0(P)$ is true in $\mathfrak{M}_0$. Since $\Sigma_0(P)$ is identical with the Ramsey-sentence as described above, $\Sigma_0(P)$ is true in $\mathfrak{M}_0$ if, and only if, $\mathfrak{M}_0$ is admitted by P. And then, clearly, there exists a model of $\mathscr{L}_1$ which satisfies both the conditions considered. In consequence, all sentences of $\mathscr{L}_1$ (and thus, all analytic sentences in L_1) are always interpreted, meaningful statements. It is evident that, in this way, the consequence (ii) of the definition I will be avoided too. No synthetic sentence in L_0 can become analytic in L_1, since, according to (b), all sentences of $\mathscr{L}_0$ which follow from $\Sigma_0(P) \to P$ belong to tautologies in L_0.

Let us call attention to the logical relation between the notions of Anal (L_1) as defined by II and III*. Clearly, the latter is applicable only to finite sets of postulates. But then turns out to be equivalent to the former:

THEOREM 1. If P is finite, then $\alpha \in \text{Anal}(L_1)$ in the sense of II if, and only if, $\alpha \in \text{Anal}(L_1)$ in the sense of III*.

Of course, II can also be employed when P is infinite. Thus, to those who are not satisfied with definition I, II may be suggested as a solution in the case of infinite sets of postulates.

Still another approach has been suggested (cf. Przełęcki [12], Wójcicki [17]), which might be regarded as a compromise between the proposals presented under Subsections A and C. It differs from all the proposals considered thus far in that it does not conform to the definitional schema (*). According to it, the semantical system L_1 is characterized as in Subsection A, but the set Anal(L_1) is defined as in Subsection C; we accept the semantical rules R_0 and R_I, and, at the same time, we adopt a definition of Anal(L_1) of the type III, e.g. the Carnap's definition III*. It is easily seen that this procedure avoids one of the shortcomings of the definition I, namely the consequence (ii). The consequence (i), of course, cannot be avoided in this way. To get rid of it, we have to loosen the semantical characteristic of the language L_1 — in one of the ways indicated above.

In conclusion, we wish to express our opinion that the only person competent to make a choice between the alternatives presented is the user of the given language. The logician has to confine himself to pointing out the consequences of every decision.

III

Let us analyze more closely the problem of 'splitting up' a set of postulates P into a synthetic (factual) component and an analytic (definitional) one. Since the considerations contained in this section will be carried out from the model-theoretic point of view, we shall refer to the notion of an S-component and that of an A-component defined here as to the semantical notions. Their syntactical (proof-theoretic) counterparts will be discussed in the next section. Once more it will be convenient to employ the subscripts 'm' and 'p' to distinguish between these two kinds of notion. Thus, apart from 'S-component', 'A-component', we shall use in the sequel the abbreviations: 'S_m-component', 'A_m-component', 'S_p-component', 'A_p-component'.

It is clear enough that in order to function as an S-component of P any set of sentences P_S must possess two properties: (i) P_S must not impose any conditions upon the way of interpreting the predicates $b_1, \ldots, b_n$, (ii) P_S and P must not differ in what they say about the reality being described by the predicates $a_1, \ldots, a_m$. It seems that the following two conditions bring an adequate and, at the same time, precise formulation of the requirements (i) and (ii) in terms of the model theory:

I. The truth-value of a sentence belonging to P_S is independent from the meanings of the predicates $b_1, \ldots, b_n$. More precisely, for any two models $\mathfrak{M}$, $\mathfrak{M}' \in \mathrm{Mod}(\mathscr{L}_1)$ such that $\mathfrak{M}|_0 = \mathfrak{M}'|_0$ the following biconditional holds:

$$P_S \subseteq \mathrm{Ver}(\mathfrak{M}) \quad \text{if and only if} \quad P_S \subseteq \mathrm{Ver}(\mathfrak{M}').$$

II. The set of models of $\mathscr{L}_0$ admitted by P_S is exactly the same as that admitted by P. In symbols: for every $\mathfrak{M} \in \mathrm{Mod}(\mathscr{L}_0)$,

$$\mathfrak{M} \in \mathrm{Adm}(P_S) \quad \text{if and only if} \quad \mathfrak{M} \in \mathrm{Adm}(P).$$

The clauses I and II taken as the clauses characterizing the notion of an S_m-component of P are equivalent to the following definition:

DEFINITION 1. A set of sentences $P_S \subseteq \mathrm{Sent}(\mathscr{L}_1^*)$ is an S_m-component of the set of postulates P if and only if for every model $\mathfrak{M} \in \mathrm{Mod}(\mathscr{L}_1)$,

$$P_S \subseteq \mathrm{Ver}(\mathfrak{M}) \quad \text{if and only if} \quad \mathfrak{M}|_0 \in \mathrm{Adm}(P).$$

The task of defining the notion of an A_m-component is somewhat more troublesome. Two properties seem to be decisive in order that a set of

sentences P_A may play the role of an A-component of P: (i) P_A, taken as a set of postulates for $b_1, \ldots, b_n$ must endow these predicates with the same meanings as they had by virtue of P, (ii) P_A must be deprived of any factual content. We propose the following interpretation of the somewhat ambiguous requirement (i):

I. For every model $\mathfrak{M} \in \text{Mod}(\mathscr{L}_1)$, if $\mathfrak{M}|_0 \in \text{Adm}(P)$ then

$$P_A \subseteq \text{Ver}(\mathfrak{M}) \quad \text{if and only if} \quad P \subseteq \text{Ver}(\mathfrak{M}).$$

The most natural interpretation of the claim (ii) in model theoretic terms seems to be given by the condition:

II. P_A is non-creative with respect to $\mathscr{L}_0$ in the semantical sense. Thus we arrive at the definition:

DEFINITION 2. A set of sentences $P_A \subseteq \text{Sent}(\mathscr{L}_1^*)$ is an A_m-component of the set of postulates P if and only if $P_A \in \text{Noncreat}_m(\mathscr{L}_0)$ and for every model $\mathfrak{M} \in \text{Mod}(\mathscr{L}_1)$, if $\mathfrak{M}|_0 \in \text{Adm}(P)$ then

$$P_A \subseteq \text{Ver}(\mathfrak{M}) \quad \text{if and only if} \quad P \subseteq \text{Ver}(\mathfrak{M}).$$

A few theorems stated below describe the most important properties of the defined notions.

THEOREM 2. If P_S and P_S' are both S_m-components of P then P_S equiv$_m^* P_S'$.

THEOREM 3. If P_S is an S_m-component of P, $\alpha \in \text{Sent}(\mathscr{L}_0^*)$, then $\alpha \in \text{Cn}_m^*(P)$ if and only if $\alpha \in \text{Cn}_m^*(P_S)$.

THEOREM 4. If P_S is an S_m-component of P, then there is $A \subseteq \text{Sent}(\mathscr{L}_0^*)$ such that P_S equiv$_m^* A$.

THEOREM 5. If P_A is an A_m-component of P, $\alpha \in \text{Sent}(\mathscr{L}_0^*)$ and $\alpha \in \text{Cn}_m^*(P_A)$, then $\alpha \in \text{Cn}_m^*(0)$.

THEOREM 6. P is an A_m-component of P if and only if 0 is an S_m-component of P.

THEOREM 7. If P_S is an S_m-component of P, P_A is an A_m-component of P, then P equiv$_m^* P_S \cup P_A$.

Turning back to Carnap's proposal we may ask now whether $\Sigma_0(P)$ and $\Sigma_0(P) \to P$ are components of P in the sense of the given definitions. It may be proved that:

THEOREM 8. If the set of postulates P is a finite set, then $\Sigma_0(P)$ is an S_m-component of P.

THEOREM 9. If the set of postulates P is a finite set, then $\Sigma_0(P) \to P$ is an A_m-component of P.

It follows from the Theorem 2 that if P is finite, the S_m-components of P are sets logically equivalent to $\Sigma_0(P)$. Let us stress however that this cannot be said about the A_m-components. They may turn out to be non-equivalent to $\Sigma_0(P) \to P$ (an example will be given in the sequel). It may be proved only that:

THEOREM 10. If the set of postulates P is a finite set, P_A is an A_m-component of P, then $\Sigma_0(P) \to P \in Cn_m^*(P_A)$.

Hence Carnap's analytic component is the weakest of all the A_m-components of P.

So far our considerations were mainly based on the results obtained by Wójcicki in [17]. Still, apart from some divergences in terminology, the present discussion differs from that contained in [17] in one important point. The analysis in [17] is carried out under the assumption that P is finite. In particular, the definitions of both components of P given in [17], though essentially akin to those formulated here, are restricted to the case of finite P. It should be noted that most authors (including Carnap) which have dealt with the problem of analyticity have not considered the infinite sets of postulates. There is nothing strange in this since in practice we mostly stipulate the meanings of terms with the help of finite sets of postulates, using infinite sets at the very special occasions only. On the other hand, the assumption stating the infiniteness of P leads too often to problems which seem to be very difficult to solve. The following open question may serve as an example:

($+$) Assuming that P is infinite, are there sets P_S, P_A being an S_m-component and an A_m-component, respectively?

The fact that we do not know how to construct an A_m-component of P, when P is an infinite set, and that we do not know even whether such a component exists, causes some troubles in defining the notion of analyticity. One of the ways in which they can be solved consists in adopting the procedure B given in the preceding section.

We shall examine now in a rather detailed way another problem. As we have mentioned before, there is no guarantee that the A_m-components of P are logically equivalent to each other. Let us produce an example. Let $r_i(x)$ and $s_j(x)$ be well-formed formulas of $\mathscr{L}_0$ containing x as the only free variable, and let b_1 be a one-place predicate. The

sentences

$$\bigwedge x\big(r_i(x) \to b_1(x)\big),$$
$$\bigwedge x\big(s_j(x) \to \sim b_1(x)\big)$$

will be called reduction sentences for b_1. Let $\{b_1, \ldots, b_n\} = \{b_1\}$ (i.e. $n=1$) and let P be a finite set of reduction sentences for b_1. Note that the finite set of reduction sentences:

$$\bigwedge x\big(r_1(x) \to b_1(x)\big),$$
$$\bigwedge x\big(r_2(x) \to b_1(x)\big),$$
$$\cdots \cdots \cdots \cdots$$
$$\bigwedge x\big(r_k(x) \to b_1(x)\big)$$

is logically equivalent to the single sentence

$$\bigwedge x\big[(r_1(x) \vee \ldots \vee r_k(x)) \to b_1(x)\big].$$

Similarly the set

$$\bigwedge x\big(s_1(x) \to \sim b_1(x)\big),$$
$$\bigwedge x\big(s_2(x) \to \sim b_1(x)\big),$$
$$\cdots \cdots \cdots \cdots$$
$$\bigwedge x\big(s_l(x) \to \sim b_1(x)\big)$$

is logically equivalent to the sentence

$$\bigwedge x\big[(s_1(x) \vee \ldots \vee s_l(x)) \to \sim b_1(x)\big].$$

Therefore, without any loss of generality, we may assume that P is a two-element set of reduction sentences:

$$P = \{\, \bigwedge x\big(r(x) \to b_1(x)\big),\ \bigwedge x\big(s(x) \to \sim b_1(x)\big)\}.$$

It is an easy matter to verify that the sentence

$$(0) \qquad \bigwedge x\big(\sim r(x) \vee \sim s(x)\big)$$

is logically equivalent to $\varSigma_0(P)$, and hence, by Theorem 8, it is an S_m-component of P. It may also be proved that the following pair of sentences:

$$(1a) \qquad \bigwedge x\big[(r(x) \wedge \sim s(x)) \to b_1(x)\big],$$
$$(1b) \qquad \bigwedge x\big[(\sim r(x) \wedge s(x)) \to \sim b_1(x)\big]$$

proposed by Przełęcki [11] as an analytic component of P is an A_m-component of P in the sense of Definition 2. We shall use the symbol P_A^1 to denote this component.

(We wish to call attention to the fact that the sentences (0), (1a), (1b) are elementary ones, i.e. they are sentences formulated in the first-order language $\mathscr{L}_1$. We should not expect that P always, no matter what its nature is, possesses elementary components. Unfortunately, the problem of existence of such components, treated in a general way, is a very difficult one.)

Note now that the sentences

(2a) $\quad \bigwedge x (r(x) \to b_1(x))$,

(2b) $\quad \bigwedge x [(\sim r(x) \wedge s(x)) \to \sim b_1(x)]$,

and similarly the sentences

(3a) $\quad \bigwedge x [(r(x) \wedge \sim s(x)) \to b_1(x)]$,

(3b) $\quad \bigwedge x (s(x) \to \sim b_1(x))$

also constitute A_m-components of P. They will be denoted by P_A^2 and P_A^3, respectively. One may easily check that any two of the components P_A^1, P_A^2, P_A^3 are not equivalent to each other. What more, none of them is equivalent to Carnap's A-component. The latter — we shall refer to it as to P_A^4 — is logically equivalent to the conjunction of the sentences:

(4a) $\quad \bigwedge x (\sim r(x) \vee \sim s(x)) \to \bigwedge x (r(x) \to b_1(x))$,

(4b) $\quad \bigwedge x (\sim r(x) \vee \sim s(x)) \to \bigwedge x (s(x) \to \sim b_1(x))$.

In order to compare P_A^1 and P_A^4 note that the sentences (1a) and (1b) are pairwise equivalent to

(1a′) $\quad \bigwedge x [(\sim r(x) \vee \sim s(x)) \to (r(x) \to b_1(x))]$,

(1b′) $\quad \bigwedge x [(\sim r(x) \vee \sim s(x)) \to (s(x) \to \sim b_1(x))]$.

Hence P_A^1 entails Carnap's component P_A^4 but, in general, the converse does not hold (Mejbaum [8]). It is almost immediately seen that both P_A^2 and P_A^3 are stronger than either of the components P_A^1 and P_A^4.

Let us turn back to the problem which already was discussed in Section II. The results we have obtained here are irrelevant to those proposals which define the set Anal(L_1) as Cn(P). They reveal however that the

definition: $\mathrm{Anal}(L_1) = \mathrm{Cn}\big(\Sigma_0(P) \to P\big)$ is not the only possible alternative. Clearly, instead of Carnap's component $\Sigma_0(P) \to P$ we may as well employ any of the A_m-components of P. Thus we are faced with the problem of choice.

Since one may be tempted to put

$$\mathrm{Anal}(L_1) = \mathrm{Cn}\big(\bigcup_{t \in T} P_A^t \big),$$

where $\{P_A^t\}_{t \in T}$ is the set of A_m-components of P, we wish to make clear that, in general, the sum $\bigcup_{t \in T} P_A^t$ is not an A_m-component of P. To realize this it is enough to notice that, for P being the set of reduction sentences for b_1, $\bigcup_{t \in T} P_A^t$ contains as a subset $\bigcup_{i=1,2,3,4} P_A^i$ and the latter set is logically equivalent to P.

There are two arguments which may be used in favour of Carnap's A-component. Firstly, it may be argued that it is the weakest component, and hence this choice is the most cautious. If we put $\mathrm{Anal}_t(L_1) = \mathrm{Cn}(P_A^t)$ and $\mathrm{Anal}_C(L_1) = \mathrm{Cn}\big(\Sigma_0(P) \to P\big)$ then $\mathrm{Anal}_C(L_1) \subseteq \mathrm{Anal}_t(L_1)$, for every t. Secondly, Carnap's component is constructed by a simple well-defined method which can be applied to any finite set of postulates. So far we have not presented any method which is as general as Carnap's one.

Though the both arguments are suggestive, none of them may be treated as entirely convincing. It seems that it is not the task of a logician to propose a particular choice of an A_m-component; that has to be undertaken by the user of the given language. The logical analysis is to make clear what are the implications of any decision.

Concluding this chapter, we shall briefly describe a method of constructing an A_m-component of P. The method has been proposed recently by Wójcicki [19]. We shall say that Q is a maximal non-creative (in the model-theoretic sense) subset of P if and only if the following conditions hold:

(i) $Q \subseteq \mathrm{Cn}_m(P)$,

(ii) $Q \in \mathrm{Noncreat}_m(\mathscr{L}_0)$,

(iii)   For every Q', if (a) $Q' \subseteq \mathrm{Cn}_m(P)$, (b) $Q' \in \mathrm{Noncreat}_m(\mathscr{L}_0)$, and (c) $Q \subseteq \mathrm{Cn}_m(Q')$, then $Q' \subseteq \mathrm{Cn}_m(Q)$.

Let $\{P_N^\mu\}_{\mu \in M}$ be the set of maximal non-creative subsets of P. It may be proved that

THEOREM 11. If P is finite, then $\bigcap_{\mu \in M} \mathrm{Cn_m}(P_N^\mu)$ is an $\mathrm{A_m}$-component of P.

The following question is open:

$(++)$ Assuming that P is infinite, is $\bigcap_{\mu \in M} \mathrm{Cn_m}(P_N^\mu)$ an $\mathrm{A_m}$-component of P?

It may be interesting to compare the components $\bigcap_{\mu \in M} \mathrm{Cn_m}(P_N^\mu)$ and $\Sigma_0(P) \to P$. By Theorem 10 we conclude that the former implies the latter. We do not know, however, whether the converse holds or not. But it seems that this question has to be answered in the negative. For instance, if P is the set of reduction sentences as considered above, we are inclined to believe that P_A^1 rather than P_A^4 is equivalent to $\bigcap_{\mu \in M} \mathrm{Cn_m}(P_N^\mu)$.

IV

The concepts of an S-component and an A-component defined in the preceding section are of a definitely semantical nature. Their definitions, 1 and 2, characterize the sets P_S and P_A by certain model theoretic conditions, such as, e.g., the condition of semantical non-creativity. And what is more, they cannot be characterized by any purely syntactical means. There are no known proof-theoretic conditions which would be logically equivalent to the model theoretic conditions used in Definition 1 or 2. In particular, we have no general syntactical criterion of the semantical non-creativity available; as we have mentioned before, the syntactical concept of non-creativity, when applied to elementary languages, proves to be essentially weaker than the semantical one. In consequence, the semantical components turn out to be, in a sense, 'non-operative' concepts. Whether or not a set of sentences of L_1 is an $\mathrm{S_m}$-component (or an $\mathrm{A_m}$-component) of a given set of postulates P may be, in some cases, hardly decidable for the user of language L_1.

Such considerations have led to the introduction of a different, more 'operative', pair of components: the syntactical (proof-theoretic) ones — $\mathrm{S_p}$-component and $\mathrm{A_p}$-component, for short. These components are conceived of as subject to the following formal requirements: (a) they must belong to the elementary language $\mathscr{L}_1$, and (b) they must be characterized by proof-theoretic conditions only. We can arrive at such concepts by the following procedure. The clause

(a) $\mathfrak{M}|_0 \in \mathrm{Adm}(P),$

which is made use of in the Definitions 1 and 2, is to be replaced throughout by the caluse

(b) $Cn_p(P) \cap Sent(\mathscr{L}_0) \subseteq Ver(\mathfrak{M})$.

So modified, and restricted to the elementary languages $\mathscr{L}_0$ and $\mathscr{L}_1$,
the Definitions 1 and 2 become equivalent to the following ones, which
have been put forward by Przełęcki [13] as definitions of an S_p-com-
ponent and an A_p-component of the set of postulates P:

DEFINITION 3. A set of sentences $P_S \subseteq Sent(\mathscr{L}_1)$ is an S_p-component of
the set of postulates P if and only if

$$P_S \; equiv_p \, Cn_p(P) \cap Sent(\mathscr{L}_0).$$

DEFINITION 4. A set of sentences $P_A \subseteq Sent(\mathscr{L}_1)$ is an A_p-component
of the set of postulates P if and only if

(i) $P_A \in Noncreat_p(\mathscr{L}_0)$,

(ii) $P_A \cup Cn_p(P) \cap Sent(\mathscr{L}_0) \; equiv_p \, P$.

The syntactical components defined above do not coincide with the
semantical ones. This is evident in view of the non-equivalence of clauses
(a) and (b): if P is non-creative in the syntactical but creative in the semantical sense, there exist models of $\mathscr{L}_1$ which fulfil (b) but not (a).

The syntactical components, being different from the semantical ones,
share with the latter several important properties. Theorems strictly
analogous to those characterizing the semantical components, i.e. to
Theorems 2–7 of the preceding section, hold for the syntactical components as well. They can be obtained from the former by substituting:
S_p for S_m, A_p for A_m, $\mathscr{L}_0$ for $\mathscr{L}_0^*$, Cn_p for Cn_m^*, and $equiv_p$ for $equiv_m^*$.

It may also be proved

THEOREM 12. If P_S is an S_m-component of P, P_A is an A_m-component of
P, $P_S \subseteq Sent(\mathscr{L}_1)$, $P_A \subseteq Sent(\mathscr{L}_1)$, then P_S is an S_p-component of P, P_A
is an A_p-component of P.

Theorem 12 shows, in view of some of the foregoing examples, that
the conditions (i) and (ii) of Definition 4 do not determine an A_p-com-
ponent in a unique way: each of the sets $P_A^1 - P_A^4$ constitutes an A_p-com-
ponent of the set P composed of two reduction sentences.

The fundamental problem of the existence – for an arbitrary set P – of
S_p- and A_p-components defined above remains open. Anyway, there is
no general method of constructing such sets available.

Having the syntactical concept of an A-component at our disposal, we can define now, with the help of it, the notion of an analytic sentence in L_1 in the way followed earlier (described in Section II as the Procedure C). The concept of Anal(L_1) obtained in this way avoids the shortcoming of the Definition I discussed under (ii). It does not, however, get rid of the consequence (i). The proper model of L_1 is here characterized as a model of $\mathscr{L}_1$ which satisfies the semantical rule $R_0: \mathfrak{M}_1|_0 = \mathfrak{M}_0$, and the semantical rule to the effect that an A_p-component of P be true in $\mathfrak{M}_1$. Now, if this A_p-component, being non-creative in the syntactical sense, is creative in the semantical sense (and, as we know, such sets do exist), there is no guarantee that a model of $\mathscr{L}_1$ so characterized will exist. Whether it exists or not, depends on what the model $\mathfrak{M}_0$ proves to be like. And this, as it has been assumed, is a matter of experience. In consequence, the meaningfulness of the analytic sentences will be dependent on experience too.

There is, however, a narrower sense of the concept of experience, which does not lead to the same conclusion. It is a sense relativized to a given language. In the case of language L_1, the so-called 'data of experience' will be restricted to those properties of the (ostensively determined) model $\mathfrak{M}_0$ which are expressible in language L_0; they may be simply identified with the sentences belonging to Ver $(\mathfrak{M}_0)$. What Ver $(\mathfrak{M}_0)$ (and not, generally, $\mathfrak{M}_0$!) proves to be like is here called a 'matter of experience'. Now, in the case considered above, the existence of the proper model of L_1 depends only on those properties of $\mathfrak{M}_0$ which are not expressible in language L_0 (as the A_p-component does not entail any non-tautological sentences of L_0). And so, the question whether such a model exists is not a question of experience — in the sense being proposed now; no empirical findings which might be obtained by the users of language L_1 can decide the question.

V

The language L_1 discussed by us thus far may be thought of as a result of one enlargement of the initial language L_0: it has been constructed from the latter by introducing into it one set of terms by means of one set of postulates. Now there certainly are languages which cannot be conceived of in this way. They should be treated rather as a result of several suc-

cessive enlargements of a given language. Languages of this kind give rise to some special problems concerning the definition of an analytic sentence as applied to them. The matter has been discussed by Nowaczyk [10] and Przełęcki [12], and some of the suggested solutions will be presented in this section, though in a very sketchy manner.

Let the language L_n be the result of n succesive enlargements of the language L_0 described above. Each language L_i in the sequence

$$L_1, ..., L_n$$

results from the preceding one, L_{i-1}, by adding some new non-logical constants, characterized by a set of postulates P_i. For the sake of simplicity, we shall here assume that all the sets P_i are finite ones (P_i will, thus, symbolize the corresponding conjunction), and that each of them introduces only one new term, b_i. Thus, corresponding to the sequence of postulates

$$P_1, ..., P_n$$

(let $P = \{P_1, ..., P_n\}$) there is a sequence of terms being introduced

$$b_1, ..., b_n.$$

We have to answer, with regard to language L_n, the same questions as before: what are the conditions which characterize the semantical system L_n, and how the analytic sentences in the system L_n are to be defined?

Two possibilities must be distinguished at the outset.

I. According to the first, each postulate P_i is being treated as a postulate for all the b-terms which appear in it; thus, the whole set P constitutes, collectively, one set of postulates for all the terms $b_1, ..., b_n$. Such an approach brings us back to the case discussed previously. From the semantical point of view, there is no difference between the present language L_n and the language L_1, dealt with in the preceding sections. In consequence, the problem of defining the set Anal(L_n) will be solved exactly as before.

II. According to the other possibility, each postulate P_i is treated as a postulate for the term b_i only; in other words, the term b_i is to be understood in such a way that the postulate P_i, in which all the remaining terms retain their established meaning, be true. It is this approach which

suggests certain new solutions to the problem of analyticity with regard to a semantical system like L_n. Generally speaking, all these solutions consist in selecting, from the whole set of postulates P, a suitable subset, and defining the analytic sentences in L_n as logical consequences of the selected postulates. Such a procedure is being adopted in order to ensure the truth of all the analytic sentences in L_n. There have been proposed different criteria of selection, corresponding to different intuitions connected with the idea of a 'truth-guarantee'. The following criterion has been formulated by Nowaczyk [10].

The subset in question — let us denote it by P_L — is defined inductively as follows:

(1)   · If K is the set of all those postulates preceding P_i in the sequence P which belong to P_L, then P_i will belong to P_L if and only if

(i) for every $j < i$: if b_j occurs in P_i, then $P_j \in P_L$,

(ii) $\mathfrak{M}_0 \in \mathrm{Adm}(K \cup P_i)$.

Now, if the semantical system L_n is characterized by the conditions

(R$_0^n$) $\mathfrak{M}_n|_0 = \mathfrak{M}_0$,

(R$_A^n$) $P_L \subseteq \mathrm{Ver}(\mathfrak{M}_n)$,

and the analytic sentences in L_n are, in accordance with the schema (∗), defined as follows:

(A) $\mathrm{Anal}(L_n) = \mathrm{Cn}(P_L)$,

it is easy to see that all the analytic sentences in L_n must certainly be true in L_n. In spite of that, they cannot be said to be completely independent of experience. As the clause (ii) clearly shows, whether or not a postulate P_i belongs to the subset P_L may depend on what $\mathfrak{M}_0$ turns out to be like; in consequence, whether a given sentence of L_n is analytic or not may be a matter of experience.

Some narrower class of analytic sentences in L_n, independent of experience in the sense being now considered, has been defined by Nowaczyk through the following procedure. Notice that associated with each model $\mathfrak{M} \in \mathrm{Mod}(\mathscr{L}_0)$ there is a subset of postulates P selected according to criterion (1) (where $\mathfrak{M}$ is substituted for $\mathfrak{M}_0$), which will be symbolized

by $P_L^{\mathfrak{M}}$, and a corresponding set of consequences $\text{Cn}\,(P_L^{\mathfrak{M}})$. Let us take the product of all such sets,

$$\bigcap_{\mathfrak{M}\,\in\,\text{Mod}(\mathscr{L}_0)} \text{Cn}\,(P_L^{\mathfrak{M}}).$$

If we identify the set of analytic sentences in L_n with this product

(B) $\text{Anal}(L_n) = \bigcap_{\mathfrak{M}\,\in\,\text{Mod}(\mathscr{L}_0)} \text{Cn}\,(P_L^{\mathfrak{M}}),$

we get a set of sentences which avoid the consequence mentioned above. Neither the membership in the set defined by B, nor the truth of its members can be a matter of experience; none of these facts depends on any particular feature of the model $\mathfrak{M}_0$.

A similar result can be obtained in another way, suggested by Prze-łęcki [12]. The procedure amounts to a suitable modification of the criterion (1). The clause (ii) is replaced by another one, which requires that P_i be non-creative with respect to K:

(ii) $P_i \in \text{Noncreat}(K).$[2]

According to whether the non-creativity in (ii) is taken in the semantical or in the syntactical sense, we get two different concepts of the set P_L, and hence, of the set $\text{Anal}(L_n)$, as defined by A: a stronger and a weaker one. The following question remains open:

(+ + +) How is the semantical concept of $\text{Anal}(L_n)$ arrived at in this way related to that defined by B?

The properties and the mutual relationships of the concepts introduced in this section are still in need of a closer examination.

BIBLIOGRAPHY

[1] K. Ajdukiewicz, 'Le problème du fondament des propositions analytiques', *Studia Logica* 8 (1958) 259–272; included in the present volume as: 'The Problem of Justifying Analytic Sentences'.

[2] L. Borkowski, 'Deductive Foundation and Analytic Propositions', *Studia Logica* 19 (1966) 59–72.

[3] Z. Czerwiński, 'Zdania analityczne, logika i doświadczenie' ('Analytic Sentences, Logic and Experience'), *Rozprawy Logiczne*, Warszawa 1964, pp. 23–30.

[4] M. Kokoszyńska, 'Deduction as a Method of Proof', *Atti del XII Congresso Internazionale di Filosofia* 5 (1960).

[5] M. Kokoszyńska, 'O dwojakim rozumieniu uzasadnienia dedukcyjnego' ('Two Concepts of Deductive Justification'), *Studia Logica* 13 (1962) 177–196.

[6] M. Kokoszyńska, 'O dyrektywach inferencji ('On the Rules of Inference'), *Rozprawy Logiczne*, Warszawa 1964, pp. 77–90.

[7] M. Kokoszyńska, *On Deduction. The Foundation of Statements and Decisions*, Warszawa 1965.

[8] W. Mejbaum, 'Wielkość fizyczna i doświadczenie' ('Physical Magnitude and Experience'), *Studia Filozoficzne* 2 (1965) 43–93.

[9] A. Nowaczyk, 'Pojęcie zdania analitycznego w problematyce teoriopoznaw-czej' ('The Notion of an Analytic Sentence in Epistemology'), *Zeszyty Naukowe Uniwersytetu Łódzkiego* I/42 (1967).

[10] A. Nowaczyk, 'Analytic Sentences in the Semantic System', this volume, pp. 457–497.

[11] M. Przełęcki, 'Pojęcia teoretyczne a doświadczenie' ('Theoretical Concepts and Experience'), *Studia Logica* 11 (1961) 91–138.

[12] M. Przełęcki, 'O pojęciu zdania analitycznego' ('On the Notion of an Analytic Sentence'), *Studia Logica* 14 (1963) 155–182.

[13] M. Przełęcki, 'Teorie empiryczne w ujęciu logiki współczesnej' ('Empirical Theories from the Logical Point of View'), *Fragmenty Filozoficzne* III, Warszawa 1967, pp. 75–102.

[14] M. Przełęcki, 'The Role of Noncreativity of Meaning Postulates in Empirical Theories'. Unpublished.

[15] M. Przełęcki, *The Logic of Empirical Theories*, London 1968.

[16] R. Suszko, 'Logika formalna a niektóre zagadnienia teorii poznania' ('Formal Logic and Some Problems of Epistemology'), *Myśl Filozoficzna* 2, 3 (1957) 27–56, 34–67.

[17] R. Wójcicki, 'Analityczne komponenty definicji arbitralnych' ('Analytic Components of Arbitrary Definitions'), *Studia Logica* 14 (1963) 119–154.

[18] R. Wójcicki, 'Analityczność, syntetyczność, empiryczna sensowność zdań' ('Analyticity, Syntheticity, Empirical Meaningfulness of Sentences'), *Studia Filozoficzne* 3 (1966) 33–65. English version appeared in *Studia Filozoficzne* 4 (1970), Suplementary Volume in English.

[19] R. Wójcicki: 'Postulates and Their Analytical Consequences'. Forthcoming.

REFERENCES

* Reprinted from *Synthese* 19 (1968–1969), D. Reidel Publishing Company, Dordrecht-Holland.

[1] A model $\mathfrak{M} = \langle U, R_1, \ldots, R_k \rangle$ is said to be an extension of a model $\mathfrak{M}'$ $= \langle U', R'_1, \ldots, R'_k \rangle$ if, and only if, $U' \subseteq U$ and each relation R_i $(i = 1, \ldots, k)$ restricted to the universe U' coincides with the relation R'_i.

[2] The notion of being non-creative with respect to a set of sentences is understood as follows. Let $\mathscr{L}' \in \text{Enlarg}\,(\mathscr{L})$, $A \subseteq \text{Sent}\,(\mathscr{L}')$, $B \subseteq \text{Sent}\,(\mathscr{L})$. $A \in \text{Noncreat}_m\,(B)$ if and only if for every $\mathfrak{M} \in \text{Mod}\,(\mathscr{L})$, if $B \subseteq \text{Ver}\,(\mathfrak{M})$, then $\mathfrak{M} \in \text{Adm}\,(A \cup B)$; $A \in \text{Noncreat}_p\,(B)$ if and only if for every $\varphi \in \text{Sent}\,(\mathscr{L})$, if $\varphi \in \text{Cn}_p\,(A \cup B)$, then $\varphi \in \text{Cn}_p\,(B)$. Thus, if $B = 0$, we get the former notion of being non-creative with respect to a given language.

KLEMENS SZANIAWSKI

A METHOD OF DECIDING BETWEEN
N STATISTICAL HYPOTHESES*

I am going to discuss briefly a possible generalization of sequential probability ratio test; this well-known test has been defined, and its properties
investigated, by Abraham Wald.[1] The (primary) purpose to be achieved
by Wald's method was: to test a simple statistical hypothesis against
one simple alternative. Some ingenious devices have been used by Wald
to extend the range of applications of the method: they all consisted in
reducing the problem to the above-mentioned form, i.e. that of the
choice between two simple hypotheses. It seems, however, that the
method itself could be slightly generalized to cover the case of choosing
between N simple hypotheses. The present paper is a preliminary discussion of such a possibility.

It is only fair to add that the proposed generalization may turn out
to have no appreciable practical significance. Even if that were so it
might be found to have some interesting properties from a more theoretical point of view: that of the logic of uncertain inference.

We are going to assume that the functional form of the distribution
of a random variable X is known to be $f(x; \theta)$,[2] where θ is a parameter
the exact value of which is to be determined on the basis of experimental
evidence. Let us assume further that one and only one of the hypotheses
$h_1, ..., h_N$ concerning the value of θ is true, where h_i states that $\theta = \theta_i$.
The problem now is how to decide between those hypotheses (i.e. which
hypothesis to accept) on the basis of an outcome of n independent
trials, the number n of trials being not fixed in advance. The desired
rule of inference is thus to be of a sequential nature.

Following Wald, by a sequential rule of inference (with respect to
the case described above) we shall mean a rule defined in this way. For
each n ($n = 1, 2, ...$) the set S_n of all possible samples $\{x_1, ..., x_n\}$ is
divided into $N+1$ disjoint subsets $R_n^1, ..., R_n^N, R_n^{N+1}$; if the sample ob-

tained belongs to the subset R_n^s $(d=1, ..., N)$ the hypothesis h_S is to be accepted; if the sample belongs to the subset R_n^{N+1} no hypothesis is accepted and $(n+1)$th trial is made. [3]

The rule of inference I am going to formulate belongs to the class of rules just defined. It can be stated as follows:

For an outcome $\{x_1, ..., x_n\}$ of n independent trials $(n=1, 2, ...)$ accept the hypothesis h_r $(1 \leqslant r \leqslant N)$ if and only if

(1) $f_r(x_1) \cdot f_r(x_2) \cdot ... \cdot f_r(x_n) \geqslant A_{rs} \cdot f_s(x_1) \cdot ... \cdot f_s(x_n)$ for all $s \neq r$,

where the numbers $|A_{rs}|$ are constants >1 and the expression $f_\alpha(x)$ stands for $f(x; \theta_\alpha)$.

If there is no such r for which the set of inequalities (1) is satisfied by the outcome $\{x_1, ..., x_n\}$, then no hypothesis is accepted and the $(n+1)$th trial should be made.

It is easy to see that the above rule (let us call it R) is a generalization of the sequential probability ratio test: we obtain this test in the case of $N=2$ by putting A for A_{21} and B for $1/A_{12}$. The rule R then reduces to the form:

if $Z_n \leqslant B$ accept h_1,

if $A \leqslant Z_n$ accept h_2,

if $B < Z_n < A$ take $(n+1)$th observation,

where $Z_n = f_2(x_1) \cdot ... \cdot f_2(x_n)/f_1(x_1) \cdot ... \cdot f_1(x_n)$.

The rule R may also be looked upon as a generalization of the much discussed maximum likelihood principle (for the case when the number of hypotheses is finite). The maximum likelihood principle recommends the acceptance of the hypothesis for which the probability (or probability density) of the outcome actually obtained is maximum. Obviously, if we put in the rule R all the constants A_{rs} equal to 1 we get the principle in question. Of course, in such a special case there is always (that is, for all n) at least one hypothesis h_r satisfying the condition (1). That is why the m.l. principle is a 'degenerate' sequential rule, i.e. such for which the number n of trials is fixed in advance; also it does not guarantee a unique solution, in view of the fact that more than one hypothesis can satisfy (1) if all the constants A_{rs} in it vanish.

Some properties of the rule R can be roughly determined in a way analogous to that followed by Wald. Let us denote by ω_r the probability that R will lead to acceptance of the hypothesis h_r when h_r is in fact true; by β_{rs} $(r \neq s)$ we are going to denote the probability that R will lead to the acceptance of h_r when h_s is true. And we shall say that $\{x_1, \ldots, x_n\}$ is a sample of the type r if the relation

$$(2) \qquad f_t(x_1) \cdot \ldots \cdot f_t(x_m) \geqslant A_{ts} f_s(x_1) \cdot \ldots \cdot f_s(x_m) \qquad \text{for all } s \neq t$$

is not satisfied for $m = 1, \ldots, n-1$, $1 \leqslant t \leqslant N$ and its is satisfied for $m = n$, $t = r$. According to the rule R, the hypothesis h_r is accepted if and only if the sample actually obtained is of the type r. But each sample of the type r satisfies the condition (1). Thus for each particular $s \neq r$, the probability (or probability density) of obtaining the sample $\{x_1, \ldots, x_n\}$ of the type r is at least A_{rs} times as great under the assumption that h_r than under the assumption that h_s. Therefore, the probability measure of all samples of the type r under the assumption that h_r must be at least A_{rs} times their probability measure under the assumption that h_s. We thus have:

$$(3) \qquad \omega_r \geqslant A_{rs} \beta_{rs} \qquad \text{for all } s \neq r.$$

Hence:

$$(4) \qquad \beta_{rs} \leqslant \frac{1}{A_{rs}} \qquad \text{for all } s \neq r.$$

The inequality (4) provides an upper bound for the probability of the error consisting in accepting the hypothesis h_r when some other hypothesis h_s is in fact true. Obviously, this upper bound is set too high: it is enough to notice that in deriving (4) from (3) we have taken advantage of the fact that ω_r cannot be greater than 1, while actually ω_r is always lower than 1. This is important from the point of view of practical applications: if we want to keep the probabilities of errors within certain limits and to have the expected number of trials not greater than it is necessary to achieve this purpose (because of the cost of experimentation), then a more exact relation between β_{rs} and A_{rs} is needed. However, for the limited purposes of this paper it is enough to remark that β_{rs} depends upon A_{rs} and by the proper choice of this constant it can be made as small as desired.

We are coming now to the question of ω_r, that is: the probability of accepting h_r when h_r is true. Let us consider first the probability, say α_r, of recjecting h_r when h_r is true. In view of (4) we have:

$$(5) \qquad \alpha_r = \sum_{s \neq r} \beta_{sr} \leqslant \sum_{s \neq r} \frac{1}{A_{sr}}.$$

Suppose now that the probability (under the assumption that h_r is true) that the procedure will terminate in a finite number of trials is 1. We would then have:

$$(6) \qquad \omega_r = 1 - \alpha_r \geqslant 1 - \sum_{s \neq r} \frac{1}{A_{sr}}$$

and by a suitable choice of the constants A_{sr}, ω_r could be made as high as desired. But this last assumption has not been proved so far. Before we proceed to the discussion of this point it should be remarked that — irrespective of the problem of controlling ω_r — something of the sort ought to be proved if the rule R is to have some (even if only theoretical) value. It ought to be shown that, whichever of the hypotheses $h_1, \ldots, h_N$ is true, the probability is 1 that the rule R will lead to the acceptance of some hypothesis in a finite number of trials. Otherwise the possibility would be open that the rule R will fail as a means of arriving at a decision.

Let us first remark that in order to prove:

(7)   The probability is 1 that there is such m that the procedure terminates after m trials,

it is sufficient to prove:

(8)   The probability that the procedure terminates after not more than n trials tends to 1 as n increases indefinitely.

In order to see this let us write A_m for the sentence 'the procedure terminates after m trials'. (7) and (8) reduce then, respectively, to

$$(9) \qquad P[(\exists_m) A_m] = 1$$

and

$$(10) \qquad \lim_{n \to \infty} P[(\exists_m)(m \leqslant n \wedge A_m)] = 1$$

or, to apply an obvious abbreviation:

(9′) $P(Z)=1$

and

(10′) $\lim\limits_{n\to\infty} P(S_n)=1$.

But obviously, for each n: S_n implies Z. Therefore, for each n:

(11) $P(Z)\geqslant P(S_n)$.

Suppose now that (9′) is not true, i.e.

(12) $P(Z)=q<1$.

But from (10′) it follows that for the some n_0

(13) $P(S_{n_0})>q$,

which in view of (11) entails:

(14) $P(Z)>q$.

We have thus arrived at a contradiction which shows that (10′) and the negation of (9′) are incompatible. Therefore (9′) follows from (10′).

It is one of our assumptions that exactly one of the hypotheses $h_1, \ldots, h_N$ is true. Without loss of generality we shall assume that the true hypothesis is h_r. By definition of the rule R a sufficient condition for the procedure to terminate in not more than n steps is that the relation

(15) $f_r(x_1)\cdot \ldots \cdot f_r(x_n)\geqslant A_{rs}\cdot f_s(x_1)\cdot \ldots \cdot f_s(x_n)$

be satisfied by the sample $\{x_1, \ldots, x_n\}$ for all $s\neq r$. If for any $s\neq r$ the probability (under the assumption that h_r) that the relation (15) is satisfied tends to 1 as n increases indefinitely, then the probability that all the relations of the type (15) — i.e. for all $s\neq r$ — are satisfied also tends to 1 as n increases indefinitely. Therefore, in order to prove (8) (and thus (7)) it is enough to show that for any $s\neq r$ and under the assumption that h_r is true:

(16) $\lim\limits_{n\to\infty} P[f_r(x_1)\cdot \ldots \cdot f_r(x_n)\geqslant A_{rs}\cdot f_s(x_1)\cdot \ldots \cdot f_s(x_n)]=1$.

I am now going to prove (16) for two, admittedly most important, cases:
(A) that of the parameter p is in binomial distribution, and (B) that
of μ in normal distribution with known variance.

(A)   Suppose that the hypotheses h_1, ..., h_N ascribe definite
        values to the parameter p in binomial distribution, i.e. hy-
        pothesis h_i states that $p=p_i$. (16) reduces now to the fol-
        lowing form. Under the assumption that $p=p_r$:

(17) $$\lim_{n\to\infty} P\left[p_r^k\cdot(1-p_r)^{n-k}\geqslant A\cdot p_s^k\cdot(1-p_s)^{n-k}\right]=1),$$

where k is the number of successes in the n independent trials, $p_s\neq p_r$
and A is a constant $\geqslant 1$ (the subscript in A_{rs} can be dropped now).
    From the law of great numbers (and in view of the asumption that
$p=p_r$) we have:

(18) $$\lim_{n\to\infty} P\left(\left|\frac{k}{n}-p_r\right|<\varepsilon\right)=1,$$

where ε is any positive constant. It is, therefore, sufficient to prove that:

(19) $$\left|\frac{k}{n}-p_r\right|<\varepsilon$$

implies

(20) $$p_r^k(1-p_r)^{n-k}\geqslant A p_s^k(1-p_s)^{n-k}.$$

But (20) is equivalent to

(21) $$\frac{k}{n}\log p_r+\left(1-\frac{k}{n}\right)\log(1-p_r)-\frac{k}{n}\log p_s-$$

$$-\left(1-\frac{k}{n}\right)\log(1-p_s)\geqslant\frac{1}{n}\log A.$$

From (19) we have:

(22) $$p_r-\varepsilon<\frac{k}{n}<p_r+\varepsilon.$$

Making appropriate substitutions in (21) we now obtain

(23) $p_r \log p_r + (1 - p_r) \log (1 - p_r) - p_r \log p_s - (1 - p_r) \log (1 - p_s) -$

$- \varepsilon [\log p_r + \log (1 - p_r) + \log p_s + \log (1 - p_s)] \geqslant \dfrac{1}{n} \log A$.

As, however, ε is as small as we wish and $\dfrac{1}{n} \log A \to 0$, it is sufficient to prove that

(24) $\varphi(p_r, p_s) = p_r \log p_r + (1 - p_r) \log (1 - p_r) - p_r \log p_s -$

$- (1 - p_r) \log (1 - p_s) > 0$.

It is easy to verify that $\varphi(p_r, p_s)$ has minimum for $p_r = p_s$, and this minimum is equal to 0. In view of the assumption that $p_r \neq p_s$ this completes the proof.

(B) Suppose now that the hypotheses $h_1, \ldots, h_N$ ascribe definite values to the mean μ in a normal distribution with known variance σ^2, i.e. a hypothesis h_i states that $\mu = \mu_i$. (16) then reduces to the following form. Under the assumption that $\mu = \mu_r$:

(25) $\lim\limits_{n \to \infty} P[n_r(x_1) \cdot \ldots \cdot n_r(x_n) \geqslant A \cdot n_s(x_1) \cdot \ldots \cdot n_s(x_n)] = 1$,

where $n_r(x)$, $n_s(x)$ are normal densities with respective means μ_r, μ_s $(\mu_r \neq \mu_s)$ and the standard deviation σ; A is a constant > 1. Taking logarithms and dividing by n we obtain:

(26) $\lim\limits_{n \to \infty} P\left(\dfrac{1}{n} \sum\limits_{i=1}^{n} \log n_r(x_i) - \dfrac{1}{n} \sum\limits_{i=1}^{n} \log n_s(x_i) \geqslant \dfrac{1}{n} \log A \right) = 1$.

But from Tchebysheff's law of great numbers we have

(27) $\lim\limits_{n \to \infty} P\left(\left| \dfrac{1}{n} \sum\limits_{i=1}^{n} \log n_r(x_i) - C_r \right| < \varepsilon \right) = 1$ for all $\varepsilon > 0$,

and

(28) $\lim\limits_{n \to \infty} P\left(\left| \dfrac{1}{n} \sum\limits_{i=1}^{n} \log n_s(x_i) - C_s \right| < \varepsilon \right) = 1$ for all $\varepsilon > 0$,

where (in view of the assumption that $\mu = \mu_r$):

$$(29) \qquad C_r = E[\log n_r(x)] = \int_{-\infty}^{\infty} n_r(x) \log n_r(x)\, dx$$

and

$$(30) \qquad C_s = E[\log n_s(x)] = \int_{-\infty}^{\infty} n_r(x) \log n_s(x)\, dx.$$

In view of the fact that ε is as small as we wish and $\dfrac{1}{n}\log A \to 0$ it is, therefore, sufficient to prove that $C_r > C_s$. But

$$(31) \qquad C_r = \int_{-\infty}^{\infty} n_r(x) \left[\log \frac{1}{\sigma\sqrt{2\pi}} - \frac{(x-\mu_r)^2}{2\sigma^2} \right] dx = \log \frac{1}{\sigma\sqrt{2\pi}} - \frac{1}{2}$$

and

$$(32) \qquad C_s = \log \frac{1}{\sigma\sqrt{2\pi}} - \frac{1}{2\sigma^2} \int_{-\infty}^{\infty} (x-\mu_s)^2 n_r(x)\, dx$$

$$= \log \frac{1}{\sigma\sqrt{2\pi}} - \frac{1}{2\sigma^2}(m_r^{(2)} - 2\mu_r\mu_s + \mu_s^2),$$

where $m_r^{(2)}$ is the second moment of $n_r(x)$. Thus

$$(33) \qquad C_r - C_s = \frac{1}{2}\left[\frac{1}{\sigma^2}(m_r^{(2)} - 2\mu_r\mu_s + \mu_s^2) - 1 \right]$$

$$= \frac{1}{2\sigma^2}[m_r^{(2)} - 2\mu_r\mu_s + \mu_s^2 - (m_r^{(2)} - \mu_r^2)] = \frac{(\mu_r - \mu_s)^2}{2\sigma^2}.$$

Therefore $C_r - C_s > 0$, because of the assumption that $\mu_r \neq \mu_s$; which completes the proof.

To sum up. A sequential rule R of inference has been formulated. The premiss of such an inference is the report on the outcome of n independent trials, the number n not being fixed in advance. The conclusion is one of N simple statistical hypotheses. The rule R is a generalization of the sequential probability ratio test, as defined by Wald. At the same time it turns out to be a generalization of maximum likelihood principle (in the case when the number of hypotheses is finite).

Some properties of the rule R have been roughly ascertained. It turns out that (as it is the case for the sequential probability ratio test) the probabilities of errors can, in principle, be made as small as desired by a suitable choice of constants in R. For two classes of hypotheses it has been proved that the rule R will lead with probability 1 to a decision in a finite number of steps.

It ought to be stressed, however, that if the rule R is to have any practical significance its properties ought to be investigated much closer than that. The expected number of trials is particularly important in this context. Its estimation seems to be a rather difficult task.

REFERENCES

* First published in *Studia Logica* **XII** (1961). A version of this paper appeared in Polish in *Studia Filozoficzne* **1** (1961).
[1] A. Wald, *Sequential Analysis*, J. Wiley, New York 1947.
[2] $f(x; \theta)$ is a probability density function if the distribution is continuous; in the case of discrete distribution it should be replaced by a probability function. A slightly different formulation is needed for the two cases, in some contexts.
[3] As Wald justly remarked (*op. cit.*, p. 35–36), a rule of inference with the number n of trials fixed can be looked upon as a special ('degenerate') case of a sequential rule. It will be characterized by the fact that for $m=1, \ldots, n-1$ all the subsets except R_m^{N+1} are empty while for $m=n$ the subset R_m^{N+1} is empty.

KLEMENS SZANIAWSKI

INTERPRETATIONS OF THE MAXIMUM LIKELIHOOD PRINCIPLE *

0. INTRODUCTORY REMARKS

The maximum likelihood principle is one of the simplest and the most intuitive rules of non-deductive inference. It was originally formulated by Ronald A. Fisher [2] for the use of mathematical statistics, and more particularly for its part concerned with point estimation, i.e. with determining the numerical value of a parameter, for example, a mean. But the idea expressed by the maximum likelihood principle has a more general character. It may be formulated as follows.

Let H be an n-element set of sentences, the exclusive disjunction of which is true. We shall call *hypotheses* the elements of the set H.

Let e be a sentence, the truth of which has been established. The sentence e (evidence) will be called the *premise* or *empirical data*.

Moreover, let there be for each $h \in H$ a given $p(e|h)$, i.e. the probability of e under the assumption that h. The chosen notation would indicate that it is a conditional probability. Indeed, the expression $p(e|h)$ can be interpreted in this way. But in the case when the meaningfulness of assigning probability to hypotheses raises doubts this expression will be considered as the probability of the sentence e, parametrical with respect to h. It is anyway substantial that e is constant while h ranges over the set H.

The question answered by the maximum likelihood principle is, of course, that of the choice of hypothesis in the light of empirical data. As it is known, the answer is: the hypothesis should be chosen which maximalizes $p(e|h)$.

In other words, the maximum likelihood principle tells us to choose the hypothesis with relation to which the empirical data are the most probable. The name of the rule originates from the fact that $p(e|h)$,

i.e. the probability of *e* with respect to *h*, was called by Fisher the like-lihood of the hypothesis *h* with relation to the data *e*. Thus the considered rule promotes the choice of the hypothesis most likely in the light of empirical data.

This view of the likelihood of a hypothesis is, of course, something different from its probability. The origin of the maximum likelihood principle amounts precisely to that: it was meant to formulate an inference rule which would not refer to probabilities of hypotheses. These probabilities – if they are to be interpreted objectively – are rarely given, and hence it would be unrealistic to make an inference rule dependent on them. Contemporary mathematical statistics is wholly based on this view. It was only a more common use of the concept of subjective probability that made it possible in the last few decades to develop the so-called Bayesian theory of probabilistic inferences, in which the maximum likelihood principle and related rules do not play any substantial role (cf. Szaniawski [7]).

It is not difficult to indicate the reasons for which the maximum likelihood principle seems intuitively convincing. Namely, it is a generalization of the commonly used procedure of elimination of hypotheses: the hypothesis is rejected in the light of which the experimentally validated sentence *e* is minimally probable. This reasoning, which Czerwiński [1] proposed to call 'weakened *modus tollens*', is somehow analogous to its deductive counterpart. Instead of the implication it contains a high conditional probability of the negation of the sentence *e*, and as the truth of *e* has been established it is the basis for the rejection of the antecedent of this quasi-implication. By repeatedly applying this elimination procedure to the hypotheses of the set *H*, we gradually reduce that set, and eventually the only hypothesis (hypotheses) left is (are) the one (those) for which $p(e|h)$ assumes the maximum value – conformingly to the maximum likelihood principle.

It naturally remains an open question whether the above-presented justification should be considered satisfactory. No matter, however, how this question is to be resolved, it is no doubt advisable to realize the possible interpretations of the maximum likelihood principle in the categories of the aim which directs the choice of a hypothesis. Such an aim should be specified in terms of a function evaluating the possible results of inference (and called, following C. Hempel, epistemic utility).

Then it should be found out in what circumstances the maximum likelihood principle leads to the maximalization of that function.

The remarks presented below concern the problem formulated as above. They are partially a repetition of already known views and partially their extension.

1. MAXIMALIZATION OF PROBABILITY OF CONCLUSION

Probably the simplest aim of an inference is to accept a true hypothesis. If the task is reduced to this, the evaluative function — let us call it u_1 — gives the same value to all correct conclusions as well as to all erroneous cases. It only differentiates error and non-error, to the advantage of the latter. This leads to the following epistemic utility:

(1) $$u_1(h, k) = \begin{cases} 1 & \text{if } h = k, \\ 0 & \text{if } h \neq k. \end{cases}$$

It is the function of two variables: the accepted hypothesis and the true one; i.e. both h and k range over the set H. In the case of identity of arguments the conclusion is true; in the opposite case one of the $n(n-1)$ possible errors occurs.

The mean value of $u_1(h, k)$, naturally equal to the probability of h, is the epistemic utility of accepting the hypothesis h. Hence the aim determined by (1) amounts to the maximalization of that probability which, according to Bayes' theorem, is given by the formula:

(2) $$p(h|e) = p(e|h) \cdot p(h) \cdot C,$$

where

(3) $$C = [\Sigma_h\, p(e|h)\, p(h)]^{-1}$$

is a constant factor normalizing the probability.

It immediately results from (2) that if for all h

(4) $$p(h) = \frac{1}{n},$$

i.e. all hypotheses are *a priori* equally probable, then the maximalization of $p(h|e)$ is equivalent to the maximalization of the likelihood. The

maximum likelihood principle is therefore optimal with respect to the aim determined by u_1 — under the assumption of equiprobability of hypotheses.

Let us remind that the so-called principle of insufficient reason, attributed to Laplace,[1] tells us to consider hypotheses equally probable if nothing indicates that one of them should be distinguished. As it can be seen, by using the principle of insufficient reason likelihood is maximalized.

2. MAXIMALIZATION OF TRANSMITTED INFORMATION

Another interpretation of the maximum likelihood principle has been given by Hintikka [3]. It is based on the concept of transmitted information. Taking as a starting point the known measure of information contained in a sentence s:

$$(5) \qquad \inf(s) = -\log p(s).$$

Hintikka defines the information transmitted by the sentence h about the subject matter of e as follows:

$$(6) \qquad \text{trans inf}(h|e) = \inf(e) - \inf(e|h) = \log[p(e|h)/p(e)].$$

Here is the justification of this definition: "$\inf(e|h)$ is the information e adds to that of h, i.e. the uncertainty that there remains concerning e even after we have learned that h is true. Hence $\inf(e) - \inf(e|h)$ measures the reduction of our uncertainty concerning e which takes place when we come to know, not e, but h." (Hintikka [3], p. 316.)

To explain the sentence e, Hintikka says, is as much as: to indicate an h such that will transmit possibly much information about what is stated by e. "What we want to do is to find an h such that the truth of e is not unexpected, given h" (p. 321). If, therefore, the aim to be achieved by choosing h is the explanation of e, then the corresponding epistemic utility, let us say u_2, is as follows:

$$(7) \qquad u_2(h) = \log[p(e|h)/p(e)].$$

The maximalization of u_2 is therefore equivalent to the maximalization of $p(e|h)$, and this precisely is the maximum likelihood principle.

"Thus we arrive at the famous maximum likelihood principle as the natural method of choosing one's explanatory hypothesis in the

kind of circumstances indicated (...). Thus the importance of this principle in statistics has an interesting general reason which can be brought out in terms of the concept of information" (p. 322).

The peculiarity of this approach lies in that it does not take into consideration the logical value of h, i.e. of the conclusion, at all. The epistemic utility is here the function of only one argument, i.e. of the accepted hypothesis, and it does not depend on which of the hypotheses $h \in H$ is true. I believe this can be explained in two ways.

The first explanation requires the change of assumptions. Let us suppose that the hypotheses of the set H are not pairwise exclusive, but conversely: that they have all been found true. The task is, then, to find the best explanation of e among those true sentences. Such a procedure may be considered, although for obvious reasons it could hardly be called inference.

Another way of understanding Hintikka's proposed maximalization of u_2 consists in considering it as the first stage of the explanation of e. It would consist in distinguishing in the set H of a hypothesis suitable to explain e—in the sense that it conveys most information about e. The hypothesis thus distinguished could then be accepted as explanation—provided its truth was somehow demonstrated, or at least made probable. Such a two-stage explanatory procedure could hardly be objected to (the first stage would consist in utilizing the maximum likelihood principle), but it should be realized that the status of the discussed principle would then undergo a basic modification.

For, according to the classical approach, the maximum likelihood principle is a rule of inference; its application therefore leads to the acceptance of the sentence h as true, h being correlated as a conclusion to the sentence e, i.e. the premise. But the principle, proposed by Hintikka, of distinguishing an element of the set H by maximalizing u_2 does not have this property. For there is no justification of a procedure which consists in accepting as true hypothesis h maximalizing u_2—because the sentence e has been accepted as true. If we were to follow Hintikka and accept that the maximum likelihood principle is justified by its function of explaining an established fact, then this same principle could no longer be considered a rule of inference—contrarily to the original intention of its author and probably contrarily to the practice of its applications.

3. MAXIMALIZATION OF PROBABILITY OF CONCLUSION AND OF TRANSMITTED INFORMATION

It results from the above presented considerations, that it is a necessary condition for the maximum likelihood principle to maintain its status of a rule of inference that the epistemic utility maximalized by that principle evaluate and differentiate truth and falsehood. In other words, epistemic utility should fulfil the following condition (for all h, k such that $h, k \in H, h \neq k$):

$$(8) \qquad u(h, h) > u(k, h).$$

If, moreover, Hintikka's idea is to be preserved (i.e. the evaluation is to take into consideration the degree to which the premise is explained by the hypothesis), then the function u should fulfil two further conditions, which we shall formulate as follows:

For all h, k and l non-identical in pairs, of the set H:

$$(9) \qquad \text{trans inf}(h|e) > \text{trans inf}(k|e)$$

entails

$$(10) \qquad u(h, h) > u(k, k)$$

and

$$(11) \qquad u(h,l) > u(k, l).$$

The condition $(9) \Rightarrow (10)$ postulates the preferential ordering of true conclusions with respect to trans inf. The condition $(9) \Rightarrow (11)$ postulates the same for false conclusions.[2]

There are of course more than one function u fulfilling the above-presented conditions. Within the limits determined by these conditions the choice of a concrete epistemic utility should probably follow the criterion of simplicity. One of the simplest is the device proposed in Szaniawski [8].

Let us suppose that t is a function which is to generate preferential orderings. Let it be a positive function. Then the fulfilment of the con-

dition (8) may be assured by defining u as follows:

(12) $u(h, k) = \begin{cases} t(h) & \text{for } h = k, \\ t(h) - \text{Sup } t(h) & \text{for } h \neq k. \end{cases}$

It is clear that the condition (8) is fulfilled as $t(h) - \text{Sup } t(h) < 0 < t(h)$. The fulfilment of the other two conditions is an immediate consequence of the definition (12).

The identification of $t(h)$ with trans inf $(h|e)$ is opposed only by the fact that the latter function may assume negative values. This is not, however, a substantial obstacle. It suffices to note that it is a sufficient condition for the positiveness of trans inf $(h|e)$ that

(13) $p(e|h) > p(e)$.

This condition states the positive probabilistic dependence between the hypothesis h and the sentence e. But only such hypotheses can pretend to the role of explanation of the sentence e. Therefore, if a certain h does not fulfil the condition (13), it can be eliminated from the set H (of course, only if the explanation of e codetermines the choice of a hypothesis). This entails that the assumption of the positiveness of trans inf is in this case no substantial limitation.

Thus, substituting in (12) trans inf $(h|e)$ for $t(h)$ and considering

(14) $\text{Sup trans inf}(h|e) = \inf(e) = -\log p(e)$

we obtain the epistemic utility u_3 fulfilling the above-formulated conditions for u:

(15) $u_3(h, k) = \begin{cases} \log[p(e|h)/p(e)] & \text{if } h = k, \\ \log p(e|h) & \text{if } h \neq k. \end{cases}$

Dtermining the mean of k in terms of the probability distribution function $p(k|e)$ we obtain the average utility connected with the acceptance of the hypothesis h:

(16) $E_{p(k|e)} u_3(h, k) = \log p(e|h) - p(h|e) \log p(e)$.

We can now answer the question concerning the assumptions under which the maximum likelihood principle is equivalent to the maximalization of transmitted information, if that maximalization is limited by the condition (8). Because of (16), the answer is the same as when

the only aim was to obtain a true conclusion: the maximum likelihood principle may be interpreted as the maximalization of u_3 if and only if the hypotheses of H are *a priori* equiprobable.

4. SEQUENTIAL GENERALIZATION

It is worth remarking at the end that the maximum likelihood principle is a special case of a procedure which assumes the possibility of accumulating empirical data. It is so, for example, when we successively examine the elements of a set (not necessarily finite), in order to decide at some moment that the accumulated data are sufficient to choose one of the hypotheses about that set. Let us denote the data accumulated after n stages of research by e_n; of course, for every n: $e_{n+1} \Rightarrow e_n$.

A sequential rule of inference (Wald [9]) is one which takes into consideration the character of the data outlined above: for every n it determines e_n for which one of the hypotheses (and which) is to be accepted, and e_n for which no choice should be made and one should proceed to e_{n+1}. There is a sequential rule of inference (Szaniawski [5]) — let us call it R — which is a generalization of the maximum likelihood principle.

The Rule R. Each pair of different hypotheses h, k of the set H has been assigned a certain constant $A_{hk} \geqslant 1$. At the nth stage of research the hypothesis h should be accepted if and only if

(17)   for every $k \neq h$: $p(e_n|h) \geqslant A_{hk} \cdot p(e_n|k)$;

in the case when (17) does not occur for any h, proceed to e_{n+1}.

It is easy to note that rule R becomes the maximum likelihood principle if we assume that

(18) $A_{hk} = 1$ for all $h, k \in H$.

The condition (18) eliminates the sequential character of rule R: as the set H is finite, there is at least one[3] hypothesis $h \in H$ fulfilling (17) for $A_{hk} = 1$, i.e. proceeding to e_{n+1} is unnecessary.

Whatever it may appear, the generalization of the maximum likelihood principle in the form of rule R is natural. It preserves what is substantial for the principle: the comparison of hypotheses in terms of their likelihood. But some elasticity is introduced into this comparison. Namely, the inference maker is given the possibility of a non-symmetrical treatment of hypotheses — by an appropriate choice of the constants A_{hk}. If,

for example, this person is particularly interested in avoiding the rejection of a distinguished hypothesis h when it is true, he will choose a relatively small A_{hk}, for all k. Different choices of numerical values of the parameters of rule R would be directed by analogical reasons. The assignment of values different from 1 to these parameters entails the sequential character of the rule R. For it is not generally excluded that no hypothesis h fulfils the condition (17), which requires to suspend the opinion and to extend the empirical basis.

A guarantee is of course necessary that this process will terminate — with a probability equal to one — in a finite number of steps, leading to the acceptance of one of the hypotheses of the set H. It is possible to demonstrate this property of the rule R only by assuming a more exact characterization of the nature of the probabilistic relation between the members of H and the successive extensions of the empirical data. In the paper mentioned above (Szaniawski [5]) the demonstration has been carried out for two rather important cases.

A similar precizing would also be required for the examination of the rule R from the point of view adopted in the Sections 1–3 of the present paper. This would naturally be a more complicated question.

6. SUMMARY

The above-presented remarks dealt with the following problem: under what assumptions is it possible to interpret the maximum likelihood principle as the maximalization of the aim of research determined as: 1) acceptance of the true conclusion, 2) acceptance of the best explanation of the empirical data. With respect to the first aim the maximum likelihood principle is optimal under the assumption that all the hypotheses are *a priori* equiprobable. With respect to the second aim (understood in agreement with Hintikka's proposal) it is optimal without any additional assumptions — but only on the condition it does not permit to accept the distinguished sentence as true, i.e. — that it is not a rule of inference. If the aims 1) and 2) are put together the optimality of the maximum likelihood principle is again conditioned by the *a priori* equiprobability of the hypotheses of the set H.

In the final section the maximum likelihood principle has been presented as a special case of a (sequential) rule of inference.

BIBLIOGRAPHY

[1] Z. Czerwiński, 'On the Relation of Statistical Inference to Traditional Induction and Deduction', *Studia Logica* **9** (1958) 243–264.

[2] R. A. Fisher, 'On the Mathematical Foundations of Theoretical Statistics', *Philosophical Transactions of the Royal Society* **222**, Series *A* (1922).

[3] J. Hintikka, 'The Varieties of Information and Scientific Explanation'. In: *Logic, Methodology and Philosophy of Science III* (ed. by van Rootselaar and Staal), North Holland (1968) 311–331.

[4] J. Levi, *Gambling with Truth. An Essay on Induction and the Aims of Science*, Alfred A. Knopf, New York 1967.

[5] K. Szaniawski, 'A Method of Deciding Between *N* Statistical Hypotheses', *Studia Logica* **12** (1961) 135–143.

[6] K. Szaniawski, 'Zasada największej wiarygodności (próba częściowej oceny)', in: *Rozprawy logiczne. Księga Pamiątkowa ku czci Kazimierza Ajdukiewicza* ('The Maximum Likelihood Principle — an Attempt at Partial Evaluation', in: *Logical Dissertations. Papers in the Honour of Kazimierz Ajdukiewicz*) PWN, Warszawa 1964.

[7] K. Szaniawski, 'Dwie koncepcje indukcji', in: *Fragmenty Filozoficzne, seria trzecia. Księga Pamiątkowa ku czci Tadeusza Kotarbińskiego* ('Two Concepts of Induction', in: *Philosophical Fragments, third series. Papers in the Honour of Tadeusz Kotarbiński*), PWN, Warszawa 1967.

[8] K. Szaniawski, 'Types of Information and Their Role in the Methodology of Science', *Proceedings of the Conference on Formal Methods in Methodology of Science*, Jabłonna 1974 (in print).

[9] A. Wald, *Sequential Analysis*, J. Wiley, New York 1947.

REFERENCES

* First published in *Rozprawy Logiczne*, Warszawa 1964. Translated by S. Wojnicki.

[1] Hence the so-called Laplace's criterion in decision-making. The paper by Szaniawski [6] has demonstrated that under certain assumptions the MLP is the optimal — in Laplace's sense — method of point estimation.

[2] The postulates adopted here are analogical to those formulated by J. Levi for other purposes (cf. Levi [4], p. 76).

[3] It is possible (which has already been mentioned) that more than one hypothesis fulfil (17), i.e. maximalize $p(e_n|h)$. The method of dealing with non-uniqueness, for example, by an arbitrary choice of the hypothesis, is not relevant here.

KLEMENS SZANIAWSKI

TWO CONCEPTS OF INFORMATION*

0. INTRODUCTORY REMARKS

Among the different meanings of the word 'information', there is at least one which makes it closely connected with decision-making. It usually is referred to as 'pragmatic information',[1] the adjective indicating that what we have in mind is information for some purpose.

There has been, so far as I know, no attempt to explore systematically the notion of pragmatic information. Elsewhere I tried to define the concept ([7], [8]) and to investigate some of its properties for the special case of categorical, i.e. non-probabilistic, information ([6], [7]). The main purpose of the present paper is to compare the concept of pragmatic information with the classical, entropy-based notion, due to Shannon. In order to be able to do this I shall have to assume that the joint probability distribution of a two-dimensional random variable is given, for this is what the statistical, i.e. Shannon, concept of information presupposes. I shall also make another special assumption; more on this later. After the relation between the two concepts has been made apparent, I will indicate some consequences of lifting the two restrictions.

1. PRAGMATIC INFORMATION, PROBABILITY FULLY SPECIFIED

Let $\langle X, S \rangle$ be a two-dimensional random variable with the joint probability distribution function

$$(1) \qquad p(x, s) = \Pr\{X=x \wedge S=s\}$$

defined for all $x \in X$ and $s \in S$, where X and S are the sets of values of the variables X and S, respectively. For the sake of simplicity, I shall assume that all the sets spoken of here are finite.

The absolute and conditional probabilities, obtainable from $p(x, s)$, will be denoted by

(2) $p(s) = \sum_x p(x, s)$,

(3) $p(x|s) = p(x, s)/p(s)$

and, analogously, by $p(x)$, $p(s|x)$.

Let us further assume that a decision problem $U = \langle A, S, u \rangle$ is given, where: $A =$ the set of possible actions, $S =$ the set of states of the world, $u =$ utility, i.e. a real function defined on $A \times S$ and representing valuation by the decision maker of all the combinations: action a taken, state s occurs $(a \in A, s \in S)$. The valuation is extended over the probability mixtures of the elements of $A \times S$, according to the principle (von Neumann and Morgenstern) that the expected value of utility is utility.

The connection between $\langle X, S \rangle$ and $\langle A, S, u \rangle$ is easily seen to consist in the fact that the set S of states of the world is identical with the set of values of the random variable S. Thus the actual value of the variable S determines the consequences of the actions in A. Obviously, if the purpose is to maximize u by a suitable choice of action from the set A, the decision could, in general, be improved by providing the decision maker with some information concerning the actual value of S.

This is where the variable X comes in. The knowledge of the value of X reduces, through $p(x, s)$, the uncertainty concerning S (on the average, of course, and if we except certain special cases which will be discussed later). Hence the possibility of improving the decision. Pragmatic information is precisely the measure of this improvement.

To make use in decision making of information provided by X means to make the choice of action depend on the value X takes. This is achieved by a decision function (the term goes back to A. Wald), i.e. by a function d from X to A. The interpretation of d is as follows: $d(x) = a$ means that the decision maker is committed to take action a in the case of $X = x$. Of course, there are many such functions (the set of all d will be denoted by D). However, in the case under consideration, it is fairly easy to determine which of them is optimal. Clearly, the best thing to do is to associate with any given x such action which maximizes expected utility under the probability distribution $p(s|x)$. Formally, an optimal decision function is defined by the following formula, valid for all x and a:

(4) $d_{opt}(x) = a \Rightarrow \sum_s u(a, s) p(s|x) = \max_a \sum_s u(a, s) p(s|x)$.

And by averaging this maximal utility over the set X we obtain

$$(5) \qquad u(d_{opt}) = \sum_x \max_a \sum_s u(a, s) p(x, s),$$

which is the utility associated with the optimal decision function.[2]

On the other hand, if information provided by X is not available, the best course of action is for the decision maker to choose this a which maximizes expected utility under the probability distribution $p(s)$. The utility associated with such a_{opt} is:[3]

$$(6) \qquad u(a_{opt}) = \max_a \sum_s u(a, s) p(s).$$

The increase (if any) of utility, due to the variable X, is equal to the difference between $u(d_{opt})$ and $u(a_{opt})$. According to the previous remarks, this measures pragmatic information.

Pragmatic information will be denoted by $C(X, S; U)$ to bring out the fact that it depends on the decision problem U.

$$(7) \qquad C(X, S; U) = u(d_{opt}) - u(a_{opt}).$$

Substituting (5) and (6) into (7) we get

$$(8) \qquad C(X, S; U) = \sum_x \max_a \sum_s u(a, s) p(x, s) -$$
$$- \max_a \sum_s u(a, s) p(s).$$

If we take (2) into account and introduce the abbreviation

$$(9) \qquad v(a, x) = \sum_s u(a, s) p(x, s),$$

C can be represented more symmetrically as

$$(10) \qquad C(X, S; U) = \sum_x \max_a v(a, x) - \max_a \sum_x v(a, x),$$

which, of course, reduces to

$$(11) \qquad C(X, S; U) = \min_a \sum_x \left(\max_a v(a, x) - v(a, x) \right).$$

As all elements in the above sum are, by the definition of maximum, non-negative, the sum itself is also non-negative and it attains its lowest value, i.e. zero, iff

$$(12) \qquad \bigvee_a \bigwedge_x \left(v(a, x) = \max_a v(a, x) \right).$$

In view of

(13) $v(a, x) = \max_a v(a, x)$

$\Leftrightarrow \sum_s u(a, s) p(s|x) = \max_a \sum_s u(a, s) p(s|x).$

the meaning of the condition (12) is fairly obvious. (12) in fact states that there exists such an action in A which, *for all x*, maximizes expected utility under the conditional probability distribution $p(s|x)$. In other words, there exists an action, optimal absolutely, i.e. optimal whichever value the variable X assumes.

It is hardly surprising that (12) is a sufficient and necessary condition for the pragmatic information to be zero. If an action is invariably adopted, no matter which statement of the form $X = x$ happens to be true, then the solution of the decision problem U does not depend on the information. Following the terminological suggestion by Mr G. Lissowski I shall say in such a case that S is independent of X according to U.

Whether the condition (12) holds or not depends both on the decision problem U and on the joint probability distribution $p(x, s)$. This may be instantiated by two special cases.

If U is such that an action dominating in utility all other actions exists, i.e.

(14) $\bigvee_{a^*} \bigwedge_a \bigvee_s (u(a^*, s) \geqslant u(a, s)),$

then (12) holds and $C(X, S; U) = 0$ for all probability distribution functions $p(x, s)$.

Also, if the variables X and S are stochastically independent, i.e.

(15) $\bigwedge_x \bigwedge_s (p(x, s) = p(x) p(s)),$

then S is independent of X according to any U, and $C(X, S; U) = 0$ for all U. This is easily seen. Assuming that (15) holds, we have

(16) $C(X, S; U) = \sum_x \max_a \sum_s u(a, s) p(x) p(s) -$

$- \max_a \sum_s u(a, s) p(s) = 0.$

2. STATISTICAL INFORMATION

The Shannon measure, often called statistical information, is defined as

(17) $I(X, S) = H(S) - H(S|X).$

In the above expression $H(S)$ is the entropy of the variable S, i.e. the

expected value of the logarithm of its probability distribution function (with the minus sign, to make it non-negative):

$$(18) \quad H(S) = -\sum_s p(s) \log p(s).$$

$H(S|X)$ is conditional entropy $H(S|X=x)$, averaged with respect to X:

$$(19) \quad H(S|X) = -\sum_x p(x) \sum_s p(s|x) \log p(s|x).$$

Entropy being a measure of uncertainty associated with a random variable, it is easily seen that $I(X, S)$ measures the average decrease of uncertainty concerning S, due to the information brought by the variable X.

The philosophical importance of this type of information concept has been stressed by Hintikka in [3]. This paper contains also an admirable discussion of possible varieties of such an approach.

3. STATISTICAL AND PRAGMATIC INFORMATION: A COMPARISON

A look at (7) and (17) is enough to notice certain similarity between the two concepts of information. In both cases, the situation where knowledge of the value assumed by the variable X is available is compared with the case where it is not. And the 'gain' made possible by the knowledge of X is measured, either in utility (for C) or in decrease of uncertainty (for I). The similarity could be made even more explicit if loss, i.e. negative utility, were used instead of utility in the definition of C.

Further comparison is made difficult by the fact that pragmatic information depends on the decision problem U, whereas statistical information does not. Perhaps the simplest way to clarify this is to assume the decision problem to be fixed and carry out the comparison, relative to this special kind of U.

Neither is the choice of a suitable U difficult to make. The decision problem, let us denote it by E, will be defined as follows. The set A of actions is assumed to be identical with S, and the utility function will be binary:

$$(20) \quad u(a, s) = \begin{cases} 1 & \text{if } a = s, \\ 0 & \text{if } a \neq s. \end{cases}$$

The interpretation of E is straightforward. The decision in this case consists in choosing one of the states of the world. If the state chosen coincides with the state which actually occurs, the outcome is 'success' and is evaluated as 1; otherwise it is 'failure' and rates 0. Such a problem (guessing the actual state of the world) could be called epistemic, following Hempel who introduced the expression 'epistemic utility'.[4]

When U has the special form of E, the definition of C simplifies considerably. All but one element of the sum over S vanish and utility no longer appears explicity in the expression. As the result we have

$$(21) \qquad C(X, S; E) = \sum_x \max_s p(x, s) - \max_s p(s)$$

$$= \sum_x \max_s p(x, s) - \max_s \sum_x p(x, s)$$

$$= \min_s \sum_x (\max_s p(x, s) - p(x, s)),$$

which depends exclusively on the joint distribution function $p(x, s)$, just as $I(X, S)$ does.

Let us now recall some well-known properties of statistical information. It satisfies the following inequalities:

$$(22) \qquad 0 \leqslant I(X, S) \leqslant H(S) \leqslant \log n,$$

where n is the number of elements in the set S.

As a counterpart of (22), we have, for pragmatic information relativized to E:

$$(23) \qquad 0 \leqslant C(X, S; E) \leqslant 1 - \max_s p(s) \leqslant 1 - n^{-1}.$$

What is more, both C and I reach their respective critical values under similar conditions. Let us examine this in more detail.

A sufficient condition for the two expressions to become equal to zero is the stochastic independence of the variables X and S:

$$(24) \qquad \text{If (15) holds, then } I(X, S) = C(X, S; E) = 0.$$

The result for I is well known (in fact, if implication is replaced by equivalence, it may be said to define I, for I is the only function which satisfies it); its validity for C has been shown above.

It may be remarked in passing that the necessary and sufficient condition for $C(X, S; E)$ to be zero, i.e. the special case of (12), is

$$(25) \qquad \bigvee_{s^*} \bigwedge_x (p(x, s^*) = \max_s p(x, s)).$$

In other words, $C(X, S; E)=0$ iff there exists a value of S which is most probable for all values of X.

The second critical point is defined by the following condition on $p(x, s)$:

(26) $\qquad \bigwedge_x \bigvee_s (p(s|x)=1).$

The condition states that if we know what value X assumes, we have the value of S uniquely determined. As this obviously represents the best possible case, such information will be called perfect. Now, if pragmatic or statistical information is perfect, it attains its highest value:

(27) $\qquad$ If (26) then $I(X, S)=H(S)$ and

$$C(X, S; E)=1-\max_s p(s).$$

Both $H(S)$ and $1-\max_s p(s)$ depend entirely on the absolute probability distribution of S. And both expressions reach their respective maximal values if the probability distribution is uniform:

(28) $\qquad \bigwedge_s (p(s)=n^{-1}),$

(29) $\qquad$ If (26) and (28) then

$$I(X, S)=\log n \text{ and } C(X, S; E)=1-n^{-1}.$$

The last expressions, $\log n$, and $1-n^{-1}$, are strictly increasing functions of n.

It thus turns out that pragmatic information, when relativized to the epistemic problem, varies in essentially the same way as statistical information. The formulae (15), (26) and (28) define sufficient conditions for the critical values of the two expressions.

The parallelism is not surprising. Statistical information consists in the reduction of uncertainty concerning the value of S; while pragmatic information with respect to the epistemic problem consists in improving the prediction of the value of S. Clearly, the probabilistic relation between the variables X and S must have the same kind of influence on the numerical value of both types of information.

Going a step further, one might suspect that under suitable assumptions concerning the form of utility functions, the two concepts can be made to coincide exactly, up to linear transformation. This possibility will not be explored here.

In terms of the present discussion, there are, of course, noticeable differences between the two concepts. We mention a few of them. Sufficient *and* necessary conditions for information to become minimum are different: stochastic independence in the case of I, independence according to the epistemic problem in the case of C. Also $I(X, S)$ is symmetric in the two variables, while $C(X, S; E)$ is not (which follows from the fact that the epistemic problem based on S is, in general, not identical with that based on X, for the two sets are not necessarily equipotent). There are, undoubtedly, several other discrepancies, not inconsistent with the essential similarity.

4. PRAGMATIC INFORMATION, PROBABILITY PARTLY SPECIFIED

It has been assumed above that the probabilistic relation between the variables X and S is given in the form of the joint probability distribution function $p(x, s)$. Without this assumption it would have been impossible to compare pragmatic information with the statistical one.

Such an approach, however, is subject to the well-known criticism of being unrealistic, unless one is prepared to adopt the subjectivic view on probability. It is argued that in the expression

$$(30) \qquad p(x, s) = p(x|s) p(s),$$

only the first factor is known (for all x and s), while the so-called *a priori* probabilities $p(s)$ are usually inaccessible and sometimes do not even make sense.

It is not my purpose here to take sides on the much debated issue whether the so-called Bayesian approach (consisting in the use of *a priori* probabilities) is the right one or not. The number of serious problems involved in this classical controversy makes it impossible to state one's own view in a few sentences. Besides, it may be doubted that a definitive solution, in any sense of the word, is possible at all.

The only point I want to make is that as long as differing views exist, it would be a sound policy not to prejudge the issue in definitions of probability-based concepts. Pragmatic information is a concept which lends itself to such treatment: it can be made elastic enough to suit people of different philosophical persuasions on the matter of *a priori* probabilities.

Let us, therefore, replace the assumption of joint probability distributions by the following one. For each $s \in S$, a probability distribution $p(x|s)$ over X is given. The distribution function $p(x|s)$ may be interpreted either as a genuine conditional distribution, or (if one does not want to assume that S varies randomly) as a parametric distribution with respect to s. Pragmatic information can then be defined in the following way.

For any decision function d and any s, the utility of adopting d when s is the actual state of the world is computed as

$$(31) \qquad u(d, s) = \sum_x u(d(x), s) p(x|s).$$

In this way, the original decision problem $\langle A, S, u \rangle$ becomes extended to $\langle D, S, u \rangle$. We use the word 'extended' because the relation $A \subset D$ may be assumed to hold, considering that any $a \in A$ is identical with a decision function associating this a with all $x \in X$. The gain in utility due to such an extension is now identified with pragmatic information. The problem arises, how to assess the gain in question.

To be able to do this, we must have a means of comparing the elements of A with the elements of $D - A$, i.e. with non-trivial decision functions. To compute the averages over S of $u(d, s)$ and $u(a, s)$ would be the obvious way of doing it if we assumed the *a priori* distribution $p(s)$ to be known. This is, in fact, what we did previously: the optimal d (in terms of average utility) was then compared with optimal a. Such a procedure remains valid and leads to a definition of pragmatic information for the case when the *a priori* probability distribution (whatever its nature, subjective or objective) is given. If it is not given, maximization of expected utility (MEU) must be replaced by some other criterion of decision-making which would then provide a standard of comparison for the elements of D.

Before we discuss this solution, let us summarize the four possible cases by means of the following table.

		A priori probabilities	
		not given	given
Information	not given	$u(a, s)$	$E_s u(a, s)$
	given	$E_x u(d(x), s)$	$E_x E_s u(d(x), s)$

The expressions in the table are those to be maximized: in the first row, by the choice of a; in the second row, by the choice of d (the letter E stands, of course, for the operator of taking expected value). The expressions in the second column have no free variables except those subject to choice. In the expressions in the first column, the free variable s appears, hence the necessity of a criterion. The discussion below is in terms of d, this case being the broader one in view of: $A \subset D$.

Generally speaking, a criterion K of decision making associates with each d the number $u_K(d)$, i.e. the K-value in utility of this decision function. Analogously, we shall have $u_K(a)$ for each a. Thus, for instance, $u_K(d)$ may be the arithmetic mean of utilities $u(d, s)$ (where K=the Laplace criterion) or their minimum (K=maximin) or, for that matter, their expected value (K=MEU, relative to subjective estimate $p(s)$ of chances), etc. For all such cases, pragmatic information is now defined [5] as

$$(32) \qquad C(X, S; U, K) = \max_d u_K(d) - \max_a u_K(a).$$

The concept is doubly relative: to the decision problem U and to the citerion K of decision-making. The last relativization is a consequence of the fact that *a priori* probabilities are no more a necessary part of the conceptual set-up. They may, however, be additionally assumed, in which case they consitute, through MEU, one specialization of K.

Thus, for instance, if K=maximin (denoted by W) we have

$$(33) \qquad C(X, S; U, W)$$

$$= \max_d \min_s \sum_x u\big(d(x), s\big)\, p(x|s) - \max_a \min_s u(a, s).$$

And similarly for other criteria of decision making.[6]

We remark, in passing, that the concept of pragmatic information may be used in assessing the criteria themselves. For, it seems sensible to adopt the following postulate, to be satisfied by any criterion K of decision-making. If U is non-trivial (i.e., there is in A no action dominating all other actions in utility) then $C(X, S; U, K) > 0$. In the light of the above, the maximum postulate, for instance, does not qualify as a reasonable principle: it can be shown [7] that pragmatic information relative to maximin is equal to zero if only U has a saddle point.

5. SOME SPECIAL CASES

Pragmatic information, as defined above, seems to exclude certain intuitively natural cases which do not involve probability and possibly do also without the variable X. For instance, the decision maker may be told simply which one is the actual state of the world or to which subset of S it belongs (the list of subsets being specified in advance). Such information may be assumed to be true or nor.

It is not difficult to see that such cases are covered by the concept introduced above. We obtain the desired results by suitably interpreting the set X and imposing strong conditions on the probability distribution $p(x|s)$. A few examples will make this assertion clear.

Let us interpret X as a partition of S, i.e. for each $x, x' \in X$:

$$(34) \qquad \emptyset = x \subset S, \quad \bigcup x = S, \quad x \neq x' \Rightarrow x \cap x' = \emptyset.$$

If, in addition, the distributions $p(x|s)$ satisfy the condition:

$$(35) \qquad \text{if } s \in x \text{ then for all } x' \neq x: \quad p(x|s) > p(x'|s),$$

then it is natural to attach a semantic interpretation to the elements of X. Each x can be thought of as the statement that the actual state of the world belongs to the subset x of S.

Condition (34) can be strengthened by the further assumption that each x contains exactly one element of S. This establishes a one-one correspondence between the two sets, hence they may be identified:

$$(36) \qquad X = S.$$

Again, (35) can be strengthened to:

$$(37) \qquad \text{if } s \in x \quad \text{then} \quad p(x|s) = 1.$$

Condition (37) ensures the truth of the statement that the actual state of the world belongs to x. Of course, the best possible situation is defined by (36) and (37).[8] In such a case, we obtain true and precise information concerning the state of the world. Information of this kind deserves the name of perfect: indeed, conditions (36) and (37) entail (26), which served for defining the notion.

The preceding remarks indicate the possibility of a typology, according to the nature of the set X and the properties of the probabilistic relation between the elements of X and S.

Obviously, if additional assumptions of the type mentioned above are made, all specializations of the general expression (32) for pragmatic information simplify considerably. The case of perfect information has been investigated in more detail in [6]. The somewhat broader class, defined by (34) and (37), was considered in [8]. In both papers conditions were looked for under which the pragmatic information (relative to different criteria) takes on extremal values. They are, of course, much simpler than those for the more complex case of $C(X, S; U)$.

BIBLIOGRAPHY

[1] C. Cherry, *On Human Communication: A Review, a Survey and a Criticism*, New York, 1957.

[2] C. G. Hempel, 'Inductive Inconsistencies', *Synthese* 12 (1960) 439–469.

[3] J. Hintikka, 'The Varieties of Information and Scientific Explanation', in *Logic, Methodology and Philosophy of Science* III (ed. by B. van Rotselaar and J. F. Staal), Amsterdam 1968.

[4] K. Szaniawski, 'Some Remarks Concerning the Criterion of Rational Decision Making', *Studia Logica* 9 (1960) 221–239.

[5] K. Szaniawski, 'A Pragmatic Justification of Rules of Statistical Inference', in: *The Foundation of Statements and Decisions* (ed. by K. Ajdukiewicz), Warszawa 1965.

[6] K. Szaniawski, 'The Value of Perfect Information', *Synthese* 17 (1967) 408–424.

[7] K. Szaniawski, 'Information and Decision' (in Polish), *Zagadnienia Naukoznawstwa* 13 (1968) 69–79.

[8] K. Szaniawski, 'Pragmatic Value of Information' (in Polish), in *Problemy Psychologii Matematycznej* (ed. by J. Kozielecki), Warszawa 1971.

REFERENCES

* Reprinted from *Theory and Decision* 1 (1974), D. Reidel Publishing Company, Dordrecht-Holland.

[1] Cf., for instance, Cherry [1]. This shows, incidentally, that the notion was loosely discussed at least some 15 years ago.

[2] Strictly speaking, d_{opt} is a representative of a (non-empty) class, as there may be more than one optimal decision function. However, if this is the case, they all satisfy (4), hence lead to the same utility (5).

[3] The same remark applies to the non-uniqueness of a_{opt} as to that of d_{opt}.

[4] See [2], cf. also [5].

[5] The definition can be justified in more detail in terms of the (non-negative) cost of obtaining information. Pragmatic information is then thought of as the highest cost, compatible with the optimality of a decision function belonging to the set $D - A$.

[6] Some of them are discussed in [4].

[7] For the proof (in the special case of perfect information), see [6].

[8] Strictly speaking, in (37) the element sign ought then to be replaced by identity.

RYSZARD WÓJCICKI

SEMANTICAL CRITERIA OF EMPIRICAL
MEANINGFULNESS *

I

This paper is an attempt to apply some methods of model theory in order to analyse the concept of empirical meaningfulness (significance), and to investigate some of its properties. In my introductory remarks I shall discuss briefly Carnap's last definition of empirical significance [3]. I have to explain why I am reopening a problem which is widely considered as satisfactorily solved (cf. [1]). Besides I want to recall some notions of methodology of empirical sciences, and some basic ideas connected with the object of the essay. In the concluding parts of the article philosophical comments will be rather short.

I shall start with an exposition of Carnap's conception. For convenience I have applied a symbolism slightly different from Carnap's. Let L_O and L_T be the observational and the theoretical parts respectively of the total language of science. The set of observational terms (non-logical constants of L_O) will be symbolized by V_O; analogously, the set of theoretical terms (non-logical constants of L_T) by V_T. Observational terms designate observable properties of events or things and relations among them. It is assumed that any sentence of L_O is understood by all members of the language community in the same sense. Thus a complete interpretation of the observational terms is given. As to the theoretical terms, they obtain their meaning through rules of correspondence $\mathfrak{S}_C$ (C-rules) and theoretical postulates $\mathfrak{S}_T$ (T-postulates). The C-rules formulated as rules of inference or as postulates (especially if the underlaying logic is strong enough) establish a connection between certain sentences of a special kind in L_O and sentences of L_T. Not every theoretical term need occur in a C-rule. Those which do not occur in a C-rule receive an indirect interpretation through the T-postulates which

connect the terms without C-rules with terms, occurring in C-rules. Empirical meaningfulness is therefore only relative, i.e. relative to theories.

Using 'significance' as a technical expression for 'empirical meaning-fulness' Carnap offers the following two definitions:

D1.   A term T_i is *significant relative to the class* K *of terms, with respect to* L_T, L_O, $\mathfrak{S}_T$, *and* $\mathfrak{S}_C =_{Df}$ the terms of K belong to V_T, T_i belongs to V_T but not to K, and there are three sentences, α_{T_i} and α_K in L_T, and α_O in L_O, such that the following conditions are fulfilled:

(a) α_{T_i} contains T_i as the only descriptive term.
(b) The descriptive terms in α_K belong to K.
(c) The conjunction $\alpha_{T_i} \wedge \alpha_K \wedge \mathfrak{S}_T \wedge \mathfrak{S}_C$ is consistent, i.e., not logically false.
(d) α_O is logically implied by the conjunction $\alpha_{T_i} \wedge \alpha_K \wedge \mathfrak{S}_T \wedge \wedge \mathfrak{S}_C$.
(e) α_O is not logically implied by $\alpha_K \wedge \mathfrak{S}_T \wedge \mathfrak{S}_C$.

D2.   A term T_k is *significant with respect to* L_T, L_O, $\mathfrak{S}_T$ *and* $\mathfrak{S}_C =_{Df}$ there is a sequence of terms $T_1, ..., T_k$ of V_T, such that every term T_i $(i=1, ..., k)$ is significant relative to the class of those terms which precede it in the sequence, with respect to L_T, L_O, $\mathfrak{S}_T$ and $\mathfrak{S}_C$.

There is no doubt that the definition of a significant term as given by Carnap is based upon a very simple and convincing idea: a theoretical term T_i is empirically meaningful if and only if there is a T_i-sentence (a sentence which does not contain non-logical terms different from T_i) playing an essential role in an argument which provides us with an observational conclusion. Nevertheless, some of its consequences are, in my opinion, strikingly counterintuitive.

Consider the following. Let T_1 be unquestionably meaningless with respect to L_T, L_O, $\mathfrak{S}_T$ and $\mathfrak{S}_C$. Let now us call the set of all logical consequences of the conjunction $\mathfrak{S}_T \wedge \mathfrak{S}_C$ the theory $\mathfrak{S}$. Assume that we add to the theory $\mathfrak{S}$ a new term T_2 with the help of the following definition (T_1, T_2 and O_1 being one-place predicates):

(1) $T_2(x) \equiv O_1(x) \wedge T_1(x)$,

where O_1 is an observational term. (1) is a new C-rule in the enlarged theory.

One can expect that everybody would admit that this operation cannot affect the meaning of T_1. It is well-known that definitions like (1) are non-creative, and therefore that they do not impose any additional conditions affecting the interpretation of the terms involved in the definiens. In the case of (1), for instance, nothing compels us to admit that the meanings of O_1 and T_1 are changed. All possible interpretations of O_1 and T_1, before the definition (1) was formulated, remain possible interpretations of O_1 and T_1 after it was formulated.

Let us denote by $\mathfrak{S}^*$ the theory $\mathfrak{S}$ enriched by the definition (1), by L_{T^*}—the theoretical language of $\mathfrak{S}^*$, and by $\mathfrak{S}_C^*$, $\mathfrak{S}_T^*$ the set of C-rules and T-postulates of $\mathfrak{S}^*$ respectively. Assume that neither the following sentence nor its negation is valid in $\mathfrak{S}$ and thus in $\mathfrak{S}^*$:

(2) $\bigvee x \bigvee y (x \neq y \wedge O_1(x) \wedge O_1(y))$.

Note that (2) is consequence of (1) and the following sentence:

(3) $\bigvee x \bigvee y (x \neq y \wedge T_2(x) \wedge T_2(y))$.

From these we may easily conclude that T_2 is significant relative to the empty class of theoretical terms with respect to L_{T^*}, L_O, $\mathfrak{S}_T^*$ and $\mathfrak{S}_C^*$ and thus T_2 is simply significant. Very similar considerations convince us that T_1 is significant relative to the class $\{T_2\}$ containing the term T_2 as its only element. The following sentence (4) can be used as α_{T_1} (compare the definition D1):

(4) $\bigwedge x T_1(x)$.

As $\alpha_{\{T_2\}}$ we can put

(5) $\sim \bigvee x \bigvee y (x \neq y \wedge T_2(x) \wedge T_2(y))$.

The required sentence α_O will be

(6) $\sim \bigvee x \bigvee y (x \neq y \wedge O_1(x) \wedge O_1(y))$.

Since T_1 is significant relative to the class $\{T_2\}$ and T_2 is significant, then, according to D2, T_1 is significant. Obviously there is something wrong. The term T_1 previously meaningless becomes significant, though its meaning is not changed.

One may be tempted to say: "Definition (1) provides an indirect connection between O_1 and T_1 through T_2 which did not exist before. Therefore we cannot maintain correctly that the meaning of T_1 remains the same." This is a confusing argument. What is the idea of meaning involved in this reasoning? Does every new definition always effect changes in the meaning of all the terms which it involves or such changes occur only sometimes? If sometimes, when? What about the definition:

$$(7) \qquad T_2(x) \equiv O_1(x) \land O_2(x)$$

assuming that O_1 and O_2 are one-place observational terms? Does (7) effect changes of the meaning of O_1 and O_2 or not? If the meanings of O_1 and O_2 are changed perhaps they are not longer observational terms. If their meanings are the same, what then is the difference between (1) and (7) which would allow us in the case of T_1 to say that its meaning is not the same as before. Take a concrete example. Suppose a sociological theory contains a notion such as 'the forces of destiny of a nation' characterized in such a way that it is evidently empirically meaningless. Let us accept the following definition: "The creative possibilites of a nation are the sum of its economic possibilities and forces its destiny". This definition can be presented in the form (1). The notion of 'economic possibilities' may possess a very good empirical characterization. But can we actually say that the notion of 'the forces of destiny' is now more understandable? [1]

We shall pass now to the second remark about Carnap's definition. Let a one-place predicate T_1 be defined in the theory $\mathfrak{S}$ simply as follows:

$$(8) \qquad T_1(x) \equiv O_1(x),$$

where O_1 is a one-place observational term. It seems that meaningfulness of T_1 is unquestionable. However, one can easily check that if the theory $\mathfrak{S}$ is such that for every sentence α involving O_1 as the only non-logical predicate, either α or the negation of α belongs to $\mathfrak{S}$ (is one of the theorems of $\mathfrak{S}$) then T_1 is empirically meaningless. The reason is that in such an event every such sentence α_{T_1} as required by the clause (a) of D1 is either a logical consequence of the T-postulates and C-rules of $\mathfrak{S}$ or is contradictory with them, and therefore α_{T_1} fails to satisfy either the clause (c) or one of the calauses (d) and (e) of D1.

One can say that the term defined by (8) is to be considered as an observational term synonymous with O_1. But we can change our example and consider more complicated definitions. In every case we shall be able to prove that the defined term can be meaningful only if the theory $\mathfrak{S}$ is not too rich in observational theorems (theorems which do not involve theoretical terms). Namely if for every observational sentence α, either α or the negation of α belongs to $\mathfrak{S}$ then every theoretical term is empirically meaningless, not only the theoretical terms defined by means of the observational terms.

It is possible that the definition of significance given by Carnap can be improved in such a way that it will not imply the undesirable consequences pointed out in my remarks. Let me state however that the criteria of empirical meaningfulness proposed in this paper have not been obtained by changing D1 and D2. This has been done by an entirely different approach to the matter.[2]

II

As the field of our considerations we shall use a language L based on the first-order logic with or without identity and involving only the predicate symbols $O_1, ..., O_m, T_1, ..., T_n$ as non-logical constants. $O_1, ..., O_m$ will be called *O-terms*, analogously $T_1, ..., T_n$ will be called *T-terms*. If it is needed the list of *T*-terms of L can be enlarged by new predicates, for instance $T'_1, ..., T'_k$. The list of *O*-terms is assumed to be closed. Obviously every such enlargement leads to a new language which will be called an *extension of* L. Let me make clear that, unless stated explicitly to the contrary, all the notions considered will be with respect to the language L.

There will be no assumption about the nature of differences between *O*- and *T*-terms which could intervene in formal issues. From the purely logical point of view this distinction is quite arbitrary. Nevertheless it is important for philosophical reasons. *O*-terms will be considered as terms which are characterized in such a way that their empirical meaningfulness provides no doubts. We can just admit that they are observational terms. *T*-terms will represent in our considerations theoretical terms.[3] We shall assume that their meanings (interpretations) are determined exclusively by a set of postulates (axioms) of an empirical

theory $\mathfrak{S}$ and consequently dependent on meanings of O-terms. We can consider the set of postulates of $\mathfrak{S}$ as a combination of T-postulates and C-rules.

Let now me introduce some notions which will be useful later. Each relational system $\mathfrak{M} = \langle U, O_1, ..., O_m, T_1, ..., T_n \rangle$ where U is a non empty set, and $O_1, ..., O_m, T_1, ..., T_n$ are relations on U (among the elements of U), having the same number of arguments as the corresponding $O_1, ..., O_m, T_1, ..., T_n$ respectively, will be called a *possible model* of L. The set U is called the *universe of* $\mathfrak{M}$, the relations $O_1, ..., O_m$, $T_1, ..., T_n$ are called the *interpretation of* $O_1, ..., O_m, T_1, ..., T_n$ in $\mathfrak{M}$ respectively. The notion of a formula (sentence) which *holds (is true) in* $\mathfrak{M}$ will be assumed as known. The set of all sentences true in $\mathfrak{M}$ will be symbolized by $E(\mathfrak{M})$. If X is a set of sentences such that $X \subseteq E(\mathfrak{M})$, $\mathfrak{M}$ will be called a model of X. The notions of a possible model of L and a model of a set of sentences of L are different and they must not be confused.[4] $D(\mathfrak{M})$ will denote the universe of $\mathfrak{M}$. By $w_{\mathfrak{M}}(O_{i_1}, ..., O_{i_k}, T_{j_1}, ..., T_{j_l})$ we symbolize a sequence of relations which are respectively the interpretations of $O_{i_1}, ..., O_{i_k}, T_{j_1}, ..., T_{j_l}$ in the model $\mathfrak{M}$. For instance, if $\mathfrak{M} = \langle U, O_1, ..., O_m, T_1, ..., T_n \rangle$ then $D(\mathfrak{M}) = U$ and $w_{\mathfrak{M}}(O_{i_1}, ..., O_{i_k}, T_{j_1}, ..., T_{j_l}) = \langle O_{i_1}, ..., O_{i_k}, T_{j_1}, ..., T_{j_l} \rangle$.

The symbols $\mathscr{W}_{\mathfrak{S}}(O_{i_1}, ..., O_{i_k}, T_{j_1}, ..., T_{j_l})$ and $\Omega_{U, O_1, ..., O_m}^{\mathfrak{S}, O_1, ..., O_m}(T_i)$, defined below, will denote certain sets. The first one is a set of relational systems, the second a set of interpretations of T_i.

DEFINITION 1. $\langle U, O_{i_1}, ..., O_{i_k}, T_{j_1}, ..., T_{j_l} \rangle \in \mathscr{W}_{\mathfrak{S}}(O_{i_1}, ..., O_{i_k}, T_{j_1}, ..., T_{j_l})$ if and only if there is a possible model $\mathfrak{M}$ such that the following conditions hold:

    i. $U = D(\mathfrak{M})$,

    ii. $\langle O_{i_1}, ..., O_{i_k}, T_{j_1}, ..., T_{j_l} \rangle = \mathscr{W}_{\mathfrak{M}}(O_{i_1}, ..., O_{i_k}, T_{j_1}, ..., T_{j_l})$,

    iii. $\mathfrak{S} \subseteq E(\mathfrak{M})$.

If a relational system $\langle U, O_{i_1}, ..., O_{i_k}, T_{j_1}, ..., T_{j_l} \rangle \in \mathscr{W}_{\mathfrak{S}}(O_{i_1}, ..., O_{i_k}, T_{j_1} ..., T_{j_l})$ we shall often say that $O_{i_1}, ..., O_{i_k}, T_{j_1}, ..., T_{j_l}$ give a $\mathfrak{S}$-*compatible* or simply compatible interpretation of $O_{i_1}, ..., O_{i_k}, T_{j_1}, ..., T_{j_l}$ in U.

DEFINITION 2. $T \in \Omega_{U, O_1, ..., O_m}^{\mathfrak{S}, O_1, ..., O_m}(T_i)$ if and only if $\langle U, O_1, ..., O_m, T \rangle \in \mathscr{W}_{\sigma}(O_1, ..., O_m, T_i)$.

Given a universe U and given relations $O_1, ..., O_m$ on U, the set $\Omega_{U, O_1, ..., O_m}^{\mathfrak{S}, O_1, ..., O_m}(T_i)$ is the set of all interpretations T of T_i such that O_1, ..., O_m, T give a $\mathfrak{S}$-compatible interpretation of $O_1, ..., O_m, T_i$ in U. For instance, let T_i be characterized only by means of the two so-called reduction sentences (see [2]), say:

(1) $\bigwedge x(O_1(x) \to T_i(x))$

(2) $\bigwedge x(O_2(x) \to \sim T_i(x))$

O_1, O_2, T_i being one-place predicates. If O_1, O_2 are the sets corresponding to predicates O_1 and O_2, then $\Omega_{U, O_1, ..., O_m}^{\mathfrak{S}, O_1, ..., O_m}(T_i)$ is the set of all interpretations T of T_i such that $O_1 \subseteq T \subseteq \bar{O}_2$ where $\bar{O}_2 = U - O_2$. $\Omega_{U, O_1, ..., O_m}^{\mathfrak{S}, O_1, ..., O_m}(T_i)$ could be called the set of all possible interpretations of T_i with respect to $\mathfrak{S}$ and interpretations $O_1, ..., O_m$ of O-terms. I have to mention here that the notion of a possible interpretation of a term was introduced to the philosophy of science by Kokoszyńska (see e.g. [11]). She used it to investigate the notions of the analytical and the synthetical. It seems that the concept of a possible interpretation can be regarded as one of the fundamental concepts of the methodology of empirical sciences.

The following notions will play an important role in our considerations: the notion of non-creative extension of a theory $\mathfrak{S}$, the notion of O-extension of a theory $\mathfrak{S}$, and the notion of O-completeness of a theory $\mathfrak{S}$.

DEFINITION 3. A theory $\mathfrak{S}'$ formulated in L or in an extension of L is a *non-creative extension of* $\mathfrak{S}$ if and only if:

i. $\mathfrak{S} \subseteq \mathfrak{S}'$,

ii. $\mathscr{W}_{\mathfrak{S}}(O_1, ..., O_m, T_1, ..., T_n) = \mathscr{W}_{\mathfrak{S}'}(O_1, ..., O_m, T_1, ..., T_n)$.

Perhaps it should be recalled that the theory $\mathfrak{S}$ is entirely formulated in L.

Let $Cn(X)$ denote the set of all logical consequences of a set of sentences X. By an O-sentence we shall understand a sentence which does not contain any non logical constants different from $O_1, ..., O_m$. A T-sentence, an O, T_i-sentence, a $T_{i_1}, ..., T_{i_k}$-sentence will be understood in a similar way.

DEFINITION 4. A theory $\mathfrak{S}'$ is an *O-extension* of a theory $\mathfrak{S}$ if and only if there is a set X of *O*-sentences such that $\mathfrak{S}'=\mathrm{Cn}(\mathfrak{S}\cup X)$.

DEFINITION 5. A theory $\mathfrak{S}$ is *O-complete* if and only if for any *O*-sentence α, α or its negation belongs to $\mathfrak{S}$.

III

The intuitive concept of empirical meaningfulness is ambiguous and elusive in such a degree that many philosophers do not believe that it is possible to find its satisfactory explication at all. The very sceptical position on this matter has been taken by Hempel ([6], cf. also [13]).[5] Actually the lack of an enough clear idea of empirical meaningfulness at least effects that the task of defining of this concept has been painful indeed. In this situation, it is desirable to approach the problem by laying down certain basic requirements to be satisfied by a correctly defined concept of an empirically meaningful term. These requirements, I will call them the conditions of adequacy, should express ideas as simple and convincing as possible.

The first condition of adequacy involves the notion of *O-isolated* term.

DEFINITION 6. T_i is an *O-isolated term* of a theory $\mathfrak{S}$ if and only if there are *T*-terms $T_{i_1}, \ldots, T_{i_k}$ such that:

i. T_i is one of $T_{i_1}, \ldots, T_{i_k}$,

ii.   for any $\mathbf{U}, \mathbf{O}_1, \ldots, \mathbf{O}_m, \mathbf{T}_1, \ldots, \mathbf{T}_n$ if

         a. $\langle \mathbf{U}, \mathbf{T}_1, \ldots, \mathbf{T}_k \rangle \in \mathscr{W}_{\mathfrak{S}}(T_{i_1}, \ldots, T_{ik})$ and

         b. $\langle \mathbf{U}, \mathbf{O}_1, \ldots, \mathbf{O}_m, \mathbf{T}_{k+1}, \ldots, \mathbf{T}_n \rangle \in \mathscr{W}_{\mathfrak{S}}(O_1, \ldots, O_m,$
         $T_{i_{k+1}}, \ldots, T_{i_n})$,

         where $T_{i_{k+1}}, \ldots, T_{i_n}$ are the remaining *T*-terms, then

         c. $\langle \mathbf{U}, \mathbf{O}_1, \ldots, \mathbf{O}_m, \mathbf{T}_1, \ldots, \mathbf{T}_n \rangle$
         $\in \mathscr{W}_{\mathfrak{S}}(O_1, \ldots, O_m, T_{i_1}, \ldots, T_{i_n})$.

Roughly speaking clause ii. reads: if the terms $T_{i_1}, \ldots, T_{i_k}$ are interpreted as $\mathbf{T}_1, \ldots, \mathbf{T}_k$ compatibly with $\mathfrak{S}$ then they can retain these meanings, no matter how one interprets the other terms.

·If $\mathfrak{S}$ is finitely first order axiomatizable theory, it can be proved by means of a theorem due to Craig ([4], Theorem 5) that Definition 6 is equivalent to the following

DEFINITION 6*. T_i is an *O-isolated term* of a theory $\mathfrak{S}$ if and only if there are T-terms $T_{i_1}, \ldots, T_{i_k}$ such that:

i. T_i is one of $T_{i_1}, \ldots, T_{i_k}$,

ii.   there is a set X_1 of $T_{i_1}, \ldots, T_{i_k}$-sentences, and a set X_2 of $O, T_{i_{k+1}}, \ldots, T_{i_n}$-sentences where $T_{i_{k+1}}, \ldots, T_{i_n}$ are the remaining T-terms such that $\mathfrak{S} = \mathrm{Cn}\,(X_1 \cup X_2)$.

Because there are no meaning connections between O-isolated terms and O-terms, it is entirely evident that the O-isolated terms cannot be regarded as empirically meaningful (cf. also [3]). Let us introduce the expression '*O-meaningful*' as a synonym of '*empirically meaningful theoretical term T-term*', and the expression '*O-meaningless*' as a synonym of '*empirically meaningless theoretical term*'. Obviously, T_i is O-meaningless if and only if T_i is not O-meaningful. The first condition of adequacy is the following:

CA1.   No O-isolated term of a theory $\mathfrak{S}$ is O-meaningful with respect to $\mathfrak{S}$.

Before passing to the second condition of adequacy we shall introduce a new notion, namely the notion of *empirical indistinguishability* of two T-terms (*O-indistinguishability*).

DEFINITION 7. Two T-terms T_i, T_j are *O-indistinguishable* (*empirically indistinguishable*) with respect to $\mathfrak{S}$, in symbols $T_i \simeq_{\mathfrak{S}} T_j$, if and only if for any $U, O_1, \ldots, O_m$:

$$\Omega^{\mathfrak{S}, O_1, \ldots, O_m}_{U, O_1, \ldots, O_m}(T_i) = \Omega^{\mathfrak{S}, O_1, \ldots, O_m}_{U, O_1, \ldots, O_m}(T_j).$$

One can easily see that:

3.1. $T_i \simeq_{\mathfrak{S}} T_i$.

3.2.   If $T_i \simeq_{\mathfrak{S}} T_j$ then $T_j \simeq_{\mathfrak{S}} T_i$.

3.3.   If $T_i \simeq_{\mathfrak{S}} T_j$ and $T_j \simeq_{\mathfrak{S}} T_k$ then $T_i \simeq_{\mathfrak{S}} T_k$.

CA2. If $T_i \simeq_{\mathfrak{S}} T_j$ and T_i is O-meaningful with respect to a theory $\mathfrak{S}$ then T_j is O-meaningful with respect to $\mathfrak{S}$.

CA2*. If $T_i \simeq_{\mathfrak{S}} T_j$ and T_i is O-meaningless with respect to a theory $\mathfrak{S}$ then T_j is O-meaningless with respect to $\mathfrak{S}$.

CA2 and **CA2*** are easily seen to be equivalent in virtue of 3.2.

Having in mind that the meanings of T-terms are determined only by the postulates of $\mathfrak{S}$, we easily see that $T_i \simeq_{\mathfrak{S}} T_j$ implies that there is no possibility to distinguish the relation symbolized by T_i from that symbolized by T_j. If, for instance, T_i and T_j are one-place predicates and a is a name of an arbitrary individual object then (still assuming that $T_i \simeq_{\mathfrak{S}} T_j$) each empirical procedure which can be used to settle whether $T_i(a)$ is valid or not, can be used to settle whether $T_j(a)$ is valid, and conversely. Moreover such a procedure can provide us only with the same result positive or negative both in the case of $T_i(a)$ and in the case of $T_j(a)$. If there are some differences between the meaning of T_i and T_j, they must be of a purely verbal, neutral with respect to the empirical evidence character.[6] We can say that all O-indistinguishable terms possess precisely the same empirical status. This seems to be an enough forceful justification of **CA2** and **CA2***.

I shall formulate now the next two conditions of adequacy.

CA3. If T_i is O-meaningful with respect to a theory $\mathfrak{S}$, and $\mathfrak{S}'$ is a non-creative extension of $\mathfrak{S}$ then T_i is O-meaningful with respect to $\mathfrak{S}'$.

CA4. If T_i is O-meaningless with respect to a theory $\mathfrak{S}$, and $\mathfrak{S}'$ is a non-creative extension of $\mathfrak{S}$ then T_i is O-meaningless with respect to $\mathfrak{S}'$.

Acceptance of **CA3** and **CA4** seems to be a natural consequence of acceptance **CA2** and **CA2***. Let $\mathfrak{S}$ be a theory, and let $\mathfrak{S}'$ be a non-creative extension of $\mathfrak{S}$. Assume that the non-logical constants of $\mathfrak{S}'$ are O_1, ..., O_m, T_1, ..., T_n, T_{n+1}, ..., T_p (the non-logical constants of $\mathfrak{S}$ being O_1, ..., O_m, T_1, ..., T_n). Let us agree to replace the terms T_1, ..., T_p of $\mathfrak{S}'$ by new symbols T_1', ..., T_p' respectively. This replacement does not essentially alter the theory. Consider now the theory $\mathfrak{S}''$ which is the

union of the theories $\mathfrak{S}$ and $\mathfrak{S}'$, the last with its T-terms renamed: $\mathfrak{S}'' = \mathrm{Cn}\,(\mathfrak{S} \cup \mathfrak{S}')$. Note that in $\mathfrak{S}''$ the meanings of $T_1, \ldots, T_n$ are determined by exactly the same postulates as in $\mathfrak{S}$, and the meanings of $T_1', \ldots, T_p'$ by exactly the same postulates as in $\mathfrak{S}'$. We can admit there-fore that all these terms have retained their meanings. It can be proved that for each i $(1 \leqslant i \leqslant n)$: $T_i \simeq_{\mathfrak{S}''} T_i'$. Hence, according to **CA2** and **CA2***, T_i and T_i' have to be either both O-meaningful or both O-meaningless with respect to $\mathfrak{S}''$. Because these terms carry over their meaning un-changed from $\mathfrak{S}$ and $\mathfrak{S}'$ into $\mathfrak{S}''$, we may say something more: namely, that if T_i is O-meaningful with respect to $\mathfrak{S}$, then T_i' is also O-meaningful with respect to $\mathfrak{S}'$, and similarly if T_i is O-meaningless with respect to $\mathfrak{S}$ then T_i' is also O-meaningless with respect to $\mathfrak{S}'$. This is exactly what is required by the conditions **CA3** and **CA4**. Obviously this argument is not a proof of **CA3** and **CA4**. I hope however, that it makes clear why I was inclined to accept these two conditions of adequacy after accepting **CA2** and **CA2***.

Let me remark that there is an obvious connection between **CA4** and the first of my remarks on Carnap's definition of significance. The definitional extension of the theory which I have discussed at the begin-ning of the paper was a non-creative extension. We can now say that Carnap's definition of significance fails with respect to **CA4**.

CA5 will be the last condition of adequacy I shall formulate.

CA5.   If T_i is O-meaningful with respect to $\mathfrak{S}$, then there is at least one consistent O-complete O-extension of $\mathfrak{S}$ such that T_i is O-meaningful with respect to $\mathfrak{S}$.

The short justification **CA5**, that I want to offer, is the following. Ac-cording to our assumptions on O-terms and T-terms, any empirical evidence can be fully expressed in O-sentences. Therefore if there are no O-complete O-extension $\mathfrak{S}'$ of $\mathfrak{S}$ such that T_i is O-meaningful with respect to $\mathfrak{S}'$ we can say in advance that any possible enrichment of our knowledge will show finally the meaninglessness of T_i. If this is the case, it would be unnatural to consider T_i as meaningful even before reaching an enlargement of $\mathfrak{S}$ which forces to admit that T_i is meaningless. Slightly modifying my second criticism of Carnap's proposal, one can show that Carnap's definition fails with respect to **CA5** too.

IV

The idea of the definition of empirical meaninfulness which will be discussed now is similar to that on which the definition of confirmability of predicates due to Carnap was based [2]. However we shall deal with the matter in a semantical way whereas Carnap's analysis was purely syntactical. Let us start with some introductory considerations that are to show the way in which the desirable definition of empirical meaningfulness can be reached.

Assume that in a theory $\mathfrak{S}$ the only postulates involving a two-place predicate T_i are:

(1) $\qquad \bigwedge x \bigwedge y \left(A_1(x, y) \to T_i(x, y) \right),$

(2) $\qquad \bigwedge x \bigwedge y \left(A_2(x, y) \to \sim T_i(x, y) \right),$

where $A_1(x, y)$, $A_2(x, y)$ are formulas of L having x and y as the only free variables, and not involving any T-terms. Let me state that (1) and (2) have been deliberately chosen as having the form of reduction sentences. In this way I have wanted to bring into relief the connection between Carnap's approach and the one presented now. Having in mind the idea of empirically meaningful term as a term whose meaning depends at least in some degree on the meanings of O-terms, we may be inclined to lay down the following criterion of O-meaningfulness of the term T_i considered.

(∗) $\qquad$ T_i is O-meaningful if and only if at least one of the formulas $A_1(x, y)$, $A_2(x, y)$ is not equivalent to a formula not involving O-terms.

Speaking about equivalence we have to distinguish three possibilities:

a. *Logical equivalence*. (Two formulas A, B of L are *logically equivalent* if and only if the formula $\ulcorner A \equiv B \urcorner$ holds in all models of L).

b. *$\mathfrak{S}$-equivalence*. (Two formulas A, B are *$\mathfrak{S}$-equivalent* if and only if the formula $\ulcorner A \equiv B \urcorner$ holds in all models of $\mathfrak{S}$).[7]

c. *Empirical equivalence*.

This last notion of equivalence will engage our attention for a while. Among all possible definitions of empirical equivalence there is one we shall mention briefly. In the case of the language of an empirical theory, say the language L just considered, we may distinguish among all pos-

sible models of L a particular model $\mathfrak{M}$ such that the language L has been specially constructed to describe this model, i.e. to describe the objects of the universe of $\mathfrak{M}$, and the relations occurring in $\mathfrak{M}$. This distinguished model can be called the *proper model of* L (cf. [10]). Formulas A, B are empirically equivalent, in the sense I have intended to present, if and only if the formula $\ulcorner A \equiv B \urcorner$ holds in the proper model $\mathfrak{M}$ of L.

There may be some reasons to admit that the equivalence mentioned in (∗) is to be understood in the sense of empirical equivalence. However philosophers have tried to define empirical meaningfulness in such a way that, given a theory $\mathfrak{S}$ and a term T_i, one can settle by using purely logical means whether T_i is meaningful or not. To continue this tendency, we shall not accept the equivalence in (∗) in the sense of empirical equivalence.

Let us now note that T_i seems to fail to be meaningful when $A_1(x, y)$ and $A_2(x, y)$ can be proved to be equivalent to some formulas $L_1(x, y)$, $L_2(x, y)$ involving only logical terms, both in the case when the pertinent proofs can be given only with the help of the postulates of $\mathfrak{S}$, and can be given without these postulates. In either of these cases we can replace the postulates (1), (2) by the following postulates

$$(3) \qquad \bigwedge x \bigwedge y (A_1(x, y) \equiv L_1(x, y)),$$

$$(4) \qquad \bigwedge x \bigwedge y (A_2(x, y) \equiv L_2(x, y)),$$

$$(1') \qquad \bigwedge x \bigwedge y (L_1(x, y) \rightarrow T_i(x, y)),$$

$$(2') \qquad \bigwedge x \bigwedge y (L_2(x, y) \rightarrow \sim T_i(x, y))$$

without affecting the theory $\mathfrak{S}$. Now the only postulates for T_i (1') and (2') in the new axiom system do not involve any non logical terms. In such an event it does not seem reasonable to regard T_i as empirically meaningful, no matter whether the postulates of $\mathfrak{S}$ are necessary or not to prove (3) and (4). By this remark I want to justify the decision to accept the equivalence in (∗) in the sense of $\mathfrak{S}$-equivalence only.

The criterion (∗), as formulated for the particular theory we are discussing now, does not admit of an immediate generalization to an arbitrary set of postulates for (T_i). Because of this, we shall try to formulate it in a different way. Let us continue to consider the theory $\mathfrak{S}$

in which the term T_i was introduced by means of postulates (1) and (2). Note that given a universe $\mathbf{U}$ and relations $\mathbf{O}_1, \ldots, \mathbf{O}_m$ on $\mathbf{U}$ corresponding to predicates $O_1, \ldots, O_m$ there can be, in general, more than one interpretation $\mathbf{T}$ of T_i such that $\mathbf{O}_1, \ldots, \mathbf{O}_m, \mathbf{T}$ give a $\mathfrak{S}$-compatible interpretation of $O_1, \ldots, O_m, T_i$. As a result of that T_i may turn out to be 'vague': there may be some ordered pairs $\langle \mathbf{u}_1, \mathbf{u}_2 \rangle$ of elements of $\mathbf{U}$ such that the question whether T_i is applicable to $\langle \mathbf{u}_1, \mathbf{u}_2 \rangle$ or not cannot be answered definitely.

Let us call (cf. [12]).

a. The set of ordered pairs $\langle u_1, u_2 \rangle$ to which T_i is definitely applicable — the *positive extension of T_i* in $\mathbf{U}$ with respect to $\mathfrak{S}$ and the interpretations $\mathbf{O}_1, \ldots, \mathbf{O}_m$ of $O_1, \ldots, O_m$ respectively, in symbols: $\mathrm{PE}_{\mathbf{U},\, \mathbf{O}_1,\, \ldots,\, \mathbf{O}_m}^{\mathfrak{S},\, O_1,\, \ldots,\, O_m}(T_i)$.

b. The set of ordered pairs $\langle \mathbf{u}_1, \mathbf{u}_2 \rangle$ to which T_i is definitely non-applicable — the *negative extension* of T_i in $\mathbf{U}$ with respect to $\mathfrak{S}$ and the interpretations $\mathbf{O}_1, \ldots, \mathbf{O}_m$ of $O_1, \ldots, O_m$ respectively, in symbols: $\mathrm{NE}_{\mathbf{U},\, \mathbf{O}_1,\, \ldots,\, \mathbf{O}_m}^{\mathfrak{S},\, O_1,\, \ldots,\, O_m}(T_i)$.

We shall not need the special name for the set of pairs which do not belong either to the positive or to the negative extension of T_i. This set could be called the *margin of vagueness* of T_i in $\mathbf{U}$ with respect to $\mathfrak{S}$ and the given interpretations of O-terms.

In the case of the particular theory $\mathfrak{S}$, we have been discussing, $\mathrm{PE}_{\mathbf{U},\, \mathbf{O}_1,\, \ldots,\, \mathbf{O}_m}^{\mathfrak{S},\, O_1,\, \ldots,\, O_m}(T_i)$ can be simply taken as the set of all ordered pairs of elements of $\mathbf{U}$ which satisfy $A_1(x, y)$ under the given interpretation $\mathbf{O}_1, \ldots, \mathbf{O}_m$ of $O_1, \ldots, O_m$. Similarly, $\mathrm{NE}_{\mathbf{U},\, \mathbf{O}_1,\, \ldots,\, \mathbf{O}_m}^{\mathfrak{S},\, O_1,\, \ldots,\, O_m}(T_i)$ can be taken as the set of all ordered pairs of elements of $\mathbf{U}$ which satisfy $A_2(x, y)$. The general definition of both $\mathrm{PE}_{\mathbf{U},\, \mathbf{O}_1,\, \ldots,\, \mathbf{O}_m}^{\mathfrak{S},\, O_1,\, \ldots,\, O_m}(T_i)$ and $\mathrm{NE}_{\mathbf{U},\, \mathbf{O}_1,\, \ldots,\, \mathbf{O}_m}^{\mathfrak{S},\, O_1,\, \ldots,\, O_m}(T_i)$ will be given a little later.

The following criterion of O-meaningfulness for the term T_i express an idea very close to the on which the criterion was based.

(∗∗) T_i is O-meaningful if and only if there is a non empty set $\mathbf{U}$ and relations $\mathbf{O}_1, \ldots, \mathbf{O}_m$ on $\mathbf{U}$ such that either $\mathrm{PE}_{\mathbf{U},\, \mathbf{O}_1,\, \ldots,\, \mathbf{O}_m}^{\mathfrak{S},\, O_1,\, \ldots,\, O_m}(T_i)$ or $\mathrm{NE}_{\mathbf{U},\, \mathbf{O}_1,\, \ldots,\, \mathbf{O}_m}^{\mathfrak{S},\, O_1,\, \ldots,\, O_m}(T_i)$ is not a logical two-place relation on $\mathbf{U}$. (The logical two-place relations on $\mathbf{U}$ are the following: the identity on $\mathbf{U}$, the diversity on $\mathbf{U}$, the empty two-place relation on $\mathbf{U}$, the full two-place relation on $\mathbf{U}$.)

It can be proved that if T_i is O-meaningful in the sense (∗∗) then it is

O-meaningful in the sense (∗) also; this proof can be easily given with the help of the Theorem 4.1 formulated later.

Note now that if T_i is O-meaningless in the sense (∗∗) we can prove that for any cardinal number $\mathfrak{m}$ there are formulas $L_1(x, y)$, $L_2(x, y)$ not involving any non logical constants such that for every model $\mathfrak{M}$ of $\mathfrak{S}$ if $\overline{\overline{D(\mathfrak{M})}} = \mathfrak{m}$ then the following formulas:

$$A_1(x, y) = L_1(x, y),$$

$$A_2(x, y) = L_2(x, y)$$

hold in $\mathfrak{S}$. Although this does not prove that O-meaningfulness in the sense (∗) implies O-meaningfulness in the sense (∗∗), however reveals some essential intuitive connections between (∗) and (∗∗).

The criterion (∗∗) suggests a way towards a general definition of O-meaningfulness. We shall formulate it after some preparations.

The following definition will be formulated as usually with respect to an arbitrary theory $\mathfrak{S}$.

DEFINITION 8. $PE^{\mathfrak{S}, O_1, ..., O_m}_{U, O_1, ..., O_m}(T_i) = \bigcap_{\mathbf{T}} (\mathbf{T} \in \Omega^{\mathfrak{S}, O_1, ..., O_m}_{U, O_1, ..., O_m}(T_i)$.

DEFINITION 9. $NE^{\mathfrak{S}, O_1, ..., O_m}_{U, O_1, ..., O_m}(T_i) = \bigcap_{\mathbf{T}} (\overline{\mathbf{T}} \in \Omega^{\mathfrak{S}, O_1, ..., O_m}_{U, O_1, ..., O_m}(T_i)$.

($\overline{\mathbf{T}}$ denotes the complementary relation with respect to $\mathbf{T}$.)

By turn we shall define the notion of logical relation. Let $\mathbf{U}$ be a non-empty set and let f be a one-one mapping of the set $\mathbf{U}$ onto itself. Let $\mathbf{U}$ be a k-place ($k \leqslant 1$) relation on $\mathbf{U}$ (if $k = 1$ $\mathbf{T}$ is simply a subset of $\mathbf{U}$). By $\mathbf{T}^f$ we shall denote the relation on $\mathbf{U}$ defined by the following condition:

$$\langle \mathbf{u}_1, ..., \mathbf{u}_k \rangle \in \mathbf{T} \text{ if and only if } \langle f(\mathbf{u}_1), ..., f(\mathbf{u}_k) \rangle \in \mathbf{T}^f.$$

DEFINITION 10. $\mathbf{T}$ is a *logical relation on a set* $\mathbf{U}$ if and only if whenever f is a one-one mapping of $\mathbf{U}$ onto itself, then $\mathbf{T} = \mathbf{T}^f$.[8]

Now we are in the position to give the announced definition of O-meaningfulness.

DEFINITION A. A term T_i is *O-meaningful with respect to a theory $\mathfrak{S}$* if and only if there is a set $\mathbf{U}$ and relations $\mathbf{O}_1, ..., \mathbf{O}_m$ on $\mathbf{U}$ such that either $PE^{\mathfrak{S}, O_1, ..., O_m}_{U, O_1, ..., O_m}(T_i)$ or $NE^{\mathfrak{S}, O_1, ..., O_m}_{U, O_1, ..., O_m}(T_i)$ (at least one of them) is not a logical relation on $\mathbf{U}$.

Definition A meets suggestions of many authors who have called attention to connections between meaningfulness and vagueness of theoretical terms (see e.g. Przełęcki [14]).

We shall prove that Definition A does indeed conform to the conditions **CA1–CA5** of adequacy. At first we are going to prove the following lemma.

LEMMA 1. If for any set **U** and relations $O_1, ..., O_m$ on **U**, and for any one-one mapping f of **U** onto itself:

$$\Omega_{U, O_1, ..., O_m}^{\mathfrak{S}, O_1, ..., O_m}(T_i) = \Omega_{U, O_1, ..., O_m}^{\mathfrak{S}, O_1, ..., O_m}(T_i),$$

then both

a. $\quad PE_{U, O_1, ..., O_m}^{\mathfrak{S}, O_1, ..., O_m}(T_i)$ is a logical relation,

and

b. $\quad NE_{U, O_1, ..., O_m}^{\mathfrak{S}, O_1, ..., O_m}(T_i)$ is a logical relation.

Proof. We shall start by showing that, under the assumption of the antecedent of Lemma 1, for any one-one mapping f of **U** onto itself:

(α) $\quad$ if $T \in \Omega_{U, O_1, ..., O_m}^{\mathfrak{S}, O_1, ..., O_m}(T_i)$ then $T^f \in \Omega_{U, O_1, ..., O_m}^{\mathfrak{S}, O, ..., O_m}(T_i)$.

Assume that

(1) $\quad T \in \Omega_{U, O_1, ..., O_m}^{\mathfrak{S}, O_1, ..., O_m}(T_i)$.

Hence

(2) $\quad \langle U, O_1, ..., O_m, T \rangle \in \mathscr{W}_{\mathfrak{S}}(O_1, ..., O_m, T_i)$.

It is well known that all models isomorphic with a given model $\mathfrak{M}$ of $\mathfrak{S}$ are models of $\mathfrak{S}$. Therefore if f is one-one mapping of **U** onto itself then, by (2), we obtain

(3) $\quad \langle U, O_1^f, ..., O_m^f, T^f \rangle \in \mathscr{W}_{\mathfrak{S}}(O_1, ..., O_m, T_i)$.

By turn the formula (3) yields

(4) $\quad T^f \in \Omega_{U, O_1^f, ..., O_m^f}^{\mathfrak{S}, O_1, ..., O_m}(T_i)$.

From the antecedent of the lemma if follows by (4) that

(5) $\quad T^f \in \Omega_{U, O_1, ..., O_m}^{\mathfrak{S}, O_1, ..., O_m}(T_i)$

which proves (α).

Let us now prove that if the antecedent of Lemma 1 holds then a. and b. are valid. We shall only prove a., the proof of b. being similar.

Suppose T_i is a k-place predicate. Let

(1) $\qquad \langle u_1, \ldots, u_k \rangle \in PE^{\mathfrak{S}, O_1, \ldots, O_m}_{U, O_1, \ldots, O_m}(T_i);$

in order to show that $PE^{\mathfrak{S}, O_1, \ldots, O_m}_{U, O_1, \ldots, O_m}(T_i)$ is a logical relation we have to show that for every one-one mapping f of U onto itself

(2) $\qquad \langle f(u_1), \ldots, f(u_k) \rangle \in PE^{\mathfrak{S}, O_1, \ldots, O_m}_{U, O_1, \ldots, O_m}(T_i).$

Let T_1 be a relation on U such that

(3) $\qquad T_1 \in \Omega^{\mathfrak{S}, O_1, \ldots, O_m}_{U, O_1, \ldots, O_m}(T_i).$

Then by (1)

(4) $\qquad \langle u_1, \ldots, u_k \rangle \in T_1.$

Using (α) we obtain from (3)

(5) $\qquad T^f_1 \in \Omega^{\mathfrak{S}, O_1, \ldots, O_m}_{U, O_1, \ldots, O_m}(T_i).$

From (4) it follows that

(6) $\qquad \langle f(u_1), \ldots, f(u_k) \rangle \in T^f_1.$

Assume now that (2) is not valid. Because of (5) and (6) there is a relation T_2 such that

(7) $\qquad T_2 \in \Omega^{\mathfrak{S}, O_1, \ldots, O_m}_{U, O_1, \ldots, O_m}(T_i)$

and

(8) $\qquad \langle f(u_1), \ldots, f(u_k) \rangle \bar{\in} T_2.$

Take now the inverse f^{-1} of f. Since f^{-1} is one-one mapping of U onto itself, by (α) we obtain from (7):

(9) $\qquad T^{f^{-1}}_2 \in \Omega^{\mathfrak{S}, O_1, \ldots, O_m}_{U, O_1, \ldots, O_m}(T_i).$

From (8) it follows

(10) $\qquad \langle f^{-1}(f(u_1)), \ldots, f^{-1}(f(u_k)) \rangle \bar{\in} T^{f^{-1}}_2$

and hence

(11) $\qquad \langle u_1, \ldots, u_k \rangle \bar{\in} T^{f^{-1}}_2.$

From (9) and (11) we should have

$$(12) \qquad \langle \mathbf{u}_1, \ldots, \mathbf{u}_k \rangle \bar{\in} \mathrm{PE}_{\mathbf{U},O_1;\ldots;O_m}^{\mathfrak{S};O_1;\ldots;O_m}(T_i)$$

contradicting the assumption (1). This proves that (2) is in fact valid, and our proof of Lemma 1 is completed.

We shall examine Definition A with regard to the conditions **CA1**–**CA5**.

4.1. Definition A implies **CA1**.

Proof. We shall show that: (β) if T_i is O-isolated term then for every set **U**, and relations $O_1, \ldots, O_m$ on **U**, and for every one-one mapping f of **U** onto itself

$$\mathbf{T}_1 \in \Omega_{\mathbf{U},O_1;\ldots;O_m}^{\mathfrak{S};O_1;\ldots;O_m}(T_i) \text{ if and only if } \mathbf{T}_1 \in \Omega_{\mathbf{U},O_1^f;\ldots;O_m^f}^{\mathfrak{S};O_1;\ldots;O_m}(T_i).$$

4.1 is an immediate consequence of (β) and Lemma 1.

In order to prove (β) suppose that the antecedent of (β) is valid, and moreover

$$(1) \qquad \mathbf{T}_1 \in \Omega_{\mathbf{U},O_1;\ldots;O_m}^{\mathfrak{S};O_1;\ldots;O_m}(T_i).$$

Let $T_{i_1}, \ldots, T_{i_k}$ be terms required by Definition 6, and let T_i be identical with T_{i_1}. It follows from (1) that there exist $\mathbf{T}_2, \ldots, \mathbf{T}_n$ such that

$$(2) \qquad \langle \mathbf{U}, O_1, \ldots, O_m, \mathbf{T}_1, \ldots, \mathbf{T}_k, \mathbf{T}_{k+1}, \ldots, \mathbf{T}_n \rangle$$
$$\in \mathscr{W}_{\mathfrak{S}}(O_1, \ldots, O_m, T_{i_1}, \ldots, T_{i_k}, T_{i_{k+1}}, \ldots, T_{i_n}),$$

$T_{i_{k+1}}, \ldots, T_{i_n}$ being the T-terms different from $T_{i_1}, \ldots, T_{i_k}$. Hence in particular

$$(3) \qquad \langle \mathbf{U}, \mathbf{T}_1, \ldots, \mathbf{T}_k \rangle \in \mathscr{W}_{\mathfrak{S}}(T_{i_1}, \ldots, T_{i_k}).$$

Let f be an arbitrary one-one mapping **U** onto itself. Then from (2) we shall also have

$$(4) \qquad \langle \mathbf{U}, O_1^f, \ldots, O_m^f, \mathbf{T}_1^f, \ldots, \mathbf{T}_n^f \rangle \in \mathscr{W}_{\mathfrak{S}}(O_1, \ldots, O_m, T_{i_1}, \ldots, T_{i_n})$$

and therefore

$$(5) \qquad \langle \mathbf{U}, O_1^f, \ldots, O_m^f, \mathbf{T}_{k+1}^f, \ldots, \mathbf{T}_n^f \rangle$$
$$\in \mathscr{W}_{\mathfrak{S}}(O_1, \ldots, O_m, T_{i_{k+1}}, \ldots, T_{i_n}).$$

By (3), (5) and Definition 6 we obtain

(6) $\qquad \langle \mathbf{U}, \mathbf{O}_1^f, \dots, \mathbf{O}_m^f, \mathbf{T}_1, \dots, \mathbf{T}_k, \mathbf{T}_{k+1}^f, \dots, \mathbf{T}_n^f \rangle$

$\qquad \in \mathscr{W}_{\mathfrak{S}}(O_1, \dots, O_m, T_{i_1}, \dots, T_{i_k}, T_{i_{k+1}}, \dots, T_{i_n})$

and hence

(7) $\qquad \mathbf{T}_1 \in \Omega_{\mathbf{U}, \mathbf{O}_1^f, \dots, \mathbf{O}_m^f}^{\mathfrak{S}, O_1, \dots, O_m}(T_i).$

In this way we have proved that

(8) $\qquad$ if $\mathbf{T}_1 \in \Omega_{\mathbf{U}, \mathbf{O}_1^f, \dots, \mathbf{O}_m^f}^{\mathfrak{S}, O_1, \dots, O_m}(T_i)$ then $\mathbf{T}_1 \in \Omega_{\mathbf{U}, \mathbf{O}_1^f, \dots, \mathbf{O}_m^f}^{\mathfrak{S}, O_1, \dots, O_m}(T_i).$

Note now that (8) is valid for every mapping f and every relations $\mathbf{O}_1, \dots$
$\dots, \mathbf{O}_m$ on $\mathbf{U}$. By substituting in formula (8) $\mathbf{O}_i^f$ in the place of $\mathbf{O}_i$ and
f^{-1} in the place of f we obtain

(9) $\qquad$ if $\mathbf{T}_1 \in \Omega_{\mathbf{U}, \mathbf{O}_1, \dots, \mathbf{O}_m}^{\mathfrak{S}, O_1, \dots, O_m}(T_i)$, then $\mathbf{T}_1 \in \Omega_{\mathbf{U}, (\mathbf{O}_1^f)^{f^{-1}}, \dots, (\mathbf{O}_m^f)^{f^{-1}}}^{\mathfrak{S}, O_1, \dots, O_m}(T_i)$

and consequently

(10) $\qquad$ if $\mathbf{T}_1 \in \Omega_{\mathbf{U}, \mathbf{O}_1^f, \dots, \mathbf{O}_m^f}^{\mathfrak{S}, O_1, \dots, O_m}(T_i)$, then $\mathbf{T}_1 \in \Omega_{\mathbf{U}, \mathbf{O}_1, \dots, \mathbf{O}_m}^{\mathfrak{S}, O_1, \dots, O_m}(T_i).$

By (8) and (10), (β) is established and with it Theorem 4.1.

4.2. Definition A implies CA2.

Proof. If $T_i \simeq_{\mathfrak{S}} T_j$ then according to Definition 7

(1) $\qquad \Omega_{\mathbf{U}, \mathbf{O}_1, \dots, \mathbf{O}_m}^{\mathfrak{S}, O_1, \dots, O_m}(T_i) = \Omega_{\mathbf{U}, \mathbf{O}_1, \dots, \mathbf{O}_m}^{\mathfrak{S}, O_1, \dots, O_m}(T_j).$

Hence by Definition 7, Definition 8 and Definition 9

(2) $\qquad \mathrm{PE}_{\mathbf{U}, \mathbf{O}_1, \dots, \mathbf{O}_m}^{\mathfrak{S}, O_1, \dots, O_m}(T_i) = \mathrm{PE}_{\mathbf{U}, \mathbf{O}_1, \dots, \mathbf{O}_m}^{\mathfrak{S}, O_1, \dots, O_m}(T_j)$

and

(3) $\qquad \mathrm{NE}_{\mathbf{U}, \mathbf{O}_1, \dots, \mathbf{O}_m}^{\mathfrak{S}, O_1, \dots, O_m}(T_i) = \mathrm{NE}_{\mathbf{U}, \mathbf{O}_1, \dots, \mathbf{O}_m}^{\mathfrak{S}, O_1, \dots, O_m}(T_j).$

From (2), (3) and Definition A it follows that if T_i is O-meaningful
with respect to $\mathfrak{S}$ then T_j is O-meaningful with respect to $\mathfrak{S}$ also. This
completes the proof of the theorem.

4.3. Definition A implies CA3.

4.4. Definition A implies CA4.

We shall prove 4.3 and 4.4 together.

Proof. It is an evident consequence of the definition of a non-creative extension of $\mathfrak{S}$ (Definition 3) that: if $\mathfrak{S}'$ is a non-creative extension of $\mathfrak{S}$ then

(1) $\qquad \Omega_{U,O_1,\ldots,O_m}^{\mathfrak{S},O_1,\ldots,O_m}(T_i) = \Omega_{U,O_1,\ldots,O_m}^{\mathfrak{S}',O_1,\ldots,O_m}(T_i)$.

It follows from (1) that

(2) $\qquad \mathrm{PE}_{U,O_1,\ldots,O_m}^{\mathfrak{S},O_1,\ldots,O_m}(T_i) = \mathrm{PE}_{U,O_1,\ldots,O_m}^{\mathfrak{S}',O_1,\ldots,O_m}(T_i)$

and

(3) $\qquad \mathrm{NE}_{U,O_1,\ldots,O_m}^{\mathfrak{S},O_1,\ldots,O_m}(T_i) = \mathrm{NE}_{U,O_1,\ldots,O_m}^{\mathfrak{S}',O_1,\ldots,O_m}(T_i)$.

CA3 and **CA4** now are seen to be immediate consequences of (2) and (3), and thus the proof is complete.

4.5. Definition A implies **CA5**.

Proof. Let T_i be O-meaningful in $\mathfrak{S}$. According to Definition A this assumption is equivalent to the following: there is a set **U** and relations $O_1, \ldots, O_m$ on **U** such that

(1) $\qquad$ either $\mathrm{PE}_{U,O_1,\ldots,O_m}^{\mathfrak{S},O_1,\ldots,O_m}(T_i)$ or $\mathrm{NE}_{U,O_1,\ldots,O_m}^{\mathfrak{S},O_1,\ldots,O_m}(T_i)$ (at least one of them) is not a logical relation.

It follows from (1) that

(2) $\qquad$ The set of relations $\Omega_{U,O_1,\ldots,O_m}^{\mathfrak{S},O_1,\ldots,O_m}(T_i)$ is not empty. For if it is empty then both $\mathrm{PE}_{U,O_1,\ldots,O_m}^{\mathfrak{S},O_1,\ldots,O_m}(T_i)$ and $\mathrm{NE}_{U,O_1,\ldots,O_m}^{\mathfrak{S},O_1,\ldots,O_m}(T_i)$ are empty, and therefore logical relations contrary to (1).

Let us denote by **M** the class of models of $\mathfrak{S}$ defined as follows:

(3) $\qquad \mathfrak{M} \in \mathbf{M}$ if and only if $\mathfrak{S} \subseteq E(\mathfrak{M})$, $D(\mathfrak{M}) = \mathbf{U}$, and

$$w_{\mathfrak{M}}(O_1, \ldots, O_m) = \langle O_1, \ldots, O_m \rangle.$$

It follows from (2) that

(4) $\qquad$ There is at least one model $\mathfrak{M}$ which belongs to **M**.

It is easily seen from (3) that

(5) $\qquad$ For every O-sentence α, α holds either in all or in none of the models belonging to **M**.

Let us indroduce the symbol X_M as follows:

(6) X_M is the set of all O-sentences which hold in all models belonging to **M**.

From (5) we we see that

(7)   Every theory $\mathfrak{S}^*$ such that $X_M \subseteq \mathfrak{S}^*$ is O-complete.

Note that X_M itself is an O-complete theory. Consider now a theory $\mathfrak{S}'$ such that

(8) $\mathfrak{S}' = \mathrm{Cn}(\mathfrak{S} \cup X_M)$.

It follows from the definition of X_M that every model of X_M is a model of $\mathfrak{S}$. Since there is at least one model of X_M as guaranteed by (4) then

(9)   The set $\mathfrak{S} \cap X_M$ is logically consistent.

By (7), (8), (9) we have:

(10) $\mathfrak{S}'$ is a consistent O-complete O-extension of $\mathfrak{S}$.

Since every model of $\mathfrak{S}'$ is a model of $\mathfrak{S}$ then, in particular, for the set **U**, and relations $O_1, \ldots, O_m$ on **U**, we have

(11) $\Omega^{\mathfrak{S}', O_1, \ldots, O_m}_{U, O_1, \ldots, O_m}(T_i) \subseteq \Omega^{\mathfrak{S}, O_1, \ldots, O_m}_{U, O_1, \ldots, O_m}(T_i)$.

To prove that the inclusion in the opposite direction is valid also, assume that $T = \Omega^{\mathfrak{S}, O_1, \ldots, O_m}_{U, O_1, \ldots, O_m}(T_i)$. This implies that there is a model $\mathfrak{M}$ such that $\mathfrak{S} \subseteq E(\mathfrak{M})$, $D(\mathfrak{M}) = U$ and therefore by (5) and (6) we have $X_M \subseteq E(\mathfrak{M})$. Now because X_M and $\mathfrak{S}$ are both included in $E(\mathfrak{M})$ hence by (8) $\mathfrak{S}' \subseteq E(\mathfrak{M})$. It proves that $T \in \Omega^{\mathfrak{S}', O_1, \ldots, O_m}_{U, O_1, \ldots, O_m}(T_i)$ and this completes the proof of the opposite inclusion. Combining this results with (11) we have

(12) $\Omega^{\mathfrak{S}', O_1, \ldots, O_m}_{U, O_1, \ldots, O_m}(T_i) = \Omega^{\mathfrak{S}, O_1, \ldots, O_m}_{U, O_1, \ldots, O_m}(T_i)$.

(1) and (12) yield

(13)   either $PE^{\mathfrak{S}', O_1, \ldots, O_m}_{U, O_1, \ldots, O_m}(T_i)$ or $NE^{\mathfrak{S}', O_1, \ldots, O_m}_{U, O_1, \ldots, O_m}(T_i)$ (at least one of them) is not logical relation on **U**.

By (8), (13) and Definition A we have:

(14) T_i is O-meaningful with respect to the theory $\mathfrak{S}'$ being a consistent O-complete O-extension of $\mathfrak{S}$.

This ends the proof of 4.5.

We have therefore proved as promised earlier that Definition A conforms to the conditions **CA1–CA5**. It is a matter of further investigations to what extent Definition A meets other possible demands of philosophers. Let me remark here that Definition A shares some consequences with the requirement of confirmability for empirically meaningful terms [2]. For certain reasons many authors are inclined to regard the requirement of confirmability as too rigorous. Similar reasons may also apply to Definition A.

Take for instance a two place predicate T_i introduced in a theory $\mathfrak{S}$ only by the following two postulates:

$$(1) \qquad \bigwedge x [O_1(x) \to \bigvee y (O_2(y) \wedge T_i(x, y))],$$

$$(2) \qquad \bigwedge x [O_3(x) \to \bigvee y (O_4(y) \wedge \sim T_i(x, y))]$$

(O_1, O_2, O_3, O_4 being one-place O-terms). Some authors are inclined to say that T_i as characterized by (1) and (2) is empirical meaningful although it is (in general) not comfirmable term. Assuming that $\mathfrak{S}$ is a consistent theory in which the following formulas are not valid

$$(3) \qquad \bigwedge x \bigwedge y (O_2(x) \wedge O_2(y) \to x = y),$$

$$(4) \qquad \bigwedge x \bigwedge y (O_4(x) \wedge O_4(y) \to x = y)$$

we can easily prove that T_i is O-meaningless in the sense of Definition A with respect to $\mathfrak{S}$. Hence we come across a controversial case here. If we decide to share the point of view of those who want to consider T_i as a meaningful term, we should try to define meaningfulness in a different way. This matter will be taken up in the next part of this paper.

*

REMARK. Let me remark here briefly that the example discussed above suggests the following change of Definition A. Let us denote by $D_p(T)$ the p-domain of a k-place relation T ($1 \leqslant p \leqslant k$): $u_p \in D_p(T)$ if and only if there are objects $\langle u_1, \ldots, u_{p-1}, u_{p+1}, \ldots, u_k \rangle$ such that $\langle u_1, \ldots, u_k \rangle \in T$. Now we could offer the following definitions.

DI. For any k-place predicate T_i and for any p ($1 \leqslant p \leqslant k$)

a. $\text{PED}^{\mathfrak{S},\ O_1, \ldots, O_m}_{p\text{U}, O_1, \ldots, O_m}(T_i) = \bigcap_{D_p(T)} (T \in \Omega^{\mathfrak{S},\ O_1, \ldots, O_m}_{\text{U}, O_1, \ldots, O_m}(T_i)),$

b. $\text{NED}_{p\mathbf{U},\,\mathbf{O}_1,\,...,\,\mathbf{O}_m}^{\mathfrak{S},\,O_1,\,...,\,O_m}(T_i) = \bigcap_{D_p(\mathbf{T})}(\bar{\mathbf{T}} \in \Omega_{\mathbf{U},\,\mathbf{O}_1,\,...,\,\mathbf{O}_m}^{\mathfrak{S},\,O_1,\,...,\,O_m}(T_i))$.

DII.   A theoretical k-place predicate T_i is *O-meaningful with respect to a theory* $\mathfrak{S}$ if and only if there is a set $\mathbf{U}$ and relations $\mathbf{O}_1,\,...,\,\mathbf{O}_m$ on $\mathbf{U}$ such that for a certain number p $(1 \leqslant p \leqslant k)$ $\text{PED}_{p\mathbf{U},\,\mathbf{O}_1,\,...,\,\mathbf{O}_m}^{\mathfrak{S},\,O_1,\,...,\,O_m}(T_i)$ or $\text{NED}_{p\mathbf{U},\,\mathbf{O}_1,\,...,\,\mathbf{O}_m}^{\mathfrak{S},\,O_1,\,...,\,O_m}(T_i)$ are not both logical relations on $\mathbf{U}$.

This definition is discussed more exactly in another paper of mine [20].

V

Consider the following situation. Let us take a set $\mathbf{U}$ and relations $\mathbf{O}_1,\,...,\,\mathbf{O}_m$ such that $\Omega_{\mathbf{U},\,\mathbf{O}_1,\,...,\,\mathbf{O}_m}^{\mathfrak{S},\,O_1,\,...,\,O_m}(T_i)$ is not empty. Let us now change the interpretations of O-terms in $\mathbf{U}$ in such a way that for the new relations $\mathbf{O}_1,\,...,\,\mathbf{O}_m$ corresponding to the O-terms, the set $\Omega_{\mathbf{U},\,\mathbf{O}_1',\,...,\,\mathbf{O}_m'}^{\mathfrak{S},\,O_1,\,...,\,O_m}(T_i)$ is not empty either. It may occur that

$$\Omega_{\mathbf{U},\,\mathbf{O}_1,\,...,\,\mathbf{O}_m}^{\mathfrak{S},\,O_1,\,...,\,O_m}(T_i) \neq \Omega_{\mathbf{U},\,\mathbf{O}_1',\,...,\,\mathbf{O}_m'}^{\mathfrak{S},\,O_1,\,...,\,O_m}(T_i).$$

In such an event we may say that the interpretation of T_i determined by $\mathfrak{S}$ is at least partly dependent on interpretations of O-terms, or to introduce a technical expression $- T_i$ is *O-dependent* with respect to $\mathfrak{S}$.

DEFINITION 10. T_i is *O-dependent with respect to* $\mathfrak{S}$ if and only if there is a set $\mathbf{U}$ and relations $\mathbf{O}_1,\,...,\,\mathbf{O}_m,\,\mathbf{O}_1',\,...,\,\mathbf{O}_m'$ on $\mathbf{U}$ such that:

i. $\Omega_{\mathbf{U},\,\mathbf{O}_1,\,...,\,\mathbf{O}_m}^{\mathfrak{S},\,O_1,\,...,\,O_m}(T_i)$ and $\Omega_{\mathbf{U},\,\mathbf{O}_1',\,...,\,\mathbf{O}_m'}^{\mathfrak{S},\,O_1,\,...,\,O_m}(T_i)$ are both non-empty,

and

ii. $\Omega_{\mathbf{U},\,\mathbf{O}_1,\,...,\,\mathbf{O}_m}^{\mathfrak{S},\,O_1,\,...,\,O_m}(T_i) \neq \Omega_{\mathbf{U},\,\mathbf{O}_1',\,...,\,\mathbf{O}_m'}^{\mathfrak{S},\,O_1,\,...,\,O_m}(T_i)$.

*

REMARK. It can be shown that there is an interesting though simple connection between the notion of O-dependence and the model-theoretical notion of definability. This last notion can be defined as follows:

DIII. T_i is *definable (in the model-theoretical sense) in O-terms with respect to* $\mathfrak{S}$ if and only if for every set $\mathbf{U}$ and relations $\mathbf{O}_1,\,...,\,\mathbf{O}_m,\,\mathbf{T}_1,\,\mathbf{T}_2$ on $\mathbf{U}$: if $\langle \mathbf{U},\,\mathbf{O}_1,\,...,\,\mathbf{O}_m,\,\mathbf{T}_1 \rangle \in \mathscr{W}_{\mathfrak{S}}(O_1,\,...,\,O_m,\,T_i)$ and $\langle \mathbf{U},\,\mathbf{O}_1,\,...,\,\mathbf{O}_m,\,\mathbf{T}_2 \rangle \in \mathscr{W}_{\mathfrak{S}}(O_1,\,...,\,O_m,\,T_i)$ then $\mathbf{T}_1 = \mathbf{T}_2$.

(Cf. Craig [4], p. 273.) Quite analogously we can define definability in logical terms.

DIV. T_i *is definable (in the model-theoretical sense) in logical terms with respect to* $\mathfrak{S}$ *if and only if for every set* $\mathbf{U}$ *and relations* $\mathbf{T}_1$, $\mathbf{T}_2$ *on* $\mathbf{U}$: *if* $\langle \mathbf{U}, \mathbf{T}_1 \rangle \in \mathscr{W}_{\mathfrak{S}}(T_1)$ *and* $\langle \mathbf{U}, \mathbf{T}_2 \rangle \in \mathscr{W}_{\mathfrak{S}}(T_i)$ *then* $\mathbf{T}_1 = \mathbf{T}_2$.

Let us introduce the notion of essential definability in O-terms.

DV. T_i *is essentially definable (in the model-theoretical sense) in* O-*terms with respect to* $\mathfrak{S}$ *if and only if* T_i *is definable (in the model-theoretical sense) in* O-*terms with respect to* $\mathfrak{S}$, *but it is not definable (in the model-theoretical sense) in logical terms with respect to* $\mathfrak{S}$.

It is not difficult to see that if T_i is esentially definable in O-terms with respect to $\mathfrak{S}$ then T_i is O-dependent term with respect to $\mathfrak{S}$. To prove it, let us assume that T_i is essentially definable in O-terms respect to $\mathfrak{S}$. Hence T_i is not definable in logical terms with respect to $\mathfrak{S}$, and therefore there are two models $\mathfrak{M}_1$ and $\mathfrak{M}_2$ of $\mathfrak{S}$ such that:

(1) $D(\mathfrak{M}_1) = D(\mathfrak{M}_2)$

and

(2) $w_{\mathfrak{M}_1}(T_i) \neq w_{\mathfrak{M}_2}(T_i)$.

Let us put

a. $D(\mathfrak{M}_1) = D(\mathfrak{M}_2) = \mathbf{U}$,

b. $w_{\mathfrak{M}_1}(O_1, \ldots, O_m, T_i) = \langle O_1, \ldots, O_m, \mathbf{T} \rangle$,

c. $w_{\mathfrak{M}_2}(O_1, \ldots, O_m, T_i) = \langle O_1', \ldots, O_m', \mathbf{T}' \rangle$.

It follows from our considerations that

(3) $\mathbf{T} \in \Omega_{\mathbf{U}, O_1, \ldots, O_m}^{\mathfrak{S}, O_1, \ldots, O_m}(T_i)$ and $\mathbf{T}' \in \Omega_{\mathbf{U}, O_1', \ldots, O_m'}^{\mathfrak{S}, O_1, \ldots, O_m}(T_i)$.

It can be easily seen that if $\mathbf{T}, \mathbf{T}' \in \Omega_{\mathbf{U}, O_1, \ldots, O_m}^{\mathfrak{S}, O_1, \ldots, O_m}(T_i)$ then our assumption that T_i is essentially definable in O-terms with respect to $\mathfrak{S}$ yields $\mathbf{T} = \mathbf{T}'$.

Since $\mathbf{T} = w_{\mathfrak{M}_1}(T_i)$ and $\mathbf{T}' = w_{\mathfrak{M}_2}(T_i)$ it would contradict (2). Therefore

(4) $\mathbf{T}' \bar{\in} \Omega_{U, O_1, ..., O_m}^{\mathfrak{S}, O_1, ..., O_m}(T_i).$

The conditions i. and ii. of Definition 10 are immediate consequences of (3) and (4). This completes the proof.

*

It seems natural to admit that O-dependence with respect to $\mathfrak{S}$ has to be a necessary condition of meaningfulness. Can we not admit that this is also a sufficient condition? Unfortunately this decision would not be correct with respect to condition **CA5** of adequacy. It is possible to find a theory $\mathfrak{S}$ and a T-term T_i such that T_i is O-dependent with respect to $\mathfrak{S}$ but is not O-dependent with respect to any O-complete, O-extension of $\mathfrak{S}$. Let for instance the only postulates for T_i in $\mathfrak{S}$ be:

(1) $\bigvee x \bigvee y \left(x \neq y \wedge O_1(x) \wedge O_1(y) \right)$

            $\rightarrow \bigvee x \bigvee y \left(x \neq y \wedge T_1(x) \wedge T_1(y) \right),$

(2) $\sim \bigvee x \bigvee y \left(x \neq y \wedge O_1(x) \wedge O_1(y) \right)$

            $\rightarrow \bigvee x \bigvee y \left(x \neq y \wedge \sim T_1(x) \wedge \sim T_1(y) \right)$

and suppose that neither the antecedent of (1) nor the antecedent of (2) is valid in $\mathfrak{S}$. In such an event we can easily prove that T_i is O-dependent with respect to $\mathfrak{S}$ but is not O-dependent with respect to any O-extension of $\mathfrak{S}$ in which either

$$\bigvee x \bigvee y \left(x \neq y \wedge O_1(x) \wedge O_1(y) \right)$$

or the negation of this formula is valid. To avoid this obstacle we shall define O-meaningfulness as follows:

DEFINITION B. A term T_i is *O-meaningful with respect to* $\mathfrak{S}$ if and only if:

i. T_i is O-dependent with respect to $\mathfrak{S}$.

ii.   There is an O-complete O-extension $\mathfrak{S}'$ of $\mathfrak{S}$ such that T_i is O-dependent with respect to $\mathfrak{S}'$.

The clause i. of this definition is superfluous and could be dropped. It follows from Lemma 3, to be formulated and proved later, that ii. entails i.

As in the case of Definition A, we have to prove that **CA1–CA5** are consequences of Definition **B**.

5.1. Definition B implies CA1.

Proof. To prove 5.1 it is sufficient to prove that if T_i is O-isolated then it is not O-dependent. Let T_i be O-isolated term and let $T_{i_1}, \ldots, T_{i_k}$ be the terms required by Definition 6. We shall assume that $T_{i_1} = T_i$. It is easily seen that if $\mathfrak{S}$ is not consistent then T_i is not O-dependent with respect to $\mathfrak{S}$. We may therefore assume that $\mathfrak{S}$ is consistent. Consider now any set **U** and relations $O_1, \ldots, O_m, O'_1, \ldots, O'_m$ on **U** such that

$$(1) \qquad \Omega_{U, O_1, \ldots, O_m}^{\mathfrak{S}, O_1, \ldots, O_m}(T_i) \text{ is not empty}$$

and also

$$(2) \qquad \Omega_{U, O'_1, \ldots, O'_m}^{\mathfrak{S}, O_1, \ldots, O_m}(T_i) \text{ is not empty.}$$

We have to prove that

$$(\gamma) \qquad \Omega_{U, O_1, \ldots, O_m}^{\mathfrak{S}, O_1, \ldots, O_m}(T_i) = \Omega_{U, O'_1, \ldots, O'_m}^{\mathfrak{S}, O_1, \ldots, O_m}(T_i).$$

Assume that

$$(3) \qquad \mathbf{T}_1 \in \Omega_{U, O_1, \ldots, O_m}^{\mathfrak{S}, O_1, \ldots, O_m}(T_i).$$

Hence there are relations $\mathbf{T}_1, \ldots, \mathbf{T}_k$ such that

$$(4) \qquad \langle \mathbf{U}, \mathbf{T}_1, \ldots, \mathbf{T}_k \rangle \in \mathscr{W}_{\mathfrak{S}}(T_{i_1}, \ldots, T_{i_k}).$$

From (2) it follows that there are relations $\mathbf{T}_{k+1}, \ldots, \mathbf{T}_n$ such that

$$(5) \qquad \langle \mathbf{U}, \mathbf{O}'_1, \ldots, \mathbf{O}'_m, \mathbf{T}_{k+1}, \ldots, \mathbf{T}_n \rangle$$
$$\in \mathscr{W}_{\mathfrak{S}}(O_1, \ldots, O_m, T_{i_{k+1}}, \ldots, T_{i_n})$$

where $T_{i_{k+1}}, \ldots, T_{i_n}$ are T-terms different from $T_{i_1}, \ldots, T_{i_n}$. Definition 6, (4) and (5) imply

$$(6) \qquad \langle \mathbf{U}, \mathbf{O}'_1, \ldots, \mathbf{O}'_m, \mathbf{T}_1, \ldots, \mathbf{T}_n \rangle \in \mathscr{W}_{\mathfrak{S}}(O_1, \ldots, O_m, T_{i_1}, \ldots, T_{i_n}),$$

Hence $\mathbf{T}_1 \in \Omega_{U, O'_1, \ldots, O'_m}^{\mathfrak{S}, O_1, \ldots, O_m}(T_i)$. In an analogous way it can be shown that if $\mathbf{T}_1 \in \Omega_{U, O'_1, \ldots, O'_m}^{\mathfrak{S}, O_1, \ldots, O_m}(T_i)$, then $\mathbf{T}_1 \in \Omega_{U, O_1, \ldots, O_m}^{\mathfrak{S}, O_1, \ldots, O_m}(T_i)$. This establishes (γ) and with it the theorem.

5.2. Definition B implies **CA2**.

Before proving 5.2 we shall prove the following lemma.

LEMMA 2. For any set **U** and relations $O_1, ..., O_m$ on **U**, if

$$\Omega_{U, O_1, ..., O_m}^{\mathfrak{S}, O_1, ..., O_m}(T_i) = \Omega_{U, O_1, ..., O_m}^{\mathfrak{S}, O_1, ..., O_m}(T_j)$$

then for every O-extension $\mathfrak{S}'$ of $\mathfrak{S}$ $\Omega_{U, O_1, ..., O_m}^{\mathfrak{S}', O_1, ..., O_m}(T_i) = \Omega_{U, O_1, ..., O_m}^{\mathfrak{S}', O_1, ..., O_m}(T_j)$.

Proof. If $\mathfrak{S}'$ is O-extension of $\mathfrak{S}$ then there is a set of O-sentences **X** such that

(1) $\qquad \mathfrak{S}' = \mathrm{Cn}(\mathfrak{S} \cup \mathbf{X})$.

Assume now that

(2) $\qquad \mathbf{T} \in \Omega_{U, O_1, ..., O_m}^{\mathfrak{S}', O_1, ..., O_m}(T_i) :$

Hence there is a possible model $\mathfrak{M}$ such that

(3) $\qquad D(\mathfrak{M}) = \mathbf{U}$,

(4) $\qquad w_{\mathfrak{M}}(O_1, ..., O_m, T_i) = \langle O_1, ..., O_m, \mathbf{T} \rangle$,

(5) $\qquad \mathfrak{S}' \subseteq E(\mathfrak{M})$.

By (1) and (5) we obtain

(6) $\qquad \mathfrak{S} \subseteq E(\mathfrak{M})$.

(3), (4) and (6) yield

(7) $\qquad \mathbf{T} \in \Omega_{U, O_1, ..., O_m}^{\mathfrak{S}, O_1, ..., O_m}(T_i)$.

Suppose that the antecedent of Lemma 2 is valid, then

(8) $\qquad \mathbf{T} \in \Omega_{U, O_1, ..., O_m}^{\mathfrak{S}, O_1, ..., O_m}(T_j)$.

Hence there is a model $\mathfrak{M}'$ such that

(9) $\qquad D(\mathfrak{M}') = \mathbf{U}$,

(10) $\qquad w_{\mathfrak{M}'}(O_1, ..., O_m, T_j) = \langle O_1, ..., O_m, \mathbf{T} \rangle$,

(11) $\qquad \mathfrak{S} \subseteq E(\mathfrak{M}')$.

Since $w_{\mathfrak{M}'}(O_1, ..., O_m) = w_{\mathfrak{M}}(O_1, ..., O_m)$ and **X** is a set of O-sentences, therefore

(12) $\qquad$ if $\mathbf{X} \subseteq E(\mathfrak{M})$, then $\mathbf{X} \subseteq E(\mathfrak{M}')$.

But it follows from (1) and (5) that $X \subseteq E(\mathfrak{M})$, then also

(13)　　　$X \subseteq E(\mathfrak{M}')$.

By (11) and (13) we have

(14)　　　$\mathfrak{S}' \subseteq E(\mathfrak{M}')$.

(9), (10) and (14) yield

(15)　　　$T \in \Omega_{U,O_1,...,O_m}^{\mathfrak{S}',O_1,...,O_m}(T_j)$.

Thus we have established that if $T \in \Omega_{U,O_1,...,O_m}^{\mathfrak{S}',O_1,...,O_m}(T_i)$, then we have $T \in \Omega_{U,O_1,...,O_m}^{\mathfrak{S}',O_1,...,O_m}(T_j)$. We can obtain the converse implication by an entirely analogous argument, and thus the proof is complete.

Let us turn to the Theorem 5.2.

Proof. Assume that for any set U and relations $O_1, ..., O_m$ on U

(1)　　　$\Omega_{U,O_1,...,O_m}^{\mathfrak{S},O_1,...,O_m}(T_i) = \Omega_{U,O_1,...,O_m}^{\mathfrak{S},O_1,...,O_m}(T_j)$.

Assume moreover that T_i is O-meaningful with respect to $\mathfrak{S}$. Hence

(2)　　　T_i is O-dependent with respect to $\mathfrak{S}$,

and

(3)　　　There is a consistent O-complete O-extension $\mathfrak{S}'$ of $\mathfrak{S}$ such that T_i is O-dependent with respect to $\mathfrak{S}$.

It is an immediate consequence of (1), (2) and Definition 10 that

(4)　　　T_j is O-dependent with respect to $\mathfrak{S}$.

By (1), (3) and Lemma 2 we have

(5)　　　There is a consistent O-complete O-extension $\mathfrak{S}''$ of $\mathfrak{S}$ such that T_j is O-dependent with respect to $\mathfrak{S}''$.

This complets the proof of the Theorem 5.2.

5.3. Definition B implies CA3.

5.4. Definition B implies CA4.

The Theorems 5.3 and 5.4 will be proved simultaneously.

Proof. Using Definitions 3, 4, 5 and Definition B we can easily establish that for any theory $\mathfrak{S}$, for any non-creative extension $\mathfrak{S}_1$ of $\mathfrak{S}$,

and for any set X of O-sentences:

(1) T_i is O-dependent with respect to $\mathfrak{S}$ if and only if T_i is O-dependent with respect to $\mathfrak{S}_1$.

(2) $\mathfrak{S}' = \mathrm{Cn}(\mathfrak{S} \cup X)$ is a consistent O-complete O-extension of $\mathfrak{S}$ if and only if $\mathfrak{S}'_1 = \mathrm{Cn}(\mathfrak{S}_1 \cup X)$ is a consistent O-complete O-extension of $\mathfrak{S}_1$.

(3) T_i is O-dependent with respect to $\mathfrak{S}' = \mathrm{Cn}(\mathfrak{S} \cup X)$ if and only if T_i is O-dependent with respect to $\mathfrak{S}'_1 = \mathrm{Cn}(\mathfrak{S}_1 \cup X)$.

CA3 and **CA4** are almost immediate consequences of (1), (2), (3).

5.5. Definition B implies **CA5**.

Proof. Let T_i be O-meaningful with respect to $\mathfrak{S}$ and let $\mathfrak{S}$ be an O-complete O-extension of $\mathfrak{S}$ such that T_i is O-dependent with respect to $\mathfrak{S}'$ (the existence of $\mathfrak{S}$ is guaranteed by the clause ii. of Definition B). It can be easily seen that T_i is O-meaningful with respect to $\mathfrak{S}'$.

VI

We shall now compare Definition A with Definition B. For this purpose let us turn for a while to the example that we have discussed on the page 668. We can easily check that, quite independent of whether the formulas (3), (4) are valid in $\mathfrak{S}$ or not, the term T_i introduced by the formulas (1) and (2) is O-meaningful in the sense of Definition B except in the case when $\mathfrak{S}$ is not consistent. This indicates that T-terms O-meaningful in the sense of Definition B are not necessarily O-menaningful in the sense of Definition A. We can prove however that:

6.1. If a T-term T_i is O-meaningful with respect to a theory $\mathfrak{S}$ in the sense of Definition A then T_i is O-meaningful with respect to $\mathfrak{S}$ in the sense of Definition B.

To prove 6.1 we shall prove the following lemma at first.

LEMMA 3. If a T-term T_i is not O-dependent with respect to a theory $\mathfrak{S}$ then T_i is not O-dependent with respect to any O-extension $\mathfrak{S}'$ of $\mathfrak{S}$.

Proof. According to the Definition 10, the atencedent of Lemma 3 is equivalent to the following.

(1)   For any set U and relations $O_1, ..., O_m, O'_1, ..., O'_m$ on U if

$$\Omega^{\mathfrak{S}, \, O_1, \, ..., \, O_m}_{U, \, O_1, \, ..., \, O_m}(T_i) \text{ and } \Omega^{\mathfrak{S}, \, O_1, \, ..., \, O_m}_{U, \, O'_1, \, ..., \, O'_m}(T_i)$$

are both non-empty, then

$$\Omega^{\mathfrak{S}, \, O_1, \, ..., \, O_m}_{U, \, O_1, \, ..., \, O_m}(T_i) = \Omega^{\mathfrak{S}, \, O_1, \, ..., \, O_m}_{U, \, O'_1, \, ..., \, O'_m}(T_i).$$

Assume now that the consequent of Lemma 3 is not valid. Then

(2)   there is a set X of O-sentences such that:

i. $\mathfrak{S}' = \text{Cn}\ (\mathfrak{S} \cup X)$ is O-extension of $\mathfrak{S}$;

ii.   for some $U, O_1, ..., O_m, O'_1, ..., O'_m$:

    a. $\Omega^{\mathfrak{S}', \, O_1, \, ..., \, O_m}_{U, \, O_1, \, ..., \, O_m}(T_i)$ and $\Omega^{\mathfrak{S}', \, O_1, \, ..., \, O_m}_{U, \, O'_1, \, ..., \, O'_m}(T_i)$ are both non-empty, and

    b. $\Omega^{\mathfrak{S}', \, O_1, \, ..., \, O_m}_{U, \, O_1, \, ..., \, O_m}(T_i) \neq \Omega^{\mathfrak{S}', \, O_1, \, ... \, O_m}_{U, \, O'_1, \, ..., \, O'_m}(T_i).$

Let us assume that there is a relation T such that

(3) $T \in \Omega^{\mathfrak{S}', \, O_1, \, ..., \, O_m}_{U, \, O_1, \, ..., \, O_m}(T_i),$

(4) $T \bar{\in} \Omega^{\mathfrak{S}', \, O_1, \, ..., \, O_m}_{U, \, O'_1, \, ..., \, O'_m}(T_i).$

We could assume as well that $T \bar{\in} \Omega^{\mathfrak{S}', \, O_1, \, ..., \, O_m}_{U, \, O_1, \, ..., \, O_m}(T_i)$ and $T \in \Omega^{\mathfrak{S}', \, O_1, \, ..., \, O_m}_{U, \, O'_1, \, ..., \, O'_m}(T_i)$; the proof would then proceed in the same way. Note now that it follows from (2)i and (2)ii.a that for the same set U, and the same relations $O_1, ..., O_m, O'_1, ..., O'_m$ for which (2)ii. a is valid

(5) $\Omega^{\mathfrak{S}, \, O_1, \, ..., \, O_m}_{U, \, O_1, \, ..., \, O_m}(T_i)$ and $\Omega^{\mathfrak{S}, \, O_1, \, ..., \, O_m}_{U, \, O'_1, \, ..., \, O'_m}(T_i)$ are both non-empty.

Hence by (1) we have

(6) $\Omega^{\mathfrak{S}, \, O_1, \, ..., \, O_m}_{U, \, O_1, \, ..., \, O_m}(T_i) = \Omega^{\mathfrak{S}, \, O_1, \, ..., \, O_m}_{U, \, O'_1, \, ..., \, O'_m}(T_i).$

(2)i and (3) yield

(7) $T \in \Omega^{\mathfrak{S}, \, O_1, \, ..., \, O_m}_{U, \, O_1, \, ..., \, O_m}(T_i).$

By (6) we obtain from (7)

(8) $T \in \Omega^{\mathfrak{S}, \, O_1, \, ..., \, O_m}_{U, \, O'_1, \, ..., \, O'_m}(T_i).$

Since according to (2)ii.a $\Omega^{\mathfrak{S}', \, O_1, \, ..., \, O_m}_{U, \, O'_1, \, ..., \, O'_m}(T_i)$ is not empty there is a possible model $\mathfrak{M}$ such that $\mathfrak{S}' \subseteq E(\mathfrak{M})$, $D(\mathfrak{M}) = U$, and $w_{\mathfrak{M}}(O_1, ..., O_m) = \langle O'_1, ..., O'_m \rangle$. Hence in virtue of (2)i also $X \subseteq E(\mathfrak{M})$. But because X

is a set of O-sentences,

(9)   For every possible model $\mathfrak{M}$, if $D(\mathfrak{M}) = U$ and $w_{\mathfrak{M}}(O_1, ..., O_m)$
$= \langle \mathbf{O}_1', ..., \mathbf{O}_m' \rangle$, then $X \subseteq E(\mathfrak{M})$.

From (8) we obtain

(10)   There is a possible model $\mathfrak{M}$ such that $\mathfrak{S} \subseteq E(\mathfrak{M})$, $D(\mathfrak{M}) = U$,
and $w_{\mathfrak{M}}(O_1, ..., O_m, T_i) = \langle \mathbf{O}_1', ..., \mathbf{O}_m', \mathbf{T} \rangle$.

It follows from (9) and (10) that

(11)   There is a model $\mathfrak{M}$ such that $\mathfrak{S}' \subseteq E(\mathfrak{M})$, $D(\mathfrak{M}) = U$, and
$w_{\mathfrak{M}}(O_1, ..., O_m, T_i) = \langle \mathbf{O}_1', ..., \mathbf{O}_m', \mathbf{T} \rangle$

which contradicts (4). This completes the proof.

We shall now present a proof of 6.1.

Proof. An indirect proof be will given. We shall show that if T_i is O-meaningless with respect to a theory $\mathfrak{S}$ in the sense of Definition B then T_i is O-meaningless with respect to $\mathfrak{S}$ in the sense of Definition A. Suppose that T_i is O-meaningless with respect to $\mathfrak{S}$ in the sense of Definition B. This assumption is equivalent to the following:

(1)   Either

i. T_i is not O-dependent with respect to $\mathfrak{S}$
or

ii. T_i is O-dependent with respect to none O-complete O-extension $\mathfrak{S}'$ of $\mathfrak{S}$.

According to Lemma 3, the clause i of (1) implies the clause ii. Therefore our proof will be complete if we show that ii implies O-meaninglesness T_i in the sense of Definition A. Note that ii yields,

(2)   For any $\mathfrak{S}'$ being an O-complete O-extension of $\mathfrak{S}$, and for any set U and relations $\mathbf{O}_1, ..., \mathbf{O}_m, \mathbf{O}_1', ..., \mathbf{O}_m'$ on U if both

$\Omega_{U, O_1, ..., O_m}^{\mathfrak{S}', O_1, ..., O_m}(T_i)$ and $\Omega_{U, O_1', ..., O_m'}^{\mathfrak{S}', O_1, ..., O_m}(T_i)$

are non-empty then

$\Omega_{U, O_1, ..., O_m}^{\mathfrak{S}', O_1, ..., O_m}(T_i) = \Omega_{U, O_1', ..., O_m'}^{\mathfrak{S}', O_1, ..., O_m}(T_i)$.

Take an arbitrary set U, relations $\mathbf{O}_1, ..., \mathbf{O}_m$ on U, and arbitrary one-one mapping f of U onto itself. Since for any theory $\mathfrak{S}$, $\Omega_{U, O_1, ..., O_m}^{\mathfrak{S}, O_1, ..., O_m}(T_i)$,

$\Omega_{U,\,O_1',\,...,\,O_m'}^{\mathfrak{S},\,O_1,\,...,\,O_m}(T_i)$ are either both empty or both non empty then by (2) we obtain in practicular

(3)   for any $\mathfrak{S}'$ being O-complete O-extension of $\mathfrak{S}$,

$$\Omega_{U,\,O_1,\,...,\,O_m}^{\mathfrak{S}',\,O_1,\,...,\,O_m}(T_i)=\Omega_{U,\,O_1',\,...,\,O_m'}^{\mathfrak{S}',\,O_1\,...,\,O_m}(T_i).$$

From (3) we obtain by Lemma 1,

(4) T_i is O-meaningless with respect to every O-complete O-extension of $\mathfrak{S}$.

The implication which we have to prove is now an immediate consequence of (4) and 4.5.

Let us make only one remark more about differences between the notion of O-meaningfulness in the sense of Definition A, and the notion of O-meaningfulness in the sense of Definition B.

The following theorem is an immediate consequence of Lemma 3.

6.2. If a T-term T_i is O-meaningless with respect to a theory $\mathfrak{S}$ in the sense of Definition B then T_i is, in the same sense, O-meaningless with respect to any theory $\mathfrak{S}'$ which is O-extension of $\mathfrak{S}$.

It is not difficult to see that the analogous theorem for O-meaninglessness in the sense of Definition A is not vaild. A term T_i which is O-meaningless in the sense of Definition A with respect to a theory $\mathfrak{S}$ can be O-meaningful with respect to same O-extensions of $\mathfrak{S}$. Let us consider again the example from the page 668. If (1) and (2) are both valid in $\mathfrak{S}$, but neither (3), (4) nor the negations of (3), (4) belong to $\mathfrak{S}$ then T_i is O-meaningless in the sense of Definition A. Let us now enrich $\mathfrak{S}$ by (3) and (4). It is easily seen that T_i is O-meaningful with respect to the new theory in the sense of Definition A.

The difference between the two notions of O-meaningfulness pointed out above can give a reason for one's preference for Definition B. According to the opinion of many philosophers, empirically meaningless terms must not occur in scientific theories. Taking this position it is safer to understand O-meaninglessness in the sense of Definition B. The removal of a term from a scientific theory may turn out to be a fault if a later empirical evidence (some O-sentences accpeted later) reveals its O-meaningfulness. There can be some doubts however, see [20], whether the program of science without O-meaningless terms is sufficiently justified indeed. Perhaps the notion of O-meaningfulness have to

be regarded as descriptive only rather than normative one. This matter will not be discussed here.

The question which of the two definitions of O-meaningfulness is better is deliberately left open. It has not been the aim of this paper to settle definitely what constitutes empirical meaningfulness. I have only tried to show some ways which may be helpful in handling the problem, and present some ideas which perhaps are capable of further development.

BIBLIOGRAPHY

[1] Y. Bar-Hillel, 'Rudolf Carnap, The Methodological Character of Theoretical Concepts' (review), *Journal of Sambolic Logic* 25 (1960).

[2] R. Carnap, 'Testability and Meaning', *Philosophy of Science* 3 and 4 (1936 –1937).

[3] R. Carnap, 'The Methodological Character of Theoretical Concepts', *Minnesota Studies in the Philosophy of Science* 1 (1956).

[4] W. Craig, 'Three Uses of the Herbrand-Gentzen Theorem in Relating Model-Theory and Proof-Theory', *Journal of Symbolic Logic* 22 (1957).

[5] H. Feigl, 'Some Major Issues and Developments in the Philosophy of Science of Logical Empiricism', *Minnesota Studies in the Philosophy of Science* 1 (1956).

[6] C. G. Hempel, 'Problems and Changes in the Empiricist Criterion of Meaning', *Revue Internationale de Philosophie* 4 (1950).

[7] C. G. Hempel, 'The Theoretician's Dilemma – A study in the Logic of Theory Construction', *Minnesota Studies in the Philosophy of Science* 2 (1958).

[8] D. Kaplan, 'Significance and Analyticity – A Comment of Some Recent Proposals of Carnap', manuscript.

[9] D. Kaplan, *On Significance*, manuscript.

[10] J. Kemeny, 'A new Approach to Semantics', *Journal of Symbolic Logic* 21 (1956).

[11] M. Kokoszyńska, 'O dwojakim rozumieniu uzasadniania dedukcyjnego' ('On Two Ways of Interpreting Deductive Justification'), *Studia Logica* XIII (1962).

[12] T. Kubiński, 'Nazwy nieostre' ('Vague Names'), *Studia Logica* VIII (1958).

[13] P. Marhenke, 'The Criterion of Significance', *Proceedings and Addresses of the American Philosophical Association* 23 (1950).

[14] M. Przełęcki, 'W sprawie terminów nieostrych' ('On vague terms'), *Studia Logica* XIII (1958).

[15] W. O. Quine, *Mathematical Logic*, Harvard 1955.

[16] R. Suszko, 'Syntactic Structure and Semantic Reference', *Studia Logica* VII, IX (1958).

[17] R. Suszko, *Wykłady z logiki formalnej* (*Lectures on Formal Logic*), Warszawa 1965.

[18] R. Wójcicki, 'Sensowność terminów teoretycznych—Krytyczne uwagi o pewnej koncepcji R. Carnapa' ('Meaningfulness of Theoretical terms—Some Critical Remarks on a Certain Carnap's Conception'), *Ruch Filozoficzny* XXII (1964).

[19] R. Wójcicki, 'O warunkach empirycznej sensowności terminów', in: *Teoria a doświadczenie* ('On the Conditions of Empirical Meaningfulness of Terms', in: *Theory and Experience*, Warszawa 1965.

[20] R. Wójcicki, 'The Relative Meaningfulness of Theoretical Terms', *Studia Logica* XIX.

REFERENCES

* First published in *Studia Logica* XIX (1966).

[1] There are no essential differences between the criticism of Carnap's definition presented here and the critical remarks on this definition I had had oportunity to present on one of meetings of Wrocław Division of Polish Philosophical Society (Wrocławski Oddział Polskiego Towarzystwa Filozoficznego) in 1962 (see [18]). I learned only much later that the same objection against Carnap's conception had been raised earlier by David Kaplan in his two non-published papers [8], [9].

[2] Perhaps it is worthwhile to mention that Carnap's definition implies some other difficulties I did not discuss here. For example, an interesting criticism of Carnap's proposal has been given by Hempel [7].

[3] For more detailed remarks on this matter see [20].

[4] The notion of a possible model of L introduced here corresponds to the notion of a model for a vocabulary of L defined in [16].

[5] Obviously this pessimistic point of view is far not unanimously accepted. One can find an interesting reply to the doubts of Hempel in an essay writen by Feigl [5].

[6] Assume that the set of the postulates of $\mathfrak{S}$ is symmetric with respect to T_i and T_j, i.e. if T_i and T_j are interchanged everywhere the set as whole remains unchanged, T_i and T_j characterized by such a set of postulates are O-indistinguishable (see [19], Theorem 2). Nevertheless our assumption does not exclude that the set of postulates entails the formula:

(1) $$\sim \bigwedge x_1 ... \bigwedge x_k (T_i(x_1,...,x_k) = T_j(x_1,...,x_k))$$

T_i, T_j being k-place predicates. Because of (1), even if **T** can serve as a interpretation both of T_i and of T_j, and T_j cannot be interpreted as **T** at the same time. This makes clear, why O-indistinguishability does not imply the identity of meaning of the pertinent terms.

[7] The explanations of a and b given in parentheses are informative only. As to 'the corners' '⌐' and '⌐' used here, see [15].

[8] The notion of a logical object was introduced by A. Tarski in a lecture given in Polish Academy of Sciences, Warsaw 1959. It corresponds to the notion of an absolute invariant, see [17].

RYSZARD WÓJCICKI

BASIC CONCEPTS OF FORMAL METHODOLOGY
OF EMPIRICAL SCIENCES*

Taking as the starting point some concepts of mathematical logic, especially those of logical semantics (theory of models), I will try to elaborate and adjust them to the needs of the methodology of empirical sciences. Proceeding along these lines, I will aim at constructing a set-theoretic model of the relationships which connect the formal components of an empirical theory (its language and the set of asserted statements) with the empirical systems that the theory is to describe. The discussion will result in producing two such models. The second one will be devised in such a manner as to conform with the approximative nature of empirical inquiry. Inside this conceptual framework there will be defined and discussed several notions, among others some types of regularities (determinism, some correlation regularities) and the notion of approximative truth.

The main ideas of this report have been outlined in two papers published in Polish [17], [18]. I should like to express may gratitude to my friend Professor Marian Przełęcki for his criticism of my earlier essays. His valuable comments certainly allowed me to present some points in a more adequate and straightforward manner.

I. INTRODUCTORY REMARKS

There are two ways of viewing a mathematical theory utilized in meta-mathematical investigations: syntactical (formal) and semantical one. From the syntactical standpoint, to define a mathematical theory we need to define its *language*, the *derivability relation* (usually be setting up a set of logical rules of inference), and finally to define the set of *axioms* of the theory. Thus a *mathematical theory* may be conceived as a triple

(A) $(\mathfrak{L}, \vdash_R, A_0)$,

where $\mathfrak{L}$ is the language, $\vdash_R$ stands for the derivability relation based on the set of inference rules R, and A_0 is the set of axioms of the theory. Writing

(1) $A \vdash_R \alpha$,

we shall state that the sentence α is derivable from the set of sentences (premises) A by means of rules in R. The elements of the set

(2) $\{\alpha : A_0 \vdash_R \alpha\}$,

i.e. all the sentences α of the language $\mathfrak{L}$ that can be derived from the axioms A_0 by means of rules in R, are called the *theorems* of the theory represented by the triple (A).

Triples like (A) are sometimes called (cf. e.g. Shoenfield [13]) *formal systems*. A mathematical theory in the semantic sense is a pair composed of a formal system and a set of its intended interpretations. The latter, at least in the case of elementary theories (theories formalized by means of the first order predicate calculus) are so-called relational structures, i.e. set-theoretic entities of the form

(3) $(X, R_1, \ldots, R_n)$,

where X is a set and $R_1, \ldots, R_n$ are relations defined on X. To illustrate this point, let us consider a couple of examples. As known, on the ground of set theory, we can define both the concept of a natural number and that of the sucessor relation, and thus we can define a certain relational structure. The arithmetic of natural numbers is often considered as the theory of this structure. As the second example, let us mention topology. It is thought of as the theory of all couples of the form (X, C) where X is an arbitrary set (space) and C is a closure operator on X, i.e. an operator that satisfies some well-known conditions.

It will be more convenient to represent the semantic notion of a mathematical theory by means of quadruples of the form

(B) $(\mathfrak{L}, \vdash_R, A_0, K)$

than by means of the pair $((\mathfrak{L}, \vdash_R, A_0), K)$. Clearly, the symbol K stands here for the set of intended interpretations of the relevant formal system. This, in particular, implies that $\mathfrak{L}$ is a language framed in such a way that it may be used to speak about all the structures in K. Con-

structing a theory we try to meet, among others, the following two requirements.

(1.1) The rules in R are infallible in the sense that if $A \vdash_R \alpha$ then, under every interpretation of extralogical concepts of $\mathfrak{L}$, α is true whenever all sentences in A are true.

(1.2) The sentences in A_0 are true under all intended interpretations.

We shall call them the *adequacy requirements*. Examining mathematical theories we need not presuppose that they are adequate in the sense that they satisfy (1.1) and (1.2).

Since the elements of the schema (B) (and thus also (A)) need not be components of a 'real life' mathematical theory, they are rather abstract entities devised so as to model the components of existing theories, we shall occasionally refer to (A) (to (B)) as a syntactic (semantic) *model of the notion of a mathematical theory*. Observe that a model of the notion of a mathematical theory should not be confused with a model of a mathematical theory. The latter notion is a synonym of the notion of an interpretation.

Might the techniques of metamathematics and its specific approach to the subject matter of investigations prove to be useful in attempting to give an account of peculiarities of empirical science? How, eventually, should the schemas (A) and (B) be modified for this purpose? Most of the problems we are going to discuss will be related to these two questions. Let us prepare the ground for subsequent considerations with some preliminary remarks stated in loose language. First of all let us notice that mathematics plays in empirical science the role entirely analogous to that which logic plays in mathematics. To put it more adequately, mathematics along with logics constitute the formal framework of empirical theories. Clearly, it can happen that an empirical theory involves no mathematical concepts. Theories of this kind, especially those of them which are based on first order predicate calculus (elementary theories), are particularly suitable for methodological investigations. The results of such investigations, being of some interest by themselves, can often be carried over to more complicated empirical theories. For this reason there is a number of treatises which deal with elementary empirical theories. Let me mention a monograph which seems to be especially valuable. I mean Przełęcki's *The Logic of Empirical*

Theories [11]. In most cases however, empirical theories make use of some mathematical tools. While dealing with theories of this kind, we must reinterpret the notion of derivability. If thus far $\vdash_R$ has been understood as logical derivability, now it ought to be understood as derivability which is defined both by logical and mathematical means available in the empirical theory under discussion.

Looking for a set-theoretic model of the notion of an empirical theory, we can try to construct it by modifying in a suitable manner either of the schemas (A) or (B). Since in this paper we shall focus our attention on some modifications of (B), now let us briefly discuss the schema (A) to examine, at least superficially, the possibilities it provides. For obvious reasons, an empirical theory could not be defined simply as a triple of the form (A). At least some statements asserted in empirical theories possess a factual content, while the criteria of assertion of mathematical statements, being entirely determined by rules of inference and axioms, are purely formal. This problem was widely discussed and settled in many different ways. For instance, in his celebrated paper 'Sprache und Sinn' [1], Ajdukiewicz admits that, besides inference rules, some languages are provided with rules allowing assertion of some statements subject to existence of some empirical facts. Ajdukiewicz calls them empirical rules of meaning. A number of philosophers sympathized with Ajdukiewicz's ideas, cf. e.g. Kokoszyńska [7]. Let us also mention, in this connection, Grzegorczyk's paper [6] which brought an exposition of a formal theory of empirical procedures of settling observational hypotheses. This line of thought gives rise to the following alteration of (A). An empirical theory is a quadruple

(A.1) $\quad (\mathfrak{L}, \vdash_R, E, A_0),$

where the symbols $\mathfrak{L}$, $\vdash_R$, A_0 keep their meaning unchanged, and E stands for a set of empirical rules of meaning (a set of empirical procedures of settling some factual hypotheses).

An alternative, but as a matter of fact inessentially different, way of defining the notion of an empirical theory is suggested by some ideas of representatives of logical empirism (cf. especially Carnap [2], [3], C. I. Lewis [9], M. Schlick [12]). Following that line, an empirical theory could be defined as a quadruple

(A.2) $\quad (\mathfrak{L}, \vdash_R, C, A_0),$

where $(\mathfrak{L}, \vdash_R, A_0)$ are to be understood once again in the formerly defined manner and C is a set of so-called correspondence rules which provide an interpretation, perhaps partial, of the primitive terms of the theory (theoretical terms) in the observation language.

I am familiar with no example of a logical reconstruction of an existing empirical theory which has been carried out after either of the patterns (A.1) and (A.2). Furthermore, I am afraid that there are some sound reasons to suspect that any attempt to produce such a reconstruction would encounter serious, if not unsurmountable, troubles (cf. in this connection, Giedymin [5]). The semantic approach seems to be much more promising. Let us turn back then to the schema (B).

From the semantic point of view, the substantial difference between mathematical and empirical theories lies in that they describe entities of different kind. While mathematics refers to abstract entities (at least so far as we have in mind the pure mathematics in contradistinction to interpreted mathematical theories such as interpreted geometry being a part of mathematical physics or theoretical mechanics), the empirical sciences refer always, in one way or another, to empirical (physical) objects, their properties and relationships they bear. Let us discuss the matter in somewhat more exhaustive manner.

The most typical entities being subjects of mathematical theories are relational structures of the form (3). Clearly, structures of that form can also be examined in empirical science. It may happen that we are interested in describing a population X in terms of the qualities $R_1, \ldots, R_n$. Any such inquiry provides us with an example of an empirical theory which refers to a structure of the form (3). Observe that such a theory, despite the qualitative character of its primitive terms, can make an extended use of mathematics, say mathematical statistics. In what follows, however, we shall be interested in those theories which describe the examined objects in terms of quantities not qualities, and thus the empirical systems they refer to differ from structures of the form (3). This matter will be discussed in detail in the next chapter. Still, for the purpose of the present discussion, we need to introduce two notions which are related to that of an empirical system (structure). Whenever an empirical theory is considered as a theory which may be applied to an empirical system X in the sense that the laws of that theory may be used in describing that system, we shall say that the system X is a *domain*

of the theory or else, is an *intended interpretation* of it. We shall avoid
using the term 'model' in the latter context, having in mind some other
applications of this word. The set of all domains of a theory will be
called its *range*. Clearly we do not exlude that the range of a theory is a
unit set. Still, it seems that whenever a theory is a theory, not only a
purely descriptive report on the results of some inquiry, and thus contains some general laws applicable to different objects, in different
areas, and in different periods of time, any attempt to give a realistic
account of what the theory is about leads to the conception of numerous
domains corresponding to numerous applications of the theory. I discussed this matter in may earlier papers [17], [18] mentioned in the
introductory note. Quite independently the issue was examined and
solved in the same manner by Sneed in his extremely valuable monograph [14].

Let us ponder how we can define the range of an empirical theory.
We certainly cannot use for this purpose merely set theory, which is so
powerful a tool in all metamathematical investigations. If the range of
an empirical theory can be defined in a linguistic manner, it can be
done only with the help of another theory sufficiently strong for this
purpose. We should not be surprised, however, if the range of an
empirical theory could not be adequately defined in a verbal manner at
all. Almost always, when an empirical law is asserted, is it provided with
some comments which are to make it clear what is the range of applicability of the law. These comments, however, putting aside their shortened and incomplete character, are usually stated in ambiguous terms having their meanings essentially dependent on the context in which they occur. The phrases such as 'natural conditions', 'characteristic turbidity of
the solution', 'full thermal isolation', 'enough small dimension' are unfortunately unavoidable components of such glosses. Studying an empirical theory one must, beyond some theoretical knowledge, afford
some practical ability to settle whether in a given situation a law may
or may not be applied. If the experts know how to select the systems
being the subject of their theory (branch of science) from all the entities
they deal with, and if their settlements on this matter evince a great
degree of unanimity, we are allowed to say that the range of this theory
is well defined, no matter if a verbal definition is available or not. Observe that to define a domain of an empirical theory, it is not enough

to define the set of its elements and the parameters it contains. We must also point out the period of time during which the defined structure is to be counted as a structure behaving in accordance with the laws of the theory. Whether or not a law provides an adequate description of an empirical system usually depends upon some conditions extraneous to this system. The planets follow the orbits determined by the laws of mechanics and gravitation and thus the solar system is a domain of particle mechanics provided, however, that there are no factors which desturb the natural course of events. Similarly, the trajectory of a meteorite which moves at a great altitude is determined by laws of mechanics and the gravitational law. The meteorite together with Earth and perhaps some other bodies which exert some gravitational attraction on it compose another domain of particle mechanics supplemented with the gravitational law under the proviso that the meteorite has not yet penetrated thick strata of Earth's atmosphere. Cuttings of tissue-paper and a small lead ball considered in the short period of time when they fall freely down in a glass tube from which the air has been pumped out compose still another domain of the theory discussed here.

Let us conclude our considerations as follows. It may happen that an adequate verbal definition of the range of an empirical theory is impossible. This, clearly, does not mean that this notion does not make sense, though it may affect prospects of investigations into some methodological problems related to a particular theory. In our subsequent considerations, however, we shall be concerned with general methodological problems. Imitating the approach applied in metamathematics we shall examine some sets of empirical theories defined by means of some conditions put upon the languages of those theories, the derivability relations, axioms and ranges. In an oblique manner, our consideration will be related to all empirical theories which might be reconstructed, without affecting their factual content, so as to satisfy those conditions.

It is assumed that the reader is familiar with rudiments of set theory and standard set-theoretic notation. However, I shall give an idea of the notation which will be applied and some highlights thereof. The symbols $\in$, $\subseteq$, $\cap$, $\cup$, will have their usual set-theoretic meanings. $\emptyset$ denotes the *empty set*. $X \times Y$ denotes the *Cartesian product* of the sets X and Y, X^n is the nth *Cartesian power* of the set X. $\bigcup X$ stands for the

union of the family of sets X, and $\bigcap X$ for its *product* (*intersection*). $\{x_1, \ldots, x_n\}$ is the set composed of the objects indicated in the brackets, $(x_1, \ldots, x_n)$ is the ordered set (*n-tuple*) of those objects. $\{x: \Phi(x)\}$ denotes the class of all the objects x which satisfy the condition $\Phi(x)$. $f: X \to Y$ reads: f is a mapping from X into the set Y. If $f: X \to Y$ and $X_0 \subseteq X$, the symbol $f/_{X_0}$ denotes the *restriction of f* to the set X_0. If x_1, x_2 are real numbers and $x_1 \leqslant x_2$ then $[x_1, x_2]$ denotes the set of all real numbers x such that $x_1 \leqslant x \leqslant x_2$. Whenever $x_1 \neq x_2$, the interval $[x_1, x_2]$ will be said to be *proper*. We shall use the symbol Re to denote the set of all real numbers. We shall also use the symbol $[Re]$ defined as follows.

(4) $$[Re] = \{[x_1, x_2] : x_1, x_2 \in Re, \text{ and } x_1 \leqslant x_2\}.$$

II. THE SET-THEORETIC MODEL OF THE FIRST KIND OF THE NOTION OF AN EMPIRICAL THEORY. REGULARITIES AND MODELS

1. *Quantitative structures.* Given any empirical object a and a time instant, we shall denote by (t, a) 'the object a taken at the time t'. Following R. Carnap we shall call every pair of the form (t, a) a *thing-slice*. Under the interpretation which we have proposed here, the symbol makes sense only if the object a exists (in the empirical sense) at the time t. It will be convenient to assume, at the very beginning, that the time instants are represented in a fixed manner by means of real numbers. For obvious reason, we shall assume that the correspondence between time instants and real numbers is one-to-one and furthermore it is fixed in such a way that the following two conditions are satisfied.

(2.1) $t_1 \leqslant t_2$ implies that the time instant corresponding to t_1 is not later (in the usual physical sense) than the time instant corresponding to t_2.

(2.2) $|t_1 - t_2| = |t_1' - t_2'|$ implies that the intervals of time determined by the time instants t_1, t_2 and t_1', t_2' respectively are equal.

Let U be a set of entirely arbitrary couples which satisfy the following conditions.

(2.3) If $(t, a) \in U$, then t is a real number.

(2.4) If $t_1 \leqslant t \leqslant t_2$ and (t_1, a), $(t_2, a) \in U$ then $(t, a) \in U$.

The aim of this construction is transparent. Write

(1) $Ob(U) = \{a : \text{for some } t, (t, a) \in U\}$.

We shall interpret the set $Ob(U)$ as the set of empirical objects and the set U as the set of thing-slices of objects in $Ob(U)$. Under this interpretation, to say that *a exists at time t*, means the same as to say that $(t, a) \in U$. The interval

(2) $\tau_U(a) = \{t : (t, a) \in U\}$

will be called the *period of existence of the object a in U*, and the union

(3) $\tau(U) = \bigcup \{\tau_U(a) : a \in Ob(U)\}$

will be called the *period of existence of U*. A set like U will be called a *universe*. If U and V are two universes and $V \subseteq U$ we shall say that V is a *subuniverse* of the universe U. Observe that if V is a subuniverse of U then $Ob(V) \subseteq Ob(U)$ but the identity $Ob(U) = Ob(V)$ need not hold. Some objects of the universe U may not exist in the universe V.

Given a universe U, we shall use the symbol $U^{(n)}$, where n is a natural number $\geqslant 1$, to denote the set defined as follows:

(4) $U^{(n)} = (t, a_1, \ldots, a_n) : (t, a_1), \ldots, (t, a_n) \in U\}$.

The set $U^{(n)}$ will be called the *nth limited Cartesian power of U*. Every mapping

(5) $F : U^{(n)} \to Re$

will be called an *n-ary numerical parameter* defined on U.

The number n will be called the *arity* of F. Whenever, $Ob(U)$ is interpreted as a set of empirical objects, F will be interpreted as a numerical time-dependent magnitude (quantity) characterizing objects in A. By writing

(6) $F(t, a_1, \ldots, a_n) = x$

we shall state that the magnitude F measured on the objects $a_1, \ldots, a_n$ at the time t takes the value x.

In this paper we shall not discuss any issues of the theory of measurement. Given any quantity F we shall always assume that F is measured in a fixed frame of reference by means of a fixed system of units. Thus

the results of measurement, the values of F, may be treated as pure numbers.

A system

$$(7) \qquad \mathfrak{X} = (X, F_1, \ldots, F_n)$$

will be said to be a *quantitative system (structure) defined on U*, if and only if U is a universe, X is a subuniverse of U and $F_1, \ldots, F_n$ are numerical parameters defined on U. Put

$$(8) \qquad Ob_t(X) = \{a: (t, a) \in X\},$$

$Ob_t(X)$ is then the set of all those objects which exist in X at time t. If

$$(9) \qquad Ob_t(X) = Ob(X)$$

for every $t \in \tau(X)$, the structure (7) will be said to be *conservative*. Roughly speaking a conservative structure is a structure which consists of the same objects at every moment of time. The set X will be called the scope of the structure $\mathfrak{X}$. Whenever we shall write $\mathfrak{X}$, $\mathfrak{Y}$, $\mathfrak{Z}$, to represent certain structures, their extensions will be denoted by X, Y, Z respectively. Throughout this chapter, whenever we shall use the term '*structure*' we shall mean a quantitative structure defined on some universe. Given two structures

$$(10) \qquad \mathfrak{X} = (X, F_1, \ldots, F_n),$$

$$(11) \qquad \mathfrak{Y} = (Y, G_1, \ldots, G_n)$$

which need not to be defined on the same universe, we shall say that they are *similar* provided that for every i, $1 \leqslant i \leqslant n$, the parameters F_i and G_i are of the same arity.

If $F_1, \ldots, F_n$ are $s_1, \ldots, s_n$-ary parameters respectively, the system of numbers

$$(12) \qquad s = (s_1, \ldots, s_n)$$

will be said to be the *similarity type* of the structure $\mathfrak{X}$. Thus the structures $\mathfrak{X}$ and $\mathfrak{Y}$ are of the same similarity type. We shall say that $\mathfrak{X}$ and $\mathfrak{Y}$ are *strictly similar* if and only if for every i, $F_i = G_i$. It follows that strictly similar structures may differ only as to the sets of objects they involve.

2. *Languages conformed to quantitative empirical systems.* The quantitative systems we have defined are obviously of a particular kind. Observe e.g. that they do not contain any vector quantities. In this case the loss of generality is an apparent one. Every vector can be represented by the system of magnitudes corresponding to the projections of the vector onto the axes of the chosen frame of reference. The structures which we are going to discuss do not contain quantities being functions of periods of time, such as e.g. work. For still another example, note that we have not taken into account the quantities defined on set of objects, say position of the centre of mass. Perhaps a more general approach would be desired. I have decided, however, to give as clear an exposition as possible of the ideas to be discussed here rather than one of maximal generality.

The next task will consist defining the notion of a language conformed to the description quantitative structures. A difficulty we shall face is the following one. It would be inconvenient to specify particular mathematical notions which are to appear in the languages we are interested in. Any decision concerning this matter would limit in an artificial way the range of our consideration. On the other hand, leaving the question of choice of a particular mathematical apparatus unsolved, we lose the possibility of defining some syntactical and semantical notions in the standard manner. We shall overcome this difficulty in the following way. The main definition of the notion of a language will be a general one. We shall illustrate it, however, by discussing some languages of a familar sort, namely the ones which are results of incorporating some mathematical terms into the language of the first order predicate calculus. It will easily be seen that these languages fall under the general definition.

Select any set K of quantitative systems. By a *language conformed to the structures in K*, briefly a *language of K*, we shall understand any pair

$$(12) \qquad \mathfrak{L} = (L, \{f_{\mathfrak{X}} : \mathfrak{X} \in K\})$$

which satisfies the following two conditions.

(2.5) L is a nonempty set.

(2.6) For every $\mathfrak{X} \in K, f_{\mathfrak{X}} : L \to \{T, F\}$.

The elements of the set $\{T, F\}$ will be interpreted as *truth* and *falsity*.

An element of L will be called a *sentence* of L and a mapping $f_{\mathfrak{X}}$ will be called an *interpretation* of L; more precisely $f_{\mathfrak{X}}$ is the interpretation of L in the structure $\mathfrak{X}$. The set of all interpretations of L will be called the *semantics* of $\mathfrak{L}$. In accordance with the meanings of the symbols T and F, we shall say that a sentence α of L is *true* in $\mathfrak{X}$, whenever $f_{\mathfrak{X}}(\alpha) = T$, otherwise we shall say that it is *false*. If α is true in $\mathfrak{X}$, $\mathfrak{X}$ will be said to be a *realization* of α. A sentence which is true in every $\mathfrak{X} \in K$ is said to be a *tautology* of $\mathfrak{L}$ and a sentence which is false in every $\mathfrak{X} \in K$ is said to be *contradictory*. If α is true in a structure $\mathfrak{X} \in K$, whenever all sentences in A are true in $\mathfrak{X}$, we shall say that α is a *semantical consequence* of A (*A entails semantically* α).

In what follows we shall deal exclusively with the languages conformed to a certain class of all conservative structures of the same similarity type. Having such a class of structures in mind, we shall examine in somewhat more exact manner a particular class of languages conformed to them. Denote by K_s the set of all conservative structures of the similarity type s. Let the structures in K_s be the structures of the form

(13) $(X, F_1, \ldots, F_n)$.

Observe that every such structure can easily be transformed into a three-sorted relational structure of the form

(14) $\big(Ob(X), \tau(X), Re, F_1/_X, \ldots, F_n/_X\big)$.

Let F_i be s_i-ary parameter. We have applied the symbol $F_i/_X$ in a slightly incorrect manner, namely $F_i/_X$ should be treated as a symbol of the restriction of F_i to $X^{(s_i)} \times Re$. Such a restriction of the parameters of the structure (13) is necessary in order to adjust them to the universes of discourse of the structure (14). As we remember, in general those parameters are defined on a superset of the set X. Observe also that assuming that the structures in K_s are conservative we have guaranteed that the restricted parameters $F_i/_X$ are defined for all objects in $Ob(X)$ and for all time instants in $\tau(X)$. Let us enlarge the relational structure (14) by adding some operation defined on the set of real numbers Re. Since our task here is to give an illustration of the general notion of a language, it does not matter which operations are selected for this purpose. Let

(15) $\big(Ob(X), \tau(X), Re, F_1/_X, \ldots, F_n/_X, E_1, \ldots, E_n\big)$

be the resultant relational structure. Any such structure may be described in the language of an applied first order calculus conformed to three-sorted relational structures. For definitions of many-sorted structures and their first order languages cf. [8]. An alternative approach is the following. Consider the one-sorted relational structure

(16) $(Ob(X) \cup Re, Ob(X), \tau(X), Re, F_1/_X, \ldots, F_n/_X, E_1, \ldots, E_m)$.

Now, all we need to speak about this structure is the standard first order language provided with some mathematical symbols (the symbols of the set Re and the symbols of operations $E_1, \ldots, E_m$). I shall assume that the reader is familiar with the techniques of forming such languages; they are discussed in every textbook of logic. A particular example of such a language was given in Montague's paper [10]. It would be then an unnecessary nuisance to describe the construction of those languages once again. Let $\mathfrak{L}_0$ be a language of the sort discussed conformed to the structure (16). Consider any structure

(17) $\mathfrak{Y} = (Y, G_1, \ldots, G_n)$

being an element of K_s. We shall say that a sentence α of the language $\mathfrak{L}_0$ is true in the structure $\mathfrak{Y}$, and the structure $\mathfrak{Y}$ is the realization of α if and only if it is true in the structure

(18) $(Ob(X) \cup Re, Ob(Y), \tau(Y), Re, G_1/_Y, \ldots, G_n/_Y, E_1, \ldots, E_m)$.

Observe that $Re, E_1, \ldots, E_n$ form the common part of structures (16) and (18). This expresses the fact that the meanings of mathematical symbols in languages of empirical theories are fixed. In the language discussed then, the variables which run over the set of real numbers, should always be interpreted as variables running over this set, and the symbols of operations $E_1, \ldots, E_n$ should always be interpreted as the symbols of those operations. Any complete description of the language $\mathfrak{L}_0$ involves both a definition of the set of sentences of this language and the set of interpretations of this language in a relational structure of the form (16). Defining the notion of truth relatively to structures in K_s we have defined in oblique manner the set of interpretations of L_0 in the structures K_s. In this way the language $\mathfrak{L}_0$ may be considered to be a language in the sense of our general definition. Observe that the notion of a tautology of $\mathfrak{L}_0$ and that of a semantic entailment depend

in an essential manner on the fixed meanings of mathematical symbols. Considering $\mathfrak{L}_0$ to be a typical representative of the languages we shall deal with throughout this paper, we shall refer to the tautologies of those languages as the *mathematical tautologies*, to the contradictions as the *mathematical contradictions*; whenever α is entailed semantically by the set of sentences A we shall often say that α is *mathematically entailed* by A. Clearly the notion of mathematical tautology (contradiction, entailment) does not coincide with that of logical tautology (contradiction, entailment). Furthermore, it should be noticed, that the considered notions in their mathematical version possess, in general, quite different properties than their logical counterparts. E.g. the notion of logical entailment can be defined in an adequate way by the notion of proof, the latter being defined in such a way that the question of whether a given sequence of sentences is a proof or not is decidable in an effective manner. This depends upon a particular choice of the mathematical operations $E_1, \ldots, E_n$, but in most cases the notion of provability cannot be used to replace the notion of mathematical entailment.

3. *The first model of the notion of an empirical theory and some of its applications.* We are now in the position to discuss the notion of an empirical theory in a more close and precise manner. Consider a quadruple

(I) $\qquad (\mathfrak{L}, \vdash, A_0, K)$

satisfying the following conditions.

(2.7) K is a set of conservative, strictly similar quantitative systems.

(2.8) $\mathfrak{L}$ is a language conformed to the set of all quantitative systems similar to systems in K.

(2.9) $\vdash$ is a derivability relation defined on the set of all sentences of the language $\mathfrak{L}$.

(2.10) A_0 is a set of sentences of $\mathfrak{L}$.

The notion of derivability which, thus far, was applied in an informal manner can be introduced by a strict definition. We shall say that $\vdash$ is a derivability relation defined on the set of sentences of the language $\mathfrak{L}$ if and only if $\vdash$ satisfies the following conditions:

(2.11) If $\alpha \in A$ then $A \vdash \alpha$.

(2.12) if $\{\beta: A \vdash \beta\} \vdash \alpha$ then $A \vdash \alpha$.

(2.13) If $A \subseteq B$ and $A \vdash \alpha$ then $B \vdash \alpha$.

(2.14) If $A \vdash \alpha$ then for some finite $B \subseteq A$, $B \vdash \alpha$.

The conditions (2.11)–(2.14) are the axioms for the consequence operation given by Tarski in one of his papers [15].

The notion of derivability once defined, all the components of the schema (I) are well defined. We shall call it the *set-theoretic model of the first kind of the notion of an empirical theory*. Given an empirical theory θ reconstructed as to fall under the schema (I) we shall denote its components with the help of symbols $\mathfrak{X}_\theta$, $\vdash_\theta$, A_θ, K_θ respectively. Thus we put

$$(1) \qquad \theta = (\mathfrak{L}_\theta, \vdash_\theta, A_\theta, K_\theta).$$

Obviously K_θ is to be interpreted as the range of θ, $\mathfrak{L}_\theta$ as its language, $\vdash_\theta$ as its derivability relation. The pair

$$(2) \qquad (\mathfrak{L}_\theta, \vdash_\theta)$$

will be called the *formalism* of θ. The set A_θ need not be interpreted as the set of the axioms of the theory θ. It may be interpreted, as well, as the set of all sentences asserted on the ground of the theory at the time at which the theory is considered. To give an example of a theory which can be viewed as a quadruple of the form (I) let us mention the classical particle mechanics *PM* in the formalization given by Montague [10]. $\mathfrak{L}_{PM}$, $\vdash_{PM}$ and A_{PM} were defined in [10]. ($\vdash_{PM}$ was defined simply as mathematical entailment. Some comment on the notion of the range K_{PM} of *PM* was given in Chapter I.) Cf. also in this respect Sneed [14].

Our considerations have reached an important stage. There exists a rather large body of methodological issues which may be examined in terms of the model (I) or some of its specifications. I would like to mention here two possibilities of utilizing this model in this respect. A fundamental task of empirical inquiry is to group observed phenomena into classes in conformity with similarities they manifest. This usually requires both specifying parameters in which observed phenomena are to be described and defining the external conditions which guarantee repeatability of phenomena. Thus, in the pretheoretical stage

of inquiry, we usually define implicitly or explicitly some classes of empirical systems and only next we try to grasp the regularities they manifest by means of relevant laws. In turn, the established theories help in discovering new phenomena and extending theoretical knowledge. This is a continuous process of development of science. I believe that some of its aspects can be described in terms of the notions defined here.

Let K be an arbitrary set of strictly similar quantitative systems. To simplify our discussion we shall assume that given any structure

(3) $\quad \mathfrak{X} = (X, F_1, \ldots, F_n)$

in K, the set $Ob(X)$ is a unit set. A structure of this kind will be called *elementary.*

Write

(4) $\quad Ob(X) = \{a\}.$

The n-tuple

(5) $\quad (x_1, \ldots, x_n)$

of real numbers selected to meet the conditions

(6) $\quad F_i(t, a) = x_i,$

will be called the state of *the structure $\mathfrak{X}$ at the time t* and will be denoted by $\mathfrak{X}(t)$.

(D1)   The class K (of strictly similar and elementary structures) will be said to be *deterministic* if and only if for every couple $\mathfrak{X}, \mathfrak{Y}$ of structures in K and for all time instants $t_1 \in \tau(X)$, $t_2 \in \tau(Y)$ and for every $\lambda \in Re$ such that $t_1 \pm \lambda \in \tau(X)$ and $t_2 \pm \lambda \in \tau(Y)$, if

(i) $\mathfrak{X}(t_1) = \mathfrak{Y}(t_2)$

then

(ii) $\mathfrak{X}(t_1 + \lambda) = \mathfrak{Y}(t_2 + \lambda).$

It is often claimed that the notion of determinism should be defined as referring to empirical theories rather than to empirical reality. Some philosophers, who advocate this point of view, argue that a sound defini-

tion of determinism of empirical systems is impossible. Montague, in his extremely incisive discussion of the notion of determinism given in [10] has also inclined toward the latter opinion. I am not going to discuss this matter from the philosophical standpoint, though as far as my philosophical credo is concerned, I should like to declare my sympathies with those who believe that it is course of events not sets of sentences which are deterministic. From a purely technical point of view, I do not see any trouble in giving an adequate definition of the notion of determinism (I believe that (D1) is such a definition), provided however that we keep in mind that determinism, like all regularities, must be defined as an attribute of sets of empirical structures not of single systems. Obviously besides the notion of a deterministic class of structures we may also define the notion of a deterministic theory. We shall give such a definition later.

There is a number of possibilities for weakening the notion of determinism. In particular we can introduce, in a rather obvious manner, the notion of probabilistic determinism. Montague's 'Deterministic Theories' [10], mentioned here several times, would be an inspiring lecture for those who are interested in further alterations of the notion under discussion.

Apart from regularities which may be termed '*of deterministic type*', we may consider regularities '*of correlational type*'. We shall define certain regularities of this sort restricting once again our considerations to elementary structures. Given a structure $\mathfrak{X}$ of the form (3), let us write

$$(7) \qquad \mathfrak{X}_k = (X, F_1, \ldots, F_k)$$

for every $k \leq n$. The structure $\mathfrak{X}_k$ will be called k-reduct of $\mathfrak{X}$. One more technical notion will be needed. Let $\mathfrak{X}$ and $\mathfrak{Y}$ be similar and let $\pi = [t_1, t_2] \subseteq \tau(X)$ and $\pi_t[t_1', t_2'] \subseteq \tau(Y)$. By writing

$$(8) \qquad \mathfrak{X}(\pi) = \mathfrak{Y}(\pi')$$

we shall mean that the following conditions are satisfied.

(2.15) $|t_1 - t_2| = |t_1' - t_2'|$,

(2.16) For every $\lambda \leq |t_1 - t_2|$,

$$\mathfrak{X}(t + \lambda) = \mathfrak{Y}(t_1' + \lambda).$$

(D2)   Let K be a class of elementary and strictly similar quantitative systems of the form (3). We shall say that *quantity* F_{k+1} $(k<n)$ *is definable in K* by means of $F_1, ..., F_k$ if and only if for every $\mathfrak{X}, \mathfrak{Y} \in K$ and every couple of proper time intervals $\pi \subseteq \tau(X)$, $\pi' \subseteq \tau(Y)$, if

   (i) $\mathfrak{X}_k(\pi) = \mathfrak{Y}_k(\pi')$

then also

   (ii) $\mathfrak{X}_{k+1}(\pi) = \mathfrak{Y}_{k+1}(\pi')$.

(D3)   Assuming that K is as in (D2), we shall say that *values of* F_{k+1} $(k<n)$ *are definable in K by means of values of* $F_1, ..., F_k$ if and only if for every $\mathfrak{X}, \mathfrak{Y} \in K$ and for every $t \in \tau(X)$, $t' \in \tau(Y)$, if

   (i) $\mathfrak{X}_k(t) = \mathfrak{Y}_k(t')$

then

   (ii) $\mathfrak{X}_{k+1}(t) = \mathfrak{Y}_{k+1}(t')$.

As in the case of determinism, we may consider various weakened versions of definability. For the purposes of this paper, multyplying of introduced notions of regularities would be unnecessary. Perhaps it is worth while to mention that the notions introduced here can be rather easily generalized so as to be applicable to sets of quantitative systems of an arbitrary kind, not only to elementary ones (cf. [18]).

Given an empirical theory θ consider a triple

(9) $T = (\mathfrak{L}_\theta, \vdash_\theta, A)$

where A is arbitrary set of sentences of $\mathfrak{L}_\theta$. The triples of this kind have been called formal systems. We shall, as hitherto, apply this terminology. Write

(10) $Cn_\theta(A) = \{\alpha : A \vdash_\theta \alpha\}$.

It is rather easily seen that the following is valid.

(2.17) The set of all realizations of A (i.e. the set of all structures in which every $\alpha \in A$ is true) coincides with that of $Cn_\theta(A)$ if and only if either A is a contradictory set (i.e. it has no realizations) or the derivabil-

ity relation is unfallible (i.e. for every set of sentences B and every sentence α of $\mathfrak{L}$, B entails semantically α whenever $B \vdash_\theta \alpha$).

The set of realizations of $Cn_\theta(A)$ will be referred to as the set of *realizations of the formal system T*. By the set of *realizations of the theory θ we shall clearly understand the set of realizations of $Cn_\theta(A_\theta)$*.

Let Λ be a regularity, say one of those we have defined. We shall say that the class of quantitative structures is Λ-*regular* whenever the regularity Λ is characteristic of this class.

(D4)   We shall say that:

(a) A formal system T is Λ-*regular* (e.g. *deterministic*) if and only if the set of realizations of T is Λ-regular (deterministic).

(b) An empirical theory θ is Λ-*regular* if and only if the formal system $T_\theta = (\mathfrak{L}_\theta, \vdash_\theta, A_\theta)$ is Λ-regular.

(D5)   We shall say that a formal system T is a *model of a set of structures K with respect to the regularity Λ* if and only if the following two conditions are satisfied.

(i) T is Λ-regular.

(ii) K is a subset of the set of realizations of T.

Constructing empirical theories we aim at making them (the formal systems being parts of them) models of their ranges with respect to certain regularities.

III. THE SECOND SET-THEORETIC MODEL OF THE NOTION OF AN EMPIRICAL THEORY AND THE NOTION OF APPROXIMATIVE TRUTH

1. *The second model of the notion of an empirical theory*. Asserting inside an empirical theory a law, or in general an empirical statement, we clearly claim that it is valid in every empirical system which belongs to the range of this theory. Thus the asserted statement is true whenever it is true in all those systems. If it fails in some applications, however, we cannot simply say that it is false, since this would imply that its negation was true and the latter need not take place. Claiming that an empirical hypothesis is false we ought to point out explicitly those systems in the range of the theory in which it is false. Thus the notion of falsity turns

out to be a relative one. Perhaps there is an inconvenience in this, but obviously it cannot be counted as a serious one.

The essential difficulty connected with the empirical notion of truth is of a quite different nature. Whenever we make use of quantities defined on a dense set of real numbers, the laws stated in terms of such quantities, and after all numerous empirical hypotheses as well, cannot be claimed to be simply true. If they are true, they are approximatively true. There is no need of an extended discussion on this point. The approximative nature of a great body of empirical claims is well-known and was raised many times by philosophers and scientists.

I do not see any possibility to define the notion of approximative truth until the notion of an empirical theory is understood in conformity with the model (I). Our next task will consist in modifying this model. The key notion which will be necessary for this purpose is that of an operational quantity. Let F be a k-ary quantity defined on a universe F. By an *operational measurement* of F we shall understand an operation p which transforms F into a function pF satisfying the following two conditions:

(3.1) $pF: U^{(k)} \to [Re]$.

(3.2) For every $(t, a_1, \ldots, a_k) \in U^{(k)}$,

$$F(t, a_1, \ldots, a_k) \in pF(t, a, \ldots, a_k).$$

Every function of the form pF, where p is an operational measurement of F will be called an operational quantity corresponding to F. We shall say that pF is a k-ary parameter, whenever F is a k-ary parameter. Given two measurements p and q we shall say that q is at *least as accurate as p*, whenever for every $(t, a_1, \ldots, a_k) \in U^{(k)}$,

(3.3) $qF(t_1, a_1, \ldots, a_k) \in pF(t_1, a_1, \ldots, a_k)$.

Consider any quantitative structure

(1) $\mathfrak{X} = (X, F_1, \ldots, F_n)$

defined on a universe U. Let $p_1, \ldots, p_n$ be operational measurements of $F_1, \ldots, F_n$ respectively. Put

(2) $p = (p_1, \ldots, p_n)$.

We shall write pF_i instead of p_iF_i, for short. Consider any ordered set of the form

(3) $X = (X, \Phi_1, \ldots, \Phi_n).$

If there exists a system p of operational measurements of the form (2) such that for every Φ_i,

(4) $\Phi_i = pF_i.$

We shall say that X is an *operational system* (*structure*) defined on U. Write

(5) $p\mathfrak{X} = (X, pF_1, \ldots, pF_n).$

The operational structure $p\mathfrak{X}$ will be called an *operational system corresponding to* $\mathfrak{X}$. By the *similarity type* of the system (5) we shall understand the similarity type of the system (1). Two operational structures will be said to be *similar*, whenever they are of the same similarity type. If they correspond to the same quantitative structure, they will be said to be *strictly similar*.

Observe that, in general, every operational structure corresponds to a number of quantitative structures. It is easily seen that the structure (3) can be regarded as corresponding to every quantitative structure

(6) $(X, G_1, \ldots, G_n)$

such that the following condition is valid.

(3.4) Let both G_i and Φ_i be s_i-ary parameters, then for every $(t, a_1, \ldots, a_{s_i}) \in A^{(s_i)}$,

$$G_i(t, a_1, \ldots, a_{s_i}) \in \Phi_i(t, a_1, \ldots, a_{s_i}).$$

Whenever an operational structure X corresponds to a quantitative structure $\mathfrak{Y}$ we shall say that $\mathfrak{Y}$ is an *idealization* of X. From the purely formal point of view, the notion of an operational measurement is rather artificial. It seems to be useful, however, in making clear what is the intuitive background of the introduced notions. Let F be a dense empirical quantity in the sense that the set of all its possible values is a dense set. E.g. we may assume that F takes all real numbers as its values. Let p be a particular empirical method of measuring F. Since F is dense, to every such method there corresponds a degree of accuracy

under which the quantity F can be measured. For this reason, the results of particular measurements are not simply numbers but certain intervals of real numbers, e.g. 0.17 ± 0.03 (i.e. $[0.14; 0.20]$). The notion of an operational measurement which has been introduced is a formal counterpart of the informal notion of a method of measurement. We may denote the quantity F measured after the method p by means of the symbol pF, thus arriving to the intuitive counterpart of the formal concept of an operational quantity corresponding to F.

2. *The notion of approximative truth.* The empirical reality we deal with is always given to us in a form of an operational system. If we agree with this point of view, we immediately see how the model (I) of the notion of an empirical theory should be modified. Consider any quadruple of the form

(II) $(\mathfrak{L}, \vdash, A_0, K)$

satisfying the following conditions.

(3.4) K is a set of strictly similar operational empirical systems.

(3.5) $\mathfrak{L}$ is a language conformed with the set of all quantitative systems similar to idealizations of the systems in K.

(3.6) $\vdash$ is a derivability relation defined on the set of sentences of $\mathfrak{L}$.

(3.7) A_0 is a set of sentences of $\mathfrak{L}$.

The schema (II) will be called the *set-theoretic model of the second kind of the notion of an empirical theory.*

Let

(1) $\varXi = (\mathfrak{L}_\varXi, \vdash_\varXi, A_\varXi, K_\varXi)$

be a particular theory reconstructed so as to fall under the schema (II). Accepting the model (II), we assume that the truth or the falsity of a sentence of the language $\mathfrak{L}_\varXi$ of the theory $\varXi$ depends on how this sentence reflects the states of affairs which take place in the domains of the theory. On the other hand, the language $\mathfrak{L}_\varXi$ is conformed to the idealizations of the domains of the theory and quantitative structures similar to the latter. Thus the semantics of the language $\mathfrak{L}_\varXi$ requires that the conditions of truth and falsity of sentences of this language in the structures being elements of K must be stated in terms of their truth or falsity relatively to some quantitative structures being ideal-

izations of the operational structures in K. From the intuitive point of view, every operational system X may be conceived a system which was devised as operational counterpart of a quantitative system $\mathfrak{X}$. Then the latter can be called the *proper idealization* of X. Perhaps one would be tempted to say that α is true in X whenever it is true in the proper idealization of X. Unfortunately, for obvious reasons the notion of the proper idealization is, at least in general, useless with respect to our present task. All that we have at our disposal when we are considering a theory like $\varXi$ is the range $K_{\varXi}$. Every idealization $\mathfrak{X}$ of an operational system X in $K_{\varXi}$ is a plausible candidate to be the proper idealization of X. This state of affairs suggests the following definitions.

(D6)   Let $\mathfrak{L}$ be a language conformed to the set K of all quantitative systems of a given similarity type, and let X be a operational system corresponding to a structure $\mathfrak{X}$ in K. We shall say that:

(a) A sentence α of the language $\mathfrak{L}$ is *true* in X whenever it is true in every idealization of X.

(b) A sentence α of L *is false* in X whenever it is false in every idealization of X.

(c) A sentence α of L is *indeterminate* in X whenever it is neither true nor false in X.

Perhaps it would be desirable to discuss an example. Assume that an unary parameter F is a parameter of a structure $\mathfrak{X} \in K$. Let the language $\mathfrak{L}$ contain familiar logical and mathematical symbols taken in their standard meanings. Consider an operational structure $p\mathfrak{X}$ corresponding to $\mathfrak{X}$. Assume that for a particular object a and a particular time t,

(2) $pF(t, a) = 7.34 \pm 0.03$.

Then, the sentence of $\mathfrak{L}$

(3) $F(t, a) = 7.34 \pm 0.03$

which clearly should be understood as a shortening of the sentence

(4) $7.31 \leqslant F(t, a) \leqslant 7.37$

is true. Every mathematical consequence of (3), e.g.

(5) $F(t, a) = 7.34 \pm 0.05$

is also true. Observe that

(6) $pF(t, a) = 7.34 \pm 0.05$

is false. (The latter sentence, which need not be a sentence of $\mathfrak{L}$ at all reads, informally, as follows: a measurement of F on a at time t performed after the method p gives the value 7.34 ± 0.05. This does not agree with our assumption (2).) The sentence

(7) $F(t, a) = 7.39$

is false, and the sentence

(8) $F(t, a) = 7.36$

is indeterminate. A negation of an indeterminate statement is indeterminate. Observe however that conjunctions, disjunctions and implications formed by indeterminate sentences, need not be indeterminate.

With the help of definitions stated earlier, we may easily prove the following.

(3.8) Assuming that $\mathfrak{L}$ and K are as in (D6) and K^* is the set of operational structures corresponding to structures in K, the following is valid:

(a) If α is a mathematical (semantic) tautology of L then α is true in every $X \in K^*$.

(b) If α is mathematically semantically contradictory then α is false in every $X \in K^*$.

(c) If A is a set of sentences of $\mathfrak{L}$, all sentences in A are true in a structure $X \in K^*$, and A entails mathematically α then α is true in X.

(d) If α is a sentence of $\mathfrak{L}$ indeterminate in X and α entails mathematically β, then β is either true or indeterminate in X.

The notion of an approximative truth, which we are going to define now, will be referred to sets of sentences, formal systems and also to theories.

(D7)   Let L and X be as in (D6). We shall say that a set of sentences A of the language $\mathfrak{L}$ is *approximatively true* in X, whenever there is an idealization $\mathfrak{X}$ of X such that every sentence α in A is true in $\mathfrak{X}$.

Observe that if A is a unit set, i.e. for some sentence α, $A=\{\alpha\}$ then A is approximatively true in X provided that α is either true or indeterminate in X, i.e. α is not false in X. If A was not a unit set the assumption that all sentences in A are not false would not allow us to conclude that A is approximatively true.

(D8)   Let $\varXi$ be as in (1). Consider a formal system $T'=(\mathfrak{L}_{\varXi},\ \vdash_{\varXi},\ A)$. We shall say that T' is *approximatively true* in an operational structure X similar to the structures in $K_{\varXi}$, provided that $Cn_{\varXi}(A)$ is approximatively true in X.

By (2.17) we easily conclude that the following is valid.

(3.9) Let T' be as in (D8). If $\vdash_{\varXi}$ is infallible then T' is approximatively true whenever A is approximatively true.

(D9)   Let $\varXi$ be as in (1). We shall say that the theory $\varXi$ is *approximatively true* whenever the formal part $(\mathfrak{L}_{\varXi},\ \vdash_{\varXi},\ A_{\varXi})$ of the theory $\varXi$ is approximatively true in every $X \in K_{\varXi}$.

The intuitive sense of this definition is, I believe, easy to grasp. Let e.g.

(9) $K^{*}=\{p_1 X,\ p_2 X,\ ...,\ p_N X\}$.

The operational measurements in the systems p_i (each p_i is to be interpreted like p in (2) in the preceding section) may differ as to their degree of accuracy. If $\varXi$ is approximatively true, none of them allows us to obtain a result of measurement that would be incompatible with the theory $\varXi$, i.e. with its laws or with the methods of proof characteristic of its formalism. Suppose now that a system of operational measurements q allows the values of some parameters of $\varXi$ to be established in a more precise manner than any of the systems p_i. It may happen that $\varXi$ is not true in qX. This simply means that qX cannot be included into the range of the theory $\varXi$, nevertheless, provided that it is approximatively true, $\varXi$ might be a valuable theory in dealing with the structures in $K_{\varXi}$.

3. *Quantities versus operational quantities.* From the radically 'operational' point of view, it is the notion of quantity which should be defined by means of the notion of operational quantity and not vice-versa. Anyway, if we like to regard operational notions as primitive and the other

as a sort of theoretic construction, then we should be able to show the road which leads from operational notions to their 'theoretical' counterparts. This issue, which I would like to discuss here briefly for its strong relevance to the problems touched upon in this chapter, is, I believe, of considerable philosophical interest.

Let X_i, $i = 1, 2, \ldots, N$ be subuniverses of a universe U, and let k be a fixed natural number. Consider the set of operational quantities

(1) $\{\Phi_i \colon X_i^{(k)} \to [Re]\}$.

(Thus to each X_i, which need not differ with each other, there corresponds an operational quantity Φ_i.)

Put

(2) $F = \{\Phi_1, \ldots, \Phi_N\}$.

We shall say that F is a *compatible* set of operational quantities if and only if for every Φ_i, Φ_j, if

(3) $(t, a_1, \ldots, a_k) \in X_i \cap X_j$

then

(4) $\Phi_i(t, a_1, \ldots, a_k) \cap \Phi_j(t, a_1, \ldots, a_k) \neq \emptyset$.

We shall say that F is U-complete provided that

(5) $\bigcup \{X_i \colon i = 1, \ldots, N\} = U$.

A quantitative parameter

(6) $F \colon U^{(k)} \to Re$

selected so as to meet the condition

(7) $F(t, a_1, \ldots, a_k) \in \Phi_i(t, a_1, \ldots, a_k)$

will be said to be a parameter approximated by the set of operational structures F.

The quantities involved in empirical theories can be conceived as quantities approximated by some compatible sets of operational quantities complete with respect to a suitable universe. When constructing a theory, as a rule we put some additional requirements on the considered quantities. E.g. in particle mechanics we require the position to be twice

differentiable. In this sense quantities may be said to be defined usually in an ambiguous way both by theory and by methods of measurement. On their side, they play a twofold role in empirical theories. They help to simplify the description of experimental data and the systematization of them. Thus they fulfil an instrumental role. At the same time, being materially related to the operational quantities they serve to state some facts on examined empirical systems.

4. *Concluding remarks*. I do believe that the notions defined in this paper and, perhaps first of all, the approach developed here are workable in dealing with the problems of methodology of science. As I tried to show, they allow us to treat some rather subtle philosophical issues with the same standard of preciseness as applied in metamathematics.

There is one paper which should be mentioned in connection with this eassy. I mean Tarski's famous 'The Concept of Truth in Formalized Languages' [16]. The ideas which had been developed there in many ways underlay the considerations which were conducted in this study. I will be happy if this paper is regarded to be a stage in realizing Tarski's program of applying strict methods in philosophy of science.

BIBLIOGRAPHY

[1] K. Ajdukiewicz, 'Sprache und Sinn', *Erkenntnis* **IV** (1934) 100–138.

[2] R. Carnap, *Introduction to Symbolic Logic and Its Applications*, Dover, New York, pp. 241.

[3] R. Carnap, 'Empiricism, Semantics and Ontology', *Rev. Int. Phil.* (1950). Reprinted in: *Semantics and the Philosophy of Language* (ed. by L. Linsky).

[4] R. Carnap, 'The Methodological Character of Theoretical Concepts', in: *Minnesota Studies in the Philosophy of Science* **I** (1959) (ed. by H. Feigl and M. Scriven).

[5] J. Giedymin, 'On the Theoretical Sense of So-Called Observational Terms and Sentences', this volume, pp. 111–134.

[6] A. Grzegorczyk, 'Próba ugruntowania semantycznej teorii wiedzy' ('An Attempt to Establish Semantical Theory of Knowledge'), *Przegląd Filozoficzny* **XLIV** (1948) 348–372.

[7] M. Kokoszyńska-Lutman, 'O pewnym warunku semantycznej teorii wiedzy' ('On a Certain Condition for Semantical Theory of Knowledge'), *Przegląd Filozoficzny* **XLIV** (1948) 373–382.

[8] G. Kreisel, and J. L. Krivine, *Elements of Mathematical Logic*, North-Holland, Amsterdam 1967.

[9] C. I. Lewis, 'Experience and Meaning', *The Philosophical Review* **43** (1934). Reprinted in: *Readings in Philosophical Analysis* (ed. by H. Feigl and W. Sellars), Appleton — Century–Crofts, New York 1949.

[10] R. Montague, 'Deterministic Theories', in: *Decisions, Values and Groups*, Pergamon Press, 1962, pp. 325–369.

[11] M. Przełęcki, *The Logic of Empirical Theories*, London 1969.

[12] M. Schlick, 'Meaning and Verification', *The Philosophical Review* **45** (1936). Reprinted in: *Readings in Philosophical Analysis* (ed. by H. Feigl and W. Sellars), Appleton–Century–Crofts, New York 1949.

[13] J. R. Shoenfield, *Mathematical Logic*, Addison–Wesley, 1967.

[14] J. D. Sneed, *The Logical Structure of Mathematical Physics*, D. Reidel, 1971.

[15] A. Tarski, 'Fundamental Concepts of the Methodology of the Deductive Sciences', in: *Logic, Semantics, Metamathematics*, Clarendon Press, Oxford 1956.

[16] A. Tarski, 'Der Wahrheitsbegriff in den formalisierten Sprachen', *Studia Philosophica* I (1936) 261–405; originally published in Polish in 1933; English translation: 'The Concept of Truth in Formalized Languages', in: *Logic, Semantics, Metamathematics*, Clarendon Press, Oxford 1956.

[17] R. Wójcicki, 'Semantyczne pojęcie prawdy w metodologii nauk empirycznych', (Semantic Notion of Truth in Methodology of Empirical Sciences), *Studia Filozoficzne* 3 (1969) 33–48.

[18] R. Wójcicki, 'Metody formalne w problematyce teoriopoznawczej' ('Formal Methods in Epistemology'), *Studia Filozoficzne* 1 (1972) 13–41.

REFERENCE

* First published by the Institute of Philosophy and Sociology of the Polish Academy of Sciences, Warsaw 1972. Appeared also in *Ajatus* **XXXV** (1973).

ZDZISŁAW ZIEMBA

RATIONAL BELIEF, PROBABILITY AND THE JUSTIFICATION OF INDUCTIVE INFERENCE *

1. INTRODUCTORY REMARKS

The argument presented in this paper is devoted to a critical analysis of a view on justification of inductive inference. This view has been first formulated by Keynes, and is represented in Polish literature by by J. Hossiason and M. Kokoszyńska. The representatives of this view see the justification of inductive inferences in the rationality of the degree of belief in the conclusion, motivated by belief in the premises, or else in the rationality of reinforcing the degree of belief in the conclusion. In the definition of the justification of inductive inference they use the concept of probability introduced axiomatically; it is said of the probability of a sentence q with respect to a sentence p that it equals the degree of rational belief in the sentence q, motivated by the belief in the sentence p. The adherents to the concept under discussion have not always formulated their views completely. In my opinion, there is a need to try to make some concepts more precise, or to fill in some gaps of the reasoning. I have attempted — without changing the main line — to fill these gaps so that my criticism does not take advantage of minor inexactitudes in the main construction.

2. INDUCTIVE INFERENCE AS THE PASSING FROM THE COMPLETE BELIEF IN THE PREMISES TO PARTIAL BELIEF IN THE CONCLUSION

Introspection supported by some linguistic habits is to supply information about the fact that, what we call conviction is subject to gradation. It is commonly said that one is 'not completely sure', 'now more convinced than before', 'absolutely sure', etc.[1] The conviction (complete

or not) with regard to a sentence q can be motivated by the complete belief in another sentence p. It is not always reflected by the common way of speaking. One might argue that the reason for it is that the belief in the sentence q is often motivated by the totality of our knowledge and that the specification of the sentences concerned by the conviction motivating the belief in the sentence q is then left out.

Inductive inference is sometimes characterized from the psychological point of view as a process of reasoning consisting in the belief in the sentence q to a degree smaller than certainty but greater than total unbelief, while this belief is motivated by the complete belief in another sentence p. In other words — as a process of reasoning consisting in *partial* acceptance of a sentence under the influence of *total* acceptance of another sentence. Hossiason says: "From our relativistic point of view probable inference, and therefore also inductive reasoning, is an inference only to the extent it allows to pass from believing in premises, the set of which we denote by Y, to *provisional* (partial) belief in conclusion x" (emphasis by Hossiason). [2]

Inference understood as the belief in the sentence q to a degree greater than un belief and smaller than certitude, under the influence of the belief in the sentence p, is not the same as the judgement "probability of the sentence q is so and so with respect to the sentence p".[3] The sentence 'probability of the sentence q is so and so with respect to sentence p' is not a conclusion here either. For in this case one (partially) accepts the sentence q, and not the sentence 'probability of the sentence q is so and so with respect to the sentence p'. In the concept of inductive inference under discussion the sentence q is the conclusion and the sentence p — the premise.

From the formal point of view inductive inference is usually characterized by the scheme of the form:

$$(2.1) \qquad f_1$$
$$\vdots$$
$$\frac{f_n}{f_{n+1}}$$

The sentential functions above the line characterize — from the point of view of logical structure — the premises, and the functions under

the line — the conclusion. Thus the following scheme is to be inference by analogy:

$$(2.2) \qquad x_1 \in X \wedge Y^4$$

$$. \; . \; . \; . \; . \; . \; .$$

$$x_{n-1} \in X \wedge Y$$

$$\frac{x_n \in X}{x_n \in Y}$$

It should be pointed out that a scheme like (2.2) describes only the formal aspect of inference and does not contain any indication concerning the psychological aspect of inference; nothing in this scheme indicates the degree to which the conclusion is accepted. In deductive schemes such an indication is not needed, as it is taken for granted that the conclusion is accepted with full certitude. If we assume that numerical values can be assigned to respective degrees of conviction, for example in the interval $[0, 1]$ then the scheme (2.1) can be completed by adding a variable ranging over the numbers of the interval established, with the exclusion of the numbers corresponding to full certitude or full disbelief. Such a completed scheme could be written in the form:

$$(2.3) \qquad f_1$$

$$\vdots$$

$$\frac{f_n}{f_{n+1}} \, (k)$$

for example

$$(2.4) \qquad x_1 \in X \wedge Y$$

$$. \; . \; . \; . \; . \; . \; .$$

$$x_{n-1} \in X \wedge Y$$

$$\frac{x_n \in X}{x_n \in Y} \, (k)$$

I shall say that a given inference conforms with the scheme of the form (2.3) if the acceptance of the sentences corresponding to the sentential formulas above the line makes us accept, to a degree different from certitude and different from total disbelief, a sentence corresponding to the sentential formula under the line (with the qualification that the same variables of a scheme are substituted with the same constants).

It seems that the role of the variable 'k' was played by the expression 'it is probable that' or 'probably', placed before the sentential formula under the line of the scheme by authors dealing with induction in traditional logic:[5]

(2.5) $x_1 \in X \wedge Y$

     . . . .

     $x_{n-1} \in X \wedge Y$

     $$\frac{x_n \in X}{\text{probable that } x_n \in Y}$$

Under no circumstances may the expression 'probable that' be interpreted here in the sense that it is probable that $x_n \in Y$ with regard to the fact $x_1, \ldots, x_{n-1} \in X \wedge Y$ and $x_n \in X$. The inscription (2.5) would then not be the scheme of inference, but that of *assertion* about the probability of certain events, unless we understand by inductive inference a judgement about probability. Placing of the expression 'probable that' before the sentential formula under the line of the scheme suggests that this expression has some sense independent of what has been written above the line. It seems that placing the expression under discussion below the line was to stress that the conclusion was accepted *partially* although thus the degree was not specified.

I suggest to differentiate between the scheme (2.3) and the inscription made out of it by replacing the variable 'k' by a number 'k_0', i.e. the inscription of the form:

(2.6) f_1
     $\vdots$
     $$\frac{f_n}{f_{n+1}} \, (k_0)$$

for example:

(2.7) $x_1 \in X \wedge Y$

     . . . .

     $x_{n-1} \in X \wedge Y$

     $$\frac{x_n \in X}{x_n \in Y} \, (\tfrac{1}{2})$$

Following K. Ajdukiewicz we shall call the inscription of the form (2.6) the *way* of inference.[6] The inscription (2.7) is one of the ways of inference by analogy. The various ways of inference by analogy differ only in number k_0.

In order to avoid misunderstandings I should like to emphasize that neither the inscription of the form notation (2.3), nor the inscription of the form (2.6) are rules of inference (following a given scheme or way) or assertions about the justification of an inference (following a given scheme or way). A *rule* of inference following a given scheme or way contains the term 'it is allowed' and its most general form is 'it is allowed to infer according to such and such a scheme (way) with regard to such and such an aim'. An *assertion* about the justification of an inference following a given scheme or way has the most general form 'the inference according to such and such a scheme (way) is justified with regard to such and such an aim'.[7]

3. THE JUSTIFICATION OF INDUCTIVE INFERENCES

If inductive inference is understood as the belief in a sentence to a degree smaller than certitude and greater than total disbelief, motivated by a total belief in another sentence, then one is inclined to consider the rationality of the degree of belief in the conclusion as the justification of the inference. It is, at the same time, said that the degree of actual belief in the sentence p, motivated by the belief in the sentence p can be *rational* or *irrational*.[8] It is assumed that for two sentences p, q there is only one rational degree of belief in the sentence q, motivated by the belief in the sentence p. It is moreover assumed that the rational degree of conviction (at least for some pairs of sentences) can be measured and expressed in numbers.[9] A relation is constructed between two sentences p, q and number k so that this relation holds when number k is the number assigned to the rational degree of belief in the sentence q, motivated by the total belief in the sentence p.[10] The fact that the relation holds is expressed in the sentence: 'the sentence q is probable to the degree k with regard to the sentence p'.[11] This relation is *objective*, i.e. it holds independently of whether anybody and to what degree believes in the sentence q under the influence of believing in sentence p.[12] The number k is called the probability of the sentence q with regard to the sentence p.

A doubt arises here, which is substantial for determining the justification of inductive inference. Authors dealing with the rational degree of belief do not precisely explain whether the rationality of the degree of belief in the sentence q, motivated by the belief in the sentence p, depends on the truth of the sentence p or not.[13] As a result it is not known whether the number k assigned to the sentences p, q as the degree of probability of the sentence q with respect to the sentence p, is the measure of the degree of conviction with which it is rational to believe the sentence q under the influence of the belief in the not necessarily true sentence p, or whether it rather is the measure of the degree of conviction with which if is rational to believe the sentence q under the influence of the belief in the sentence p, when p is *true*. Thus it is not exactly known whether if the probability of the sentence q with regard to the sentence p is k, the degree of belief in the sentence q, equal to k, is rational, if this belief is motivated by the belief in the sentence p, even should the latter be false, or whether the degree of belief in the sentence q, equal to k is rational, when the belief in the sentence q is motivated by the belief in the sentence p, and the latter is true. For instance, it is usually accepted that if the sentence q is logically implied by the sentence p, then the probability of the sentence q with regard to the sentence p equals 1, and the number 1 is assigned to total certitude. The question then arises whether the degree of belief in the sentence q equal to certitude is rational if this belief is motivated by the belief in the sentence p, even should the latter be false, or whether rather the degree of belief in the sentence q equal to certitude is rational if this belief is motivated by the belief in the sentence p and the latter is true. I am rather inclined to favour the second alternative. It seems that the situation is analogical in the case of rationality of degrees of conviction other than certitude.

Summing up the above-presented argument I should like to point out the distinction observed in this paper between *conviction* (belief), which is a psychological phenomenon, and *objective relation* between sentences and numbers, i.e. the relation of probability. In some cases there can be a sort of correspondence between this relation and someone's actual conviction: the degree of belief in the sentence q, resulting from the belief in the sentence p, equals the probability of the sentence q with respect to the sentence p and then the degree of belief in the sentence q is rational, possibly under the condition of the sentence p being true.

Returning to the problem of justification of inductive inference, consisting in a partial acceptance of the sentence q under the influence of the total belief in the sentence p, we can say that:

1. If the rationality of the degree of belief in a sentence q, motivated by the belief in a sentence p, does not depend on the truth of the sentence p, and if the corresponding number k assigned to the sentences p, q as degree of probability of the sentence q with respect to the sentence p equals the degree to which it is reasonable to believe the sentence q under the influence of the total belief in the sentence p, then inductive inference is justified if the degree of belief in the conclusion q under the influence of the belief in the premise p equals the probability of the sentence q with respect to the sentence p, i.e. if the degree of this belief equals k.

2. If the rationality of the degree of belief in a sentence q, motivated by the belief in a sentence p, *depends* on the truth of the sentence p, and an appropriate number k assigned to the sentences p, q as the degree of probability of the sentence q with respect to the sentence p equals the degree of conviction with which it is rational to believe the sentence q under the influence of the belief in the *true* sentence p, then inductive inference is justified if the degree of belief in the sentence q under the influence of the belief in the sentence p equals the probability of the sentence q with respect to the sentence p, i.e. when the degree of this belief equals k and the premise p is true. Similarly to the case of deductive inferences, we might speak here of formal justification and material justification of inductive inference. Namely we might say that inductive inference is *formally* justified if the degree of belief in the conclusion under the influence of the belief in the premise equals the probability of the conclusion with respect to the premise, and that it is *materially* justified in the premise is true.

Correspondingly to the concept of justification of a given inference, the concept of justification of a scheme or way of inductive inference might be constructed. One is inclined to understand the justification of a scheme (way) of inference so that *every* inference following the given scheme (way) is justified (if justification of inference is understood in the same way as in (1)), or formally justified (if justification of inference is understood in the same way as in (2)). Then the way of induc-

tive inference of the form:

$$(3.1) \qquad \begin{array}{c} f_1 \\ \vdots \\ \dfrac{f_n}{f_{n+1}} \; (k_0) \end{array}$$

would be justified if for every pair of sentences p, q such that the sentence p is a substitution instance of the formula '$f_1 \wedge \ldots \wedge f_n$', and the sentence q—a substitution instance of the formula 'f_{n+1}', there was:

$$(3.2) \qquad P(p, q) = k_0 \;^{14}$$

with the reservation that the same constants are substituted for the same variables in the formulas 'f_1', ..., 'f_{n+1}'. On the other hand, the scheme of inductive inference of the form:

$$(3.3) \qquad \begin{array}{c} f_1 \\ \vdots \\ \dfrac{f_n}{f_{n+1}} \; (k) \end{array}$$

would be justified for every pair of sentences p, q such that sentence p is a substitution instance of the formula '$f_1 \wedge \ldots \wedge f_n$', and sentence q—a substitution instance of the formula 'f_{n+1}', and for every k, there was:

$$(3.4) \qquad P(p, q) = k,$$

with the reservation that in the scheme the same constants are substituted for the same variables.

It is obvious that none of the traditional schemes of inductive inference, completed with the variable 'k' in the way indicated in (2.3), can be justified in the sense suggested above. It would thus be an absurdity to say that for every pair of sentences '$(a_1 \in A \wedge B) \wedge \ldots \wedge (a_{n-1} \in A \wedge B) \wedge \wedge (a_n \in A)$', '$a_n \in B$', and for every k, the probability of the sentence '$a_n \in B$' with respect to the sentence '$(a_1 \in A \wedge B) \wedge \ldots \wedge (a_{n-1} \in A \wedge B) \wedge \wedge (a_n \in A)$' equals k. Cf. the scheme (2.4). As a result the same sentence '$a_n \in B$' would have various degrees of probability with respect to the same sentence '$(a_1 \in A \wedge B) \wedge \ldots \wedge (a_{n-1} \in A \wedge B) \wedge (a_n \in A)$'. It is therefore necessary either to agree that the scheme (2.4) is justified in the sense that every inference according to it is justified, or to say that this

scheme needs completing. If we wanted to complete a given traditional scheme of inductive inference of the form (3.3) so as to obtain a justified *way* of inference, there would have to exist and be possible to define a *unique* value of the probability of a sentence q with respect to a sentence p, where the sentences p, q are *any* pair of sentences such that p is a substitution instance of the formula '$f_1 \wedge \ldots \wedge f_n$', and q — a substitution instance of the 'f_{n+1}', with the reservation that for the same variables in the formulas 'f_1', ..., 'f_{n+1}' the same constants are substituted. For example, in order to construct a justified way of inference by analogy there would have to exist and be definable a *unique* value of probability of *any* sentence being a substitution instance of the formula '$x_n \in Y$' with respect to *any* sentence being a substitution instance of the formula '$(x_1 \in X \wedge Y) \wedge \ldots \wedge (x_{n-1} \in X \wedge Y) \wedge (x_n \in X)$', provided the same constants were substituted for t he same variables. If such a number existed and was definable, it should be substituted for '$\frac{1}{2}$' in the inscription (2.7) in order to obtain a justified way of inference by analogy. Naturally, the question remains whether such a number exists and whether it is definable. We shall deal with this problem in Section 5.

The necessity of demonstrating the existence and of calculating the mentioned value can be avoided due to a certain modification of the traditional inductive schemes, for instance that of analogy, in the following way:

(3.5) $\quad P[(x_1 \in X \wedge Y) \wedge \ldots \wedge (x_{n-1} \in X \wedge Y) \wedge (x_n \in X), x_n \in Y] = k$ [15]

$\quad x_1 \in X \wedge Y$

$\quad \cdot \ \cdot \ \cdot \ \cdot \ \cdot \ \cdot \ \cdot$

$\quad x_{n-1} \in X \wedge Y$

$\quad \dfrac{x_n \in X}{x_n \in Y} \ (k)$

For the sake of further discussion we shall generalize this scheme to the form:

(3.6) $\quad P(p, q) = k$ [16]

$\quad \dfrac{p}{q} \ (k)$

It seems that the schemes (3.5) and respectively (3.6) are valid on the grounds of the theorem:

$$(3.7) \qquad P\{[P(p, q)=k] \wedge p, q\}=k.$$

It seems that the theorem (3.7) can be accepted on the basis of the meaning assigned to the concept of probability if by probability of the sentence q with respect to the sentence p we understand the degree of belief, expressed by a number, to which it is rational to believe the sentence q under the influence of the total belief in the *true* sentence p. The theorem (3.7) then says that the number k is assigned to the rational degree of belief in the sentence q, motivated by the belief in the sentence p and in the sentence asserting that the probability of the sentence q with respect to the sentence p equals k if the sentences p, '$P(p, q)=k$' are true. In other words, the theorem (3.7) says that the number k is assigned to the rational degree of belief in the sentence q if this belief is motivated by the belief in the true sentence p and in the true sentence asserting that it is the rational degree of belief in the sentence q, motivated by the belief in the sentence p, that equals k. On the other hand, the theorem (3.7) does not seem to be true if by probability we understand the degree of belief which is rational regarding some sentence under the influence of the belief in another sentence, not necessarily true. It may then be rational to believe the sentence q in a degree greater than k if one believes in the sentence p and in the sentence asserting that the degree k of belief in the sentence q under the influence of the belief in the sentence p is rational. We shall, therefore, in discussing the justification of the schemes (3.5), (3.6), understand by the term 'probability' the degree of the rational belief in a sentence if this belief is motivated by the total belief in another *true* sentence (cf. Section 2, p. 11).

It is particularly important that in the schemes (3.5), (3.6) 'k' is a variable and not a symbol representing a definite number, and it is not necessary to specify this variable in order to demonstrate the *validity of these schemes*. Naturally, in order to use such a scheme in practice one would have to know how great k is in given substitutions for 'x', 'Y', 'X'. This, however, would already be the question of the truth of premises, of a concrete inference, and not the question of validity of the scheme.

Maybe Keynes considered the scheme (3.6) the most general scheme of inductive inference, as he wrote: "If h is the factual evidence on which we base our conviction, then there exists our knowledge in the form of the sentence q that sentence p bears the relation of probability in the degree α to h, and this knowledge justifies our rational belief in the sentence p in the degree α".[17] The author speaks here of the knowledge 'in the form of the sentence q', the knowledge that '$P(h, p)' = \alpha$. Therefore, our partial conviction in the sentence p is motivated not only by 'factual evidence h', but also by the conviction that '$P(h, p) = \alpha$', and the whole inference follows the scheme (3.6). But the justification of such an inference is based not only on the truth of the premises, among others of the premise '$P(h, p) = \alpha$', but also on the theorem (3.7).

Naturally, the scheme (3.5) is not a traditional scheme of analogy. But it may be assumed that the tasks of logic do not consist in attempts at demonstrating the validity of every scheme, should it even be actually used. The tasks of logic may be regarded as the construction of definite relations between sentences resp. the construction of schemes of inference which are the *reconstruction* of schemes actually used. In carrying out such a reconstruction one can take into consideration the existing scheme of inference and use as many as possible elements of the scheme actually used. i.e., if possible, leave the logical structure of the conclusion and of the premises unchanged by constructing a scheme only completed with premises lacking in the scheme actually used.

On the other hand one might try to defend the view that traditional schemes of inductive inferences are enthymematic schemes in the sense that they contain an incomplete description of the premises actually taken into consideration in the inference.

Still another attempt at reconstructing traditional schemes of inductive inference is possible.[18] It constists in substituting in the scheme of the type:

$$(3.8) \qquad f_1$$
$$\vdots$$
$$\frac{f_n}{f_{n+1}} \ (k)$$

the variable 'k' by an expression denoting the value of the probability

of the sentence obtained by substitution in the formula 'f_{n+1}' with respect to the sentence obtained by substitution in the formula '$f_1 \wedge \ldots \wedge f_n$', for example:

(3.9) $x_1 \in X \wedge Y$ [19]

. . . .

$x_{n-1} \in X \wedge Y$

$$\frac{x_n \in X}{x_n \in Y} \quad P[(x_1 \in X \wedge Y) \wedge \ldots \wedge (x_{n-1} \in X \wedge Y) \wedge (x_n \in X), x_n \in Y]$$

Every inference following this scheme is justified (in the sense defined above), as the conclusion is accepted in the degree equal to its probability with respect to the premises.

4. INDUCTIVE INFERENCE AS REINFORCEMENT OF BELIEF IN THE CONCLUSION AND ITS JUSTIFICATION

We shall now consider some problems connected with defining inductive inference as consisting in the reinforcement of partial belief in the conclusion under the influence of the total belief in the premises.

First of all, let us consider the following question: in comparison to what the conviction is reinforced here? It is usually understood so that before the inference is begun one already has a belief in a sentence q, in a degree smaller than certitude, a belief motivated by the belief in sentence w. The process of inductive inference is to consist in the fact that one accepts q in a degree greater than before under the influence of the belief in the sentence w and another sentence p. It is then considered that such an inference is justified if:

(4.1) $P(w, q) < P[(w \wedge p), q]$.

In her paper 'Induction et Analogie' Hossiason has generally defined induction as reasoning in which, starting from a thesis f_1, one reaches reinforcement of the conviction (degree of belief, degree of certitude) concerning a sentence h, where h is the logical reason of the sentence f_1. The author has further stated that such reasoning is justified if the premises reinforce the degree of rational belief in the conclusion, i.e. if $P[(f_1 \wedge s), h] > P(s, h)$, where the sentence f_1 results from the conjunction of sentences h, s.[20] Similarly, Kokoszyńska says that "the in-

ference about the premise (conjunction of premises) 'f' and the conclusion 'h' is correct if and only if h has with respect to f and w a greater probability than with respect to w (the assumed knowledge)". The authoress also explains that "it seems acceptable to assume that an inference providing the justification of its conclusion should give an argument motivating the *reinforcement* of our *belief* in the conclusion and make our knowledge progress in some way in the case of the truth of the premises" (italics mine $-$ Z.Z.).[21]

Serious doubts arise as to whether what has been said above on the rational degree of belief agrees with the requirement (4.1), and whether satisfying (4.1) is at all necessary for the justification of the inference just described.

First, it seems that it is *not sufficient* to say only that $P[(w \wedge p), q]$ is greater than $P(w, q)$ in order to consider *any* reinforcement of the belief in the sentence q justified. At the first glance already the reservation seems necessary that such a reinforcement cannot reach the level of certitude unless the sentence q results from the conjunction of sentences w, p. It seems necessary to determine to what level can the belief in the sentence q be reinforced. For a case seems possible that someone will believe in the sentence p motivated by the knowledge w, a belief the level of which exceeds the rational degree of belief and where any further reinforcement would be totally unjustified.

Second, it seems that it is *not necessary* that (4.1) occurs to make the reinforcement of the belief in the sentence q under the influence of the belief in the conjunction of sentences w, p rational, and thereby justified. Even if the situation were just the opposite:

(4.2) $P(w, q) > P[(w \wedge p), q]$

then if only x believes in the sentence q, motivated by the belief in sentence w, in a degree smaller than rational, he can justifiably reinforce that belief to the rational level.

It seems that authors dealing with inference as reinforcement of the degree of conviction in the conclusion sometimes mix up the concept of the *actual* degree of belief with that of *rational* degree of belief. Obviously, in an inference consisting in the reinforcement of the belief in the sentence q, we have:

(4.3) $W(w, q) < [W(w \wedge p), q]$

where '$W(w, q)$', '$W[(w \wedge p), q]$' denote respectively the degree of actual belief in the sentence q, motivated by the belief in the sentence w or in the conjunction of sentences w, p. But this condition is a *definitional* one: we speak of the reinforcement of the degree of actual belief in the definition of inference itself. If we were to understand inductive inference as reinforcement of conviction, then non-fulfilment of the condition (4.3) would mean that in the given case we would not deal with inductive inference at all. Of course, if inductive inference were to be defined as reinforcement of the *rational* degree, i.e. as passing from the belief in the sentence q motivated by the sentence w, of the degree $P(w, q)$, to the belief in the sentence q motivated by the belief in the conjunction of sentences w, p, of the degree $P[(w \wedge p), q]$, then it would be natural to require that:

$$(4.4) \qquad P(w, q) < P[(w \wedge p), q]$$

but again it would be a definitional condition of inference and not of its justification. The mental processes for which this condition would not be fulfilled would not be inductive inferences at all.

It also seems to me that those who speak of inductive inferences as of reinforcement of the belief in a given sentence forget that the reinforcement of a conviction consists in *replacing* one conviction by another, the strength of which is greater than that of the former one. It is relevant here that the final inference consists in *believing* the sentence q, and this belief remains on a determined level of intensity. If we assumed the most primitive measurability of levels of conviction, following the principle 'lower levels — smaller numbers, higher levels — greater numbers' (as it has been done with respect to temperature) then we might try to present the inference consisting in reinforcing the belief in the conclusion by the following scheme:

$$(4.5) \qquad \frac{w}{q}(k) \Longrightarrow \frac{w \wedge p}{q}(t),$$

where:

$k < 1$,

k different from the number assigned to total unbelief,

t different from the number assigned to total certitude.

Naturally, the sign '$\sqsubset\!\!\triangleright$' is neither the sign of implication nor of consequence here. By means of this sign we denote the 'passing' from the belief in the sentence q in the degree k to the belief in the sentence q in the degree t greater than k, motivated by the belief in the sentences w, p. In other words, it means that the belief in the sentence q in the degree k *precedes* the belief in sentence q in the degree t.

Thus the whole difference between the inductive inference following the scheme (4.5) and the one we were considering in Section 2 consists in the fact that here the previous belief in the sentence q is taken into consideration while there it did not matter whether the belief in the sentence q motivated by the belief in the sentences w, p had been preceded by the belief in the sentence w, and particularly whether the level of the belief had been smaller. According to the former definition of inductive inference the belief in the sentence q motivated by the belief in the sentence w could well have been *greater* than the belief in sentence q motivated by the belief in the sentences w, p. According to the former definition inductive inference could also have been the *weakening* of the belief in the sentence q. An inference consisting in reinforcing the belief in a sentence is, as a matter of fact, only a special case of an inference consisting in believing a sentence to a degree smaller than certitude and greater than total unbelief, motivated by the belief in another sentence. There is then no need to define inductive inference alternatively, as consisting in believing a sentence to a degree smaller than certitude and greater than total unbelief, motivated by the belief in another sentence, or in the reinforcement of the conviction (not attaining certitude) motivated by the belief in the sentences w, p, because such a definition is a pleonasm. The alternative definition of inductive inference has been introduced by Hossiason in the paper 'O prawomocności indukcji hipotetycznej' ('On the Justification of Hypothetical Induction'): "In these reasonings (i.e. inductive inferences – Z.Z.) we do not accept the conclusion with full certitude, we only *suppose* or *reinforce the supposition* that the conclusion is true" (italics by Hossiason).[22]

In the light of these remarks it seems that the justification of an inference consisting in the reinforcement of the belief in the sentence q should rather be defined in accordance with the view presented before. Thus an inference consisting in the reinforcement of the belief in the sentence q under the influence of the belief in the (previously unknown)

sentence p should be considered as justified only if the degree of belief in the sentence q motivated by the total belief in the sentence $w \wedge p$ is rational, i.e. equals $P[(w \wedge p), q]$.

5. PROBABILITY AS RATIONAL DEGREE OF BELIEF

The above discussion was to show how the representatives of some views on inductive inference attacked the problem of justification of such inference. In particular the role played by the concept of probability as rational degree of belief has been shown. I attempted to show how — in my opinion — schemes of inductive inference should be constructed, and particularly those of inference by analogy, in order to be adapted to the accepted definition of the psychological aspect of inductive inference and to be justified schemes. I have employed the term 'probability' without a detailed analysis of its meaning. Now the concept of probability as rational degree of belief will be considered, and the difficulties will be presented which have their roots in the very concept of probability as rational degree of belief but also concern the problem of justification of inductive inference in the approach presented above.

"The history of the theory of probability is that of the attempts to find an explanation of the pre-scientific notion of probability".[23] According to Carnap, the various definitions were different not only in their definienses, but also in the definienda. For in extra-scientific language there are at least two terms identical in their material aspect but different in meaning. Besides the term 'probability' in the sense of relative frequency there exists the term 'probability' in the sense of degree of confirmation, or else logical probability. This second term was subject to explanation in discussions devoted to probability as degree of rational belief.

"The definition of probability is not possible" — says Keynes, and he advises to restrict it to the explanation that the degree of probability is the degree of rational belief.[24] Similarly, Jeffreys thinks that it is possible to say that probability expresses the degree of rational conviction, but that such a statement cannot be considered as explaining the meaning of the term 'probability'. "Actually this concept can be properly described by reference to cases of its use".[25] Thus, instead of a definition, the representatives of the view of probability as rational degree of belief

introduce a system of axioms which – in their opinion – express their intuitions about rational belief. The systems of axioms accepted by those who understand probability as a degree of rational belief are, in principle, similar to each other. I shall quote here as an example Hossiason's axiomatics.[26] The authoress says: "Let us assume, ..., that some non-negative numbers can be assigned to the degrees of belief which we have towards some sentences on the bas s of some others – at least for sentences with which these considerations are concerned, i.e., sentences occurring in inductive reasonings. Let us agree that the number 1 is assigned to certitude. Let us call the degree of belief in the sentence x on the grounds of the set of sentences Y – if this set is not contradictory – *the likelihood of x with respect to Y*, and let us denote the corresponding number by the symbol '$w(x, Y)$'" (italics by *J.H.*). This concerns the degree of belief which we are permitted to have towards the sentence x, i.e. not the actual belief but the *rational* belief. The degree of likelihood of sentence x with respect to the sentence Y is then here the rational degree of belief in sentence x, motivated by the belief in sentences belonging to the set Y. What Hossiason calls likelihood other authors call probability. Further on I shall continue to use the term 'probability' instead of 'likelihood', and I shall write '$P(Y, x)$' where Hossiason writes '$w(x, Y)$'.

Hossiason adopts the following axioms:

I: if x is implied by Y, then $P(Y, x)=1$,

II: if $\sim(x \cdot y)$ is implied by Y, then $P(Y, x+y)=P(Y, x)+P(Y, y)$,

III: $P(Y, x \cdot y)=P(Y, x) P(Y+\{x\}, y)$,

IV: if the sets X, Z are equivalent, then $P(X, y)=P(Z, x)$.

Hossiason explains that '$x \cdot y$' is the name of the product having as components the sentences x, y; '$x+y$' is the name of the logical sum of the components x, y; the symbol '$\{x, y, ...\}$' denotes the set composed of the members $x, y, ...$ The equivalence of two sets X, Z takes place when every sentence of the set X is implied by the set of sentences Z and conversely.[27]

As we have seen, the problem of justification of inductive inference concerns the question of measurability of probability. The authors considering probability as rational degree of belief and introducing this

concept axiomatically refer to the principle of *indifference*. It is possible to make use of the introduced axioms and conventions concerning some numerical values of probability only if it is known that definite sentences make an exclusive alternative, the probability of which with respect to a given sentence equals one and the probabilities of each of its terms are equal. A criterion is therefore needed which would allow to state the equality of probabilities of elements of exclusive alternative. Among others, Keynes and Jeffreys advise to use the principle of indifference. Jeffreys says: "If there is no reason to believe one hypothesis rather than another – the probabilities are equal. In terms of our fundamental notions of the nature of inductive inference, to say that the probabilities are equal is a precise way of saying that we have no ground for choosing between the alternatives" (more accurately: between elements of a disjoint alternative – *Z.Z.*). And: "The rule that we should then take them (the probabilities – *Z.Z.*) equal is not a statement of any belief about the actual composition of the world, nor is it an inference from previous experience; it is merely the formal way of expressing ignorance".[28]

These statements are not very clear. It is said that probabilities are equal when there is *no reason* to choose one of the hypotheses. The expression 'there is no reason to choose' may be understood in various ways. It cannot mean, of course, that one's knowledge does not *actually* motivate the belief in one hypothesis to a greater degree than in another. Such an understanding of the expression 'there is no reason to choose' would simply mean that the *actual* but not necessarily rational degree of belief in one hypothesis is equal to the degree of belief in another. The equality of probabilities would then denote the equality of the *actual* degree of conviction, which has been strongly opposed by Jeffreys.

But the expression 'there is no reason to choose' can be understood otherwise. Namely, it may be considered that, if we do not know which of the two hypotheses is true, the only reason to choose or to prefer one of them is the knowledge that the probability of the chosen hypothesis is greater than that of the other. Therefore the lack of such a reason means the absence of knowledge that the probability of one hypothesis is greater than that of another. To be more precise: someone does not know the degree of probability of the components of an exclusive alternative, and he believes these hypotheses to the same degree or not. We then deal with the ignorance of the degrees of probabilities

of definite sentences. However, considering in this situation these probabilities equal may be objected to on the grounds that knowledge is arbitrarily substituted for ignorance. That the degrees of probability of some sentences *are not known* is not sufficient to assume that the probabilities are equal. There appears the danger of failure of the theory of probability as theory of rational degree of belief. For, it would be difficult to accept that if the rational degree of belief in a given sentence is unknown, then the rational degree of belief is such and such.

These two 'subjectivist' interpretations of the expression 'there is no reason to choose' can be opposed by an 'objectivist' interpretation. Kneale says: "Instead of *knowledge of absence*, Laplace and those who agree with him accept the absence of knowledge as a sufficient ground for judgements of probability" (emphasis by the author). And: "I have argued that we are entitled to treat alternatives (more exactly: elements of an exclusive alternative — Z.Z.) as equiprobable if and only if we know that the available evidence does not provide a reason for preferring anyone to the other".[29]

However, the question arises when our knowledge does not give any reason to prefer one alternative to the other. If we answer that it is so when the probabilities of the respective hypotheses are equal with respect to our knowledge — we shall enter a vicious circle. If, on the other hand, the reasons are precisely described without referring to the probability of various hypotheses, new difficulties will arise, varying depending on whether the principle of indifference is considered as the *definition* of the equality of the probabilities (in the sense of a sentence true on the grounds that it has been decided to understand the terms contained in it so that the sentence be true) or whether it is considered as a *theorem* about the equality of the given probabilities.

If the principle of indifference, appropriately precised, were to be the definition, or an additional axiom, of the equality of given probabilities, it would thereby be the definition — at least to some extent — of the concept of probability itself. For example, the establishment of the definition of the equality of two weights is at the same time the establishment to some extent of the contents of the concept of weight. Of course, one may decide to understand the term 'probability' so that in some cases probabilities be equal. But it should not be forgotten that the concept of probability has been introduced among other things in

order to formulate the criterion of justification of inferences of a given type, and the introduction of this concept was accompanied by the comment that probability is a numerical value characterizing the rational degree of belief in one sentence, motivated by the belief in another sentence (possibly true). The question then arises whether by adding the principle of indifference as partial definition of the term 'probability' (or as an additional axiom), we shall obtain the solution of the problem of inductive inferences we are interested in. The acceptance of the principle of indifference as definition equals to a decision as to the way of understanding the term 'rational degree of belief'. The concept of rational belief is arbitrarily petrified to such an extent that it entails the establishment of rational degrees of belief for several pairs of sentences other than those dealt with by the principle of indifference. On the other hand, the concept of rational degree of belief has not been definitely explained and it is not really clear why it is rational to believe a given sentence under the influence of the belief in another sentence to such and such degree. The only answer which can then be given is: this is how it has been decided to understand the term 'rational degree of belief'. However, one would prefer that the concept of rational degree of belief be first wholly explained, and that only then it be demonstrated that the degree of belief for given pairs of sentences equals so and so. In other words, one would like to know in what sense and why is such and such degree of belief in a given sentence rational, and the expected answer is not that it results from the arbitrary decision on the grounds of which a definite degree of belief in a sentence, motivated by the belief in another sentence, has been called rational.

If the principle of indifference is to be a definition, then of course it does not need any justification. A *decision* is sufficient, concerning a definite understanding of the terms contained in the principle. If, however, the principle of indifference were to be the *theorem* that in such and such cases the respective probabilities are equal, then it would obviously need a justification. It would then be necessary to show that, in the case of the established meaning of the term 'probability' the probabilities of appropriate sentences are equal. But then the equality of some probabilities would have to be possible to be established without reference to the principle of indifference. Should this be done, it would then be necessary to demonstrate that they are according to the require-

ments of the principle of indifference. But the principle of indifference is referred to precisely because there are no other ways of establishing the equality of probabilities.

The whole difficulty in determining the degree of probability without referring to other probabilities results from the vagueness of the concept of probability. Axiomatic approach to that concept leaves it somewhat unspecified, and as far as its contents is concerned, we only know what has been said in the axioms and in theorems derived from these axioms. If there is no precise definition of probability, the search for a justification of the principle of indifference must remain unsolved. Keynes has even clearly given up such a justification. In his opinion, in many cases it is possible to find out directly (*direkt erkennen*) that a conclusion is probable with respect to its premises. "Without questioning the objective character of the relation of probability we must agree that the search for a way of determining probability without intuition or a direct judgement (*direktes Urteil*) is unlikely to be successful. There is general agreement that we often directly see that a conclusion is entailed by a premise. It is not an important extension (of his opinion) to assume that we often see that a conclusion is *partly entailed by a premise*, or that it is probabilistically related to it. Moreover, the lack of results in the attempts at defining probability in terms of other logical concepts makes it reasonable to presume that some probabilistic relations can be known directly and are not obtained, by means of some rule, from data containing no reference to probability". And: "The logical system of the theory of probability is to enable us to know relations difficult to observe by using easier observable relations, that is to change uncertain knowledge into a more distinct one".[30]

This specific apriorism, represented by Keynes, is hardly defendable. For who can guarantee that we are not misled by our intuitions? Probability is the rational degree of conviction, something objective — as the author assures us — and independent of what one might suppose. On the other hand, it is not very accurate to compare an intuitive grasping of ('complete') entailment with an intuitive grasping of the equality of specified probabilities, as first, our intuitions are often misleading as far as inference is concerned, and second — they can be tested due to a sufficiently clear determination of the concept of itself. But it is impossible to test whether our intuition does not fail us with respect to the

finding of definite probabilities, as it is not very well known what is the probability of a sentence with regard to another sentence.

Another critical thought should be added to these remarks about the principle of indifference as way of stating the equality of definite probabilities which is to be further used in calculating other probabilities. Namely, what are we to do when the principle of indifference cannot be applied because of the lack of corresponding sentences of the same probability as the one dealt with? Such is for example the case when the die is not symmetrical—what is then the probability of the sentence 'I shall get a one in the next throw' with respect to the sentence 'I throw this die'? Or if it has been found out that among n objects being A there are m objects being B and that an object x is A, what is the probability of the sentence 'x is B'? In such cases we usually assume that the probability equals m/n, but, as Ramsey has remarked, nobody can say why. Ramsey finds it strange that if n is large there are no doubts that the probability equals m/n. Bur if n is small one is inclined to assert that no number can be assigned to the probability of the sentence 'x is B'. It might be said that, if n is small, the relation of probability is such that the value of the probability is not measurable and if n is large—such that the value of the probability is measurable. Such a view is paradoxical in Ramsey's opinion; it should be agreed then that there is an n_0 starting from which the relation of probability undergoes a change of character.[31]

6. PROBABILITY AND THE DEGREE OF CONVICTION

The concept of justification of inductive inferences, consisting in believing a sentence to a degree smaller than certitude and greater than unbelief, has broken down on the question of calculability of the rational degree of belief, i.e. of logical probability. Although the conditions of justification of inductive inference have been determined we cannot state about any concrete inference whether it is justified or not.

We shall now formulate another objection to that conception. Let us assume for a while that we have sucessfully measured the degrees of probability for the particular pairs of sentences. The following doubts arise: are we able to 'coordinate' the degree of conviction in the conclusion q, if we believe a premise p, with the number k, which is the probability of a sentence q with respect to the sentence p? For instance,

if by using the principle of indifference it is calculated that the probability of the sentence 'I shall throw a one' with respect to the sentence 'I throw a symmetrical die' equals 1/6, how is the degree of conviction to be chosen relatively to that number so that the degree of this conviction be rational? Similarly, if one wanted to test whether someone's actual conviction concerning the sentence q under the influence of the belief in the sentence p is a rational conviction, how would one have to proceed? And what does it mean 'to be convinced in the degree 1/6'? If someone is told that the temperature of a given object is 100°C, he can — at least to some extent — imagine the impression of heat corresponding to that temperature, and he knows, among other things, that if he touched the object with a bare hand he would burn himself. The reason is that one has had the opportunity of learning — within the limits of the exactitude of his sense — that such and such degrees on Celsius' scale correspond to such and such impression of heat. Although we cannot tell the difference between the heat of 5°C and that of 6°C, we can nevertheless differentiate the feeling of heat in the interval 0°C–5°C from that of the interval 10°C–15°C. Respective feelings of heat correspond not so much to exact numbers as to numerical intervals on Celsius' scale. Therefore, conversely, if we are told that the temperature is such and such (contained in a given interval on Celsius' scale), we expect to perceive such and such heat. The situation should be analogical with convictions. A scale of convictions should be established (for instance, the numerical interval [0, 1]), and then respective numbers or numerical subintervals of this scale should be assigned to different degrees of conviction. If numbers were coordinated — at least within some determined limits of exactitude — with degrees of conviction, then the information that the probability equals so and so much would be of some value, and the degree of conviction could be chosen more or less precisely with respect to the value of the probability. Similarly, it would then be possible to test if the degree of one's conviction is rational. Until this is done — the information that the probability of a sentence q with respect to a sentence p equals, for instance, 1/2, only permits to suppose that the rational degree of belief in the sentence q, motivated by the belief in the sentence p, is greater than the rational degree of belief in a sentence r, motivated by the belief in a sentence s, when the probability of the sentence r with respect to the sentence s

equals, for example, 1/6. Thus, if for instance the probability of the sentence 'I shall throw a one' with respect to the sentence 'I throw a symmetrical die' equals 1/6, and if the probability of the sentence 'I shall throw an even number' with respect to the sentence 'I throw a symmetrical die' equals 1/2, then until numbers of the interval [0, 1] are assigned to the degrees of conviction—it is only known that the rational degree of the conviction that one will cast an even number, which is motivated by the belief in the sentence 'I throw a symmetrical die', is *greater* than the rational degree of the belief that one will throw a one, which is motivated by the belief in the sentence 'I throw a symmetrical die'. Numbers are given here, but the information that the probability of the sentence q with respect to the sentence p equals k is valuable only to the extent that we can say on its grounds that we are allowed to believe in the sentence q more, or less or equally to the degree of conviction in the sentence s, but we do not know in what degree we can believe the sentence q. The information about the degree of probability is thus not an independent one, as on the grounds of the frequency definition of probability. This information has got a value only if connected with another information about the degree of probability, and allows us only to *compare* rational degrees of belief.

Another objection which can be made to the considered conception of justification of inductive inference is that we are not interested in the rationality of the degree of the conviction in the conclusion in itself, but in the rationality of partial belief as motive of definite action. What we are aiming at is to act rationally on the grounds of our convictions. But it is not known what to do with partial belief of rational degree if we are to restrict ourselves to the explanations presented above. Is it rational to act as if the sentence p was true if one has a partial belief in a rational degree in the sentence p? This is not a merely practical problem, devoid of value for theoretical activities. Also in the course of theoretical activity one often faces the decision: to accept a given sentence and act as if it was true or not?

REFERENCES

* First published in *Studia Logica* 12 (1961). Translated by S. Wojnicki.
[1] C. J. Ducasse says: "The fact that inclination to believe is not only subject to gradation, but also allows for at least approximate quantification, is confirmed

by the use of such expressions as 'I am *very* inclined to believe', 'I am *little* inclined to believe', 'I am inclined to believe *more* in *P* than in *Q*...', etc., or by the common use of generally respected adverbs as 'probably', 'very probably', 'possible' etc., as only modalizers of assertion" (emphasis by the author). J. C. Ducasse, 'Some Observations Concerning the Nature of Probability', *The Journal of Philosophy* **XXXVIII** (1941) 397.

[2] J. Hossiason, 'O prawomocności indukcji hipotetycznej' ('The Problem of Justification of Hypothetical Induction'), *Fragmenty Filozoficzne*, Warszawa 1934, p. 18.

[3] This is pointed out by Hossiason, *op. cit.*, p. 20.

[4] The inscription '$x_j \in X \wedge Y$' reads 'x_j is *X* and *Y*'.

[5] Cf. for example W. Biegański: *Teorya Logiki* (*Theory of Logic*), Warszawa 1912, p. 594; W. Wundt: *Logik*, Vol. 1, 3rd ed., Stuttgart, p. 329.

[6] By the way of inference K. Ajdukiewicz means "a class of concrete inferences characterized not only by a certain scheme of inference but also by the degree of certitude with which the premises and conclusions are accepted", 'O racjonalności zawodnych sposobów wnioskowania' ('On the Rationality of Fallible Ways of Inference'), *Studia Filozoficzne* **4** (1958) 22. I have given Ajdukiewicz's term another, although related meaning, calling so the *notation* of the form (2.6).

[7] Reichenbach was probably the first to draw attention to the fact that justification of inference, as it is the case with any other human act, consists in showing that it leads to the realization of a specified goal (H. Reichenbach, *Wahrscheinlichkeitslehre*, Leiden 1935, p. 416. Cf. also by the same author, *Experience and Prediction*, Chicago 1949, pp. 349–350 and 'On the Justification of Induction' *The Journal of Philosophy* **XXVII** (1940) 97). Similarly, cf. K. Ajdukiewicz, *op. cit.*, p. 15. It can be supposed that the representatives of the concept of justification of inductive inferences analyzed here assume that the aim of inference is to arrive at rational convictions.

[8] "We first of all distinguish rational convictions from irrational ones". J. M. Keynes, *Über Wahrscheinlichkeit*, Leipzig 1939, p. 2. "We distinguish sure convictions from probable ones. The latter can be rational or not and is subject to gradation". *Ibid*, p. 7.

[9] "Let us assume that the degrees of belief we may assign to certain sentences on the basis of other sentences defined non-negative numbers can be assigned – at least for the sentences discussed here, i.e. those connected with inductive reasonings". J. Hossiason, 'On the Justification of Hypothetical Induction', *op. cit.*, p. 18.

[10] "If our knowledge *h* justifies rational belief in (a sentence) *a* to a degree α, then we say that between *a* and *h* there holds a relation of probability to the degree α". Keynes, *op. cit.*, p. 2.

[11] A three-argument relation of probability between two sentences and a number should be distinguished from a two-argument relation of probability which is stated in the utterance 'a sentence *q* is probable with regard to a sentence *p*'. As T. Kubiński has rightly pointed out, Keynes is rather unprecise in distinguishing these two relations. (Cf. T. Kubiński: 'O metodzie tworzenia logik modalnych'. ('On the Method of Constructing Modal Logics'), *Studia Logica* **IV** (1956) 227–228.)

[12] "The theory of probability is of logical nature since it concerns degrees of convictions which one can rationally hold under given conditions, and not the convictions themselves actually held by a given person, which can be rational or not". Keynes, *op. cit.*, p. 2.

[13] We usually speak of the rational degree of belief in a sentence p, motivated by a *knowledge* w. But the question arises whether this knowledge is a conjunction (a set) of sentences in which a given person believes, or a conjunction (a set) of *true* sentences in which this person believes.

[14] The inscription '$P(p, q)$' reads 'The probability of the sentence q with respect to the sentence p'.

[15] The relation of probability as degree of rational belief is usually constructed in metalanguage. Thus, in the inscription '$P(p, q)$', the variables 'p', 'q' range over the names of sentences. Correspondingly, we shall interpret the symbols 'f_1', 'f_2' in the expression '$P(f_1, f_2)$' as *variables* ranging over the names of sentences. We moreover agree that, if the scheme contains the symbol '$P(f_1, f_2)$' beside the symbols 'f_1', 'f_2' then we substitute for the variable 'f_1' in the inscription '$P(f_1, f_2)$' the name of the sentence obtained by the substitution for the variables in the formula 'f_1'. We proceed accordingly with the symbol 'f_2'. That is, some inference follows the scheme (3.5) if and only if the conclusion '$a_n \in B$' is accepted to the degree k under the influence of the belief in the sentence '$a_1 \in A \wedge B$', ..., '$a_{n-1} \in A \wedge B$', '$a_n \in A$' and in the sentence asserting that the probability of the sentence '$a_n \in B$' with respect to the sentence '$(a_1 \in A \wedge B) \wedge ... \wedge (a_{n-1} \in A \wedge B) \wedge \wedge (a_n \in A)$' is k.

[16] The variables 'p', 'q' in the inscription '$P(p, q)$' range over names of sentences, and the variables 'p', 'q' beside this notation are ordinary sentential variables here. Cf. Note 15.

[17] Keynes, *op. cit.*, p. 7.

[18] This possibility was pointed out to me by Professor K. Ajdukiewicz.

[19] An inference follows the scheme (3.9) if and only if the conclusion '$a_n \in B$' is accepted under the influence of the belief in the sentences '$a_1 \in A \wedge B$', ..., '$a_{n-1} \in A \wedge B$', '$a_n \in A$' in the degree equal to the probability of the sentence '$a_n \in B$' with respect to the sentence '$(a_1 \in A \wedge B) \wedge ... \wedge (a_{n-1} \in A \wedge B) \wedge (a_n \in A)$'.

[20] J. Hossiason, 'Induction et analogie', *Mind* **200** (1941) 351 ff.

[21] M. Kokoszyńska, 'On 'good' and 'bad' induction', *Studia Logica* **5** (1957) 53.

[22] J. Hossiason, 'O prawomocności indukcji hipotetycznej' ('On the Justification of Hypothetical Induction'), *op. cit.*, pp. 16–17. Cf. also, by the same authoress: 'Definicje rozumowania indukcyjnego' ('The Definitions of Inductive Reasoning'), *Przegląd Filozoficzny* **XXXI** (1928) 363–364.

[23] R. Carnap, *Logical Foundations of Probability*, London 1951, p. 23. Cf. also his 'The Two Concepts of Probability', *Philosophy and Phenomenological Research* **IV** (1954) 513–532.

[24] J. M. Keynes, *op. cit.*, p. 5.

[25] H. Jeffreys, *Scientific Inference*, Cambridge 1957, pp. 22–23.

[26] J. Hossiason, 'O prawomocności indukcji hipotetycznej' ('On the Justification of Hypothetical Induction'), *op. cit.*, pp. 20–23.

[27] Cf. also: J. M. Keynes, *op. cit.*, pp. 105–110; M. Kokoszyńska, 'O 'dobrej' i 'złej' indukcji' ('On 'Good' and 'Bad' Induction'), *Studia Logica* V (1957) 50–51; H. Jeffreys, *Theory of Probability*, Oxford 1950, pp. 17–27; H. Jeffreys, *Scientific Inference*, Cambridge 1947, pp. 24ff; A. Shimony, 'Coherence and the Axioms of Confirmation', *Journal of Symbolic Logic* **20**, p. 3.

[28] H. Jeffreys, *Theory of Probability*, p. 34.

[29] W. C. Kneale, *Probability and Induction*, Oxford 1952, pp. 172–173.

[30] Keynes, *op. cit.*, p. 40.

[31] F. P. Ramsey, *The Foundations of Mathematics*, London 1954, p. 162.